AF879476

SOTHIS PRESS

Sothis Press aims to completely reimagine the public domain book with an emphasis on universal access to captivatingly designed print editions. We believe that technology will enable every single book ever published to be perpetually accessible in beautiful print form. Sothis is beauty in ubiquity.

www.sothis.press
1 Waterhouse Square, London, ECIN 2ST

The Thirteen Books of Euclid's Elements

Translated From the Text of Heiberg,
With Introduction and Commentary;
Books X-XIII and Appendix

Euclid

SOTHIS PRESS

ISBN 978-1-0365-2424-1

This edition is published by
Sothis Press © 2023. The company
is registered in England and Wales.
The Company Number is 14900389.

THE THIRTEEN BOOKS
OF
EUCLID'S ELEMENTS

THE THIRTEEN BOOKS
OF
EUCLID'S ELEMENTS

TRANSLATED FROM THE TEXT OF HEIBERG

WITH INTRODUCTION AND COMMENTARY

BY

T. L. HEATH, C.B., Sc.D.,

SOMETIME FELLOW OF TRINITY COLLEGE, CAMBRIDGE

VOLUME III

BOOKS X—XIII AND APPENDIX

CAMBRIDGE:
at the University Press
1908

CONTENTS OF VOLUME III

BOOK X.

INTRODUCTORY NOTE.

The discovery of the doctrine of incommensurables is attributed to Pythagoras. Thus Proclus says (*Comm. on Eucl.* I. p. 65, 19) that Pythagoras "discovered the theory of irrationals[1]"; and, again, the scholium on the beginning of Book x., also attributed to Proclus, states that the Pythagoreans were the first to address themselves to the investigation of commensurability, having discovered it by means of their observation of numbers. They discovered, the scholium continues, that not all magnitudes have a common measure. "They called all magnitudes measurable by the same measure commensurable, but those which are not subject to the same measure incommensurable, and again such of these as are measured by some other common measure commensurable with one another, and such as are not, incommensurable with the others. And thus by *assuming* their measures they referred everything to *different* commensurabilities, but, though they were different, even so (they proved that) not all magnitudes are commensurable with any. (They showed that) all magnitudes can be rational (ῥητά) and all irrational (ἄλογα) in a relative sense (ὡς πρός τι); hence the commensurable and the incommensurable would be for them *natural* (kinds) (φύσει), while the rational and irrational would rest on *assumption* or *convention* (θέσει)." The scholium quotes further the legend according to which "the first of the Pythagoreans who made public the investigation of these matters perished in a shipwreck," conjecturing that the authors of this story "perhaps spoke allegorically, hinting that everything irrational and formless is properly concealed, and, if any soul should rashly invade this region of life and lay it open, it would be carried away into the sea of becoming and be overwhelmed by its unresting currents." There would be a reason also for keeping the discovery of irrationals secret for the time in the fact that it rendered unstable so much of the groundwork of geometry as the Pythagoreans had based upon the imperfect theory of proportions which applied only to numbers. We have already, after Tannery, referred to the probability that the discovery of incommensurability must have necessitated a great recasting of the whole fabric of elementary geometry, pending the discovery of the general theory of proportion applicable to incommensurable as well as to commensurable magnitudes.

It seems certain that it was with reference to the length of the diagonal of a square or the hypotenuse of an isosceles right-angled triangle that Pythagoras made his discovery. Plato (*Theaetetus*, 147 D) tells us that Theodorus of Cyrene wrote about square roots (δυνάμεις), proving that the square roots of

[1] I have already noted (Vol. I. p. 351) that G. Junge (*Wann haben die Griechen das Irrationale entdeckt?*) disputes this, maintaining that it was the Pythagoreans, but not Pythagoras, who made the discovery. Junge is obliged to alter the reading of the passage of Proclus, on what seems to be quite insufficient evidence; and in any case I doubt whether the point is worth so much labouring.

three square feet and five square feet are not commensurable with that of one square foot, and so on, selecting each such square root up to that of 17 square feet, at which for some reason he stopped. No mention is here made of $\sqrt{2}$, doubtless for the reason that its incommensurability had been proved before, i.e. by Pythagoras. We know that Pythagoras invented a formula for finding right-angled triangles in rational numbers, and in connexion with this it was inevitable that he should investigate the relations between sides and hypotenuse in other right-angled triangles. He would naturally give special attention to the isosceles right-angled triangle; he would try to measure the diagonal, he would arrive at successive approximations, in rational fractions, to the value of $\sqrt{2}$; he would find that successive efforts to obtain an exact expression for it failed. It was however an enormous step to conclude that such exact expression was *impossible*, and it was this step which Pythagoras (or the Pythagoreans) made. We now know that the formation of the *side-* and *diagonal-*numbers explained by Theon of Smyrna and others was Pythagorean, and also that the theorems of Eucl. II. 9, 10 were used by the Pythagoreans in direct connexion with this method of approximating to the value of $\sqrt{2}$. The very method by which Euclid proves these propositions is itself an indication of their connexion with the investigation of $\sqrt{2}$, since he uses a figure made up of two isosceles right-angled triangles.

The actual method by which the Pythagoreans proved the incommensurability of $\sqrt{2}$ with unity was no doubt that referred to by Aristotle (*Anal. prior.* I. 23, 41 a 26—7), a *reductio ad absurdum* by which it is proved that, if the diagonal is commensurable with the side, it will follow that the same number is both odd and even. The proof formerly appeared in the texts of Euclid as X. 117, but it is undoubtedly an interpolation, and August and Heiberg accordingly relegate it to an Appendix. It is in substance as follows.

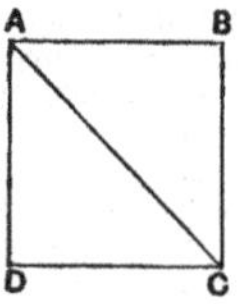

Suppose AC, the diagonal of a square, to be commensurable with AB, its side. Let $\alpha : \beta$ be their ratio expressed in the smallest numbers.

Then $\alpha > \beta$ and therefore necessarily > 1.

Now $$AC^2 : AB^2 = \alpha^2 : \beta^2,$$

and, since $$AC^2 = 2AB^2, \quad \text{[Eucl. I. 47]}$$

$$\alpha^2 = 2\beta^2.$$

Therefore α^2 is even, and therefore α is even.

Since $\alpha : \beta$ is in its lowest terms, it follows that β must be *odd*.

Put $$\alpha = 2\gamma;$$

therefore $$4\gamma^2 = 2\beta^2,$$

or $$\beta^2 = 2\gamma^2,$$

so that β^2, and therefore β, must be *even*.

But β was also odd:

which is impossible.

This proof only enables us to prove the incommensurability of the diagonal of a square with its side, or of $\sqrt{2}$ with unity. In order to prove the incommensurability of the sides of squares, one of which has *three* times the area of another, an entirely different procedure is necessary; and we find in fact that, even a century after Pythagoras' time, it was still necessary to use *separate* proofs (as the passage of the *Theaetetus* shows that Theodorus did) to establish the incommensurability with unity of $\sqrt{3}$, $\sqrt{5}$, ... up to $\sqrt{17}$.

This fact indicates clearly that the general theorem in Eucl. x. 9 that *squares which have not to one another the ratio of a square number to a square number have their sides incommensurable in length* was not arrived at all at once, but was, in the manner of the time, developed out of the separate consideration of special cases (Hankel, p. 103).

The proposition x. 9 of Euclid is definitely ascribed by the scholiast to Theaetetus. Theaetetus was a pupil of Theodorus, and it would seem clear that the theorem was not known to Theodorus. Moreover the Platonic passage itself (*Theaet.* 147 D sqq.) represents the young Theaetetus as striving after a general conception of what we call a *surd.* "The idea occurred to me, seeing that *square roots* (δυνάμεις) appeared to be unlimited in multitude, to try to arrive at one collective term by which we could designate all these square roots.... I divided number in general into two classes. The number which can be expressed as equal multiplied by equal (ἴσον ἰσάκις) I likened to a square in form, and I called it square and equilateral....The intermediate number, such as three, five, and any number which cannot be expressed as equal multiplied by equal, but is either less times more or more times less, so that it is always contained by a greater and less side, I likened to an oblong figure and called an oblong number.... Such straight lines then as square the equilateral and plane number I defined as length (μῆκος), and such as square the oblong *square roots* (δυνάμεις), as not being commensurable with the others in length but only in the plane areas to which their squares are equal."

There is further evidence of the contributions of Theaetetus to the theory of incommensurables in a commentary on Eucl. x. discovered, in an Arabic translation, by Woepcke (*Mémoires présentés à l'Académie des Sciences*, XIV., 1856, pp. 658—720). It is certain that this commentary is of Greek origin. Woepcke conjectures that it was by Vettius Valens, an astronomer, apparently of Antioch, and a contemporary of Claudius Ptolemy (2nd cent. A.D.). Heiberg, with greater probability, thinks that we have here a fragment of the commentary of Pappus (*Euklid-studien*, pp. 169—71), and this is rendered practically certain by Suter (*Die Mathematiker und Astronomen der Araber und ihre Werke*, pp. 49 and 211). This commentary states that the theory of irrational magnitudes "had its origin in the school of Pythagoras. It was considerably developed by Theaetetus the Athenian, who gave proof, in this part of mathematics, as in others, of ability which has been justly admired. He was one of the most happily endowed of men, and gave himself up, with a fine enthusiasm, to the investigation of the truths contained in these sciences, as Plato bears witness for him in the work which he called after his name. As for the exact distinctions of the above-named magnitudes and the rigorous demonstrations of the propositions to which this theory gives rise, I believe that they were chiefly established by this mathematician; and, later, the great Apollonius, whose genius touched the highest point of excellence in mathematics, added to these discoveries a number of remarkable theories after many efforts and much labour.

"For Theaetetus had distinguished square roots [*puissances* must be the δυνάμεις of the Platonic passage] commensurable in length from those which are incommensurable, and had divided the well-known species of irrational lines after the different means, assigning the *medial* to geometry, the *binomial* to arithmetic, and the *apotome* to harmony, as is stated by Eudemus the Peripatetic.

"As for Euclid, he set himself to give rigorous rules, which he established, relative to commensurability and incommensurability in general; he made precise the definitions and the distinctions between rational and irrational magnitudes, he set out a great number of orders of irrational magnitudes, and finally he clearly showed their whole extent."

The allusion in the last words must be apparently to x. 115, where it is proved that from the *medial* straight line an unlimited number of other irrationals can be derived all different from it and from one another.

The connexion between the *medial* straight line and the geometric mean is obvious, because it is in fact the mean proportional between two rational straight lines "commensurable in square only." Since $\frac{1}{2}(x+y)$ is the *arithmetic* mean between x, y, the reference to it of the binomial can be understood. The connexion between the apotome and the harmonic mean is explained by some propositions in the second book of the Arabic commentary. The harmonic mean between x, y is $\frac{2xy}{x+y}$, and propositions of which Woepcke quotes the enunciations prove that, if a rational or a medial area has for one of its sides a *binomial* straight line, the other side will be an *apotome* of corresponding order (these propositions are generalised from Eucl. x. 111—4); the fact is that $\frac{2xy}{x+y}=\frac{2xy}{x^2-y^2}\cdot(x-y)$.

One other predecessor of Euclid appears to have written on irrationals, though we know no more of the work than its title as handed down by Diogenes Laertius[1]. According to this tradition, Democritus wrote περὶ ἀλόγων γραμμῶν καὶ ναστῶν β′, *two Books on irrational straight lines and solids* (apparently). Hultsch (*Neue Jahrbücher für Philologie und Pädagogik*, 1881, pp. 578—9) conjectures that the true reading may be περὶ ἀλόγων γραμμῶν κλαστῶν, "on irrational broken lines." Hultsch seems to have in mind *straight* lines divided into two parts one of which is rational and the other irrational ("Aus einer Art von Umkehr des Pythagoreischen Lehrsatzes über das rechtwinklige Dreieck gieng zunächst mit Leichtigkeit hervor, dass man eine Linie construiren könne, welche als irrational zu bezeichnen ist, aber durch Brechung sich darstellen lässt als die Summe einer rationalen und einer irrationalen Linie"). But I doubt the use of κλαστός in the sense of breaking one straight line into parts; it should properly mean a bent line, i.e. two straight lines forming an angle or *broken short off* at their point of meeting. It is also to be observed that ναστόν is quoted as a Democritean word (opposite to κενόν) in a fragment of Aristotle (202). I see therefore no reason for questioning the correctness of the title of Democritus' book as above quoted.

I will here quote a valuable remark of Zeuthen's relating to the classification of irrationals. He says (*Geschichte der Mathematik im Altertum und Mittelalter*, p. 56) "Since such roots of equations of the second degree as are incommensurable with the given magnitudes cannot be expressed by means of the latter and of numbers, it is conceivable that the Greeks, in exact investigations, introduced no approximate values but worked on with the magnitudes they had found, which were represented by straight lines obtained by the construction corresponding to the solution of the equation. That is exactly the same thing which happens when we do not evaluate roots but content ourselves with expressing them by radical signs and other algebraical symbols. But, inasmuch as one straight line looks like another, the Greeks did not get

[1] Diog. Laert. IX. 47, p. 239 (ed. Cobet).

the same clear view of what they denoted (i.e. by simple inspection) as our system of symbols assures to us. For this reason it was necessary to undertake a classification of the irrational magnitudes which had been arrived at by successive solution of equations of the second degree." To much the same effect Tannery wrote in 1882 (*De la solution géométrique des problèmes du second degré avant Euclide* in *Mémoires de la Société des sciences physiques et naturelles de Bordeaux*, 2e Série, IV. pp. 395—416). Accordingly Book X. formed a repository of results to which could be referred problems which depended on the solution of certain types of equations, quadratic and biquadratic but reducible to quadratics.

Consider the quadratic equations

$$x^2 \pm 2\alpha x \cdot \rho \pm \beta \cdot \rho^2 = 0,$$

where ρ is a rational straight line, and α, β are coefficients. Our quadratic equations in algebra leave out the ρ; but I put it in, because it has always to be remembered that Euclid's x is a *straight line*, not an algebraical quantity, and is therefore to be found in terms of, or in relation to, a certain assumed *rational straight line*, and also because with Euclid ρ may be not only of the form a, where a represents a units of length, but also of the form $\sqrt{\frac{m}{n}} \cdot a$, which represents a length "commensurable in square only" with the unit of length, or $\sqrt{A}$ where A represents a number (not square) of units of *area*. The use therefore of ρ in our equations makes it unnecessary to multiply different *cases* according to the relation of ρ to the unit of length, and has the further advantage that, e.g., the expression $\rho \pm \sqrt{k} \cdot \rho$ is just as general as the expression $\sqrt{k} \cdot \rho \pm \sqrt{\lambda} \cdot \rho$, since ρ covers the form $\sqrt{k} \cdot \rho$, both expressions covering a length either commensurable in length, or "commensurable in square only," with the unit of length.

Now the *positive* roots of the quadratic equations

$$x^2 \pm 2\alpha x \cdot \rho \pm \beta \cdot \rho^2 = 0$$

can only have the following forms

$$\left. \begin{array}{ll} x_1 = \rho(\alpha + \sqrt{\alpha^2 - \beta}), & x_1' = \rho(\alpha - \sqrt{\alpha^2 - \beta}) \\ x_2 = \rho(\sqrt{\alpha^2 + \beta} + \alpha), & x_2' = \rho(\sqrt{\alpha^2 + \beta} - \alpha) \end{array} \right\}.$$

The negative roots do not come in, since x must be a *straight line*. The omission however to bring in negative roots constitutes no loss of generality, since the Greeks would write the equation leading to negative roots in another form so as to make them positive, i.e. they would change the sign of x in the equation.

Now the positive roots x_1, x_1', x_2, x_2' may be classified according to the character of the coefficents α, β and their relation to one another.

I. Suppose that α, β do not contain any surds, i.e. are either integers or of the form m/n, where m, n are integers.

Now in the expressions for x_1, x_1' it may be that

(1) β is of the form $\frac{m^2}{n^2}\alpha^2$.

Euclid expresses this by saying that the square on $\alpha\rho$ exceeds the square on $\rho\sqrt{\alpha^2 - \beta}$ by the square on a straight line commensurable in length with $\alpha\rho$.

In this case x_1 is, in Euclid's terminology, a *first binomial* straight line,

and x_1' a *first apotome*.

(2) In general, β not being of the form $\frac{m^2}{n^2}\alpha^2$,

x_1 is a *fourth binomial*,
x_1' a *fourth apotome*.

Next, in the expressions for x_2, x_2' it may be that

(1) β is equal to $\frac{m^2}{n^2}(\alpha^2+\beta)$, where m, n are integers, i.e. β is of the form

$$\frac{m^2}{n^2-m^2}\alpha^2.$$

Euclid expresses this by saying that the square on $\rho\sqrt{\alpha^2+\beta}$ exceeds the square on $\alpha\rho$ by the square on a straight line commensurable in length with $\rho\sqrt{\alpha^2+\beta}$.

In this case x_2 is, in Euclid's terminology, a *second binomial*,
x_2' a *second apotome*.

(2) In general, β not being of the form $\frac{m^2}{n^2-m^2}\alpha^2$,

x_2 is a *fifth binomial*,
x_2' a *fifth apotome*.

II. Now suppose that α is of the form $\sqrt{\frac{m}{n}}$, where m, n are integers, and let us denote it by $\sqrt{\lambda}$.

Then in this case

$$x_1=\rho(\sqrt{\lambda}+\sqrt{\lambda-\beta}),\quad x_1'=\rho(\sqrt{\lambda}-\sqrt{\lambda-\beta}),$$
$$x_2=\rho(\sqrt{\lambda+\beta}+\sqrt{\lambda}),\quad x_2'=\rho(\sqrt{\lambda+\beta}-\sqrt{\lambda}).$$

Thus x_1, x_1' are of the same form as x_2, x_2'.

If $\sqrt{\lambda-\beta}$ in x_1, x_1' is not surd but of the form m/n, and if $\sqrt{\lambda+\beta}$ in x_2, x_2' is not surd but of the form m/n, the roots are comprised among the forms already shown, the first, second, fourth and fifth binomials and apotomes.

If $\sqrt{\lambda-\beta}$ in x_1, x_1' is surd, then

(1) we may have β of the form $\frac{m^2}{n^2}\lambda$, and in this case

x_1 is a *third binomial* straight line,
x_1' a *third apotome*;

(2) in general, β not being of the form $\frac{m^2}{n^2}\lambda$,

x_1 is a *sixth binomial* straight line,
x_1' a *sixth apotome*.

With the expressions for x_2, x_2' the distinction between the third and sixth binomials and apotomes is of course the distinction between the cases

(1) in which $\beta=\frac{m^2}{n^2}(\lambda+\beta)$, or β is of the form $\frac{m^2}{n^2-m^2}\lambda$,

and (2) in which β is not of this form.

If we take the square root of the product of ρ and each of the six binomials and six apotomes just classified, i.e.

$$\rho^2(\alpha\pm\sqrt{\alpha^2-\beta}),\quad \rho^2(\sqrt{\alpha^2+\beta}\pm\alpha),$$

in the six different forms that each may take, we find six new irrationals with a positive sign separating the two terms, and six corresponding irrationals with a negative sign. These are of course roots of the equations

$$x^4 \pm 2\alpha x^2 . \rho^2 \pm \beta . \rho^4 = 0.$$

These irrationals really come before the others in Euclid's order (x. 36—41 for the positive sign and x. 73—78 for the negative sign). As we shall see in due course, the straight lines actually found by Euclid are

1. $\rho \pm \sqrt{k} . \rho$, the *binomial* (ἡ ἐκ δύο ὀνομάτων) and the *apotome* (ἀποτομή),

which are the positive roots of the biquadratic (reducible to a quadratic)

$$x^4 - 2(1 + k)\rho^2 . x^2 + (1 - k)^2 \rho^4 = 0.$$

2. $k^{\frac{1}{4}}\rho \pm k^{\frac{3}{4}}\rho$, the *first bimedial* (ἐκ δύο μέσων πρώτη) and the *first apotome of a medial* (μέσης ἀποτομὴ πρώτη),

which are the positive roots of

$$x^4 - 2\sqrt{k}(1 + k)\rho^2 . x^2 + k(1 - k)^2 \rho^4 = 0.$$

3. $k^{\frac{1}{4}}\rho \pm \dfrac{\sqrt{\lambda}}{k^{\frac{1}{4}}}\rho$, the *second bimedial* (ἐκ δύο μέσων δευτέρα) and the *second apotome of a medial* (μέσης ἀποτομὴ δευτέρα),

which are the positive roots of the equation

$$x^4 - 2\frac{k + \lambda}{\sqrt{k}}\rho^2 . x^2 + \frac{(k - \lambda)^2}{k}\rho^4 = 0.$$

4. $\dfrac{\rho}{\sqrt{2}}\sqrt{1 + \dfrac{k}{\sqrt{1 + k^2}}} \pm \dfrac{\rho}{\sqrt{2}}\sqrt{1 - \dfrac{k}{\sqrt{1 + k^2}}}$,

the *major* (irrational straight line) (μείζων) and the *minor* (irrational straight line) (ἐλάσσων),

which are the positive roots of the equation

$$x^4 - 2\rho^2 . x^2 + \frac{k^2}{1 + k^2}\rho^4 = 0.$$

5. $\dfrac{\rho}{\sqrt{2(1 + k^2)}}\sqrt{\sqrt{1 + k^2} + k} \pm \dfrac{\rho}{\sqrt{2(1 + k^2)}}\sqrt{\sqrt{1 + k^2} - k}$,

the "*side*" *of a rational plus a medial* (area) (ῥητὸν καὶ μέσον δυναμένη) and the "*side*" *of a medial minus a rational area* (in the Greek ἡ μετὰ ῥητοῦ μέσον τὸ ὅλον ποιοῦσα),

which are the positive roots of the equation

$$x^4 - \frac{2}{\sqrt{1 + k^2}}\rho^2 . x^2 + \frac{k^2}{(1 + k^2)^2}\rho^4 = 0,$$

6. $\dfrac{\lambda^{\frac{1}{4}}\rho}{\sqrt{2}}\sqrt{1 + \dfrac{k}{\sqrt{1 + k^2}}} \pm \dfrac{\lambda^{\frac{1}{4}}\rho}{\sqrt{2}}\sqrt{1 - \dfrac{k}{\sqrt{1 + k^2}}}$,

the "*side*" *of the sum of two medial areas* (ἡ δύο μέσα δυναμένη) and the "*side*" *of a medial minus a medial area* (in the Greek ἡ μετὰ μέσου μέσον τὸ ὅλον ποιοῦσα),

which are the positive roots of the equation

$$x^4 - 2\sqrt{\lambda} . x^2\rho^2 + \lambda\frac{k^2}{1 + k^2}\rho^4 = 0.$$

The above facts and formulae admit of being stated in a great variety of ways according to the notation and the particular letters used. Consequently the summaries which have been given of Eucl. x. by various writers differ much in appearance while expressing the same thing in substance. The first summary in algebraical form (and a very elaborate one) seems to have been that of Cossali (*Origine, trasporto in Italia, primi progressi in essa dell' Algebra*, Vol. II. pp. 242—65) who takes credit accordingly (p. 265). In 1794 Meier Hirsch published at Berlin an *Algebraischer Commentar über das zehente Buch der Elemente des Euklides* which gives the *contents* in algebraical form but fails to give any indication of Euclid's methods, using modern forms of proof only. In 1834 Poselger wrote a paper, *Ueber das zehnte Buch der Elemente des Euklides*, in which he pointed out the defects of Hirsch's reproduction and gave a summary of his own, which however, though nearer to Euclid's form, is difficult to follow in consequence of an elaborate system of abbreviations, and is open to the objection that it is not algebraical enough to enable the character of Euclid's irrationals to be seen at a glance. Other summaries will be found (1) in Nesselmann, *Die Algebra der Griechen*, pp. 165—84; (2) in Loria, *Il periodo aureo della geometria greca*, Modena, 1895, pp. 40—9; (3) in Christensen's article "Ueber Gleichungen vierten Grades im zehnten Buch der Elemente Euklids" in the *Zeitschrift für Math. u. Physik* (*Historisch-literarische Abtheilung*), XXXIV. (1889), pp. 201—17. The only summary in English that I know is that in the *Penny Cyclopaedia*, under "Irrational quantity," by De Morgan, who yielded to none in his admiration of Book x. "Euclid investigates," says De Morgan, "every possible variety of lines which can be represented by $\sqrt{(\sqrt{a} \pm \sqrt{b})}$, a and b representing two commensurable lines....This book has a completeness which none of the others (not even the fifth) can boast of: and we could almost suspect that Euclid, having arranged his materials in his own mind, and having completely elaborated the 10th Book, wrote the preceding books after it and did not live to revise them thoroughly."

Much attention was given to Book x. by the early algebraists. Thus Leonardo of Pisa (fl. about 1200 A.D.) wrote in the 14th section of his *Liber Abaci* on the theory of irrationalities (*de tractatu binomiorum et recisorum*), without however (except in treating of irrational trinomials and cubic irrationalities) adding much to the substance of Book x.; and, in investigating the equation

$$x^3 + 2x^2 + 10x = 20,$$

propounded by Johannes of Palermo, he proved that none of the irrationals in Eucl. x. would satisfy it (Hankel, pp. 344—6, Cantor, II_1, p. 43). Luca Paciuolo (about 1445—1514 A.D.) in his algebra based himself largely, as he himself expressly says, on Euclid x. (Cantor, II_1, p. 293). Michael Stifel (1486 or 1487 to 1567) wrote on irrational numbers in the second Book of his *Arithmetica integra*, which Book may be regarded, says Cantor (II_1, p. 402), as an elucidation of Eucl. x. The works of Cardano (1501—76) abound in speculations regarding the irrationals of Euclid, as may be seen by reference to Cossali (Vol. II., especially pp. 268—78 and 382—99); the character of the various odd and even powers of the binomials and apotomes is therein investigated, and Cardano considers in detail of what particular forms of equations, quadratic, cubic, and biquadratic, each class of Euclidean irrationals can be roots. Simon Stevin (1548—1620) wrote a *Traité des incommensurables grandeurs en laquelle est sommairement déclaré le contenu du Dixiesme Livre d'Euclide* (*Oeuvres mathématiques*, Leyde, 1634, pp. 219 sqq.); he speaks thus of the book: "La difficulté du dixiesme Livre d'Euclide est à plusieurs devenue en horreur, voire jusque à l'appeler la croix des mathématiciens, matière trop dure à digérer, et en la quelle n'aperçoivent aucune utilité," a passage quoted by Loria (*Il periodo aureo della geometria greca*, p. 41).

It will naturally be asked, what use did the Greek geometers actually make of the theory of irrationals developed at such length in Book x.? The answer is that Euclid himself, in Book XIII., makes considerable use of the second portion of Book x. dealing with the irrationals affected with a negative sign, the *apotomes* etc. One object of Book XIII. is to investigate the relation of the sides of a pentagon inscribed in a circle and of an icosahedron and dodecahedron inscribed in a sphere to the diameter of the circle or sphere respectively, supposed rational. The connexion with the regular pentagon of a straight line cut in extreme and mean ratio is well known, and Euclid first proves (XIII. 6) that, if a *rational* straight line is so divided, the parts are the irrationals called *apotomes*, the lesser part being a *first apotome*. Then, on the assumption that the diameters of a circle and sphere respectively are rational, he proves (XIII. 11) that the side of the inscribed regular pentagon is the irrational straight line called *minor*, as is also the side of the inscribed icosahedron (XIII. 16), while the side of the inscribed dodecahedron is the irrational called an *apotome* (XIII. 17).

Of course the investigation in Book x. would not have been complete if it had dealt only with the irrationals affected with a *negative* sign. Those affected with the positive sign, the *binomials* etc., had also to be discussed, and we find both portions of Book x., with its nomenclature, made use of by Pappus in two propositions, of which it may be of interest to give the enunciations here.

If, says Pappus (IV. p. 178), AB be the rational diameter of a semicircle, and if AB be produced to C so that BC is equal to the radius, if CD be a tangent,

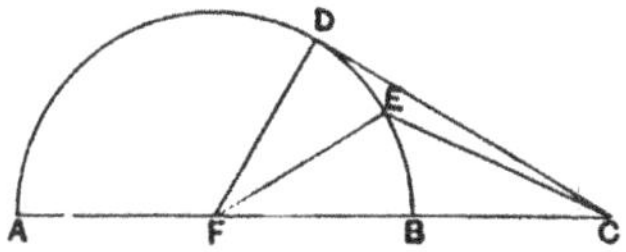

if E be the middle point of the arc BD, and if CE be joined, then CE is the irrational straight line called *minor*. As a matter of fact, if ρ is the radius,

$$CE^2 = \rho^2 (5 - 2\sqrt{3}) \text{ and } CE = \sqrt{\frac{5+\sqrt{13}}{2}} - \sqrt{\frac{5-\sqrt{13}}{2}}.$$

If, again (p. 182), CD be equal to the radius of a semicircle supposed

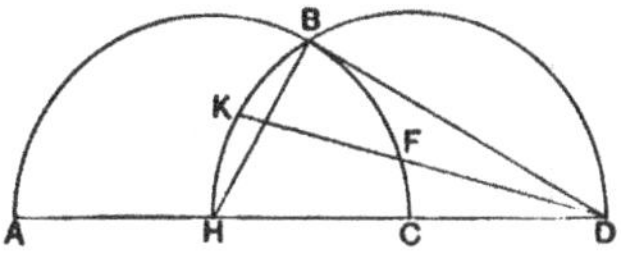

rational, and if the tangent DB be drawn and the angle ADB be bisected by DF meeting the circumference in F, then DF is the excess by which the *binomial* exceeds the *straight line which produces with a rational area a medial whole* (see Eucl. x. 77). (In the figure DK is the ***binomial*** and KF the other irrational straight line.) As a matter of fact, if ρ be the radius,

$$KD = \rho \cdot \frac{\sqrt{3}+1}{\sqrt{2}}, \text{ and } KF = \rho \cdot \sqrt{\sqrt{3}-1} = \rho \cdot \left(\sqrt{\frac{\sqrt{3}+\sqrt{2}}{2}} - \sqrt{\frac{\sqrt{3}-\sqrt{2}}{2}}\right).$$

Proclus tells us that Euclid left out, as alien to a selection of *elements*, the discussion of the more complicated irrationals, "the unordered irrationals which Apollonius worked out more fully" (Proclus, p. 74, 23), while the scholiast to Book x. remarks that Euclid does not deal with all rationals and irrationals but only the simplest kinds by the combination of which an infinite number of irrationals are obtained, of which Apollonius also gave some. The author of the commentary on Book x. found by Woepcke in an Arabic translation, and above alluded to, also says that "it was Apollonius who, beside the *ordered* irrational magnitudes, showed the existence of the *unordered* and by accurate methods set forth a great number of them." It can only be vaguely gathered, from such hints as the commentator proceeds to give, what the character of the extension of the subject given by Apollonius may have been. See note at end of Book.

DEFINITIONS.

1. Those magnitudes are said to be **commensurable** which are measured by the same measure, and those **incommensurable** which cannot have any common measure.

2. Straight lines are **commensurable in square** when the squares on them are measured by the same area, and **incommensurable in square** when the squares on them cannot possibly have any area as a common measure.

3. With these hypotheses, it is proved that there exist straight lines infinite in multitude which are commensurable and incommensurable respectively, some in length only, and others in square also, with an assigned straight line. Let then the assigned straight line be called **rational**, and those straight lines which are commensurable with it, whether in length and in square or in square only, **rational**, but those which are incommensurable with it **irrational**.

4. And let the square on the assigned straight line be called **rational** and those areas which are commensurable with it **rational**, but those which are incommensurable with it **irrational**, and the straight lines which produce them **irrational**, that is, in case the areas are squares, the sides themselves, but in case they are any other rectilineal figures, the straight lines on which are described squares equal to them.

DEFINITION 1.

Σύμμετρα μεγέθη λέγεται τὰ τῷ αὐτῷ μέτρῳ μετρούμενα, ἀσύμμετρα δέ, ὧν μηδὲν ἐνδέχεται κοινὸν μέτρον γενέσθαι.

DEFINITION 2.

Εὐθεῖαι δυνάμει σύμμετροί εἰσιν, ὅταν τὰ ἀπ' αὐτῶν τετράγωνα τῷ αὐτῷ χωρίῳ μετρῆται, ἀσύμμετροι δέ, ὅταν τοῖς ἀπ' αὐτῶν τετραγώνοις μηδὲν ἐνδέχηται χωρίον κοινὸν μέτρον γενέσθαι.

Commensurable in square is in the Greek δυνάμει σύμμετρος. In earlier translations (e.g. Williamson's) δυνάμει has been translated "in power," but, as the particular *power* represented by δύναμις in Greek geometry is *square*, I have thought it best to use the latter word throughout. It will be observed that Euclid's expression *commensurable in square only* (used in Def. 3 and constantly) corresponds to what Plato makes Theaetetus call a *square root* (δύναμις) in the sense of a *surd*. If a is any straight line, a and $a\sqrt{m}$, or $a\sqrt{m}$ and $a\sqrt{n}$ (where m, n are integers or arithmetical fractions in their lowest terms, proper or improper, but not square) are *commensurable in square only*. Of course (as explained in the Porism to X. 10) all straight lines *commensurable in length* (μήκει), in Euclid's phrase, are commensurable *in square* also; but not all straight lines which are commensurable *in square* are commensurable *in length* as well. On the other hand, straight lines *incommensurable in square* are necessarily incommensurable *in length* also; but not all straight lines which are incommensurable *in length* are incommensurable *in square*. In fact, straight lines which are *commensurable in square only* are incommensurable *in length*, but obviously not incommensurable in square.

DEFINITION 3.

Τούτων ὑποκειμένων δείκνυται, ὅτι τῇ προτεθείσῃ εὐθείᾳ ὑπάρχουσιν εὐθεῖαι πλήθει ἄπειροι σύμμετροί τε καὶ ἀσύμμετροι αἱ μὲν μήκει μόνον, αἱ δὲ καὶ δυνάμει. καλείσθω οὖν ἡ μὲν προτεθεῖσα εὐθεῖα ῥητή, καὶ αἱ ταύτῃ σύμμετροι εἴτε μήκει καὶ δυνάμει εἴτε δυνάμει μόνον ῥηταί, αἱ δὲ ταύτῃ ἀσύμμετροι ἄλογοι καλείσθωσαν.

The first sentence of the definition is decidedly elliptical. It should, strictly speaking, assert that "with a given straight line there are an infinite number of straight lines which are (1) commensurable either (a) in square only or (b) in square and in length also, and (2) incommensurable, either (a) in length only or (b) in length and in square also."

The relativity of the terms *rational* and *irrational* is well brought out in this definition. We may set out *any straight line* and call it rational, and it is then with reference to this assumed rational straight line that others are called *rational* or *irrational*.

We should carefully note that the signification of *rational* in Euclid is wider than in our terminology. With him, not only is a straight line commensurable *in length* with a rational straight line rational, but a straight line is rational which is commensurable with a rational straight line *in square only*. That is, if ρ is a rational straight line, not only is $\frac{m}{n}\rho$ rational, where m, n are integers and

m/n in its lowest terms is not square, but $\sqrt{\frac{m}{n}} \cdot \rho$ is *rational* also. We should in this case call $\sqrt{\frac{m}{n}} \cdot \rho$ irrational. It would appear that Euclid's terminology here differed as much from that of his predecessors as it does from ours. We are familiar with the phrase ἄρρητος διάμετρος τῆς πεμπάδος by which Plato (evidently after the Pythagoreans) describes the diagonal of a square on a straight line containing 5 units of length. This "inexpressible diameter of five (squared)" means $\sqrt{50}$, in contrast to the ῥητὴ διάμετρος, the "expressible diameter" of the same square, by which is meant the approximation $\sqrt{50 - 1}$, or 7. Thus for Euclid's predecessors $\sqrt{\frac{m}{n}} \cdot \rho$ would apparently not have been rational but ἄρρητος, "inexpressible," i.e. irrational.

I shall throughout my notes on this Book denote a *rational* straight line in Euclid's sense by ρ, and by ρ and σ when two different rational straight lines are required. Wherever then I use ρ or σ, it must be remembered that ρ, σ may have either of the forms a, $\sqrt{k} \cdot a$, where a represents a units of length, a being either an integer or of the form m/n, where m, n are both integers, and k is an integer or of the form m/n (where both m, n are integers) but not square. In other words, ρ, σ may have either of the forms a or $\sqrt{A}$, where A represents A units of *area* and A is integral or of the form m/n, where m, n are both integers. It has been the habit of writers to give a and $\sqrt{a}$ as the alternative forms of ρ, but I shall always use $\sqrt{A}$ for the second in order to keep the dimensions right, because it must be borne in mind throughout that ρ is an irrational *straight line*.

As Euclid extends the signification of *rational* (ῥητός, literally *expressible*), so he limits the scope of the term ἄλογος (literally *having no ratio*) as applied to straight lines. That this limitation was started by himself may perhaps be inferred from the form of words "*let* straight lines incommensurable with it *be called* irrational." Irrational straight lines then are with Euclid straight lines commensurable *neither in length nor in square* with the assumed rational straight line. $\sqrt{k} \cdot a$ where k is not square is not irrational; $\sqrt[4]{k} \cdot a$ is irrational, and so (as we shall see later on) is $(\sqrt{k} \pm \sqrt{\lambda})\, a$.

DEFINITION 4.

Καὶ τὸ μὲν ἀπὸ τῆς προτεθείσης εὐθείας τετράγωνον ῥητόν, καὶ τὰ τούτῳ σύμμετρα ῥητά, τὰ δὲ τούτῳ ἀσύμμετρα ἄλογα καλείσθω, καὶ αἱ δυνάμεναι αὐτὰ ἄλογοι, εἰ μὲν τετράγωνα εἴη, αὐταὶ αἱ πλευραί, εἰ δὲ ἕτερά τινα εὐθύγραμμα, αἱ ἴσα αὐτοῖς τετράγωνα ἀναγράφουσαι.

As applied to *areas*, the terms *rational* and *irrational* have, on the other hand, the same sense with Euclid as we should attach to them. According to Euclid, if ρ is a rational straight line in *his* sense, ρ^2 is *rational* and any area commensurable with it, i.e. of the form $k\rho^2$ (where k is an integer, or of the form m/n, where m, n are integers), is rational; but any area of the form $\sqrt{k} \cdot \rho^2$ is *irrational*. Euclid's *rational area* thus contains A *units of area*, where A is an integer or of the form m/n, where m, n are integers; and his *irrational area* is of the form $\sqrt{k} \cdot A$. His irrational *area* is then connected with his irrational *straight line* by making the latter the square root of the former. This would give us for the irrational *straight line* $\sqrt[4]{k} \,.\, \sqrt{A}$, which of course includes $\sqrt[4]{k} \,.\, a$.

αἱ δυνάμεναι αὐτά are the straight lines the squares on which are equal to the areas, in accordance with the regular meaning of δύνασθαι. It is scarcely possible, in a book written in geometrical language, to translate δυναμένη as the *square root* (of an *area*) and δύνασθαι as *to be the square root* (of an *area*), although I can use the term "square root" when in my notes I am using an algebraical expression to represent an area; I shall therefore hereafter use the word "side" for δυναμένη and "to be the side of" for δύνασθαι, so that "side" will in such expressions be a short way of expressing the "side of a *square equal to* (*an area*)." In this particular passage it is not quite practicable to use the words "side of" or "straight line the square on which is equal to," for these expressions occur just afterwards for two alternatives which the word δυναμένη covers. I have therefore exceptionally translated "the straight lines which produce them" (i.e. if squares are described upon them as sides).

αἱ ἴσα αὐτοῖς τετράγωνα ἀναγράφουσαι, literally "the (straight lines) which *describe* squares equal to them": a peculiar use of the active of ἀναγράφειν, the meaning being of course "the straight lines on which *are described* the squares" which are equal to the rectilineal figures.

BOOK X. PROPOSITIONS.

Proposition 1.

Two unequal magnitudes being set out, if from the greater there be subtracted a magnitude greater than its half, and from that which is left a magnitude greater than its half, and if this process be repeated continually, there will be left some magnitude which will be less than the lesser magnitude set out.

Let AB, C be two unequal magnitudes of which AB is the greater:

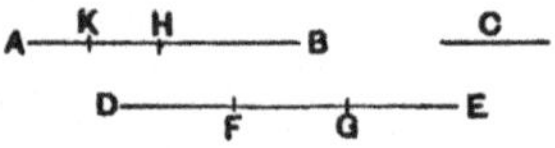

I say that, if from AB there be subtracted a magnitude greater than its half, and from that which is left a magnitude greater than its half, and if this process be repeated continually, there will be left some magnitude which will be less than the magnitude C.

For C if multiplied will sometime be greater than AB. [cf. v. Def. 4]

Let it be multiplied, and let DE be a multiple of C, and greater than AB;

let DE be divided into the parts DF, FG, GE equal to C,

from AB let there be subtracted BH greater than its half,

and, from AH, HK greater than its half,

and let this process be repeated continually until the divisions in AB are equal in multitude with the divisions in DE.

Let, then, AK, KH, HB be divisions which are equal in multitude with DF, FG, GE.

Now, since DE is greater than AB,

and from DE there has been subtracted EG less than its half,

and, from AB, BH greater than its half,

therefore the remainder GD is greater than the remainder HA.

And, since GD is greater than HA,
and there has been subtracted, from GD, the half GF,
and, from HA, HK greater than its half,
therefore the remainder DF is greater than the remainder AK.

But DF is equal to C;
therefore C is also greater than AK.

Therefore AK is less than C.

Therefore there is left of the magnitude AB the magnitude AK which is less than the lesser magnitude set out, namely C.

Q. E. D.

And the theorem can be similarly proved even if the parts subtracted be halves.

This proposition will be remembered because it is the lemma required in Euclid's proof of XII. 2 to the effect that circles are to one another as the squares on their diameters. Some writers appear to be under the impression that XII. 2 and the other propositions in Book XII. in which the method of exhaustion is used are the only places where Euclid makes use of X. 1; and it is commonly remarked that X. 1 might just as well have been deferred till the beginning of Book XII. Even Cantor (*Gesch. d. Math.* I_3, p. 269) remarks that "Euclid draws no inference from it [X. 1], not even that which we should more than anything else expect, namely that, if two magnitudes are incommensurable, we can always form a magnitude commensurable with the first which shall differ from the second magnitude by as little as we please." But, so far from making no use of X. 1 before XII. 2, Euclid actually uses it in the very next proposition, X. 2. This being so, as the next note will show, it follows that, since X. 2 gives the criterion for the incommensurability of two magnitudes (a very necessary preliminary to the study of incommensurables), X. 1 comes exactly where it should be.

Euclid uses X. 1 to prove not only XII. 2 but XII. 5 (that pyramids with the same height and triangular bases are to one another as their bases), by means of which he proves (XII. 7 and Por.) that any pyramid is a third part of the prism which has the same base and equal height, and XII. 10 (that any cone is a third part of the cylinder which has the same base and equal height), besides other similar propositions. Now XII. 7 Por. and XII. 10 are theorems specifically attributed to Eudoxus by Archimedes (*On the Sphere and Cylinder*, Preface), who says in another place (*Quadrature of the Parabola*, Preface) that the first of the two, and the theorem that circles are to one another as the squares on their diameters, were proved by means of a certain lemma which he states as follows: "Of unequal lines, unequal surfaces, or unequal solids, the greater exceeds the less by such a magnitude as is capable, if added [continually] to itself, of exceeding any magnitude of those which are comparable with one another," i.e. of magnitudes of the same kind as the original magnitudes. Archimedes also says (*loc. cit.*) that the second of the two theorems which he attributes to Eudoxus (Eucl. XII. 10) was proved by means of "a lemma similar to the aforesaid." The lemma stated thus by Archimedes is decidedly different from X. 1, which, however, Archimedes himself uses several times, while he refers to the use of it in XII. 2 (*On the Sphere and Cylinder*, I. 6). As I have before suggested (*The Works of Archimedes*, p. xlviii), the apparent difficulty caused by the mention of *two* lemmas in connexion with the theorem of Eucl. XII. 2 may be explained by reference to the proof of X. 1. Euclid there takes the lesser magnitude and says that it is possible, by multiplying it, to make it some time exceed the greater, and this statement he clearly bases on the 4th definition of Book V., to the effect that "magnitudes are said to bear a ratio to one another which can, if multiplied, exceed one another." Since then the smaller magnitude in X. 1 may be regarded as the difference between some two unequal magnitudes, it is clear that the lemma stated by Archimedes is in substance used to prove the lemma in X. 1, which appears to play so much larger a part in the investigations of quadrature and cubature which have come down to us.

Besides being employed in Eucl. X. 1, the "Axiom of Archimedes" appears in Aristotle, who also practically quotes the result of X. 1 itself. Thus he says, *Physics* VIII. 10, 266 b 2, "By continually adding to a finite (magnitude) I shall exceed any definite (magnitude), and similarly by continually subtracting from it I shall arrive at something less than it," and *ibid.* III. 7, 207 b 10 "For bisections of a magnitude are endless." It is thus somewhat misleading to use the term "Archimedes' Axiom" for the "lemma" quoted by him, since he makes no claim to be the discoverer of it, and it was obviously much earlier.

Stolz (quoted by G. Vitali in *Questioni riguardanti la geometria elementare*, pp. 91—2) showed how to prove the so-called Axiom or Postulate of Archimedes by means of the Postulate of Dedekind, thus. Suppose the two magnitudes to be straight lines. It is required to prove that, *given two straight lines, there always exists a multiple of the smaller which is greater than the other.*

Let the straight lines be so placed that they have a common extremity and the smaller lies along the other on the same side of the common extremity.

If AC be the greater and AB the smaller, we have to prove that there exists an integral number n such that $n \cdot AB > AC$.

Suppose that this is not true but that there are some points, like B, not coincident with the extremity A, and such that, n being any integer however great, $n \cdot AB < AC$; and we have to prove that this assumption leads to an absurdity.

The points of AC may be regarded as distributed into two "parts," namely

(1) points H for which there exists no integer n such that $n \cdot AH > AC$,

(2) points K for which an integer n does exist such that $n \cdot AK > AC$.

This division into parts satisfies the conditions for the application of Dedekind's Postulate, and therefore there exists a point M such that the points of AM belong to the first part and those of MC to the second part.

Take now a point Y on MC such that $MY < AM$. The middle point (X) of AY will fall between A and M and will therefore belong to the first part; but, since there exists an integer n such that $n \cdot AY > AC$, it follows that $2n \cdot AX > AC$: which is contrary to the hypothesis.

PROPOSITION 2.

If, when the less of two unequal magnitudes is continually subtracted in turn from the greater, that which is left never measures the one before it, the magnitudes will be incommensurable.

For, there being two unequal magnitudes AB, CD, and AB being the less, when the less is continually subtracted in turn from the greater, let that which is left over never measure the one before it;

I say that the magnitudes AB, CD are incommensurable.

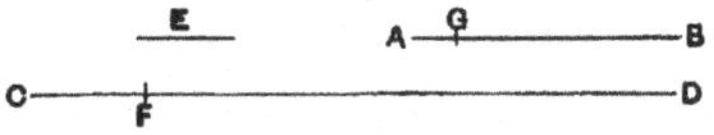

For, if they are commensurable, some magnitude will measure them.

Let a magnitude measure them, if possible, and let it be E;

let AB, measuring FD, leave CF less than itself,

let CF measuring BG, leave AG less than itself,

and let this process be repeated continually, until there is left some magnitude which is less than E.

Suppose this done, and let there be left AG less than E.

Then, since E measures AB,

while AB measures DF,

therefore E will also measure FD.

But it measures the whole CD also;

therefore it will also measure the remainder CF.

But CF measures BG;

therefore E also measures BG.

But it measures the whole AB also;

therefore it will also measure the remainder AG, the greater the less:

which is impossible.

Therefore no magnitude will measure the magnitudes AB, CD;

therefore the magnitudes AB, CD are incommensurable.

[X. Def. 1]

Therefore etc.

This proposition states the test for incommensurable magnitudes, founded on the usual operation for finding the greatest common measure. The sign of the incommensurability of two magnitudes is that this operation never comes to an end, while the successive remainders become smaller and smaller until they are less than any assigned magnitude.

Observe that Euclid says "let this process be repeated continually until there is left some magnitude which is less than E." Here he evidently assumes that the process *will* some time produce a remainder less than any assigned magnitude E. Now this is by no means self-evident, and yet Heiberg (though so careful to supply references) and Lorenz do not refer to the basis of the assumption, which is in reality x. 1, as Billingsley and Williamson were shrewd enough to see. The fact is that, if we set off a smaller magnitude once or oftener along a greater which it does not exactly measure, until the remainder is less than the smaller magnitude, we take away from the greater *more than its half*. Thus, in the figure, FD is more than the half of CD, and BG more than the half of AB. If we continued the process, AG marked off along CF as many times as possible would cut off more than its half; next, more than half AG would be cut off, and so on. Hence along CD, AB *alternately* the process would cut off more than half, then more than half the remainder and so on, so that on *both* lines we should ultimately arrive at a remainder less than any assigned length.

The method of finding the greatest common measure exhibited in this proposition and the next is of course again the same as that which we use and which may be shown thus:

$$\begin{array}{l} b)\,a\,(p \\ \quad \underline{pb} \\ \quad c)\,b\,(q \\ \qquad \underline{qc} \\ \qquad d)\,c\,(r \\ \qquad\quad \underline{rd} \\ \qquad\quad e \end{array}$$

The proof too is the same as ours, taking just the same form, as shown in the notes to the similar propositions VII. 1, 2 above. In the present case the hypothesis is that the process never stops, and it is required to prove that a, b cannot in that case have any common measure, as f. For suppose that f is a common measure, and suppose the process to be continued until the remainder e, say, is less than f.

Then, since f measures a, b, it measures $a - pb$, or c.

Since f measures b, c, it measures $b - qc$, or d; and, since f measures c, d, it measures $c - rd$, or e: which is impossible, since $e < f$.

Euclid assumes as axiomatic that, if f measures a, b, it measures $ma \pm nb$.

In practice, of course, it is often unnecessary to carry the process far in order to see that it will never stop, and consequently that the magnitudes are incommensurable. A good instance is pointed out by Allman (*Greek Geometry from Thales to Euclid*, pp. 42, 137—8). Euclid proves in XIII. 5 that, if AB be cut in extreme and mean ratio at C, and if DA equal to AC be added, then DB is also cut in extreme and mean ratio at A. This is indeed

obvious from the proof of II. 11. It follows conversely that, if BD is cut into extreme and mean ratio at A, and AC, equal to the lesser segment AD, be subtracted from the greater AB, AB is similarly divided at C. We can then mark off from AC a portion equal to CB, and AC will then be similarly divided, and so on. Now the greater segment in a line thus divided is greater than half the line, but it follows from XIII. 3 that it is less than twice the lesser segment, i.e. the lesser segment can never be marked off more than *once* from the greater. Our process of marking off the lesser segment from the greater continually is thus exactly that of finding the greatest common measure. If, therefore, the segments were commensurable, the process would stop. But it clearly does not; therefore the segments are incommensurable.

Allman expresses the opinion that it was rather in connexion with the line cut in extreme and mean ratio than with reference to the diagonal and side of a square that Pythagoras discovered incommensurable magnitudes. But the evidence seems to put it beyond doubt that the Pythagoreans did discover the incommensurability of $\sqrt{2}$ and devoted much attention to this particular case. The view of Allman does not therefore commend itself to me, though it is likely enough that the Pythagoreans were aware of the incommensurability of the segments of a line cut in extreme and mean ratio. At all events the Pythagoreans could hardly have carried their investigations into the incommensurability of the segments of this line very far, since Theaetetus is said to have made the first classification of irrationals, and to him is also, with reasonable probability, attributed the substance of the first part of Eucl. XIII., in the sixth proposition of which occurs the proof that the segments of a rational straight line cut into extreme and mean ratio are *apotomes*.

Again, the incommensurability of $\sqrt{2}$ can be proved by a method practically equivalent to that of X. 2, and without carrying the process very far. This method is given in Chrystal's *Text-book of Algebra* (I. p. 270). Let d, a be the diagonal and side respectively of a square $ABCD$. Mark off AF along AC equal to a. Draw FE at right angles to AC meeting BC in E.

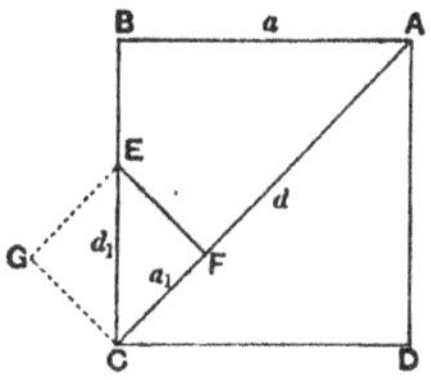

It is easily proved that

$$BE = EF = FC,$$

$$CF = AC - AB = d - a \quad \ldots\ldots\ldots(1).$$

$$CE = CB - CF = a - (d - a)$$

$$= 2a - d \quad \ldots\ldots\ldots(2).$$

Suppose, if possible, that d, a are commensurable. If d, a are both commensurably expressible in terms of any finite unit, each must be an integral multiple of a certain finite unit.

But from (1) it follows that CF, and from (2) it follows that CE, is an integral multiple of the same unit.

And CF, CE are the side and diagonal of a square $CFEG$, the side of which is *less than half the side of the original square*. If a_1, d_1 are the side and diagonal of this square,

$$\left.\begin{array}{l} a_1 = d - a \\ d_1 = 2a - d \end{array}\right\}.$$

Similarly we can form a square with side a_2 and diagonal d_2 which are less than half a_1, d_1 respectively, and a_2, d_2 must be integral multiples of the same unit, where

$$a_2 = d_1 - a_1,$$

$$d_2 = 2a_1 - d_1;$$

and this process may be continued indefinitely until (X. 1) we have a square as small as we please, the side and diagonal of which are integral multiples of a finite unit: which is absurd.

Therefore *a*, *d* are incommensurable.

It will be observed that this method is the opposite of that shown in the Pythagorean series of side- and diagonal-numbers, the squares being successively smaller instead of larger.

Proposition 3.

Given two commensurable magnitudes, to find their greatest common measure.

Let the two given commensurable magnitudes be *AB*, *CD* of which *AB* is the less;

thus it is required to find the greatest common measure of *AB*, *CD*.

Now the magnitude *AB* either measures *CD* or it does not.

If then it measures it—and it measures itself also—*AB* is a common measure of *AB*, *CD*.

And it is manifest that it is also the greatest;

for a greater magnitude than the magnitude *AB* will not measure *AB*.

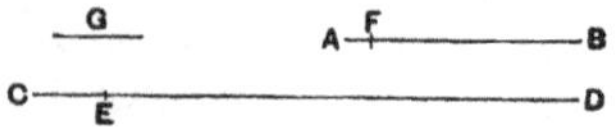

Next, let *AB* not measure *CD*.

Then, if the less be continually subtracted in turn from the greater, that which is left over will sometime measure the one before it, because *AB*, *CD* are not incommensurable; [cf. X. 2]

let *AB*, measuring *ED*, leave *EC* less than itself,

let *EC*, measuring *FB*, leave *AF* less than itself,

and let *AF* measure *CE*.

Since, then, *AF* measures *CE*,

while *CE* measures *FB*,

therefore *AF* will also measure *FB*.

But it measures itself also;

therefore *AF* will also measure the whole *AB*.

But *AB* measures *DE*;
therefore *AF* will also measure *ED*.

But it measures *CE* also;
therefore it also measures the whole *CD*.

Therefore *AF* is a common measure of *AB*, *CD*.

I say next that it is also the greatest.

For, if not, there will be some magnitude greater than *AF* which will measure *AB*, *CD*.

Let it be *G*.

Since then *G* measures *AB*,
while *AB* measures *ED*,
therefore *G* will also measure *ED*.

But it measures the whole *CD* also;
therefore *G* will also measure the remainder *CE*.

But *CE* measures *FB*;
therefore *G* will also measure *FB*.

But it measures the whole *AB* also,
and it will therefore measure the remainder *AF*, the greater the less:
which is impossible.

Therefore no magnitude greater than *AF* will measure *AB*, *CD*;
therefore *AF* is the greatest common measure of *AB*, *CD*.

Therefore the greatest common measure of the two given commensurable magnitudes *AB*, *CD* has been found.

Q. E. D.

PORISM. From this it is manifest that, if a magnitude measure two magnitudes, it will also measure their greatest common measure.

This proposition for two commensurable *magnitudes* is, *mutatis mutandis*, exactly the same as VII. 2 for numbers. We have the process

$$\begin{array}{l} b)\,a\,(p \\ \underline{pb} \\ \quad c)\,b\,(q \\ \quad \underline{qc} \\ \qquad d)\,c\,(r \\ \qquad \underline{rd} \end{array}$$

where c is equal to rd and therefore there is no remainder.

It is then proved that *d* is a common measure of *a*, *b*; and next, by a *reductio ad absurdum*, that it is the *greatest* common measure, since any common measure must measure *d*, and no magnitude greater than *d* can measure *d*. The *reductio ad absurdum* is of course one of form only.

The Porism corresponds exactly to the Porism to VII. 2.

The process of finding the greatest common measure is probably given in this Book, not only for the sake of completeness, but because in x. 5 a common measure of two magnitudes *A*, *B* is assumed and used, and therefore it is important to show that such a measure can be *found* if not already known.

PROPOSITION 4.

Given three commensurable magnitudes, to find their greatest common measure.

Let *A*, *B*, *C* be the three given commensurable magnitudes; thus it is required to find the greatest common measure of *A*, *B*, *C*.

Let the greatest common measure of the two magnitudes *A*, *B* be taken, and let it be *D*; [x. 3]

then *D* either measures *C*, or does not measure it.

First, let it measure it.

Since then *D* measures *C*,

while it also measures *A*, *B*,

therefore *D* is a common measure of *A*, *B*, *C*.

And it is manifest that it is also the greatest;

for a greater magnitude than the magnitude *D* does not measure *A*, *B*.

Next, let *D* not measure *C*.

I say first that *C*, *D* are commensurable.

For, since *A*, *B*, *C* are commensurable,

some magnitude will measure them,

and this will of course measure *A*, *B* also;

so that it will also measure the greatest common measure of *A*, *B*, namely *D*. [x. 3, Por.]

But it also measures *C*;

so that the said magnitude will measure *C*, *D*;

therefore *C*, *D* are commensurable.

Now let their greatest common measure be taken, and let it be E. [x. 3]

Since then E measures D,

while D measures A, B,

therefore E will also measure A, B.

But it measures C also;

therefore E measures A, B, C;

therefore E is a common measure of A, B, C.

I say next that it is also the greatest.

For, if possible, let there be some magnitude F greater than E, and let it measure A, B, C.

Now, since F measures A, B, C,

it will also measure A, B,

and will measure the greatest common measure of A, B. [x. 3, Por.]

But the greatest common measure of A, B is D;

therefore F measures D.

But it measures C also;

therefore F measures C, D;

therefore F will also measure the greatest common measure of C, D. [x. 3, Por.]

But that is E;

therefore F will measure E, the greater the less: which is impossible.

Therefore no magnitude greater than the magnitude E will measure A, B, C;

therefore E is the greatest common measure of A, B, C if D do not measure C,

and, if it measure it, D is itself the greatest common measure.

Therefore the greatest common measure of the three given commensurable magnitudes has been found.

PORISM. From this it is manifest that, if a magnitude measure three magnitudes, it will also measure their greatest common measure.

Similarly too, with more magnitudes, the greatest common measure can be found, and the porism can be extended.

Q. E. D.

This proposition again corresponds exactly to VII. 3 for numbers. As there Euclid thinks it necessary to prove that, a, b, c not being prime to one another, d and c are also not prime to one another, so here he thinks it necessary to prove that d, c are commensurable, as they must be since any common measure of a, b must be a measure of their greatest common measure d (x. 3, Por.).

The argument in the proof that e, the greatest common measure of d, c, is the greatest common measure of a, b, c, is the same as that in VII. 3 and x. 3.

The Porism contains the extension of the process to the case of four or more magnitudes, corresponding to Heron's remark with regard to the similar extension of VII. 3 to the case of four or more *numbers*.

PROPOSITION 5.

Commensurable magnitudes have to one another the ratio which a number has to a number.

Let A, B be commensurable magnitudes;

I say that A has to B the ratio which a number has to a number.

For, since A, B are commensurable, some magnitude will measure them.

Let it measure them, and let it be C.

And, as many times as C measures A, so many units let there be in D;

and, as many times as C measures B, so many units let there be in E.

Since then C measures A according to the units in D,

while the unit also measures D according to the units in it,

therefore the unit measures the number D the same number of times as the magnitude C measures A;

therefore, as C is to A, so is the unit to D; [VII. Def. 20]

therefore, inversely, as A is to C, so is D to the unit. [cf. v. 7, Por.]

Again, since C measures B according to the units in E,

while the unit also measures E according to the units in it, therefore the unit measures E the same number of times as C measures B;

therefore, as C is to B, so is the unit to E.

But it was also proved that,

as A is to C, so is D to the unit;

therefore, *ex aequali*,

as A is to B, so is the number D to E. [v. 22]

Therefore the commensurable magnitudes A, B have to one another the ratio which the number D has to the number E.

Q. E. D.

The argument is as follows. If a, b be commensurable magnitudes, they have some common measure c, and

$$a = mc,$$
$$b = nc,$$

where m, n are integers.

It follows that $c : a = 1 : m$ ....(1),

or, inversely, $a : c = m : 1$;

and also that $c : b = 1 : n$,

so that, *ex aequali*, $a : b = m : n$.

It will be observed that, in stating the proportion (1), Euclid is merely expressing the fact that a is the same multiple of c that m is of 1. In other words, he rests the statement on the definition of proportion in VII. Def. 20. This, however, is applicable only to four *numbers*, and c, a are not numbers but magnitudes. Hence the statement of the proportion is not legitimate unless it is proved that it is true in the sense of v. Def. 5 with regard to magnitudes in general, the numbers 1, m being *magnitudes*. Similarly with regard to the other proportions in the proposition.

There is, therefore, a hiatus. Euclid ought to have proved that magnitudes which are proportional in the sense of VII. Def. 20 are also proportional in the sense of v. Def. 5, or that the proportion of numbers is included in the proportion of magnitudes as a particular case. Simson has proved this in his Proposition C inserted in Book v. (see Vol. II. pp. 126—8). The portion of that proposition which is required here is the proof that,

if
$$\left.\begin{aligned} a &= mb \\ c &= md \end{aligned}\right\},$$

then $a : b = c : d$, in the sense of v. Def. 5.

Take any equimultiples pa, pc of a, c and any equimultiples qb, qd of b, d.

Now
$$\left.\begin{aligned} pa &= pmb \\ pc &= pmd \end{aligned}\right\}.$$

But, according as $pmb >=< qb$, $pmd >=< qd$.

Therefore, according as $pa >=< qb$, $pa >=< qd$.

And pa, pc are *any* equimultiples of a, c, and qb, qd *any* equimultiples of b, d.

Therefore $a : b = c : d$. [v. Def. 5.]

Proposition 6.

If two magnitudes have to one another the ratio which a number has to a number, the magnitudes will be commensurable.

For let the two magnitudes A, B have to one another the ratio which the number D has to the number E;

I say that the magnitudes A, B are commensurable.

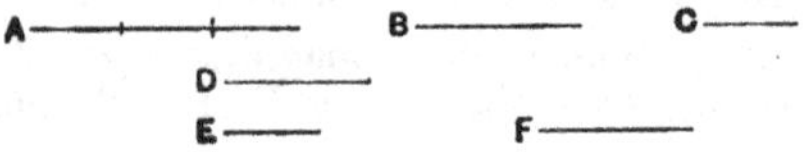

For let A be divided into as many equal parts as there are units in D,

and let C be equal to one of them;

and let F be made up of as many magnitudes equal to C as there are units in E.

Since then there are in A as many magnitudes equal to C as there are units in D,

whatever part the unit is of D, the same part is C of A also;

therefore, as C is to A, so is the unit to D. [VII. Def. 20]

But the unit measures the number D;

therefore C also measures A.

And since, as C is to A, so is the unit to D,

therefore, inversely, as A is to C, so is the number D to the unit. [cf. v. 7, Por.]

Again, since there are in F as many magnitudes equal to C as there are units in E,

therefore, as C is to F, so is the unit to E. [VII. Def. 20]

But it was also proved that,

as A is to C, so is D to the unit;

therefore, *ex aequali*, as A is to F, so is D to E. [v. 22]

But, as D is to E, so is A to B;

therefore also, as A is to B, so is it to F also. [v. 11]

Therefore A has the same ratio to each of the magnitudes B, F;

therefore B is equal to F. [v. 9]

But C measures F;

therefore it measures B also.

Further it measures A also;

therefore C measures A, B.

Therefore A is commensurable with B.

Therefore etc.

PORISM. From this it is manifest that, if there be two numbers, as D, E, and a straight line, as A, it is possible to make a straight line [F] such that the given straight line is to it as the number D is to the number E.

And, if a mean proportional be also taken between A, F, as B,

as A is to F, so will the square on A be to the square on B, that is, as the first is to the third, so is the figure on the first to that which is similar and similarly described on the second. [VI. 19, Por.]

But, as A is to F, so is the number D to the number E; therefore it has been contrived that, as the number D is to the number E, so also is the figure on the straight line A to the figure on the straight line B. Q. E. D.

15. **But the unit measures the number D; therefore C also measures A.** These words are redundant, though they are apparently found in all the MSS.

The same link to connect the proportion of numbers with the proportion of magnitudes as was necessary in the last proposition is necessary here. This being premised, the argument is as follows.

Suppose $$a : b = m : n,$$
where m, n are (integral) numbers.

Divide a into m parts, each equal to c, say,
so that $$a = mc.$$

Now take d such that $$d = nc.$$

Therefore we have $$a : c = m : 1,$$
and $$c : d = 1 : n,$$
so that, *ex aequali*, $$a : d = m : n$$
$$= a : b, \text{ by hypothesis.}$$

Therefore $b = d = nc$,

so that c measures b n times, and a, b are commensurable.

The Porism is often used in the later propositions. It follows (1) that, if a be a given straight line, and m, n any numbers, a straight line x can be found such that
$$a : x = m : n.$$

(2) We can find a straight line y such that
$$a^2 : y^2 = m : n.$$

For we have only to take y, a mean proportional between a and x, as previously found, in which case a, y, x are in continued proportion and [V. Def. 9]
$$a^2 : y^2 = a : x$$
$$= m : n.$$

Proposition 7.

Incommensurable magnitudes have not to one another the ratio which a number has to a number.

Let A, B be incommensurable magnitudes;
I say that A has not to B the ratio which a number has to a number.

For, if A has to B the ratio which a number has to a number, A will be commensurable with B. [x. 6]

But it is not;

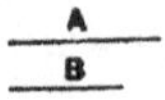

therefore A has not to B the ratio which a number has to a number.

Therefore etc.

Proposition 8.

If two magnitudes have not to one another the ratio which a number has to a number, the magnitudes will be incommensurable.

For let the two magnitudes A, B not have to one another the ratio which a number has to a number;

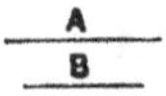

I say that the magnitudes A, B are incommensurable.

For, if they are commensurable, A will have to B the ratio which a number has to a number. [x. 5]

But it has not;
therefore the magnitudes A, B are incommensurable.

Therefore etc.

Proposition 9.

The squares on straight lines commensurable in length have to one another the ratio which a square number has to a square number; and squares which have to one another the ratio which a square number has to a square number will also have their sides commensurable in length. But the squares on straight lines incommensurable in length have not to one another the ratio which a square number has to a square number; and squares which have not to one another the ratio which a square number has to a square number will not have their sides commensurable in length either.

For let A, B be commensurable in length;

I say that the square on A has to the square on B the ratio which a square number has to a square number.

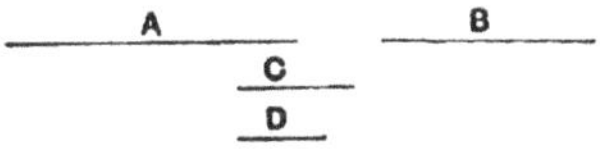

For, since A is commensurable in length with B, therefore A has to B the ratio which a number has to a number. [x. 5]

Let it have to it the ratio which C has to D.

Since then, as A is to B, so is C to D,

while the ratio of the square on A to the square on B is duplicate of the ratio of A to B,

for similar figures are in the duplicate ratio of their corresponding sides; [vi. 20, Por.]

and the ratio of the square on C to the square on D is duplicate of the ratio of C to D,

for between two square numbers there is one mean proportional number, and the square number has to the square number the ratio duplicate of that which the side has to the side; [viii. 11]

therefore also, as the square on A is to the square on B, so is the square on C to the square on D.

Next, as the square on A is to the square on B, so let the square on C be to the square on D;

I say that A is commensurable in length with B.

For since, as the square on A is to the square on B, so is the square on C to the square on D,

while the ratio of the square on A to the square on B is duplicate of the ratio of A to B,

and the ratio of the square on C to the square on D is duplicate of the ratio of C to D,

therefore also, as A is to B, so is C to D.

Therefore A has to B the ratio which the number C has to the number D;

therefore A is commensurable in length with B. [x. 6]

Next, let A be incommensurable in length with B;

I say that the square on A has not to the square on B the ratio which a square number has to a square number.

For, if the square on A has to the square on B the ratio which a square number has to a square number, A will be commensurable with B.

But it is not;

therefore the square on A has not to the square on B the ratio which a square number has to a square number.

Again, let the square on A not have to the square on B the ratio which a square number has to a square number;

I say that A is incommensurable in length with B.

For, if A is commensurable with B, the square on A will have to the square on B the ratio which a square number has to a square number.

But it has not;

therefore A is not commensurable in length with B.

Therefore etc.

PORISM. And it is manifest from what has been proved that straight lines commensurable in length are always commensurable in square also, but those commensurable in square are not always commensurable in length also.

[LEMMA. It has been proved in the arithmetical books that similar plane numbers have to one another the ratio which a square number has to a square number, [VIII. 26]

and that, if two numbers have to one another the ratio which a square number has to a square number, they are similar plane numbers. [Converse of VIII. 26]

And it is manifest from these propositions that numbers which are not similar plane numbers, that is, those which have not their sides proportional, have not to one another the ratio which a square number has to a square number.

For, if they have, they will be similar plane numbers: which is contrary to the hypothesis.

Therefore numbers which are not similar plane numbers have not to one another the ratio which a square number has to a square number.]

A scholium to this proposition (Schol. x. No. 62) says categorically that the theorem proved in it was the discovery of Theaetetus.

If a, b be straight lines, and

$$a : b = m : n,$$

where m, n are numbers,

then

$$a^2 : b^2 = m^2 : n^2;$$

and conversely.

This inference, which looks so easy when thus symbolically expressed, was by no means so easy for Euclid owing to the fact that a, b are straight lines, and m, n numbers. He has to pass from $a : b$ to $a^2 : b^2$ by means of VI. 20, Por. through the duplicate ratio; the square on a is to the square on b in the duplicate ratio of the corresponding sides a, b. On the other hand, m, n being *numbers*, it is VIII. 11 which has to be used to show that $m^2 : n^2$ is the ratio duplicate of $m : n$.

Then, in order to establish his result, Euclid *assumes* that, *if two ratios are equal, the ratios which are their duplicates are also equal.* This is nowhere proved in Euclid, but it is an easy inference from V. 22, as shown in my note on VI. 22.

The converse has to be established in the same careful way, and Euclid *assumes* that *ratios the duplicates of which are equal are themselves equal.* This is much more troublesome to prove than the converse; for proofs I refer to the same note on VI. 22.

The second part of the theorem, deduced by *reductio ad absurdum* from the first, requires no remark.

In the Greek text there is an addition to the Porism which Heiberg brackets as superfluous and not in Euclid's manner. It consists (1) of a sort of proof, or rather explanation, of the Porism and (2) of a statement and explanation to the effect that straight lines incommensurable in length are not necessarily incommensurable in square also, and that straight lines incommensurable in square are, on the other hand, always incommensurable in length also.

The Lemma gives expressions for two numbers which have to one another the ratio of a square number to a square number. *Similar plane numbers* are of the form $pm \cdot pn$ and $qm \cdot qn$, or mnp^2 and mnq^2, the ratio of which is of course the ratio of p^2 to q^2.

The converse theorem that, if two numbers have to one another the ratio of a square number to a square number, the numbers are similar plane numbers is not, as a matter of fact, proved in the arithmetical Books. It is the converse of VIII. 26 and is used in IX. 10. Heron gave it (see note on VIII. 27 above).

Heiberg however gives strong reason for supposing the Lemma to be an interpolation. It has reference to the next proposition, X. 10, and, as we shall see, there are so many objections to X. 10 that it can hardly be accepted as genuine. Moreover there is no reason why, in the Lemma itself, numbers which are *not* similar plane numbers should be brought in as they are.

[Proposition 10.

To find two straight lines incommensurable, the one in length only, and the other in square also, with an assigned straight line.

Let A be the assigned straight line;

thus it is required to find two straight lines incommensurable, the one in length only, and the other in square also, with A.

Let two numbers B, C be set out which have not to one

another the ratio which a square number has to a square number, that is, which are not similar plane numbers;

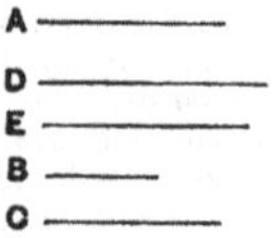

and let it be contrived that,

as B is to C, so is the square on A to the square on D

—for we have learnt how to do this— [x. 6, Por.]

therefore the square on A is commensurable with the square on D. [x. 6]

And, since B has not to C the ratio which a square number has to a square number,
therefore neither has the square on A to the square on D the ratio which a square number has to a square number;
therefore A is incommensurable in length with D. [x. 9]

Let E be taken a mean proportional between A, D;
therefore, as A is to D, so is the square on A to the square on E. [v. Def. 9]

But A is incommensurable in length with D;
therefore the square on A is also incommensurable with the square on E; [x. 11]
therefore A is incommensurable in square with E.

Therefore two straight lines D, E have been found incommensurable, D in length only, and E in square and of course in length also, with the assigned straight line A.]

It would appear as though this proposition was intended to supply a justification for the statement in x. Def. 3 that *it is proved* that there are an infinite number of straight lines (*a*) incommensurable in length only, or commensurable in square only, and (*b*) incommensurable in square, with any given straight line.

But in truth the proposition could well be dispensed with; and the positive objections to its genuineness are considerable.

In the first place, it depends on the following proposition, x. 11; for the last step concludes that, since

$$a^2 : y^2 = a : x,$$

and a, x are incommensurable in length, therefore a^2, y^2 are incommensurable. But Euclid never commits the irregularity of proving a theorem by means of a later one. Gregory sought to get over the difficulty by putting x. 10 after x. 11; but of course, if the order were so inverted, the Lemma would still be in the wrong place.

Further, the expression ἐμάθομεν γάρ, "for we have learnt (how to do this)," is not in Euclid's manner and betrays the hand of a learner (though the same expression is found in the *Sectio Canonis* of Euclid, where the reference is to the *Elements*).

Lastly the manuscript P has the number 10, in the first hand, at the top of x. 11, from which it may perhaps be concluded that x. 10 had at first no number.

It seems best therefore to reject as spurious both the Lemma and x. 10.

The argument of x. 10 is simple. If a be a given straight line and m, n numbers which have not to one another the ratio of square to square, take x such that

$$a^2 : x^2 = m : n, \qquad \text{[x. 6, Por.]}$$

whence a, x are incommensurable in length. [x. 9]

Then take y a mean proportional between a, x, whence

$$a^2 : y^2 = a : x \qquad \text{[v. Def. 9]}$$

$$[= \sqrt{m} : \sqrt{n}],$$

and x is incommensurable in length only, while y is incommensurable in square as well as in length, with a.

Proposition 11.

If four magnitudes be proportional, and the first be commensurable with the second, the third will also be commensurable with the fourth; and, if the first be incommensurable with the second, the third will also be incommensurable with the fourth.

Let A, B, C, D be four magnitudes in proportion, so that, as A is to B, so is C to D,

and let A be commensurable with B;

I say that C will also be commensurable with D.

For, since A is commensurable with B,

therefore A has to B the ratio which a number has to a number. [x. 5]

And, as A is to B, so is C to D;

therefore C also has to D the ratio which a number has to a number;

therefore C is commensurable with D. [x. 6]

Next, let A be incommensurable with B;

I say that C will also be incommensurable with D.

For, since A is incommensurable with B,

therefore A has not to B the ratio which a number has to a number. [x. 7]

And, as A is to B, so is C to D;
therefore neither has C to D the ratio which a number has to a number;
therefore C is incommensurable with D. [x. 8]

Therefore etc.

I shall henceforth, for the sake of brevity, use symbols for the terms "commensurable (with)" and "incommensurable (with)" according to the varieties described in x. Deff. 1—4. The symbols are taken from Lorenz and seem convenient.

Commensurable and *commensurable with*, in relation to areas, and *commensurable in length* and *commensurable in length with*, in relation to straight lines, will be denoted by $\frown$.

Commensurable in square only or *commensurable in square only with* (terms applicable only to straight lines) will be denoted by $\frown\!\!-$.

Incommensurable (*with*), of areas, and *incommensurable* (*with*), of straight lines will be denoted by $\smile$.

Incommensurable in square (*with*) (a term applicable to straight lines only) will be denoted by $\smile\!\!-$.

Suppose a, b, c, d to be four magnitudes such that

$$a : b = c : d.$$

Then (1), if $a \frown b$, $a : b = m : n$, where m, n are integers, [x. 5]
whence $c : d = m : n$,
and therefore $c \frown d$. [x. 6]

(2) If $a \smile b$, $a : b \neq m : n$, [x. 7]
so that $c : d \neq m : n$,
whence $c \smile d$. [x. 8]

Proposition 12.

Magnitudes commensurable with the same magnitude are commensurable with one another also.

For let each of the magnitudes A, B be commensurable with C;
I say that A is also commensurable with B.

For, since A is commensurable with C,
therefore A has to C the ratio which a number has to a number. [x. 5]

Let it have the ratio which D has to E.

Again, since C is commensurable with B,

therefore C has to B the ratio which a number has to a number. [X. 5]

Let it have the ratio which F has to G.

And, given any number of ratios we please, namely the ratio which D has to E and that which F has to G,

let the numbers H, K, L be taken continuously in the given ratios; [cf. VIII. 4]

so that, as D is to E, so is H to K,

and, as F is to G, so is K to L.

Since, then, as A is to C, so is D to E,

while, as D is to E, so is H to K,

therefore also, as A is to C, so is H to K. [V. 11]

Again, since, as C is to B, so is F to G,

while, as F is to G, so is K to L,

therefore also, as C is to B, so is K to L. [V. 11]

But also, as A is to C, so is H to K;

therefore, *ex aequali*, as A is to B, so is H to L. [V. 22]

Therefore A has to B the ratio which a number has to a number;

therefore A is commensurable with B. [X. 6]

Therefore etc.

Q. E. D.

We have merely to go through the process of compounding two ratios in numbers.

Suppose	a, b each $\frown c$.	
Therefore	$a : c = m : n$, say,	[X. 5]
	$c : b = p : q$, say.	
Now	$m : n = mp : np$,	
and	$p : q = np : nq$.	
Therefore	$a : c = mp : np$,	
	$c : b = np : nq$,	
whence, *ex aequali*,	$a : b = mp : nq$,	
so that	$a \frown b$.	[X. 6]

Proposition 13.

If two magnitudes be commensurable, and the one of them be incommensurable with any magnitude, the remaining one will also be incommensurable with the same.

Let A, B be two commensurable magnitudes, and let one of them, A, be incommensurable with any other magnitude C;

I say that the remaining one, B, will also be incommensurable with C.

For, if B is commensurable with C, while A is also commensurable with B, A is also commensurable with C. [x. 12]

But it is also incommensurable with it: which is impossible.

Therefore B is not commensurable with C; therefore it is incommensurable with it.

Therefore etc.

Lemma.

Given two unequal straight lines, to find by what square the square on the greater is greater than the square on the less.

Let AB, C be the given two unequal straight lines, and let AB be the greater of them;

thus it is required to find by what square the square on AB is greater than the square on C.

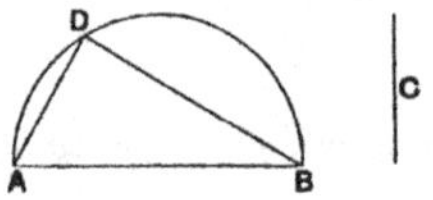

Let the semicircle ADB be described on AB, and let AD be fitted into it equal to C; [iv. 1]

let DB be joined.

It is then manifest that the angle ADB is right, [iii. 31] and that the square on AB is greater than the square on AD, that is, C, by the square on DB. [i. 47]

Similarly also, if two straight lines be given, the straight line the square on which is equal to the sum of the squares on them is found in this manner.

Let AD, DB be the given two straight lines, and let it be required to find the straight line the square on which is equal to the sum of the squares on them.

Let them be placed so as to contain a right angle, that formed by AD, DB;

and let AB be joined.

It is again manifest that the straight line the square on which is equal to the sum of the squares on AD, DB is AB. [I. 47]

Q. E. D.

The lemma gives an obvious method of finding a straight line (c) equal to $\sqrt{a^2 \mp b^2}$, where a, b are given straight lines of which a is the greater.

Proposition 14.

If four straight lines be proportional, and the square on the first be greater than the square on the second by the square on a straight line commensurable with the first, the square on the third will also be greater than the square on the fourth by the square on a straight line commensurable with the third.

And, if the square on the first be greater than the square on the second by the square on a straight line incommensurable with the first, the square on the third will also be greater than the square on the fourth by the square on a straight line incommensurable with the third.

Let A, B, C, D be four straight lines in proportion, so that, as A is to B, so is C to D;

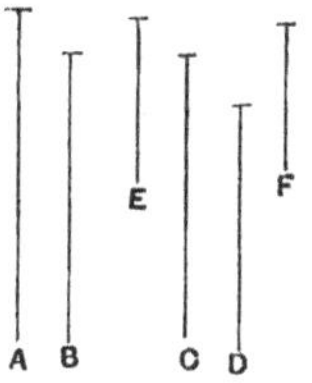

and let the square on A be greater than the square on B by the square on E, and let the square on C be greater than the square on D by the square on F;

I say that, if A is commensurable with E, C is also commensurable with F,

and, if A is incommensurable with E, C is also incommensurable with F.

For since, as A is to B, so is C to D,

therefore also, as the square on A is to the square on B, so is the square on C to the square on D. [VI. 22]

But the squares on E, B are equal to the square on A, and the squares on D, F are equal to the square on C.

Therefore, as the squares on E, B are to the square on B, so are the squares on D, F to the square on D;

therefore, *separando*, as the square on E is to the square on B, so is the square on F to the square on D; [v. 17]

therefore also, as E is to B, so is F to D; [vi. 22]

therefore, inversely, as B is to E, so is D to F.

But, as A is to B, so also is C to D;

therefore, *ex aequali*, as A is to E, so is C to F. [v. 22]

Therefore, if A is commensurable with E, C is also commensurable with F,

and, if A is incommensurable with E, C is also incommensurable with F. [x. 11]

Therefore etc.

3, 5, 8, 10. Euclid speaks of the square on the first (third) being greater than the square on the second (fourth) by the square on a straight line commensurable (incommensurable) "with *itself* (ἑαυτῇ)," and similarly in all like phrases throughout the Book. For clearness' sake I substitute "the first," "the third," or whatever it may be, for "itself" in these cases.

Suppose a, b, c, d to be straight lines such that

$$a : b = c : d \quad \text{.............................(1).}$$

It follows [vi. 22] that $\quad a^2 : b^2 = c^2 : d^2 \quad$ ....(2).

In order to prove that, *convertendo*,

$$a^2 : (a^2 - b^2) = c^2 : (c^2 - d^2)$$

Euclid has to use a somewhat roundabout method owing to the absence of a *convertendo* proposition in his Book v. (which omission Simson supplied by his Prop. E).

It follows from (2) that

$$\{(a^2 - b^2) + b^2\} : b^2 = \{(c^2 - d^2) + d^2\} : d^2,$$

whence, *separando*, $\quad (a^2 - b^2) : b^2 = (c^2 - d^2) : d^2,$ [v. 17]

and, inversely, $\quad b^2 : (a^2 - b^2) = d^2 : (c^2 - d^2).$

From this and (2), *ex aequali*,

$$a^2 : (a^2 - b^2) = c^2 : (c^2 - d^2). \quad \text{[v. 22]}$$

Hence $\quad a : \sqrt{a^2 - b^2} = c : \sqrt{c^2 - d^2}.$ [vi. 22]

According therefore as $\quad a \frown \text{ or } \smile \sqrt{a^2 - b^2},$

$$c \frown \text{ or } \smile \sqrt{c^2 - d^2}. \quad \text{[x. 11]}$$

If $a \frown \sqrt{a^2 - b^2}$, we may put $\sqrt{a^2 - b^2} = ka$, where k is of the form m/n and m, n are integers. And if $\sqrt{a^2 - b^2} = ka$, it follows in this case that $\sqrt{c^2 - d^2} = kc$.

PROPOSITION 15.

If two commensurable magnitudes be added together, the whole will also be commensurable with each of them; and, if the whole be commensurable with one of them, the original magnitudes will also be commensurable.

For let the two commensurable magnitudes AB, BC be added together;

I say that the whole AC is also commensurable with each of the magnitudes AB, BC.

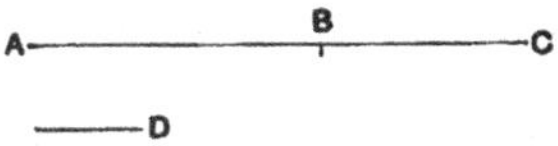

For, since AB, BC are commensurable, some magnitude will measure them.

Let it measure them, and let it be D.

Since then D measures AB, BC, it will also measure the whole AC.

But it measures AB, BC also;

therefore D measures AB, BC, AC;

therefore AC is commensurable with each of the magnitudes AB, BC. [X. Def. 1]

Next, let AC be commensurable with AB;

I say that AB, BC are also commensurable.

For, since AC, AB are commensurable, some magnitude will measure them.

Let it measure them, and let it be D.

Since then D measures CA, AB, it will also measure the remainder BC.

But it measures AB also;

therefore D will measure AB, BC;

therefore AB, BC are commensurable. [X. Def. 1]

Therefore etc.

(1) If a, b be any two commensurable magnitudes, they are of the form mc, nc, where c is a common measure of a, b and m, n some integers.

It follows that $$a + b = (m + n)c;$$
therefore $(a + b)$, being measured by c, is commensurable with both a and b.

(2) If $a + b$ is commensurable with either a or b, say a, we may put $a + b = mc$, $a = nc$, where c is a common measure of $(a + b)$, a, and m, n are integers.

Subtracting, we have $$b = (m - n)c,$$
whence $b \frown a$.

Proposition 16.

If two incommensurable magnitudes be added together, the whole will also be incommensurable with each of them; and, if the whole be incommensurable with one of them, the original magnitudes will also be incommensurable.

For let the two incommensurable magnitudes *AB*, *BC* be added together;

I say that the whole *AC* is also incommensurable with each of the magnitudes *AB*, *BC*.

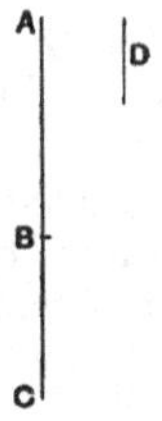

For, if *CA*, *AB* are not incommensurable, some magnitude will measure them.

Let it measure them, if possible, and let it be *D*.

Since then *D* measures *CA*, *AB*,
therefore it will also measure the remainder *BC*.

But it measures *AB* also;
therefore *D* measures *AB*, *BC*.

Therefore *AB*, *BC* are commensurable;
but they were also, by hypothesis, incommensurable:
which is impossible.

Therefore no magnitude will measure *CA*, *AB*;
therefore *CA*, *AB* are incommensurable. [x. Def. 1]

Similarly we can prove that *AC*, *CB* are also incommensurable.

Therefore *AC* is incommensurable with each of the magnitudes *AB*, *BC*.

Next, let *AC* be incommensurable with one of the magnitudes *AB*, *BC*.

First, let it be incommensurable with *AB*;
I say that *AB*, *BC* are also incommensurable.

For, if they are commensurable, some magnitude will measure them.

Let it measure them, and let it be *D*.

Since then *D* measures *AB*, *BC*,
therefore it will also measure the whole *AC*.

But it measures *AB* also:
therefore *D* measures *CA*, *AB*.

Therefore CA, AB are commensurable;
but they were also, by hypothesis, incommensurable:
which is impossible.

Therefore no magnitude will measure AB, BC;
therefore AB, BC are incommensurable. [x. Def. 1]

Therefore etc.

LEMMA.

If to any straight line there be applied a parallelogram deficient by a square figure, the applied parallelogram is equal to the rectangle contained by the segments of the straight line resulting from the application.

For let there be applied to the straight line AB the parallelogram AD deficient by the square figure DB;

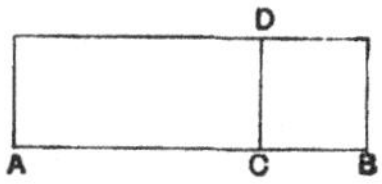

I say that AD is equal to the rectangle contained by AC, CB.

This is indeed at once manifest;
for, since DB is a square,
DC is equal to CB;
and AD is the rectangle AC, CD, that is, the rectangle AC, CB.

Therefore etc.

If a be the given straight line, and x the side of the square by which the applied rectangle is to be deficient, the rectangle is equal to $ax - x^2$, which is of course equal to $x(a - x)$. The rectangle may be written xy, where $x + y = a$. Given the area $x(a - x)$, or xy (where $x + y = a$), two different *applications* will give rectangles equal to this area, the sides of the defect being x or $a - x$ (x or y) respectively; but the second mode of expression shows that the rectangles do not differ in form but only in position.

PROPOSITION 17.

If there be two unequal straight lines, and to the greater there be applied a parallelogram equal to the fourth part of the square on the less and deficient by a square figure, and if it divide it into parts which are commensurable in length, then the square on the greater will be greater than the square on the less by the square on a straight line commensurable with the greater.

And, if the square on the greater be greater than the square on the less by the square on a straight line commensurable with the greater, and if there be applied to the greater a parallelogram equal to the fourth part of the square on the less and deficient by a square figure, it will divide it into parts which are commensurable in length.

Let A, BC be two unequal straight lines, of which BC is the greater,

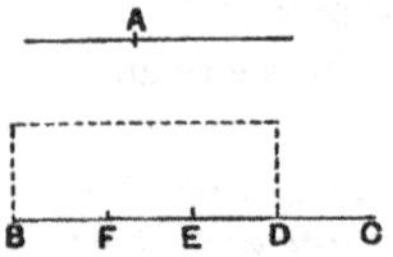

and let there be applied to BC a parallelogram equal to the fourth part of the square on the less, A, that is, equal to the square on the half of A, and deficient by a square figure. Let this be the rectangle BD, DC, [cf. Lemma]

and let BD be commensurable in length with DC;

I say that the square on BC is greater than the square on A by the square on a straight line commensurable with BC.

For let BC be bisected at the point E,

and let EF be made equal to DE.

Therefore the remainder DC is equal to BF.

And, since the straight line BC has been cut into equal parts at E, and into unequal parts at D,

therefore the rectangle contained by BD, DC, together with the square on ED, is equal to the square on EC; [II. 5]

And the same is true of their quadruples;

therefore four times the rectangle BD, DC, together with four times the square on DE, is equal to four times the square on EC.

But the square on A is equal to four times the rectangle BD, DC;

and the square on DF is equal to four times the square on DE, for DF is double of DE.

And the square on BC is equal to four times the square on EC, for again BC is double of CE.

Therefore the squares on A, DF are equal to the square on BC,

so that the square on BC is greater than the square on A by the square on DF.

It is to be proved that BC is also commensurable with DF.

Since BD is commensurable in length with DC,

therefore BC is also commensurable in length with CD. [x. 15]

But *CD* is commensurable in length with *CD*, *BF*, for *CD* is equal to *BF*. [X. 6]

Therefore *BC* is also commensurable in length with *BF*, *CD*, [X. 12]

so that *BC* is also commensurable in length with the remainder *FD*; [X. 15]

therefore the square on *BC* is greater than the square on *A* by the square on a straight line commensurable with *BC*.

Next, let the square on *BC* be greater than the square on *A* by the square on a straight line commensurable with *BC*,

let a parallelogram be applied to *BC* equal to the fourth part of the square on *A* and deficient by a square figure, and let it be the rectangle *BD*, *DC*.

It is to be proved that *BD* is commensurable in length with *DC*.

With the same construction, we can prove similarly that the square on *BC* is greater than the square on *A* by the square on *FD*.

But the square on *BC* is greater than the square on *A* by the square on a straight line commensurable with *BC*.

Therefore *BC* is commensurable in length with *FD*,

so that *BC* is also commensurable in length with the remainder, the sum of *BF*, *DC*. [X. 15]

But the sum of *BF*, *DC* is commensurable with *DC*, [X. 6]

so that *BC* is also commensurable in length with *CD*; [X. 12]

and therefore, *separando*, *BD* is commensurable in length with *DC*. [X. 15]

Therefore etc.

45. After saying literally that "the square on *BC* is greater than the square on *A* by the square on *DF*," Euclid adds the equivalent expression with δύναται in its technical sense, ἡ ΒΓ ἄρα τῆς Α μεῖζον δύναται τῇ ΔΖ. As this is untranslatable in English except by a paraphrase in practically the same words as have preceded, I have not attempted to reproduce it.

This proposition gives the condition that the roots of the equation in x,

$$ax - x^2 = \beta\left(= \frac{b^2}{4}, \text{ say}\right),$$

are commensurable with a, or that x is expressible in terms of a and integral numbers, i.e. is of the form $\frac{m}{n}a$. No better proof can be found for the fact that Euclid and the Greeks used their solutions of quadratic equations for *numerical* problems. On no other assumption could an elaborate discussion of the conditions of incommensurability of the roots with given lengths or with a given number of units of length be explained. In a purely *geometrical* solution the distinction between commensurable and incommensurable roots has no point, because each can equally easily be represented by straight lines. On the other hand, on the assumption that the *numerical* solution of quadratic equations was an important part of the system of the Greek geometers, the distinction between the cases where the roots are commensurable and incommensurable respectively with a given length or unit becomes of great importance. Since the Greeks had no means of *expressing* what we call an irrational number, the case of an equation with incommensurable roots could *only* be represented by them geometrically; and the geometrical representations had to serve instead of what we can express by formulae involving surds.

Euclid proves in this proposition and the next that, x being determined from the equation

$$x(a-x)=\frac{b^2}{4} \quad \ldots\ldots\ldots\ldots\ldots\ldots\ldots\ldots\ldots\ldots (1),$$

x, $(a-x)$ are commensurable in length when $\sqrt{a^2-b^2}$, a are so, and incommensurable in length when $\sqrt{a^2-b^2}$, a are incommensurable; and conversely.

Observe the similarity of his proof to our algebraical method of solving the equation. a being represented in the figure by BC, and x by CD,

$$EF=ED=\frac{a}{2}-x$$

and $$x(a-x)+\left(\frac{a}{2}-x\right)^2=\frac{a^2}{4}, \qquad \text{by Eucl. II. 5.}$$

If we multiply throughout by 4,

$$4x(a-x)+4\left(\frac{a}{2}-x\right)^2=a^2,$$

whence, by (1), $$b^2+(a-2x)^2=a^2,$$

or $$a^2-b^2=(a-2x)^2,$$

and $$\sqrt{a^2-b^2}=a-2x.$$

We have to prove in this proposition

(1) that, if x, $(a-x)$ are commensurable in length, so are a, $\sqrt{a^2-b^2}$,

(2) that, if a, $\sqrt{a^2-b^2}$ are commensurable in length, so are x, $(a-x)$.

(1) To prove that a, $a-2x$ are commensurable in length Euclid employs several successive steps, thus.

Since $(a-x) \frown x$,	$a \frown x$.	[X. 15]
But	$x \frown 2x$.	[X. 6]
Therefore	$a \frown 2x$	[X. 12]
	$\frown (a-2x)$.	[X. 15]
That is,	$a \frown \sqrt{a^2-b^2}$.	

(2) Since $a \frown \sqrt{a^2-b^2}$,	$a \frown a-2x$,	
whence	$a \frown 2x$.	[X. 15]
But	$2x \frown x$;	[X. 6]
therefore	$a \frown x$,	[X. 12]
and hence	$(a-x) \frown x$.	[X. 15]

It is often more convenient to use the symmetrical form of equation in this and similar cases, viz.

$$\left.\begin{aligned} xy &= \frac{b^2}{4} \\ x + y &= a \end{aligned}\right\}.$$

The result with this mode of expression is that

(1) if $x \frown y$, then $a \frown \sqrt{a^2 - b^2}$; and

(2) if $a \frown \sqrt{a^2 - b^2}$, then $x \frown y$.

The truth of the proposition is even easier to see in this case, since $(x - y)^2 = (a^2 - b^2)$.

Proposition 18.

If there be two unequal straight lines, and to the greater there be applied a parallelogram equal to the fourth part of the square on the less and deficient by a square figure, and if it divide it into parts which are incommensurable, the square on the greater will be greater than the square on the less by the square on a straight line incommensurable with the greater.

And, if the square on the greater be greater than the square on the less by the square on a straight line incommensurable with the greater, and if there be applied to the greater a parallelogram equal to the fourth part of the square on the less and deficient by a square figure, it divides it into parts which are incommensurable.

Let *A*, *BC* be two unequal straight lines, of which *BC* is the greater,

and to *BC* let there be applied a parallelogram equal to the fourth part of the square on the less, *A*, and deficient by a square figure. Let this be the rectangle *BD*, *DC*, [cf. Lemma before x. 17]

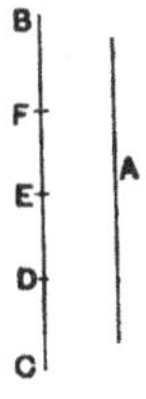

and let *BD* be incommensurable in length with *DC*;

I say that the square on *BC* is greater than the square on *A* by the square on a straight line incommensurable with *BC*.

For, with the same construction as before, we can prove similarly that the square on *BC* is greater than the square on *A* by the square on *FD*.

It is to be proved that *BC* is incommensurable in length with *DF*.

Since BD is incommensurable in length with DC, therefore BC is also incommensurable in length with CD. [x. 16]

But DC is commensurable with the sum of BF, DC; [x. 6] therefore BC is also incommensurable with the sum of BF, DC; [x. 13]
so that BC is also incommensurable in length with the remainder FD. [x. 16]

And the square on BC is greater than the square on A by the square on FD;
therefore the square on BC is greater than the square on A by the square on a straight line incommensurable with BC.

Again, let the square on BC be greater than the square on A by the square on a straight line incommensurable with BC, and let there be applied to BC a parallelogram equal to the fourth part of the square on A and deficient by a square figure. Let this be the rectangle BD, DC.

It is to be proved that BD is incommensurable in length with DC.

For, with the same construction, we can prove similarly that the square on BC is greater than the square on A by the square on FD.

But the square on BC is greater than the square on A by the square on a straight line incommensurable with BC;
therefore BC is incommensurable in length with FD,
so that BC is also commensurable with the remainder, the sum of BF, DC. [x. 16]

But the sum of BF, DC is commensurable in length with DC; [x. 6]
therefore BC is also incommensurable in length with DC, [x. 13]
so that, *separando*, BD is also incommensurable in length with DC. [x. 16]

Therefore etc.

With the same notation as before, we have to prove in this proposition that
(1) if $(a-x)$, x are incommensurable in length, so are a, $\sqrt{a^2-b^2}$, and
(2) if a, $\sqrt{a^2-b^2}$ are incommensurable in length, so are $(a-x)$, x.

Or, with the equations

$$\left.\begin{aligned} xy &= \frac{b^2}{4} \\ x+y &= a \end{aligned}\right\},$$

(1) if $x \smile y$, then $a \smile \sqrt{a^2 - b^2}$, and

(2) if $a \smile \sqrt{a^2 - b^2}$, then $x \smile y$.

The steps are exactly the same as shown under (1) and (2) of the last note, with $\smile$ instead of $\frown$, except only in the lines "$x \frown 2x$" and "$2x \frown x$" which are unaltered, while, in the references, x. 13, 16 take the place of x. 12, 15 respectively.

[Lemma.

Since it has been proved that straight lines commensurable in length are always commensurable in square also, while those commensurable in square are not always commensurable in length also, but can of course be either commensurable or incommensurable in length, it is manifest that, if any straight line be commensurable in length with a given rational straight line, it is called rational and commensurable with the other not only in length but in square also, since straight lines commensurable in length are always commensurable in square also.

But, if any straight line be commensurable in square with a given rational straight line, then, if it is also commensurable in length with it, it is called in this case also rational and commensurable with it both in length and in square; but, if again any straight line, being commensurable in square with a given rational straight line, be incommensurable in length with it, it is called in this case also rational but commensurable in square only.]

Proposition 19.

The rectangle contained by rational straight lines commensurable in length is rational.

For let the rectangle AC be contained by the rational straight lines AB, BC commensurable in length;

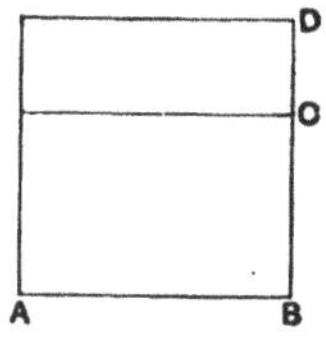

I say that AC is rational.

For on AB let the square AD be described;

therefore AD is rational. [x. Def. 4]

And, since AB is commensurable in length with BC,

while AB is equal to BD,

therefore BD is commensurable in length with BC.

And, as BD is to BC, so is DA to AC. [vi. 1]

Therefore DA is commensurable with AC. [x. 11]

But DA is rational;

therefore AC is also rational. [x. Def. 4]

Therefore etc.

There is a difficulty in the text of the enunciation of this proposition. The Greek runs τὸ ὑπὸ ῥητῶν μήκει συμμέτρων κατά τινα τῶν προειρημένων τρόπων εὐθειῶν περιεχόμενον ὀρθογώνιον ῥητόν ἐστιν, where the rectangle is said to be contained by "rational straight lines commensurable in length *in any of the aforesaid ways*." Now straight lines can only be commensurable *in length* in *one* way, the degrees of commensurability being commensurability in length and commensurability in square only. But a straight line may be *rational* in two ways in relation to a *given* rational straight line, since it may be either commensurable *in length*, or commensurable *in square only*, with the latter. Hence Billingsley takes κατά τινα τῶν προειρημένων τρόπων with ῥητῶν, translating "straight lines commensurable in length and rational in any of the aforesaid ways," and this agrees with the expression in the next proposition "a straight line once more rational in any of the aforesaid ways"; but the order of words in the Greek seems to be fatal to this way of translating the passage.

The best solution of the difficulty seems to be to reject the words "in any of the aforesaid ways" altogether. They have reference to the Lemma which immediately precedes and which is itself open to the gravest suspicion. It is very prolix, and cannot be called necessary; it appears moreover in connexion with an addition clearly spurious and therefore relegated by Heiberg to the Appendix. The addition does not even pretend to be Euclid's, for it begins with the words "for *he calls* rational straight lines those...." Hence we should no doubt relegate the Lemma itself to the Appendix. August does so and leaves out the suspected words in the enunciation, as I have done.

Exactly the same arguments apply to the Lemma added (without the heading "Lemma") to x. 23 and the same words "in any of the aforesaid ways" used with "medial straight lines commensurable in length" in the enunciation of x. 24. The said Lemma must stand or fall with that now in question, since it refers to it in terms: "And in the same way as was explained in the case of rationals...."

Hence I have bracketed the Lemma added to x. 23 and left out the objectionable words in the enunciation of x. 24.

If ρ be one of the given rational straight lines (rational of course in the sense of x. Def. 3), the other can be denoted by $k\rho$, where k is, as usual, of the form m/n (where m, n are integers). Thus the rectangle is $k\rho^2$, which is obviously rational since it is commensurable with ρ^2. [x. Def. 4.]

A rational rectangle may have any of the forms ab, ka^2, kA or A, where a, b are commensurable with the unit of length, and A with the unit of area.

Since Euclid is not able to use $k\rho$ as a symbol for a straight line commensurable in length with ρ, he has to put his proof in a form corresponding to

$$\rho^2 : k\rho^2 = \rho : k\rho,$$

whence, ρ, $k\rho$ being commensurable, ρ^2, $k\rho^2$ are so also. [x. 11]

PROPOSITION 20.

If a rational area be applied to a rational straight line, it produces as breadth a straight line rational and commensurable in length with the straight line to which it is applied.

For let the rational area AC be applied to AB, a straight line once more rational in any of the aforesaid ways, producing BC as breadth;

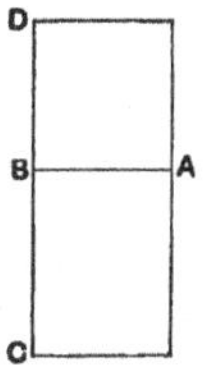

I say that BC is rational and commensurable in length with BA.

For on AB let the square AD be described;
therefore AD is rational. [x. Def. 4]

But AC is also rational;
therefore DA is commensurable with AC.

And, as DA is to AC, so is DB to BC. [VI. 1]

Therefore DB is also commensurable with BC; [x. 11]
and DB is equal to BA;
therefore AB is also commensurable with BC.

But AB is rational;
therefore BC is also rational and commensurable in length with AB.

Therefore etc.

The converse of the last. If ρ is a rational straight line, any rational area is of the form $k\rho^2$. If this be "applied" to ρ, the breadth is $k\rho$ commensurable in length with ρ and therefore rational. We should reach the same result if we applied the area to *another* rational straight line σ. The breadth is then $\frac{k\rho^2}{\sigma} = \frac{k\rho^2}{\sigma^2} \cdot \sigma = \frac{m}{n} k \cdot \sigma$ or $k'\sigma$, say.

PROPOSITION 21.

The rectangle contained by rational straight lines commensurable in square only is irrational, and the side of the square equal to it is irrational. Let the latter be called **medial**.

For let the rectangle AC be contained by the rational straight lines AB, BC commensurable in square only;

I say that AC is irrational, and the side of the square equal to it is irrational;

and let the latter be called **medial**.

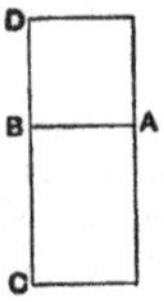

For on AB let the square AD be described;

therefore AD is rational. [x. Def. 4]

And, since AB is incommensurable in length with BC,

for by hypothesis they are commensurable in square only,

while AB is equal to BD,

therefore DB is also incommensurable in length with BC.

And, as DB is to BC, so is AD to AC; [vi. 1]

therefore DA is incommensurable with AC. [x. 11]

But DA is rational;

therefore AC is irrational,

so that the side of the square equal to AC is also irrational. [x. Def. 4]

And let the latter be called **medial**.

Q. E. D.

A *medial* straight line, now defined for the first time, is so called because it is a mean proportional between two rational straight lines commensurable in square only. Such straight lines can be denoted by ρ, $\rho\sqrt{k}$. A medial straight line is therefore of the form $\sqrt{\rho^2\sqrt{k}}$ or $k^{\frac{1}{4}}\rho$. Euclid's proof that this is irrational is equivalent to the following. Take ρ, $\rho\sqrt{k}$ commensurable in square only, so that they are incommensurable in length.

Now $$\rho : \rho\sqrt{k} = \rho^2 : \rho^2\sqrt{k},$$

whence [x. 11] $\rho^2\sqrt{k}$ is incommensurable with ρ^2 and therefore irrational [x. Def. 4], so that $\sqrt{\rho^2\sqrt{k}}$ is also irrational [*ibid.*].

A medial straight line may evidently take either of the forms $\sqrt{a\sqrt{B}}$ or $\sqrt[4]{AB}$, where of course B is not of the form k^2A.

Lemma.

If there be two straight lines, then, as the first is to the second, so is the square on the first to the rectangle contained by the two straight lines.

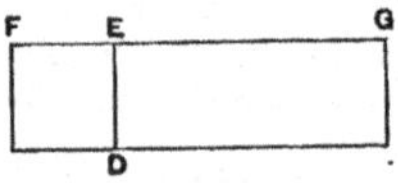

Let FE, EG be two straight lines.

I say that, as FE is to EG, so is the square on FE to the rectangle FE, EG.

For on *FE* let the square *DF* be described, and let *GD* be completed.

Since then, as *FE* is to *EG*, so is *FD* to *DG*, [VI. 1]
and *FD* is the square on *FE*,
and *DG* the rectangle *DE*, *EG*, that is, the rectangle *FE*, *EG*,
therefore, as *FE* is to *EG*, so is the square on *FE* to the rectangle *FE*, *EG*.

Similarly also, as the rectangle *GE*, *EF* is to the square on *EF*, that is, as *GD* is to *FD*, so is *GE* to *EF*.

Q. E. D.

If a, b be two straight lines,

$$a : b = a^2 : ab.$$

PROPOSITION 22.

The square on a medial straight line, if applied to a rational straight line, produces as breadth a straight line rational and incommensurable in length with that to which it is applied.

Let *A* be medial and *CB* rational,
and let a rectangular area *BD* equal to the square on *A* be applied to *BC*, producing *CD* as breadth;

I say that *CD* is rational and incommensurable in length with *CB*.

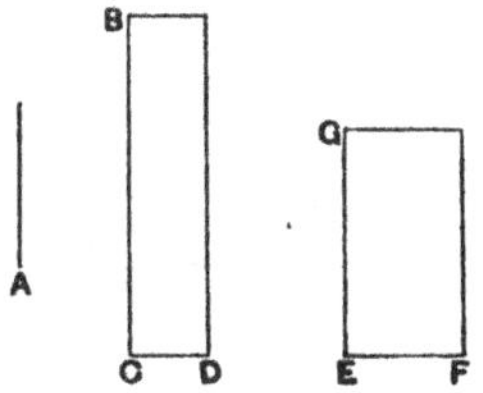

For, since *A* is medial, the square on it is equal to a rectangular area contained by rational straight lines commensurable in square only. [X. 21]

Let the square on it be equal to *GF*.

But the square on it is also equal to *BD*;
therefore *BD* is equal to *GF*.

But it is also equiangular with it;
and in equal and equiangular parallelograms the sides about the equal angles are reciprocally proportional; [VI. 14]
therefore, proportionally, as *BC* is to *EG*, so is *EF* to *CD*.

Therefore also, as the square on *BC* is to the square on *EG*, so is the square on *EF* to the square on *CD*. [VI. 22]

But the square on CB is commensurable with the square on EG, for each of these straight lines is rational;

therefore the square on EF is also commensurable with the square on CD. [X. 11]

But the square on EF is rational;

therefore the square on CD is also rational; [X. Def. 4]

therefore CD is rational.

And, since EF is incommensurable in length with EG,

for they are commensurable in square only,

and, as EF is to EG, so is the square on EF to the rectangle FE, EG, [Lemma]

therefore the square on EF is incommensurable with the rectangle FE, EG. [X. 11]

But the square on CD is commensurable with the square on EF, for the straight lines are rational in square;

and the rectangle DC, CB is commensurable with the rectangle FE, EG, for they are equal to the square on A;

therefore the square on CD is also incommensurable with the rectangle DC, CB. [X. 13]

But, as the square on CD is to the rectangle DC, CB, so is DC to CB; [Lemma]

therefore DC is incommensurable in length with CB. [X. 11]

Therefore CD is rational and incommensurable in length with CB.

Q. E. D.

Our algebraical notation makes the result of this proposition almost self-evident. We have seen that the square of a medial straight line is of the form $\sqrt{k}\cdot\rho^2$. If we "apply" this area to another rational straight line σ, the breadth is $\frac{\sqrt{k}\cdot\rho^2}{\sigma}$.

This is equal to $\frac{\sqrt{k}\cdot\rho^2}{\sigma^2}\cdot\sigma = \sqrt{k}\cdot\frac{m}{n}\sigma$, where m, n are integers. The latter straight line, which we may express, if we please, in the form $\sqrt{k'}\cdot\sigma$, is clearly commensurable with σ in square only, and therefore rational but incommensurable in length with σ.

Euclid's proof, necessarily longer, is in two parts.

Suppose that the rectangle $\sqrt{k}\cdot\rho^2 = \sigma\cdot x$.

Then (1) $\sigma:\rho = \sqrt{k}\cdot\rho : x$, [VI. 14]

whence $\sigma^2:\rho^2 = k\rho^2 : x^2$. [VI. 22]

But $\sigma^2 \frown \rho^2$, and therefore $k\rho^2 \frown x^2$. [X. 11]

And $k\rho^2$ is rational;
therefore x^2, and therefore x, is rational. [x. Def. 4]

(2) Since $\sqrt{k}\,.\,\rho \frown \rho$, $\sqrt{k}\,.\,\rho \smile \rho$.

But [Lemma] $\quad \sqrt{k}\,.\,\rho : \rho = k\rho^2 : \sqrt{k}\,.\,\rho^2$,

whence $\quad k\rho^2 \smile \sqrt{k}\,.\,\rho^2$. [x. 11]

But $\sqrt{k}\,.\,\rho^2 = \sigma x$, and $k\rho^2 \frown x^2$ (from above);

therefore $\quad x^2 \smile \sigma x$; [x. 13]

and, since $x^2 : \sigma x = x : \sigma$, [Lemma]

$$x \smile \sigma.$$

Proposition 23.

A straight line commensurable with a medial straight line is medial.

Let A be medial, and let B be commensurable with A;
I say that B is also medial.

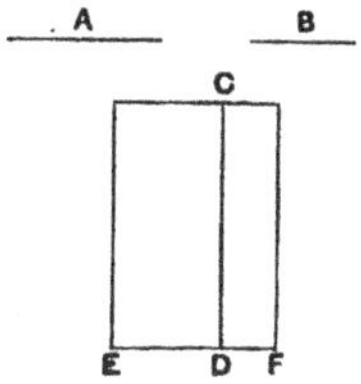

For let a rational straight line CD be set out,
and to CD let the rectangular area CE equal to the square on A be applied, producing ED as breadth;
therefore ED is rational and incommensurable in length with CD. [x. 22]

And let the rectangular area CF equal to the square on B be applied to CD, producing DF as breadth.

Since then A is commensurable with B,
the square on A is also commensurable with the square on B.

But EC is equal to the square on A,
and CF is equal to the square on B;
therefore EC is commensurable with CF.

And, as EC is to CF, so is ED to DF; [vi. 1]
therefore ED is commensurable in length with DF. [x. 11]

But ED is rational and incommensurable in length with DC;
therefore DF is also rational [x. Def. 3] and incommensurable in length with DC. [x. 13]

Therefore CD, DF are rational and commensurable in square only.

But the straight line the square on which is equal to the rectangle contained by rational straight lines commensurable in square only is medial; [x. 21]
therefore the side of the square equal to the rectangle *CD*, *DF* is medial.

And *B* is the side of the square equal to the rectangle *CD*, *DF*;
therefore *B* is medial.

PORISM. From this it is manifest that an area commensurable with a medial area is medial.

[And in the same way as was explained in the case of rationals [Lemma following x. 18] it follows, as regards medials, that a straight line commensurable in length with a medial straight line is called **medial** and commensurable with it not only in length but in square also, since, in general, straight lines commensurable in length are always commensurable in square also.

But, if any straight line be commensurable in square with a medial straight line, then, if it is also commensurable in length with it, the straight lines are called, in this case too, medial and commensurable in length and in square, but, if in square only, they are called medial straight lines commensurable in square only.]

As explained in the bracketed passage following this proposition, a straight line commensurable with a medial straight line *in square only*, as well as a straight line commensurable with it in length, is medial.

Algebraical notation shows this easily.

If $k^{\frac{1}{4}}\rho$ be the given straight line, $\lambda k^{\frac{1}{4}}\rho$ is a straight line commensurable in length with it and $\sqrt{\lambda} \, . \, k^{\frac{1}{4}}\rho$ a straight line commensurable with it in square only.

But $\lambda\rho$ and $\sqrt{\lambda} \, . \, \rho$ are both rational [x. Def. 3] and therefore can be expressed by ρ', and we thus arrive at $k^{\frac{1}{4}}\rho'$, which is clearly medial.

Euclid's proof amounts to the following.

Apply both the areas $\sqrt{k} \, . \, \rho^2$ and $\lambda^2\sqrt{k} \, . \, \rho^2$ (or $\lambda\sqrt{k} \, . \, \rho^2$) to a rational straight line σ.

The breadths $\sqrt{k} \, . \, \frac{\rho^2}{\sigma}$ and $\lambda^2\sqrt{k} \, . \, \frac{\rho^2}{\sigma}$ $\left(\text{or } \lambda\sqrt{k} \, . \, \frac{\rho^2}{\sigma}\right)$ are in the ratio of the areas $\sqrt{k} \, . \, \rho^2$ and $\lambda^2\sqrt{k} \, . \, \rho^2$ (or $\lambda\sqrt{k} \, . \, \rho^2$) themselves and are therefore commensurable.

Now [x. 22] $\sqrt{k} \, . \, \frac{\rho^2}{\sigma}$ is rational but incommensurable with σ.

Therefore $\lambda^2\sqrt{k} \, . \, \frac{\rho^2}{\sigma}$ $\left(\text{or } \lambda\sqrt{k} \, . \, \frac{\rho^2}{\sigma}\right)$ is so also;

whence the area $\lambda^2\sqrt{k}\,.\,\rho^2$ (or $\lambda\sqrt{k}\,.\,\rho^2$) is contained by two rational straight lines commensurable in square only, so that $\lambda k^{\frac{1}{4}}\rho$ (or $\sqrt{\lambda}\,.\,k^{\frac{1}{4}}\rho$) is a medial straight line.

It is in the Porism that we have the first mention of a medial *area*. It is the area which is equal to the square on a medial straight line, an area, therefore, of the form $k^{\frac{1}{2}}\rho^2$, which is, as a matter of fact, arrived at, though not named, before the medial *straight line* itself (x. 21).

The Porism states that $\lambda k^{\frac{1}{2}}\rho^2$ is a medial area, which is indeed obvious.

Proposition 24.

The rectangle contained by medial straight lines commensurable in length is medial.

For let the rectangle *AC* be contained by the medial straight lines *AB*, *BC* which are commensurable in length;

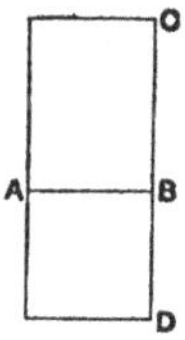

I say that *AC* is medial.

For on *AB* let the square *AD* be described;

therefore *AD* is medial.

And, since *AB* is commensurable in length with *BC*,

while *AB* is equal to *BD*,

therefore *DB* is also commensurable in length with *BC*;

so that *DA* is also commensurable with *AC*. [vi. 1, x. 11]

But *DA* is medial;

therefore *AC* is also medial. [x. 23, Por.]

Q. E. D.

There is the same difficulty in the text of this enunciation as in that of x. 19. The Greek says "medial straight lines commensurable in length in any of the aforesaid ways"; but straight lines can only be *commensurable in length* in one way, though they can be medial in two ways, as explained in the addition to the preceding proposition, i.e. they can be either commensurable in length or commensurable in square only with a *given* medial straight line. For the same reason as that explained in the note on x. 19 I have omitted "in any of the aforesaid ways" in the enunciation and bracketed the addition to x. 23 to which it refers.

$k^{\frac{1}{4}}\rho$ and $\lambda k^{\frac{1}{4}}\rho$ are medial straight lines commensurable in length. The rectangle contained by them is $\lambda k^{\frac{1}{2}}\rho^2$, which may be written $k^{\frac{1}{2}}\rho'^2$ and is therefore clearly medial.

Euclid's proof proceeds thus. Let x, λx be the two medial straight lines commensurable in length.

Therefore $$x^2 : x\,.\,\lambda x = x : \lambda x.$$

But $x \frown \lambda x$, so that $x^2 \frown x \,.\, \lambda x$. [x. 11]

Now x^2 is medial [x. 21];

therefore $x \,.\, \lambda x$ is also medial. [x. 23, Por.]

We may of course write two medial straight lines commensurable in length in the forms $mk^{\frac{1}{4}}\rho$, $nk^{\frac{1}{4}}\rho$; and these may either be $m\sqrt{a\sqrt{B}}$, $n\sqrt{a\sqrt{B}}$, or $m\sqrt[4]{AB}$, $n\sqrt[4]{AB}$.

Proposition 25.

The rectangle contained by medial straight lines commensurable in square only is either rational or medial.

For let the rectangle AC be contained by the medial straight lines AB, BC which are commensurable in square only;

I say that AC is either rational or medial.

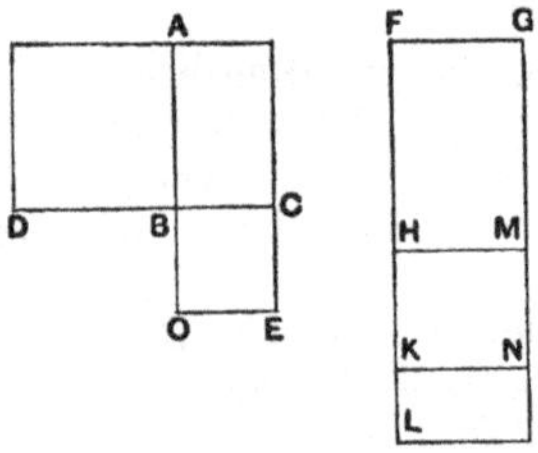

For on AB, BC let the squares AD, BE be described;

therefore each of the squares AD, BE is medial.

Let a rational straight line FG be set out,

to FG let there be applied the rectangular parallelogram GH equal to AD, producing FH as breadth,

to HM let there be applied the rectangular parallelogram MK equal to AC, producing HK as breadth,

and further to KN let there be similarly applied NL equal to BE, producing KL as breadth;

therefore FH, HK, KL are in a straight line.

Since then each of the squares AD, BE is medial,

and AD is equal to GH, and BE to NL,

therefore each of the rectangles GH, NL is also medial.

And they are applied to the rational straight line FG;

therefore each of the straight lines FH, KL is rational and incommensurable in length with FG. [x. 22]

And, since AD is commensurable with BE,

therefore GH is also commensurable with NL.

And, as GH is to NL, so is FH to KL; [vi. 1]

therefore FH is commensurable in length with KL. [x. 11]

Therefore FH, KL are rational straight lines commensurable in length;
therefore the rectangle FH, KL is rational. [X. 19]

And, since DB is equal to BA, and OB to BC,
therefore, as DB is to BC, so is AB to BO.

But, as DB is to BC, so is DA to AC, [VI. 1]
and, as AB is to BO, so is AC to CO; [*id.*]
therefore, as DA is to AC, so is AC to CO.

But AD is equal to GH, AC to MK and CO to NL;
therefore, as GH is to MK, so is MK to NL;
therefore also, as FH is to HK, so is HK to KL; [VI. 1, V. 11]
therefore the rectangle FH, KL is equal to the square on HK. [VI. 17]

But the rectangle FH, KL is rational;
therefore the square on HK is also rational.

Therefore HK is rational.

And, if it is commensurable in length with FG,
HN is rational; [X. 19]
but, if it is incommensurable in length with FG,
KH, HM are rational straight lines commensurable in square only, and therefore HN is medial. [X. 21]

Therefore HN is either rational or medial.

But HN is equal to AC;
therefore AC is either rational or medial.

Therefore etc.

Two medial straight lines commensurable in square only are of the form $k^{\frac{1}{4}}\rho$, $\sqrt{\lambda} \cdot k^{\frac{1}{4}}\rho$.

The rectangle contained by them is $\sqrt{\lambda} \cdot k^{\frac{1}{2}}\rho^2$. Now this is in general *medial*; but, if $\sqrt{\lambda} = k' \sqrt{k}$, the rectangle is $kk'\rho^2$, which is *rational*.

Euclid's argument is as follows. Let us, for convenience, put x for $k^{\frac{1}{4}}\rho$, so that the medial straight lines are x, $\sqrt{\lambda} \cdot x$.

Form the areas x^2, $x \cdot \sqrt{\lambda} \cdot x$, λx^2,
and let these be respectively equal to σu, σv, σw, where σ is a rational straight line.

Since x^2, λx^2 are medial *areas*,
so are σu, σw,
whence u, w are respectively rational and $\frown \sigma$.

But $x^2 \frown \lambda x^2$,

so that $\sigma u \frown \sigma w$,

or $u \frown w$ ....(1).

Therefore, u, w being both rational, uw is rational .. ..(2).

Now $x^2 : \sqrt{\lambda} \cdot x^2 = \sqrt{\lambda} \cdot x^2 : \lambda x^2$

or $\sigma u : \sigma v = \sigma v : \sigma w$,

so that $u : v = v : w$,

and $uw = v^2$.

Hence, by (2), v^2, and therefore v, is rational ....(3).

Now (α) if $v \frown \sigma$, σv or $\sqrt{\lambda} \cdot x^2$ is *rational*;

(β) if $v \smile \sigma$, so that $v \sim \sigma$, σv or $\sqrt{\lambda} \cdot x^2$ is *medial*.

PROPOSITION 26.

A medial area does not exceed a medial area by a rational area.

For, if possible, let the medial area *AB* exceed the medial area *AC* by the rational area *DB*,

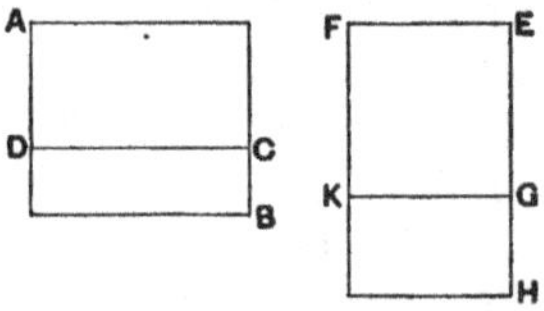

and let a rational straight line *EF* be set out;

to *EF* let there be applied the rectangular parallelogram *FH* equal to *AB*, producing *EH* as breadth,

and let the rectangle *FG* equal to *AC* be subtracted;

therefore the remainder *BD* is equal to the remainder *KH*.

But *DB* is rational;

therefore *KH* is also rational.

Since, then, each of the rectangles *AB*, *AC* is medial, and *AB* is equal to *FH*, and *AC* to *FG*,

therefore each of the rectangles *FH*, *FG* is also medial.

And they are applied to the rational straight line *EF*;

therefore each of the straight lines *HE*, *EG* is rational and incommensurable in length with *EF*. [x. 22]

And, since [*DB* is rational and is equal to *KH*,

therefore] *KH* is [also] rational;

and it is applied to the rational straight line *EF*;

therefore *GH* is rational and commensurable in length with *EF*. [x. 20]

But *EG* is also rational, and is incommensurable in length with *EF*;

therefore *EG* is incommensurable in length with *GH*. [x. 13]

And, as *EG* is to *GH*, so is the square on *EG* to the rectangle *EG*, *GH*;

therefore the square on *EG* is incommensurable with the rectangle *EG*, *GH*. [x. 11]

But the squares on *EG*, *GH* are commensurable with the square on *EG*, for both are rational;

and twice the rectangle *EG*, *GH* is commensurable with the rectangle *EG*, *GH*, for it is double of it; [x. 6]

therefore the squares on *EG*, *GH* are incommensurable with twice the rectangle *EG*, *GH*; [x. 13]

therefore also the sum of the squares on *EG*, *GH* and twice the rectangle *EG*, *GH*, that is, the square on *EH* [II. 4], is incommensurable with the squares on *EG*, *GH*. [x. 16]

But the squares on *EG*, *GH* are rational;

therefore the square on *EH* is irrational. [x. Def. 4]

Therefore *EH* is irrational.

But it is also rational:

which is impossible.

Therefore etc.

Q. E. D.

"Apply" the two given medial areas to one and the same rational straight line ρ. They can then be written in the form $\rho \,.\, k^{\frac{1}{2}}\rho$, $\rho \,.\, \lambda^{\frac{1}{2}}\rho$.

The difference is then $(\sqrt{k} - \sqrt{\lambda})\,\rho^2$; and the proposition asserts that this cannot be rational, i.e. $(\sqrt{k} - \sqrt{\lambda})$ cannot be equal to k'. Cf. the proposition corresponding to this in algebraical text-books.

To make Euclid's proof clear we will put x for $k^{\frac{1}{2}}\rho$ and y for $\lambda^{\frac{1}{2}}\rho$.

Suppose $$\rho(x - y) = \rho z,$$

and, if possible, let ρz be rational, so that z must be rational and $\frown \rho$ ...(1).

Since ρx, ρy are medial,

x and y are respectively rational and $\smile \rho$ ....(2).

From (1) and (2), $$y \smile z.$$

Now $$y : z = y^2 : yz,$$

so that $$y^2 \smile yz.$$

But $y^2 + z^2 \frown y^2$,
and $2yz \frown yz$.

Therefore $y^2 + z^2 \smile 2yz$,
whence $(y+z)^2 \smile (y^2 + z^2)$,
or $x^2 \smile (y^2 + z^2)$.

And $(y^2 + z^2)$ is rational;
therefore x^2, and consequently x, is irrational.

But, by (2), x is rational:
which is impossible.

Therefore ρz is *not* rational.

PROPOSITION 27.

To find medial straight lines commensurable in square only which contain a rational rectangle.

Let two rational straight lines *A*, *B* commensurable in square only be set out;

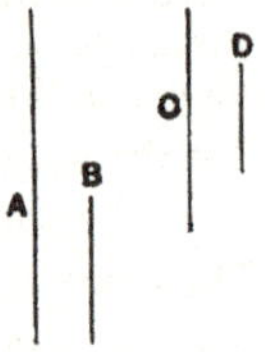

let *C* be taken a mean proportional between *A*, *B*, [VI. 13]
and let it be contrived that,

as *A* is to *B*, so is *C* to *D*. [VI. 12]

Then, since *A*, *B* are rational and commensurable in square only,
the rectangle *A*, *B*, that is, the square on *C* [VI. 17], is medial. [X. 21]

Therefore *C* is medial. [X. 21]

And since, as *A* is to *B*, so is *C* to *D*,
and *A*, *B* are commensurable in square only,
therefore *C*, *D* are also commensurable in square only. [X. 11]

And *C* is medial;
therefore *D* is also medial. [X. 23, addition]

Therefore *C*, *D* are medial and commensurable in square only.

I say that they also contain a rational rectangle.

For since, as *A* is to *B*, so is *C* to *D*,
therefore, alternately, as *A* is to *C*, so is *B* to *D*. [V. 16]

But, as *A* is to *C*, so is *C* to *B*;
therefore also, as *C* is to *B*, so is *B* to *D*;
therefore the rectangle *C*, *D* is equal to the square on *B*.

But the square on B is rational;
therefore the rectangle C, D is also rational.

Therefore medial straight lines commensurable in square only have been found which contain a rational rectangle.

Q. E. D.

Euclid takes two rational straight lines commensurable in square only, say ρ, $k^{\frac{1}{2}}\rho$.

Find the mean proportional, i.e. $k^{\frac{1}{4}}\rho$.

Take x such that $\rho : k^{\frac{1}{2}}\rho = k^{\frac{1}{4}}\rho : x$ ....(1).

This gives $x = k^{\frac{3}{4}}\rho$,
and the lines required are $k^{\frac{1}{4}}\rho$, $k^{\frac{3}{4}}\rho$.

For (α) $k^{\frac{1}{4}}\rho$ is medial.

And (β), by (1), since $\rho \frown k^{\frac{1}{2}}\rho$,

$$k^{\frac{1}{4}}\rho \frown k^{\frac{3}{4}}\rho,$$

whence [addition to X. 23], since $k^{\frac{1}{4}}\rho$ is medial,

$$k^{\frac{3}{4}}\rho \text{ is also medial.}$$

The medial straight lines thus found may take either of the forms

$$(1)\quad \sqrt{a\sqrt{B}},\quad \sqrt{\frac{B\sqrt{B}}{a}}\quad \text{or}\ (2)\quad \sqrt[4]{AB},\quad \sqrt{B\frac{\sqrt{B}}{\sqrt{A}}}.$$

Proposition 28.

To find medial straight lines commensurable in square only which contain a medial rectangle.

Let the rational straight lines A, B, C commensurable in square only be set out;
let D be taken a mean proportional between A, B, [VI. 13]
and let it be contrived that,

as B is to C, so is D to E. [VI. 12]

Since A, B are rational straight lines commensurable in square only,
therefore the rectangle A, B, that is, the square on D [VI. 17], is medial. [X. 21]

Therefore D is medial. [X. 21]

And since B, C are commensurable in square only, and, as B is to C, so is D to E, therefore D, E are also commensurable in square only. [X. 11]

But D is medial; therefore E is also medial. [X. 23, addition]

Therefore D, E are medial straight lines commensurable in square only.

I say next that they also contain a medial rectangle.

For since, as B is to C, so is D to E, therefore, alternately, as B is to D, so is C to E. [V. 16]

But, as B is to D, so is D to A; therefore also, as D is to A, so is C to E; therefore the rectangle A, C is equal to the rectangle D, E. [VI. 16]

But the rectangle A, C is medial; [X. 21] therefore the rectangle D, E is also medial.

Therefore medial straight lines commensurable in square only have been found which contain a medial rectangle.

Q. E. D.

Euclid takes three straight lines commensurable in square only, i.e. of the form ρ, $k^{\frac{1}{2}}\rho$, $\lambda^{\frac{1}{2}}\rho$, and proceeds as follows.

Take the mean proportional to ρ, $k^{\frac{1}{2}}\rho$, i.e. $k^{\frac{1}{4}}\rho$.

Then take x such that

$$k^{\frac{1}{2}}\rho : \lambda^{\frac{1}{2}}\rho = k^{\frac{1}{4}}\rho : x \quad \text{.............................(1)},$$

so that $x = \lambda^{\frac{1}{2}}\rho/k^{\frac{1}{4}}$.

$k^{\frac{1}{4}}\rho$, $\lambda^{\frac{1}{2}}\rho/k^{\frac{1}{4}}$ are the required medial straight lines.

For $k^{\frac{1}{4}}\rho$ is medial.

Now, by (1), since $k^{\frac{1}{2}}\rho \frown \lambda^{\frac{1}{2}}\rho$,

$$k^{\frac{1}{4}}\rho \frown x,$$

whence x is also medial [X. 23, addition], while $\frown k^{\frac{1}{4}}\rho$.

Next, by (1), $\lambda^{\frac{1}{2}}\rho : x = k^{\frac{1}{2}}\rho : k^{\frac{1}{4}}\rho$

$$= k^{\frac{1}{4}}\rho : \rho,$$

whence $x \cdot k^{\frac{1}{4}}\rho = \lambda^{\frac{1}{2}}\rho^2$, which is medial.

The straight lines $k^{\frac{1}{4}}\rho$, $\lambda^{\frac{1}{2}}\rho/k^{\frac{1}{4}}$ of course take different forms according as the original straight lines are of the forms (1) a, $\sqrt{B}$, $\sqrt{C}$, (2) $\sqrt{A}$, $\sqrt{B}$, $\sqrt{C}$, (3) $\sqrt{A}$, b, $\sqrt{C}$, and (4) $\sqrt{A}$, $\sqrt{B}$, c.

E.g. in case (1) they are $\sqrt{a\sqrt{B}}$, $\sqrt{\dfrac{aC}{\sqrt{B}}}$,

in case (2) they are $\sqrt[4]{AB}$, $\sqrt{\dfrac{C\sqrt{A}}{\sqrt{B}}}$,

and so on.

LEMMA 1.

To find two square numbers such that their sum is also square.

Let two numbers AB, BC be set out, and let them be either both even or both odd.

Then since, whether an even number is subtracted from an even number, or an odd number from an odd number, the remainder is even, [IX. 24, 26]

therefore the remainder AC is even.

Let AC be bisected at D.

Let AB, BC also be either similar plane numbers, or square numbers, which are themselves also similar plane numbers.

Now the product of AB, BC together with the square on CD is equal to the square on BD. [II. 6]

And the product of AB, BC is square, inasmuch as it was proved that, if two similar plane numbers by multiplying one another make some number, the product is square. [IX. 1]

Therefore two square numbers, the product of AB, BC, and the square on CD, have been found which, when added together, make the square on BD.

And it is manifest that two square numbers, the square on BD and the square on CD, have again been found such that their difference, the product of AB, BC, is a square, whenever AB, BC are similar plane numbers.

But when they are not similar plane numbers, two square numbers, the square on BD and the square on DC, have been found such that their difference, the product of AB, BC, is not square.

Q. E. D.

Euclid's method of forming right-angled triangles in integral numbers, already alluded to in the note on I. 47, is as follows.

Take two similar plane numbers, e.g. mnp^2, mnq^2, *which are either both even or both odd*, so that their difference is divisible by 2.

Now the product of the two numbers, or $m^2n^2p^2q^2$, is square, [IX. 1]
and, by II. 6,

$$mnp^2 \cdot mnq^2 + \left(\frac{mnp^2 - mnq^2}{2}\right)^2 = \left(\frac{mnp^2 + mnq^2}{2}\right)^2,$$

so that the numbers $mnpq$, $\frac{1}{2}(mnp^2 - mnq^2)$ satisfy the condition that the sum of their squares is also a square number.

It is also clear that $\frac{1}{2}(mnp^2 + mnq^2)$, $mnpq$ are numbers such that the *difference* of their squares is also square.

Lemma 2.

To find two square numbers such that their sum is not square.

For let the product of *AB*, *BC*, as we said, be square, and *CA* even,
and let *CA* be bisected by *D*.

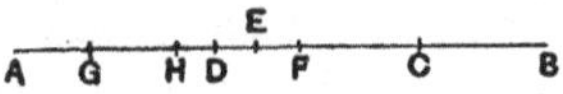

It is then manifest that the square product of *AB*, *BC* together with the square on *CD* is equal to the square on *BD*. [See Lemma 1]

Let the unit *DE* be subtracted;
therefore the product of *AB*, *BC* together with the square on *CE* is less than the square on *BD*.

I say then that the square product of *AB*, *BC* together with the square on *CE* will not be square.

For, if it is square, it is either equal to the square on *BE*, or less than the square on *BE*, but cannot any more be greater, lest the unit be divided.

First, if possible, let the product of *AB*, *BC* together with the square on *CE* be equal to the square on *BE*,
and let *GA* be double of the unit *DE*.

Since then the whole *AC* is double of the whole *CD*,
and in them *AG* is double of *DE*,
therefore the remainder *GC* is also double of the remainder *EC*;
therefore *GC* is bisected by *E*.

Therefore the product of *GB*, *BC* together with the square on *CE* is equal to the square on *BE*. [II. 6]

But the product of *AB*, *BC* together with the square on *CE* is also, by hypothesis, equal to the square on *BE*;

therefore the product of *GB*, *BC* together with the square on *CE* is equal to the product of *AB*, *BC* together with the square on *CE*.

And, if the common square on *CE* be subtracted,
it follows that *AB* is equal to *GB*:
which is absurd.

Therefore the product of *AB*, *BC* together with the square on *CE* is not equal to the square on *BE*.

I say next that neither is it less than the square on *BE*.

For, if possible, let it be equal to the square on *BF*,
and let *HA* be double of *DF*.

Now it will again follow that *HC* is double of *CF*;
so that *CH* has also been bisected at *F*,
and for this reason the product of *HB*, *BC* together with the square on *FC* is equal to the square on *BF*. [II. 6]

But, by hypothesis, the product of *AB*, *BC* together with the square on *CE* is also equal to the square on *BF*.

Thus the product of *HB*, *BC* together with the square on *CF* will also be equal to the product of *AB*, *BC* together with the square on *CE*:
which is absurd.

Therefore the product of *AB*, *BC* together with the square on *CE* is not less than the square on *BE*.

And it was proved that neither is it equal to the square on *BE*.

Therefore the product of *AB*, *BC* together with the square on *CE* is not square.

Q. E. D.

We can, of course, write the identity in the note on Lemma 1 above (p. 64) in the simpler form

$$mp^2 \cdot mq^2 + \left(\frac{mp^2 - mq^2}{2}\right)^2 = \left(\frac{mp^2 + mq^2}{2}\right)^2,$$

where, as before, mp^2, mq^2 are both odd or both even.

Now, says Euclid,

$$mp^2 \cdot mq^2 + \left(\frac{mp^2 - mq^2}{2} - 1\right)^2 \text{ is not a square number.}$$

This is proved by *reductio ad absurdum*.

The number is clearly less than $mp^2 . mq^2 + \left(\frac{mp^2 - mq^2}{2}\right)^2$, i.e. less than $\left(\frac{mp^2 + mq^2}{2}\right)^2$.

If then the number is square, its side must be greater than, equal to, or less than $\left(\frac{mp^2 + mq^2}{2} - 1\right)$, the number next less than $\frac{mp^2 + mq^2}{2}$.

But (1) the side cannot be > $\left(\frac{mp^2 + mq^2}{2} - 1\right)$ without being equal to $\frac{mp^2 + mq^2}{2}$, since they are consecutive numbers.

(2) $$(mp^2 - 2)\, mq^2 + \left(\frac{mp^2 - mq^2}{2} - 1\right)^2 = \left(\frac{mp^2 + mq^2}{2} - 1\right)^2. \qquad \text{[II. 6]}$$

If then $mp^2 . mq^2 + \left(\frac{mp^2 - mq^2}{2} - 1\right)^2$ is *also* equal to $\left(\frac{mp^2 + mq^2}{2} - 1\right)^2$,

we must have $$(mp^2 - 2)\, mq^2 = mp^2 . mq^2,$$
or $$mp^2 - 2 = mp^2:$$
which is impossible.

(3) If $$mp^2 . mq^2 + \left(\frac{mp^2 - mq^2}{2} - 1\right)^2 < \left(\frac{mp^2 + mq^2}{2} - 1\right)^2,$$

suppose it equal to $\left(\frac{mp^2 + mq^2}{2} - r\right)^2$.

But [II. 6] $(mp^2 - 2r)\, mq^2 + \left(\frac{mp^2 - mq^2}{2} - r\right)^2 = \left(\frac{mp^2 + mq^2}{2} - r\right)^2$.

Therefore
$$(mp^2 - 2r)\, mq^2 + \left(\frac{mp^2 - mq^2}{2} - r\right)^2 = mp^2 . mq^2 + \left(\frac{mp^2 - mq^2}{2} - 1\right)^2:$$
which is impossible.

Hence all three hypotheses are false, and the sum of the squares $mp^2 . mq^2$ and $\left(\frac{mp^2 - mq^2}{2} - 1\right)^2$ is *not* square.

PROPOSITION 29.

To find two rational straight lines commensurable in square only and such that the square on the greater is greater than the square on the less by the square on a straight line commensurable in length with the greater.

For let there be set out any rational straight line *AB*, and two square numbers *CD*, *DE* such that their difference *CE* is not square; [Lemma 1]

let there be described on *AB* the semicircle *AFB*,

and let it be contrived that,

as DC is to CE, so is the square on BA to the square on AF. [x. 6, Por.]

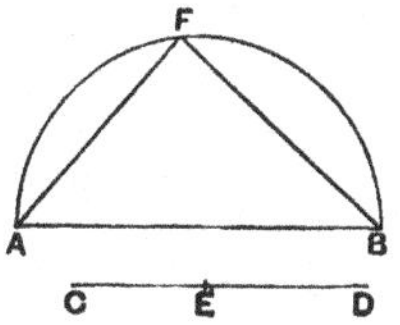

Let FB be joined.

Since, as the square on BA is to the square on AF, so is DC to CE,

therefore the square on BA has to the square on AF the ratio which the number DC has to the number CE;

therefore the square on BA is commensurable with the square on AF. [x. 6]

But the square on AB is rational; [x. Def. 4]

therefore the square on AF is also rational; [*id.*]

therefore AF is also rational.

And, since DC has not to CE the ratio which a square number has to a square number,

neither has the square on BA to the square on AF the ratio which a square number has to a square number;

therefore AB is incommensurable in length with AF. [x. 9]

Therefore BA, AF are rational straight lines commensurable in square only.

And since, as DC is to CE, so is the square on BA to the square on AF,

therefore, *convertendo*, as CD is to DE, so is the square on AB to the square on BF. [v. 19, Por., III. 31, I. 47]

But CD has to DE the ratio which a square number has to a square number;

therefore also the square on AB has to the square on BF the ratio which a square number has to a square number;

therefore AB is commensurable in length with BF. [x. 9]

And the square on AB is equal to the squares on AF, FB;

therefore the square on AB is greater than the square on AF by the square on BF commensurable with AB.

Therefore there have been found two rational straight lines BA, AF commensurable in square only and such that the square on the greater AB is greater than the square on the less AF by the square on BF commensurable in length with AB.

Q. E. D.

Take a rational straight line ρ and two numbers m^2, n^2 such that $(m^2 - n^2)$ is not a square.

Take a straight line x such that

$$m^2 : m^2 - n^2 = \rho^2 : x^2 \quad \text{..........................} (1),$$

whence $$x^2 = \frac{m^2 - n^2}{m^2} \rho^2,$$

and $$x = \rho\sqrt{1 - k^2}, \qquad \text{where } k = \frac{n}{m}.$$

Then ρ, $\rho\sqrt{1 - k^2}$ are the straight lines required.

It follows from (1) that $x^2 \frown \rho^2$,

and x is rational, but $x \smile \rho$.

By (1), *convertendo*, $m^2 : n^2 = \rho^2 : \rho^2 - x^2$,

so that $\sqrt{\rho^2 - x^2} \frown \rho$, and in fact $= k\rho$.

According as ρ is of the form a or $\sqrt{A}$, the straight lines are (1) a, $\sqrt{a^2 - b^2}$ or (2) $\sqrt{A}$, $\sqrt{A - k^2 A}$.

Proposition 30.

To find two rational straight lines commensurable in square only and such that the square on the greater is greater than the square on the less by the square on a straight line incommensurable in length with the greater.

Let there be set out a rational straight line AB, and two square numbers CE, ED such that their sum CD is not square; [Lemma 2]

let there be described on AB the semicircle AFB,

let it be contrived that,

as DC is to CE, so is the square on BA to the square on AF, [x. 6, Por.]

and let FB be joined.

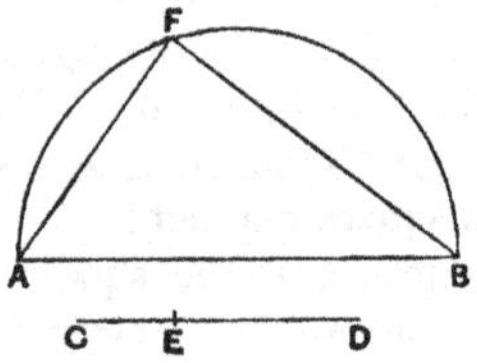

Then, in a similar manner to the preceding, we can prove that BA, AF are rational straight lines commensurable in square only.

And since, as DC is to CE, so is the square on BA to the square on AF,

therefore, *convertendo*, as CD is to DE, so is the square on AB to the square on BF. [v. 19, Por., III. 31, I. 47]

But CD has not to DE the ratio which a square number has to a square number;

therefore neither has the square on *AB* to the square on *BF* the ratio which a square number has to a square number;
therefore *AB* is incommensurable in length with *BF*. [x. 9]

And the square on *AB* is greater than the square on *AF* by the square on *FB* incommensurable with *AB*.

Therefore *AB*, *AF* are rational straight lines commensurable in square only, and the square on *AB* is greater than the square on *AF* by the square on *FB* incommensurable in length with *AB*.

Q. E. D.

In this case we take m^2, n^2 such that $m^2 + n^2$ is not square.

Find x such that $m^2 + n^2 : m^2 = \rho^2 : x^2$,

whence $$x^2 = \frac{m^2}{m^2 + n^2}\rho^2,$$

or $$x = \frac{\rho}{\sqrt{1 + k^2}}, \qquad \text{where } k = \frac{n}{m}.$$

Then ρ, $\frac{\rho}{\sqrt{1+k^2}}$ satisfy the condition.

The proof is after the manner of the proof of the preceding proposition and need not be repeated.

According as ρ is of the form a or $\sqrt{A}$, the straight lines take the form (1) a, $\sqrt{a^2 - \frac{k^2a^2}{1+k^2}}$, that is, a, $\sqrt{a^2 - B}$, or (2) $\sqrt{A}$, $\sqrt{A - B}$ and $\sqrt{A}$, $\sqrt{A - b^2}$.

Proposition 31.

To find two medial straight lines commensurable in square only, containing a rational rectangle, and such that the square on the greater is greater than the square on the less by the square on a straight line commensurable in length with the greater.

Let there be set out two rational straight lines *A*, *B* commensurable in square only and such that the square on *A*, being the greater, is greater than the square on *B* the less by the square on a straight line commensurable in length with *A*. [x. 29]

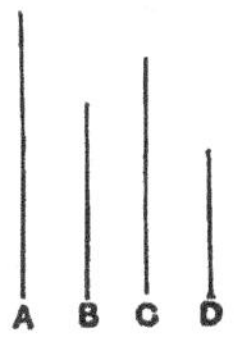

And let the square on *C* be equal to the rectangle *A*, *B*.

Now the rectangle *A*, *B* is medial; [x. 21]
therefore the square on *C* is also medial;
therefore *C* is also medial. [x. 21]

Let the rectangle C, D be equal to the square on B.

Now the square on B is rational;
therefore the rectangle C, D is also rational.

And since, as A is to B, so is the rectangle A, B to the square on B,
while the square on C is equal to the rectangle A, B,
and the rectangle C, D is equal to the square on B,
therefore, as A is to B, so is the square on C to the rectangle C, D.

But, as the square on C is to the rectangle C, D, so is C to D;
therefore also, as A is to B, so is C to D.

But A is commensurable with B in square only;
therefore C is also commensurable with D in square only. [x. 11]

And C is medial;
therefore D is also medial. [x. 23, addition]

And since, as A is to B, so is C to D,
and the square on A is greater than the square on B by the square on a straight line commensurable with A,
therefore also the square on C is greater than the square on D by the square on a straight line commensurable with C. [x. 14]

Therefore two medial straight lines C, D, commensurable in square only and containing a rational rectangle, have been found, and the square on C is greater than the square on D by the square on a straight line commensurable in length with C.

Similarly also it can be proved that the square on C exceeds the square on D by the square on a straight line incommensurable with C, when the square on A is greater than the square on B by the square on a straight line incommensurable with A. [x. 30]

I. Take the rational straight lines commensurable in square only found in x. 29, i.e. ρ, $\rho\sqrt{1-k^2}$.

Take the mean proportional $\rho(1-k^2)^{\frac{1}{4}}$ and x such that

$$\rho(1-k^2)^{\frac{1}{4}} : \rho\sqrt{1-k^2} = \rho\sqrt{1-k^2} : x.$$

Then $\rho(1-k^2)^{\frac{1}{4}}$, x, or $\rho(1-k^2)^{\frac{1}{4}}$, $\rho(1-k^2)^{\frac{3}{4}}$ are straight lines satisfying the given conditions.

For (α) $\rho^2\sqrt{1-k^2}$ is a medial area, and therefore $\rho(1-k^2)^{\frac{1}{4}}$ is a medial straight line ....(1);
and $x\,.\,\rho(1-k^2)^{\frac{1}{4}} = \rho^2(1-k^2)$ and is therefore a rational area.

(β) ρ, $\rho(1-k^2)^{\frac{1}{4}}$, $\rho\sqrt{1-k^2}$, x are straight lines in continued proportion, by construction.

Therefore $\rho : \rho\sqrt{1-k^2} = \rho(1-k^2)^{\frac{1}{4}} : x$ ....(2).

(This Euclid has to prove in a somewhat roundabout way by means of the lemma after X. 21 to the effect that $a : b = ab : b^2$.)

From (2) it follows [X. 11] that $x \frown \rho(1-k^2)^{\frac{1}{4}}$; whence, since $\rho(1-k^2)^{\frac{1}{4}}$ is medial, x or $\rho(1-k^2)^{\frac{3}{4}}$ is medial also.

(γ) From (2), since ρ, $\rho\sqrt{1-k^2}$ satisfy the remaining condition of the problem, $\rho(1-k^2)^{\frac{1}{4}}$, $\rho(1-k^2)^{\frac{3}{4}}$ do so also [X. 14].

According as ρ is of the form a or $\sqrt{A}$, the straight lines take the forms

$$(1)\qquad \sqrt{a\sqrt{a^2-b^2}},\quad \frac{a^2-b^2}{\sqrt{a\sqrt{a^2-b^2}}},$$

or

$$(2)\qquad \sqrt[4]{A(A-k^2A)},\quad \frac{A-k^2A}{\sqrt[4]{A(A-k^2A)}}.$$

II. To find medial straight lines commensurable in square only containing a rational rectangle, and such that the square on one exceeds the square on the other by the square on a straight line *incommensurable* with the former, we simply begin with the rational straight lines having the corresponding property [X. 30], viz. ρ, $\frac{\rho}{\sqrt{1+k^2}}$, and we arrive at the straight lines

$$\frac{\rho}{(1+k^2)^{\frac{1}{4}}},\quad \frac{\rho}{(1+k^2)^{\frac{3}{4}}}.$$

According as ρ is of the form a or $\sqrt{A}$, these (if we use the same transformation as at the end of the note on X. 30) may take any of the forms

$$(1)\qquad \sqrt{a\sqrt{a^2-B}},\quad \frac{a^2-B}{\sqrt{a\sqrt{a^2-B}}},$$

or

$$(2)\qquad \sqrt[4]{A(A-B)},\quad \frac{A-B}{\sqrt[4]{A(A-B)}},$$

$$\text{or}\quad \sqrt[4]{A(A-b^2)},\quad \frac{A-b^2}{\sqrt[4]{A(A-b^2)}}.$$

PROPOSITION 32.

To find two medial straight lines commensurable in square only, containing a medial rectangle, and such that the square on the greater is greater than the square on the less by the square on a straight line commensurable with the greater.

Let there be set out three rational straight lines A, B, C commensurable in square only, and such that the square on A is greater than the square on C by the square on a straight line commensurable with A, [x. 29]
and let the square on D be equal to the rectangle A, B.

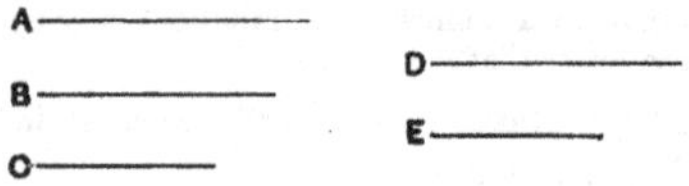

Therefore the square on D is medial;
therefore D is also medial. [x. 21]

Let the rectangle D, E be equal to the rectangle B, C.

Then since, as the rectangle A, B is to the rectangle B, C, so is A to C,
while the square on D is equal to the rectangle A, B,
and the rectangle D, E is equal to the rectangle B, C,
therefore, as A is to C, so is the square on D to the rectangle D, E.

But, as the square on D is to the rectangle D, E, so is D to E;
therefore also, as A is to C, so is D to E.

But A is commensurable with C in square only;
therefore D is also commensurable with E in square only. [x. 11]

But D is medial;
therefore E is also medial. [x. 23, addition]

And, since, as A is to C, so is D to E,
while the square on A is greater than the square on C by the square on a straight line commensurable with A,
therefore also the square on D will be greater than the square on E by the square on a straight line commensurable with D. [x. 14]

I say next that the rectangle D, E is also medial.

For, since the rectangle B, C is equal to the rectangle D, E, while the rectangle B, C is medial, [x. 21]
therefore the rectangle D, E is also medial.

Therefore two medial straight lines D, E, commensurable in square only, and containing a medial rectangle, have been found such that the square on the greater is greater than the square on the less by the square on a straight line commensurable with the greater.

Similarly again it can be proved that the square on D is greater than the square on E by the square on a straight line incommensurable with D, when the square on A is greater than the square on C by the square on a straight line incommensurable with A. [X. 30]

I. Euclid takes three straight lines of the form ρ, $\rho\sqrt{\lambda}$, $\rho\sqrt{1-k^2}$, takes the mean proportional $\rho\lambda^{\frac{1}{4}}$ between the first two ....(1), and then finds x such that

$$\rho\lambda^{\frac{1}{4}} : \rho\lambda^{\frac{1}{2}} = \rho\sqrt{1-k^2} : x \quad \text{..........................(2)},$$

whence $x = \rho\lambda^{\frac{1}{4}}\sqrt{1-k^2}$,

and the straight lines $\rho\lambda^{\frac{1}{4}}$, $\rho\lambda^{\frac{1}{4}}\sqrt{1-k^2}$ satisfy the given conditions.

Now (α) $\rho\lambda^{\frac{1}{4}}$ is medial.

(β) We have, from (1) and (2),

$$\rho : \rho\sqrt{1-k^2} = \rho\lambda^{\frac{1}{4}} : x \quad \text{..........................(3)},$$

whence $x \frown \rho\lambda^{\frac{1}{4}}$; and x is therefore medial and $\frown \rho\lambda^{\frac{1}{4}}$.

(γ) $x \,.\, \rho\lambda^{\frac{1}{4}} = \rho\sqrt{\lambda} \,.\, \rho\sqrt{1-k^2}$.

But the latter is medial; [X. 21]

therefore $x \,.\, \rho\lambda^{\frac{1}{4}}$, or $\rho\lambda^{\frac{1}{4}} \,.\, \rho\lambda^{\frac{1}{4}}\sqrt{1-k^2}$, is medial.

Lastly (δ) ρ, $\rho\sqrt{1-k^2}$ have the remaining property in the enunciation; therefore $\rho\lambda^{\frac{1}{4}}$, $\rho\lambda^{\frac{1}{4}}\sqrt{1-k^2}$ have it also. [X. 14]

(Euclid has not the assistance of symbols to prove the proportion (3) above. He therefore uses the lemmas $ab : bc = a : c$ and $d^2 : de = d : e$ to deduce from the relations

$$\left.\begin{array}{r} ab = d^2 \\ \text{and} \quad d : b = c : e \end{array}\right\}$$

that $a : c = d : e$.)

The straight lines $\rho\lambda^{\frac{1}{4}}$, $\rho\lambda^{\frac{1}{4}}\sqrt{1-k^2}$ may take any of the following forms according as the straight lines first taken are

(1) a, $\sqrt{B}$, $\sqrt{a^2-c^2}$, (2) $\sqrt{A}$, $\sqrt{B}$, $\sqrt{A-k^2A}$, (3) $\sqrt{A}$, b, $\sqrt{A-k^2A}$.

$$(1) \quad \sqrt{a\sqrt{B}}, \quad \frac{\sqrt{B(a^2-c^2)}}{\sqrt{a\sqrt{B}}};$$

$$(2) \quad \sqrt[4]{AB}, \quad \frac{\sqrt{B(A-k^2A)}}{\sqrt[4]{AB}};$$

$$(3) \quad \sqrt{b\sqrt{A}}, \quad \frac{b\sqrt{A-k^2A}}{\sqrt{b\sqrt{A}}}.$$

II. If the other conditions are the same, but the square on the first medial straight line is to exceed the square on the second by the square on a straight line *incommensurable* with the first, we begin with the three straight lines ρ, $\rho\sqrt{\lambda}$, $\frac{\rho}{\sqrt{1+k^2}}$, and the medial straight lines are

$$\rho\lambda^{\frac{1}{4}}, \quad \frac{\rho\lambda^{\frac{1}{4}}}{\sqrt{1+k^2}}.$$

The possible forms are even more various in this case owing to the more various forms that the original lines may take, e.g.

$$\begin{array}{llll} (1) & a, & \sqrt{B}, & \sqrt{a^2-C}; \\ (2) & \sqrt{A}, & b, & \sqrt{A-c^2}; \\ (3) & \sqrt{A}, & b, & \sqrt{A-C}; \\ (4) & \sqrt{A}, & \sqrt{B}, & \sqrt{A-c^2}; \\ (5) & \sqrt{A}, & \sqrt{B}, & \sqrt{A-C}; \end{array}$$

the medial straight lines corresponding to these being

$$(1) \quad \sqrt{a\sqrt{B}}, \quad \frac{\sqrt{B(a^2-C)}}{\sqrt{a\sqrt{B}}};$$

$$(2) \quad \sqrt{b\sqrt{A}}, \quad \frac{b\sqrt{A-c^2}}{\sqrt{b\sqrt{A}}};$$

$$(3) \quad \sqrt{b\sqrt{A}}, \quad \frac{b\sqrt{A-C}}{\sqrt{b\sqrt{A}}};$$

$$(4) \quad \sqrt[4]{AB}, \quad \frac{\sqrt{B(A-c^2)}}{\sqrt[4]{AB}};$$

$$(5) \quad \sqrt[4]{AB}, \quad \frac{\sqrt{B(A-C)}}{\sqrt[4]{AB}}.$$

Lemma.

Let *ABC* be a right-angled triangle having the angle *A* right, and let the perpendicular *AD* be drawn;

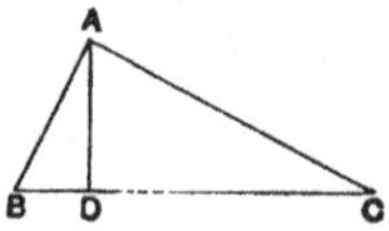

I say that the rectangle *CB*, *BD* is equal to the square on *BA*,

the rectangle *BC*, *CD* equal to the square on *CA*,

the rectangle *BD*, *DC* equal to the square on *AD*,

and, further, the rectangle *BC*, *AD* equal to the rectangle *BA*, *AC*.

And first that the rectangle *CB*, *BD* is equal to the square on *BA*.

For, since in a right-angled triangle *AD* has been drawn from the right angle perpendicular to the base,
therefore the triangles *ABD*, *ADC* are similar both to the whole *ABC* and to one another. [vi. 8]

And since the triangle *ABC* is similar to the triangle *ABD*,
therefore, as *CB* is to *BA*, so is *BA* to *BD*; [vi. 4]
therefore the rectangle *CB*, *BD* is equal to the square on *AB*. [vi. 17]

For the same reason the rectangle *BC*, *CD* is also equal to the square on *AC*.

And since, if in a right-angled triangle a perpendicular be drawn from the right angle to the base, the perpendicular so drawn is a mean proportional between the segments of the base, [vi. 8, Por.]
therefore, as *BD* is to *DA*, so is *AD* to *DC*;
therefore the rectangle *BD*, *DC* is equal to the square on *AD*. [vi. 17]

I say that the rectangle *BC*, *AD* is also equal to the rectangle *BA*, *AC*.

For since, as we said, *ABC* is similar to *ABD*,
therefore, as *BC* is to *CA*, so is *BA* to *AD*. [vi. 4]

Therefore the rectangle *BC*, *AD* is equal to the rectangle *BA*, *AC*. [vi. 16]

Q. E. D.

Proposition 33.

To find two straight lines incommensurable in square which make the sum of the squares on them rational but the rectangle contained by them medial.

Let there be set out two rational straight lines *AB*, *BC* commensurable in square only and such that the square on the greater *AB* is greater than the square on the less *BC* by the square on a straight line incommensurable with *AB*, [x. 30]

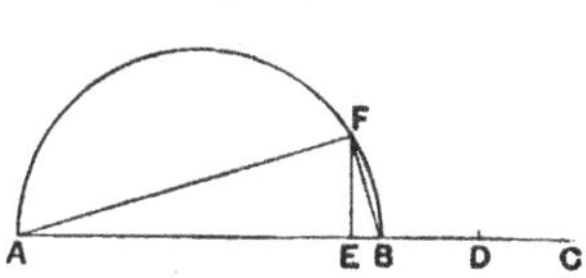

let BC be bisected at D,

let there be applied to AB a parallelogram equal to the square on either of the straight lines BD, DC and deficient by a square figure, and let it be the rectangle AE, EB; [vi. 28]

let the semicircle AFB be described on AB,

let EF be drawn at right angles to AB,

and let AF, FB be joined.

Then, since AB, BC are unequal straight lines,

and the square on AB is greater than the square on BC by the square on a straight line incommensurable with AB,

while there has been applied to AB a parallelogram equal to the fourth part of the square on BC, that is, to the square on half of it, and deficient by a square figure, making the rectangle AE, EB,

therefore AE is incommensurable with EB. [x. 18]

And, as AE is to EB, so is the rectangle BA, AE to the rectangle AB, BE,

while the rectangle BA, AE is equal to the square on AF,

and the rectangle AB, BE to the square on BF;

therefore the square on AF is incommensurable with the square on FB;

therefore AF, FB are incommensurable in square.

And, since AB is rational,

therefore the square on AB is also rational;

so that the sum of the squares on AF, FB is also rational. [i. 47]

And since, again, the rectangle AE, EB is equal to the square on EF,

and, by hypothesis, the rectangle AE, EB is also equal to the square on BD,

therefore FE is equal to BD;

therefore BC is double of FE,

so that the rectangle AB, BC is also commensurable with the rectangle AB, EF.

But the rectangle AB, BC is medial; [x. 21]

therefore the rectangle AB, EF is also medial. [x. 23, Por.]

But the rectangle AB, EF is equal to the rectangle AF, FB; [Lemma]
therefore the rectangle AF, FB is also medial.

But it was also proved that the sum of the squares on these straight lines is rational.

Therefore two straight lines AF, FB incommensurable in square have been found which make the sum of the squares on them rational, but the rectangle contained by them medial.

Q. E. D.

Euclid takes the straight lines found in x. 30, viz. ρ, $\frac{\rho}{\sqrt{1+k^2}}$.

He then solves *geometrically* the equations

$$\left.\begin{aligned} x+y&=\rho \\ xy&=\frac{\rho^2}{4(1+k^2)} \end{aligned}\right\} \quad \ldots\ldots\ldots\ldots(1).$$

If x, y are the values found, he takes u, v such that

$$\left.\begin{aligned} u^2&=\rho x \\ v^2&=\rho y \end{aligned}\right\} \quad \ldots\ldots\ldots\ldots(2),$$

and u, v are straight lines satisfying the conditions of the problem.

Solving algebraically, we get (if $x>y$)

$$x=\frac{\rho}{2}\left(1+\frac{k}{\sqrt{1+k^2}}\right),\quad y=\frac{\rho}{2}\left(1-\frac{k}{\sqrt{1+k^2}}\right),$$

whence

$$\left.\begin{aligned} u&=\frac{\rho}{\sqrt{2}}\sqrt{1+\frac{k}{\sqrt{1+k^2}}} \\ v&=\frac{\rho}{\sqrt{2}}\sqrt{1-\frac{k}{\sqrt{1+k^2}}} \end{aligned}\right\} \quad \ldots\ldots\ldots\ldots(3).$$

Euclid's proof that these straight lines fulfil the requirements is as follows.

(α) The constants in the equations (1) satisfy the conditions of x. 18;

therefore $\qquad x \smile y.$

But $\qquad x:y=u^2:v^2.$

Therefore $\qquad u^2 \smile v^2,$

and u, v are thus *incommensurable in square*.

(β) $u^2+v^2=\rho^2$, which is *rational*.

(γ) By (1), $\qquad \sqrt{xy}=\frac{\rho}{2\sqrt{1+k^2}}.$

By (2),

$$uv=\rho\,.\sqrt{xy}$$
$$=\frac{\rho^2}{2\sqrt{1+k^2}}.$$

But $\frac{\rho^2}{\sqrt{1+k^2}}$ is a medial area,

therefore uv is medial.

Since ρ, $\frac{\rho}{\sqrt{1+k^2}}$ may have any of the three forms

$$(1)\ a,\ \sqrt{a^2-B},\quad (2)\ \sqrt{A},\ \sqrt{A-B},\quad (3)\ \sqrt{A},\ \sqrt{A-b^2},$$

u, v may have any of the forms

$$(1)\quad \sqrt{\frac{a^2+a\sqrt{B}}{2}},\quad \sqrt{\frac{a^2-a\sqrt{B}}{2}};$$

$$(2)\quad \sqrt{\frac{A+\sqrt{AB}}{2}},\quad \sqrt{\frac{A-\sqrt{AB}}{2}};$$

$$(3)\quad \sqrt{\frac{A+b\sqrt{A}}{2}},\quad \sqrt{\frac{A-b\sqrt{A}}{2}}.$$

PROPOSITION 34.

To find two straight lines incommensurable in square which make the sum of the squares on them medial but the rectangle contained by them rational.

Let there be set out two medial straight lines *AB*, *BC*, commensurable in square only, such that the rectangle which they contain is rational, and the square on *AB* is greater than the square on *BC* by the square on a straight line incommensurable with *AB*; [x. 31, *ad fin.*]

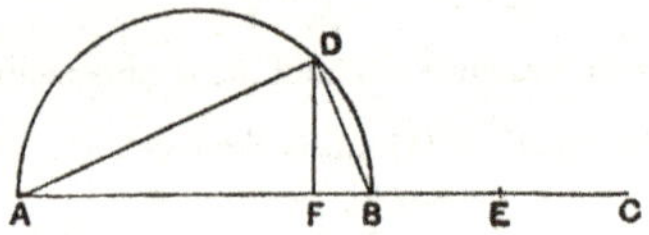

let the semicircle *ADB* be described on *AB*,

let *BC* be bisected at *E*,

let there be applied to *AB* a parallelogram equal to the square on *BE* and deficient by a square figure, namely the rectangle *AF*, *FB*; [vi. 28]

therefore *AF* is incommensurable in length with *FB*. [x. 18]

Let *FD* be drawn from *F* at right angles to *AB*, and let *AD*, *DB* be joined.

Since AF is incommensurable in length with FB, therefore the rectangle BA, AF is also incommensurable with the rectangle AB, BF. [x. 11]

But the rectangle BA, AF is equal to the square on AD, and the rectangle AB, BF to the square on DB; therefore the square on AD is also incommensurable with the square on DB.

And, since the square on AB is medial, therefore the sum of the squares on AD, DB is also medial. [III. 31, I. 47]

And, since BC is double of DF, therefore the rectangle AB, BC is also double of the rectangle AB, FD.

But the rectangle AB, BC is rational; therefore the rectangle AB, FD is also rational. [x. 6]

But the rectangle AB, FD is equal to the rectangle AD, DB; [Lemma]
so that the rectangle AD, DB is also rational.

Therefore two straight lines AD, DB incommensurable in square have been found which make the sum of the squares on them medial, but the rectangle contained by them rational.

Q. E. D.

In this case we take [x. 31, 2nd part] the medial straight lines

$$\frac{\rho}{(1+k^2)^{\frac{1}{4}}}, \quad \frac{\rho}{(1+k^2)^{\frac{3}{4}}}.$$

Solve the equations

$$\left.\begin{aligned} x + y &= \frac{\rho}{(1+k^2)^{\frac{1}{4}}} \\ xy &= \frac{\rho^2}{4(1+k^2)^{\frac{3}{2}}} \end{aligned}\right\} \quad \ldots\ldots\ldots\ldots(1).$$

Take u, v such that, if x, y be the result of the solution,

$$\left.\begin{aligned} u^2 &= \frac{\rho}{(1+k^2)^{\frac{1}{4}}} \cdot x \\ v^2 &= \frac{\rho}{(1+k^2)^{\frac{1}{4}}} \cdot y \end{aligned}\right\} \quad \ldots\ldots\ldots\ldots(2),$$

and u, v are straight lines satisfying the given conditions.

Euclid's proof is similar to the preceding.

(α) From (1) it follows [x. 18] that

$$x \smile y,$$

whence

$$u^2 \smile v^2,$$

and u, v are thus incommensurable in square.

(β) $u^2 + v^2 = \dfrac{\rho^2}{\sqrt{1+k^2}}$, which is a medial area.

(γ)
$$uv = \frac{\rho}{(1+k^2)^{\frac{1}{4}}} \cdot \sqrt{xy}$$
$$= \frac{1}{2} \cdot \frac{\rho^2}{1+k^2}, \text{ which is a rational area.}$$

Therefore uv is rational.

To find the actual form of u, v, we have, by solving the equations (1) (if $x > y$),

$$x = \frac{\rho}{2(1+k^2)^{\frac{3}{4}}}(\sqrt{1+k^2}+k),$$

$$y = \frac{\rho}{2(1+k^2)^{\frac{3}{4}}}(\sqrt{1+k^2}-k);$$

and hence
$$u = \frac{\rho}{\sqrt{2(1+k^2)}}\sqrt{\sqrt{1+k^2}+k},$$

$$v = \frac{\rho}{\sqrt{2(1+k^2)}}\sqrt{\sqrt{1+k^2}-k}.$$

Bearing in mind the forms which $\dfrac{\rho}{(1+k^2)^{\frac{1}{4}}}$, $\dfrac{\rho}{(1+k^2)^{\frac{3}{4}}}$ may take (see note on x. 31), we shall find that u, v may have any of the forms

(1) $\sqrt{\dfrac{(a+\sqrt{B})\sqrt{a^2-B}}{2}}$, $\sqrt{\dfrac{(a-\sqrt{B})\sqrt{a^2-B}}{2}}$;

(2) $\sqrt{\dfrac{(\sqrt{A}+\sqrt{B})\sqrt{A-B}}{2}}$, $\sqrt{\dfrac{(\sqrt{A}-\sqrt{B})\sqrt{A-B}}{2}}$;

(3) $\sqrt{\dfrac{(\sqrt{A}+b)\sqrt{A-b^2}}{2}}$, $\sqrt{\dfrac{(\sqrt{A}-b)\sqrt{A-b^2}}{2}}$.

Proposition 35.

To find two straight lines incommensurable in square which make the sum of the squares on them medial and the rectangle contained by them medial and moreover incommensurable with the sum of the squares on them.

Let there be set out two medial straight lines *AB*, *BC* commensurable in square only, containing a medial rectangle, and such that the square on *AB* is greater than the square on *BC* by the square on a straight line incommensurable with *AB*; [x. 32, *ad fin.*]

let the semicircle ADB be described on AB,
and let the rest of the construction be as above.

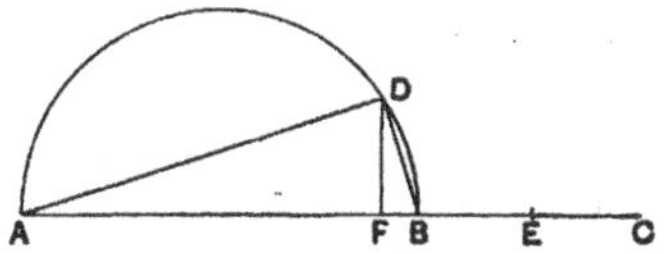

Then, since AF is incommensurable in length with FB, [x. 18]

AD is also incommensurable in square with DB. [x. 11]

And, since the square on AB is medial,
therefore the sum of the squares on AD, DB is also medial. [III. 31, I. 47]

And, since the rectangle AF, FB is equal to the square on each of the straight lines BE, DF,
therefore BE is equal to DF;
therefore BC is double of FD,
so that the rectangle AB, BC is also double of the rectangle AB, FD.

But the rectangle AB, BC is medial;
therefore the rectangle AB, FD is also medial. [x. 32, Por.]

And it is equal to the rectangle AD, DB; [Lemma after x. 32]

therefore the rectangle AD, DB is also medial.

And, since AB is incommensurable in length with BC,
while CB is commensurable with BE,
therefore AB is also incommensurable in length with BE, [x. 13]

so that the square on AB is also incommensurable with the rectangle AB, BE. [x. 11]

But the squares on AD, DB are equal to the square on AB, [I. 47]

and the rectangle AB, FD, that is, the rectangle AD, DB, is equal to the rectangle AB, BE;
therefore the sum of the squares on AD, DB is incommensurable with the rectangle AD, DB.

Therefore two straight lines *AD*, *DB* incommensurable in square have been found which make the sum of the squares on them medial and the rectangle contained by them medial and moreover incommensurable with the sum of the squares on them.

Q. E. D.

Take the medial straight lines found in X. 32 (2nd part), viz.

$$\rho\lambda^{\frac{1}{4}},\quad \rho\lambda^{\frac{1}{4}}/\sqrt{1+k^2}.$$

Solve the equations

$$\left.\begin{aligned} x+y&=\rho\lambda^{\frac{1}{4}} \\ xy&=\frac{\rho^2\sqrt{\lambda}}{4(1+k^2)} \end{aligned}\right\} \qquad \ldots\ldots(1),$$

and then put

$$\left.\begin{aligned} u^2&=\rho\lambda^{\frac{1}{4}}.x \\ v^2&=\rho\lambda^{\frac{1}{4}}.y \end{aligned}\right\} \qquad \ldots\ldots(2),$$

where x, y are the ascertained values of x, y.

Then u, v are straight lines satisfying the given conditions.

Euclid proves this as follows.

(α) From (1) it follows [X. 18] that $x \smile y$.

Therefore $$u^2 \smile v^2,$$

and $$u \smile\!\!-\, v.$$

(β) $$u^2+v^2=\rho^2\sqrt{\lambda}, \text{ which is a medial area } \ldots\ldots(3).$$

(γ) $$uv=\rho\lambda^{\frac{1}{4}}.\sqrt{xy}$$

$$=\frac{1}{2}\frac{\rho^2\sqrt{\lambda}}{\sqrt{1+k^2}}, \text{ which is a medial area } \ldots\ldots(4);$$

therefore uv is medial.

(δ) $$\rho\lambda^{\frac{1}{4}} \smile \frac{1}{2}\frac{\rho\lambda^{\frac{1}{4}}}{\sqrt{1+k^2}},$$

whence $$\rho^2\sqrt{\lambda} \smile \frac{1}{2}\frac{\rho^2\sqrt{\lambda}}{\sqrt{1+k^2}}.$$

That is, by (3) and (4),

$$(u^2+v^2) \smile uv.$$

The actual values are found thus. Solving the equations (1), we have

$$x=\frac{\rho\lambda^{\frac{1}{4}}}{2}\left(1+\frac{k}{\sqrt{1+k^2}}\right),$$

$$y=\frac{\rho\lambda^{\frac{1}{4}}}{2}\left(1-\frac{k}{\sqrt{1+k^2}}\right),$$

whence $$u=\frac{\rho\lambda^{\frac{1}{4}}}{\sqrt{2}}\sqrt{1+\frac{k}{\sqrt{1+k^2}}},$$

$$v=\frac{\rho\lambda^{\frac{1}{4}}}{\sqrt{2}}\sqrt{1-\frac{k}{\sqrt{1+k^2}}}.$$

According as ρ is of the form a or $\sqrt{A}$, we have a variety of forms for u, v, arrived at by using the same transformations as in the notes on x. 30 and x. 32 (second part), e.g.

$$(1) \quad \sqrt{\frac{(a+\sqrt{C})\sqrt{B}}{2}}, \quad \sqrt{\frac{(a-\sqrt{C})\sqrt{B}}{2}};$$

$$(2) \quad \sqrt{\frac{(\sqrt{A}+\sqrt{C})\sqrt{B}}{2}}, \quad \sqrt{\frac{(\sqrt{A}-\sqrt{C})\sqrt{B}}{2}};$$

$$(3) \quad \sqrt{\frac{(\sqrt{A}+c)\sqrt{B}}{2}}, \quad \sqrt{\frac{(\sqrt{A}-c)\sqrt{B}}{2}};$$

and the expressions in (2), (3) with b in place of $\sqrt{B}$.

PROPOSITION 36.

If two rational straight lines commensurable in square only be added together, the whole is irrational; and let it be called **binomial.**

For let two rational straight lines *AB*, *BC* commensurable in square only be added together;

I say that the whole *AC* is irrational.

For, since *AB* is incommensurable in length with *BC*—for they are commensurable in square only—

and, as *AB* is to *BC*, so is the rectangle *AB*, *BC* to the square on *BC*,

therefore the rectangle *AB*, *BC* is incommensurable with the square on *BC*. [x. 11]

But twice the rectangle *AB*, *BC* is commensurable with the rectangle *AB*, *BC* [x. 6], and the squares on *AB*, *BC* are commensurable with the square on *BC*—for *AB*, *BC* are rational straight lines commensurable in square only— [x. 15]

therefore twice the rectangle *AB*, *BC* is incommensurable with the squares on *AB*, *BC*. [x. 13]

And, *componendo*, twice the rectangle *AB*, *BC* together with the squares on *AB*, *BC*, that is, the square on *AC* [II. 4], is incommensurable with the sum of the squares on *AB*, *BC*. [x. 16]

But the sum of the squares on *AB*, *BC* is rational;

therefore the square on *AC* is irrational,

so that *AC* is also irrational. [x. Def. 4]

And let it be called **binomial.** Q. E. D.

Here begins the first hexad of propositions relating to compound irrational straight lines. The six compound irrational straight lines are formed by *adding* two parts, as the corresponding six in Props. 73—78 are formed by *subtraction*. The relation between the six irrational straight lines in this and the next five propositions with those described in Definitions II. and the Props. 48—53 following thereon (the *first*, *second*, *third*, *fourth*, *fifth* and *sixth binomials*) will be seen when we come to Props. 54—59; but it may be stated here that the six compound irrationals in Props. 36—41 can be found by means of the equivalent of *extracting the square root* of the compound irrationals in x. 48—53 (the process being, strictly speaking, the finding of the sides of the squares equal to the rectangles contained by the latter irrationals respectively and a rational straight line as the other side), and it is therefore the further removed compound irrational, so to speak, which is treated first.

In reproducing the proofs of the propositions, I shall for the sake of simplicity call the two parts of the compound irrational straight line x, y, explaining at the outset the forms which x, y really have in each case; x will always be supposed to be the greater segment.

In this proposition x, y are of the form ρ, $\sqrt{k}\,.\,\rho$, and $(x+y)$ is proved to be irrational thus.

$x \frown\!\!-\, y$, so that $x \smile y$.

Now $$x : y = x^2 : xy,$$

so that $$x^2 \smile xy.$$

But $x^2 \frown (x^2+y^2)$, and $xy \frown 2xy$;

therefore $$(x^2+y^2) \smile 2xy,$$

and hence $$(x^2+y^2+2xy) \smile (x^2+y^2).$$

But (x^2+y^2) is rational;

therefore $(x+y)^2$, and therefore $(x+y)$, is *irrational*.

This irrational straight line, $\rho+\sqrt{k}\,.\,\rho$, is called a *binomial* straight line.

This and the corresponding *apotome* $(\rho-\sqrt{k}\,.\,\rho)$ found in x. 73 are the positive roots of the equation

$$x^4 - 2(1+k)\rho^2\,.\,x^2 + (1-k)^2\rho^4 = 0.$$

Proposition 37.

If two medial straight lines commensurable in square only and containing a rational rectangle be added together, the whole is irrational; and let it be called a **first bimedial** *straight line.*

For let two medial straight lines AB, BC commensurable in square only and containing a rational rectangle be added together;

I say that the whole AC is irrational.

For, since AB is incommensurable in length with BC, therefore the squares on AB, BC are also incommensurable with twice the rectangle AB, BC; [cf. x. 36, ll. 9—20] and, *componendo*, the squares on AB, BC together with twice the rectangle AB, BC, that is, the square on AC [II. 4], is incommensurable with the rectangle AB, BC. [x. 16]

But the rectangle AB, BC is rational, for, by hypothesis, AB, BC are straight lines containing a rational rectangle;

therefore the square on AC is irrational;

therefore AC is irrational. [x. Def. 4]

And let it be called a **first bimedial** straight line.

Q. E. D.

Here x, y have the forms $k^{\frac{1}{4}}\rho$, $k^{\frac{3}{4}}\rho$ respectively, as found in x. 27.

Exactly as in the last case we prove that

$$x^2 + y^2 \smile 2xy,$$

whence $$(x + y)^2 \smile 2xy.$$

But xy is rational;

therefore $(x + y)^2$, and consequently $(x + y)$, is *irrational*.

The irrational straight line $k^{\frac{1}{4}}\rho + k^{\frac{3}{4}}\rho$ is called a *first bimedial* straight line.

This and the corresponding first *apotome of a medial* $(k^{\frac{1}{4}}\rho - k^{\frac{3}{4}}\rho)$ found in x. 74 are the positive roots of the equation

$$x^4 - 2\sqrt{k}(1 + k)\rho^2 . x^2 + k(1 - k)^2 \rho^4 = 0.$$

Proposition 38.

If two medial straight lines commensurable in square only and containing a medial rectangle be added together, the whole is irrational; and let it be called a **second bimedial** *straight line.*

For let two medial straight lines AB, BC commensurable in square only and containing a medial rectangle be added together;

I say that AC is irrational.

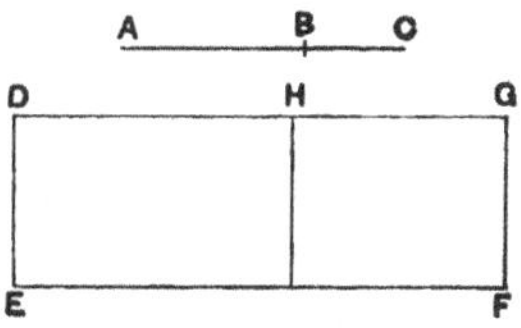

For let a rational straight line DE be set out, and let the parallelogram DF equal to the square on AC be applied to DE, producing DG as breadth. [I. 44]

Then, since the square on AC is equal to the squares on AB, BC and twice the rectangle AB, BC, [II. 4]

let EH, equal to the squares on AB, BC, be applied to DE;

therefore the remainder *HF* is equal to twice the rectangle *AB*, *BC*.

And, since each of the straight lines *AB*, *BC* is medial, therefore the squares on *AB*, *BC* are also medial.

But, by hypothesis, twice the rectangle *AB*, *BC* is also medial.

And *EH* is equal to the squares on *AB*, *BC*, while *FH* is equal to twice the rectangle *AB*, *BC*; therefore each of the rectangles *EH*, *HF* is medial.

And they are applied to the rational straight line *DE*; therefore each of the straight lines *DH*, *HG* is rational and incommensurable in length with *DE*. [x. 22]

Since then *AB* is incommensurable in length with *BC*, and, as *AB* is to *BC*, so is the square on *AB* to the rectangle *AB*, *BC*, therefore the square on *AB* is incommensurable with the rectangle *AB*, *BC*. [x. 11]

But the sum of the squares on *AB*, *BC* is commensurable with the square on *AB*, [x. 15] and twice the rectangle *AB*, *BC* is commensurable with the rectangle *AB*, *BC*. [x. 6]

Therefore the sum of the squares on *AB*, *BC* is incommensurable with twice the rectangle *AB*, *BC*. [x. 13]

But *EH* is equal to the squares on *AB*, *BC*, and *HF* is equal to twice the rectangle *AB*, *BC*.

Therefore *EH* is incommensurable with *HF*, so that *DH* is also incommensurable in length with *HG*. [vi. 1, x. 11]

Therefore *DH*, *HG* are rational straight lines commensurable in square only; so that *DG* is irrational. [x. 36]

But *DE* is rational; and the rectangle contained by an irrational and a rational straight line is irrational; [cf. x. 20] therefore the area *DF* is irrational, and the side of the square equal to it is irrational. [x. Def. 4]

But AC is the side of the square equal to DF; therefore AC is irrational.

And let it be called a **second bimedial** straight line.

Q. E. D.

After proving (l. 21) that *each* of the squares on AB, BC is medial, Euclid states (ll. 24, 26) that EH, which is equal to the *sum* of the squares, is a medial area, but does not explain why. It is because, by hypothesis, the squares on AB, BC are commensurable, so that the sum of the squares is commensurable with either [x. 15] and is therefore a medial area [x. 23, Por.].

In this case [x. 28, note] x, y are of the forms $k^{\frac{1}{4}}\rho$, $\lambda^{\frac{1}{2}}\rho/k^{\frac{1}{4}}$ respectively.

Apply each of the areas (x^2+y^2) and $2xy$ to a rational straight line σ, i.e. suppose

$$x^2+y^2=\sigma u,$$
$$2xy=\sigma v.$$

Now it follows from the hypothesis, x. 15 and x. 23, Por. that (x^2+y^2) is a medial area; and so is $2xy$, by hypothesis;
therefore σu, σv are medial areas.

Therefore each of the straight lines u, v is rational and $\smile \sigma$ .... (1).

Again $x \smile y$;
therefore $x^2 \smile xy$.
But $x^2 \frown x^2+y^2$ and $xy \frown 2xy$;
therefore $x^2+y^2 \smile 2xy$,
or $\sigma u \smile \sigma v$,
whence $u \smile v$ ....(2).

Therefore, by (1), (2), u, v are rational and $\frown$.
It follows, by x. 36, that $(u+v)$ is irrational.
Therefore $(u+v)\,\sigma$ is an irrational area [this can be deduced from x. 20 by *reductio ad absurdum*],
whence $(x+y)^2$, and consequently $(x+y)$, is irrational.

The irrational straight line $k^{\frac{1}{4}}\rho+\dfrac{\lambda^{\frac{1}{2}}\rho}{k^{\frac{1}{4}}}$ is called a *second bimedial* straight line.

This and the corresponding *second apotome of a medial* $\left(k^{\frac{1}{4}}\rho-\dfrac{\sqrt{\lambda}}{k^{\frac{1}{4}}}\rho\right)$ found in x. 75 are the positive roots of the equation

$$x^4-2\frac{k+\lambda}{\sqrt{k}}\rho^2\,.\,x^2+\frac{(k-\lambda)^2}{k}\rho^4=0.$$

Proposition 39.

If two straight lines incommensurable in square which make the sum of the squares on them rational, but the rectangle contained by them medial, be added together, the whole straight line is irrational: and let it be called **major**.

For let two straight lines AB, BC incommensurable in square, and fulfilling the given conditions [x. 33], be added together;

I say that AC is irrational.

For, since the rectangle AB, BC is medial, twice the rectangle AB, BC is also medial. [x. 6 and 23, Por.]

But the sum of the squares on AB, BC is rational; therefore twice the rectangle AB, BC is incommensurable with the sum of the squares on AB, BC, so that the squares on AB, BC together with twice the rectangle AB, BC, that is, the square on AC, is also incommensurable with the sum of the squares on AB, BC; [x. 16] therefore the square on AC is irrational, so that AC is also irrational. [x. Def. 4]

And let it be called **major**.

Q. E. D.

Here x, y are of the form found in x. 33, viz.

$$\frac{\rho}{\sqrt{2}}\sqrt{1+\frac{k}{\sqrt{1+k^2}}},\quad \frac{\rho}{\sqrt{2}}\sqrt{1-\frac{k}{\sqrt{1+k^2}}}.$$

By hypothesis, the rectangle xy is medial; therefore $2xy$ is medial.

Also (x^2+y^2) is a rational area.

Therefore $$x^2+y^2 \smile 2xy,$$

whence $$(x+y)^2 \smile (x^2+y^2),$$

so that $(x+y)^2$, and therefore $(x+y)$, is irrational.

The irrational straight line $\frac{\rho}{\sqrt{2}}\sqrt{1+\frac{k}{\sqrt{1+k^2}}}+\frac{\rho}{\sqrt{2}}\sqrt{1-\frac{k}{\sqrt{1+k^2}}}$ is called a *major* (irrational) straight line.

This and the corresponding *minor* irrational found in x. 76 are the positive roots of the equation

$$x^4-2\rho^2 . x^2+\frac{k^2}{1+k^2}\rho^4=0.$$

PROPOSITION 40.

If two straight lines incommensurable in square which make the sum of the squares on them medial, but the rectangle contained by them rational, be added together, the whole straight line is irrational; and let it be called the **side of a rational plus a medial area.**

For let two straight lines *AB*, *BC* incommensurable in square, and fulfilling the given conditions [x. 34], be added together;

I say that *AC* is irrational.

For, since the sum of the squares on *AB*, *BC* is medial, while twice the rectangle *AB*, *BC* is rational,
therefore the sum of the squares on *AB*, *BC* is incommensurable with twice the rectangle *AB*, *BC*;
so that the square on *AC* is also incommensurable with twice the rectangle *AB*, *BC*. [x. 16]

But twice the rectangle *AB*, *BC* is rational;
therefore the square on *AC* is irrational.

Therefore *AC* is irrational. [x. Def. 4]

And let it be called the **side of a rational plus a medial area**.

Q. E. D.

Here x, y have [x. 34] the forms

$$\frac{\rho}{\sqrt{2(1+k^2)}}\sqrt{\sqrt{1+k^2}+k}, \quad \frac{\rho}{\sqrt{2(1+k^2)}}\sqrt{\sqrt{1+k^2}-k}.$$

In this case (x^2+y^2) is a medial, and $2xy$ a rational, area; thus

$$x^2+y^2 \smile 2xy.$$

Therefore $$(x+y)^2 \smile 2xy,$$

whence, since $2xy$ is rational,

$(x+y)^2$, and consequently $(x+y)$, is irrational.

The irrational straight line

$$\frac{\rho}{\sqrt{2(1+k^2)}}\sqrt{\sqrt{1+k^2}+k}+\frac{\rho}{\sqrt{2(1+k^2)}}\sqrt{\sqrt{1+k^2}-k}$$

is called (for an obvious reason) the "*side*" *of a rational plus a medial* (area).

This and the corresponding irrational with a minus sign found in x. 77 are the positive roots of the equation

$$x^4-\frac{2}{\sqrt{1+k^2}}\rho^2 . x^2+\frac{k^2}{(1+k^2)^2}\rho^4=0.$$

Proposition 41.

If two straight lines incommensurable in square which make the sum of the squares on them medial, and the rectangle contained by them medial and also incommensurable with the sum of the squares on them, be added together, the whole straight line is irrational; and let it be called the **side of the sum of two medial areas.**

For let two straight lines *AB*, *BC* incommensurable in square and satisfying the given conditions [x. 35] be added together;

I say that *AC* is irrational.

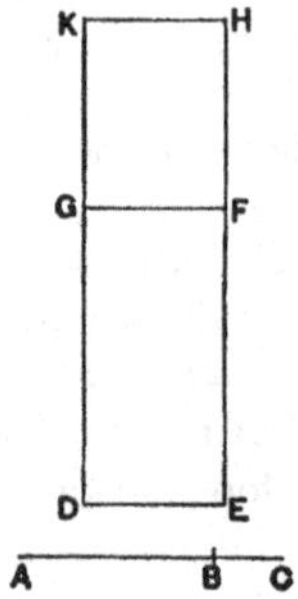

Let a rational straight line *DE* be set out, and let there be applied to *DE* the rectangle *DF* equal to the squares on *AB*, *BC*, and the rectangle *GH* equal to twice the rectangle *AB*, *BC*;

therefore the whole *DH* is equal to the square on *AC*. [II. 4]

Now, since the sum of the squares on *AB*, *BC* is medial,

and is equal to *DF*,

therefore *DF* is also medial.

And it is applied to the rational straight line *DE*;

therefore *DG* is rational and incommensurable in length with *DE*. [x. 22]

For the same reason *GK* is also rational and incommensurable in length with *GF*, that is, *DE*.

And, since the squares on *AB*, *BC* are incommensurable with twice the rectangle *AB*, *BC*,

DF is incommensurable with *GH*;

so that *DG* is also incommensurable with *GK*. [VI. 1, x. 11]

And they are rational;

therefore *DG*, *GK* are rational straight lines commensurable in square only;

therefore *DK* is irrational and what is called binomial. [x. 36]

But *DE* is rational;

therefore *DH* is irrational, and the side of the square which is equal to it is irrational. [x. Def. 4]

But *AC* is the side of the square equal to *HD*;

therefore *AC* is irrational.

And let it be called the **side of the sum of two medial areas.**

Q. E. D.

In this case x, y are of the form

$$\frac{\rho\lambda^{\frac{1}{4}}}{\sqrt{2}}\sqrt{1+\frac{k}{\sqrt{1+k^2}}},\quad \frac{\rho\lambda^{\frac{1}{4}}}{\sqrt{2}}\sqrt{1-\frac{k}{\sqrt{1+k^2}}}.$$

By hypothesis, (x^2+y^2) and $2xy$ are medial areas, and

$$x^2+y^2 \smile 2xy \quad \text{..........................(1).}$$

'Apply' these areas respectively to a rational straight line σ, and suppose

$$\left.\begin{array}{r} x^2+y^2=\sigma u \\ 2xy=\sigma v \end{array}\right\} \quad \text{..........................(2).}$$

Since then σu and σv are both medial areas, u, v are rational and both are $\smile \sigma$ ....(3).

Now, by (1) and (2),

$$\sigma u \smile \sigma v,$$

so that

$$u \smile v.$$

By this and (3), u, v are rational and $\frown$.

Therefore [X. 36] $(u+v)$ is irrational.

Hence $\sigma(u+v)$ is irrational [deduction from X. 20].

Thus $(x+y)^2$, and therefore $(x+y)$, is irrational.

The irrational straight line

$$\frac{\rho\lambda^{\frac{1}{4}}}{\sqrt{2}}\sqrt{1+\frac{k}{\sqrt{1+k^2}}}+\frac{\rho\lambda^{\frac{1}{4}}}{\sqrt{2}}\sqrt{1-\frac{k}{\sqrt{1+k^2}}}$$

is called (again for an obvious reason) the "*side*" *of the sum of two medials* (medial areas).

This and the corresponding irrational with a minus sign found in X. 78 are the positive roots of the equation

$$x^4-2\sqrt{\lambda}\,.\,x^2\rho^2+\lambda\frac{k^2}{1+k^2}\rho^4=0.$$

LEMMA.

And that the aforesaid irrational straight lines are divided only in one way into the straight lines of which they are the sum and which produce the types in question, we will now prove after premising the following lemma.

Let the straight line AB be set out, let the whole be cut into unequal parts at each of the points C, D, and let AC be supposed greater than DB;

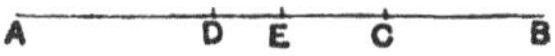

I say that the squares on AC, CB are greater than the squares on AD, DB.

For let AB be bisected at E.

Then, since AC is greater than DB,

let DC be subtracted from each;

therefore the remainder AD is greater than the remainder CB.

But AE is equal to EB;

therefore DE is less than EC;

therefore the points C, D are not equidistant from the point of bisection.

And, since the rectangle AC, CB together with the square on EC is equal to the square on EB, [II. 5]

and, further, the rectangle AD, DB together with the square on DE is equal to the square on EB, [*id.*]

therefore the rectangle AC, CB together with the square on EC is equal to the rectangle AD, DB together with the square on DE.

And of these the square on DE is less than the square on EC;

therefore the remainder, the rectangle AC, CB, is also less than the rectangle AD, DB,

so that twice the rectangle AC, CB is also less than twice the rectangle AD, DB.

Therefore also the remainder, the sum of the squares on AC, CB, is greater than the sum of the squares on AD, DB.

Q. E. D.

3. **and which produce the types in question.** The Greek is ποιουσῶν τὰ προκείμενα εἴδη, and I have taken εἴδη to mean "types (of irrational straight lines)," though the expression might perhaps mean "satisfying the *conditions* in question."

This proves that, if $x + y = u + v$, and if u, v are more nearly equal than x, y (i.e. if the straight line is divided in the second case nearer to the point of bisection), then

$$(x^2 + y^2) > (u^2 + v^2).$$

It is first proved by means of II. 5 that

$$2xy < 2uv,$$

whence, since $(x + y)^2 = (u + v)^2$, the required result follows.

PROPOSITION 42.

A binomial straight line is divided into its terms at one point only.

Let AB be a binomial straight line divided into its terms at C;

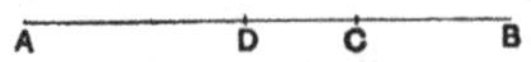

therefore AC, CB are rational straight lines commensurable in square only. [x. 36]

I say that AB is not divided at another point into two rational straight lines commensurable in square only.

For, if possible, let it be divided at D also, so that AD, DB are also rational straight lines commensurable in square only.

It is then manifest that AC is not the same with DB.

For, if possible, let it be so.

Then AD will also be the same as CB,

and, as AC is to CB, so will BD be to DA;

thus AB will be divided at D also in the same way as by the division at C:

which is contrary to the hypothesis.

Therefore AC is not the same with DB.

For this reason also the points C, D are not equidistant from the point of bisection.

Therefore that by which the squares on AC, CB differ from the squares on AD, DB is also that by which twice the rectangle AD, DB differs from twice the rectangle AC, CB,

because both the squares on AC, CB together with twice the rectangle AC, CB, and the squares on AD, DB together with twice the rectangle AD, DB, are equal to the square on AB. [II. 4]

But the squares on AC, CB differ from the squares on AD, DB by a rational area,

for both are rational;

therefore twice the rectangle AD, DB also differs from twice the rectangle AC, CB by a rational area, though they are medial [x. 21]:

which is absurd, for a medial area does not exceed a medial by a rational area. [x. 26]

Therefore a binomial straight line is not divided at different points;

therefore it is divided at one point only.

Q. E. D.

This proposition proves the equivalent of the well-known theorem in surds that,

if $a + \sqrt{b} = x + \sqrt{y}$,

then $a = x, \quad b = y$,

and if $\sqrt{a} + \sqrt{b} = \sqrt{x} + \sqrt{y}$,

then $a = x, \quad b = y \quad (\text{or } a = y, \quad b = x)$.

The proposition states that a *binomial* straight line cannot be split up into *terms* (ὀνόματα) in two ways. For, if possible, let

$$x + y = x' + y',$$

where x, y, and also x', y', are the *terms* of a binomial straight line, x', y' being different from x, y (or y, x).

One pair is necessarily more nearly equal than the other. Let x', y' be more nearly equal than x, y.

Then $$(x^2 + y^2) - (x'^2 + y'^2) = 2x'y' - 2xy.$$

Now by hypothesis $(x^2 + y^2)$, $(x'^2 + y'^2)$ are *rational* areas, being of the form $\rho^2 + k\rho^2$;

but $2x'y'$, $2xy$ are *medial* areas, being of the form $\sqrt{k} \cdot \rho^2$;

therefore the difference of two medial areas is rational:

which is impossible. [x. 26]

Therefore x', y' cannot be different from x, y (or y, x).

Proposition 43.

A first bimedial straight line is divided at one point only.

Let AB be a first bimedial straight line divided at C, so that AC, CB are medial straight lines commensurable in square only and containing a rational rectangle; [x. 37]

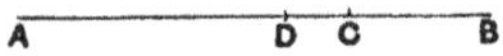

I say that AB is not so divided at another point.

For, if possible, let it be divided at D also, so that AD, DB are also medial straight lines commensurable in square only and containing a rational rectangle.

Since, then, that by which twice the rectangle AD, DB differs from twice the rectangle AC, CB is that by which the squares on AC, CB differ from the squares on AD, DB,

while twice the rectangle AD, DB differs from twice the rectangle AC, CB by a rational area—for both are rational—

therefore the squares on AC, CB also differ from the squares on AD, DB by a rational area, though they are medial:

which is absurd. [x. 26]

Therefore a first bimedial straight line is not divided into its terms at different points;

therefore it is so divided at one point only.

Q. E. D.

In this case, with the same hypothesis, viz. that

$$x + y = x' + y',$$

and x', y' are more nearly equal than x, y,

we have as before $\quad (x^2 + y^2) - (x'^2 + y'^2) = 2x'y' - 2xy.$

But, from the given properties of x, y, and x', y', it follows that $2xy$, $2x'y'$ are *rational*, and $(x^2 + y^2)$, $(x'^2 + y'^2)$ *medial*, areas.

Therefore the difference between two medial areas is rational: which is impossible. [X. 26]

PROPOSITION 44.

A second bimedial straight line is divided at one point only.

Let AB be a second bimedial straight line divided at C, so that AC, CB are medial straight lines commensurable in square only and containing a medial rectangle; [X. 38]

it is then manifest that C is not at the point of bisection, because the segments are not commensurable in length.

I say that AB is not so divided at another point.

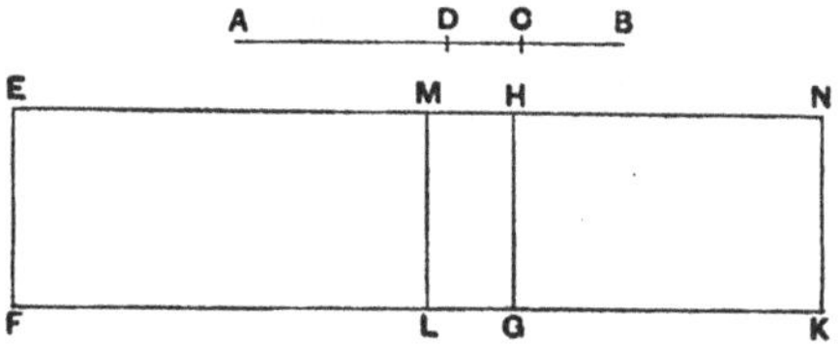

For, if possible, let it be divided at D also, so that AC is not the same with DB, but AC is supposed greater;

it is then clear that the squares on AD, DB are also, as we proved above [Lemma], less than the squares on AC, CB;

and suppose that AD, DB are medial straight lines commensurable in square only and containing a medial rectangle.

Now let a rational straight line EF be set out,

let there be applied to EF the rectangular parallelogram EK equal to the square on AB,

and let EG equal to the squares on AC, CB be subtracted;

therefore the remainder HK is equal to twice the rectangle AC, CB. [II. 4]

Again, let there be subtracted EL, equal to the squares on AD, DB, which were proved less than the squares on AC, CB [Lemma];

therefore the remainder MK is also equal to twice the rectangle AD, DB.

Now, since the squares on AC, CB are medial, therefore EG is medial.

And it is applied to the rational straight line EF; therefore EH is rational and incommensurable in length with EF. [x. 22]

For the same reason HN is also rational and incommensurable in length with EF.

And, since AC, CB are medial straight lines commensurable in square only, therefore AC is incommensurable in length with CB.

But, as AC is to CB, so is the square on AC to the rectangle AC, CB; therefore the square on AC is incommensurable with the rectangle AC, CB. [x. 11]

But the squares on AC, CB are commensurable with the square on AC; for AC, CB are commensurable in square. [x. 15]

And twice the rectangle AC, CB is commensurable with the rectangle AC, CB. [x. 6]

Therefore the squares on AC, CB are also incommensurable with twice the rectangle AC, CB. [x. 13]

But EG is equal to the squares on AC, CB, and HK is equal to twice the rectangle AC, CB; therefore EG is incommensurable with HK, so that EH is also incommensurable in length with HN. [vi. 1, x. 11]

And they are rational; therefore EH, HN are rational straight lines commensurable in square only.

But, if two rational straight lines commensurable in square only be added together, the whole is the irrational which is called binomial. [x. 36]

Therefore EN is a binomial straight line divided at H.

In the same way EM, MN will also be proved to be rational straight lines commensurable in square only; and EN will be a binomial straight line divided at different points, H and M.

And EH is not the same with MN.

For the squares on AC, CB are greater than the squares on AD, DB.

But the squares on AD, DB are greater than twice the rectangle AD, DB;
therefore also the squares on AC, CB, that is, EG, are much greater than twice the rectangle AD, DB, that is, MK,
so that EH is also greater than MN.

Therefore EH is not the same with MN.

Q. E. D.

As the irrationality of the *second bimedial* straight line [x. 38] is proved by means of the irrationality of the binomial straight line [x. 36], so the present theorem is reduced to that of x. 42.

Suppose, if possible, that the second bimedial straight line can be divided into its terms as such in two ways, i.e. that

$$x + y = x' + y',$$

where x', y' are nearer equality than x, y.

Apply $x^2 + y^2$, $2xy$ to a rational straight line σ, i.e. let

$$x^2 + y^2 = \sigma u,$$
$$2xy = \sigma v.$$

Then, as in x. 38, the areas $x^2 + y^2$, $2xy$ are medial, so that σu, σv are medial;
therefore u, v are both rational and $\smile \sigma$ ....(1).

Again, by hypothesis, x, y are medial straight lines commensurable in square only;

therefore $x \smile y$.

Hence $x^2 \smile xy$.

And $x^2 \frown (x^2 + y^2)$, while $xy \frown 2xy$;

therefore $(x^2 + y^2) \smile 2xy$,

or $\sigma u \smile \sigma v$,

and hence $u \smile v$ ....(2).

Therefore, by (1) and (2), u, v are rational straight lines commensurable in square only;
therefore $u + v$ is a *binomial* straight line.

Similarly, if $x'^2 + y'^2 = \sigma u'$ and $2x'y' = \sigma v'$,
$u' + v'$ will be proved to be a binomial straight line.

And, since $(x + y)^2 = (x' + y')^2$, and therefore $(u + v) = (u' + v')$, it follows that a binomial straight line is divided as such in two ways:
which is impossible. [x. 42]

Therefore $x + y$, the given second bimedial straight line, can only be so divided in one way.

In order to prove that $u + v$, $u' + v'$ represent a *different* division of the same straight line, Euclid assumes that $x^2 + y^2 > 2xy$. This is of course an easy inference from II. 7; but the assumption of it here renders it probable that the Lemma after x. 59 is interpolated.

Proposition 45.

A major straight line is divided at one and the same point only.

Let AB be a major straight line divided at C, so that AC, CB are incommensurable in square and make the sum of the squares on AC, CB rational, but the rectangle AC, CB medial; [x. 39]

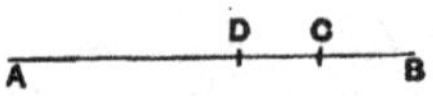

I say that AB is not so divided at another point.

For, if possible, let it be divided at D also, so that AD, DB are also incommensurable in square and make the sum of the squares on AD, DB rational, but the rectangle contained by them medial.

Then, since that by which the squares on AC, CB differ from the squares on AD, DB is also that by which twice the rectangle AD, DB differs from twice the rectangle AC, CB, while the squares on AC, CB exceed the squares on AD, DB by a rational area—for both are rational—

therefore twice the rectangle AD, DB also exceeds twice the rectangle AC, CB by a rational area, though they are medial:

which is impossible. [x. 26]

Therefore a major straight line is not divided at different points;

therefore it is only divided at one and the same point.

Q. E. D.

If possible, let the *major* irrational straight line be divided into terms in two ways, viz. as $(x+y)$ and $(x'+y')$, where x', y' are supposed to be nearer equality than x, y.

We have then, as in x. 42, 43,

$$(x^2+y^2)-(x'^2+y'^2)=2x'y'-2xy.$$

But, by hypothesis, (x^2+y^2), $(x'^2+y'^2)$ are both *rational*, so that their difference is rational.

Also, by hypothesis, $2x'y'$, $2xy$ are both *medial* areas;

therefore the difference of two medial areas is a rational area:

which is impossible. [x. 26]

Therefore etc.

Proposition 46.

The side of a rational plus a medial area is divided at one point only.

Let AB be the side of a rational plus a medial area divided at C, so that AC, CB are incommensurable in square and make the sum of the squares on AC, CB medial, but twice the rectangle AC, CB rational; [x. 40]

I say that AB is not so divided at another point.

For, if possible, let it be divided at D also, so that AD, DB are also incommensurable in square and make the sum of the squares on AD, DB medial, but twice the rectangle AD, DB rational.

Since then that by which twice the rectangle AC, CB differs from twice the rectangle AD, DB is also that by which the squares on AD, DB differ from the squares on AC, CB,

while twice the rectangle AC, CB exceeds twice the rectangle AD, DB by a rational area,

therefore the squares on AD, DB also exceed the squares on AC, CB by a rational area, though they are medial:

which is impossible. [x. 26]

Therefore the side of a rational plus a medial area is not divided at different points;

therefore it is divided at one point only.

Q. E. D.

Here, as before, if we use the same notation,

$$(x^2+y^2)-(x'^2+y'^2)=2x'y'-2xy,$$

and the areas on the left side are, by hypothesis, both medial, while the areas on the right side are both rational.

Thus the result of x. 26 is contradicted, as before.

Therefore etc.

Proposition 47.

The side of the sum of two medial areas is divided at one point only.

Let AB be divided at C, so that AC, CB are incommensurable in square and make the sum of the squares on AC,

CB medial, and the rectangle *AC*, *CB* medial and also incommensurable with the sum of the squares on them;
I say that *AB* is not divided at another point so as to fulfil the given conditions.

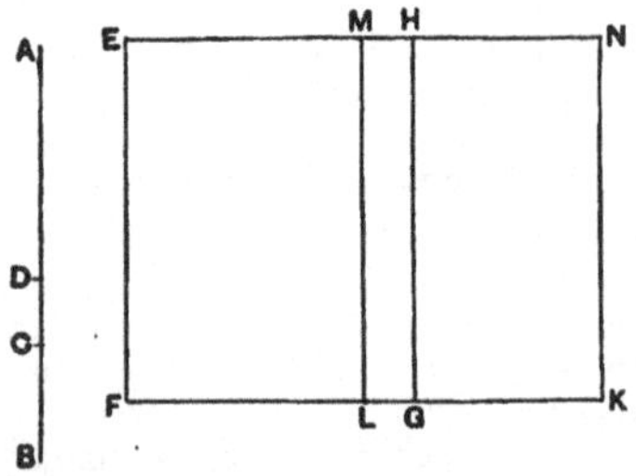

For, if possible, let it be divided at *D*, so that again *AC* is of course not the same as *BD*, but *AC* is supposed greater;
let a rational straight line *EF* be set out,
and let there be applied to *EF* the rectangle *EG* equal to the squares on *AC*, *CB*,
and the rectangle *HK* equal to twice the rectangle *AC*, *CB*;
therefore the whole *EK* is equal to the square on *AB*. [II. 4]

Again, let *EL*, equal to the squares on *AD*, *DB*, be applied to *EF*;
therefore the remainder, twice the rectangle *AD*, *DB*, is equal to the remainder *MK*.

And since, by hypothesis, the sum of the squares on *AC*, *CB* is medial,
therefore *EG* is also medial.

And it is applied to the rational straight line *EF*;
therefore *HE* is rational and incommensurable in length with *EF*. [X. 22]

For the same reason
HN is also rational and incommensurable in length with *EF*.

And, since the sum of the squares on *AC*, *CB* is incommensurable with twice the rectangle *AC*, *CB*,
therefore *EG* is also incommensurable with *GN*,
so that *EH* is also incommensurable with *HN*. [VI. 1, X. 11]

And they are rational;

therefore EH, HN are rational straight lines commensurable in square only;

therefore EN is a binomial straight line divided at H. [x. 36]

Similarly we can prove that it is also divided at M.

And EH is not the same with MN;

therefore a binomial has been divided at different points: which is absurd. [x. 42]

Therefore a side of the sum of two medial areas is not divided at different points;

therefore it is divided at one point only.

Using the same notation as in the note on x. 44, we suppose that, if possible,

$$x + y = x' + y',$$

and we put

$$\left.\begin{aligned} x^2 + y^2 &= \sigma u \\ 2xy &= \sigma v \end{aligned}\right\} \text{ and } \left.\begin{aligned} x'^2 + y'^2 &= \sigma u' \\ 2x'y' &= \sigma v' \end{aligned}\right\}.$$

Then, since $x^2 + y^2$, $2xy$ are medial areas, and σ rational,

u, v are both rational and $\smile \sigma$ ....(1).

Also, by hypothesis, $x^2 + y^2 \smile 2xy$,

whence $u \smile v$ ....(2).

Therefore, by (1) and (2), u, v are rational and $\frown$.

Hence $u + v$ is a *binomial* straight line. [x. 36]

Similarly $u' + v'$ is a binomial straight line.

But $u + v = u' + v'$;

therefore a binomial straight line is divided into terms in two ways: which is impossible. [x. 42]

Therefore etc.

DEFINITIONS II.

1. Given a rational straight line and a binomial, divided into its terms, such that the square on the greater term is greater than the square on the lesser by the square on a straight line commensurable in length with the greater, then, if the greater term be commensurable in length with the rational straight line set out, let the whole be called a **first binomial** straight line;

2. but if the lesser term be commensurable in length with the rational straight line set out, let the whole be called a **second binomial**;

3. and if neither of the terms be commensurable in length with the rational straight line set out, let the whole be called a **third binomial.**

4. Again, if the square on the greater term be greater than the square on the lesser by the square on a straight line incommensurable in length with the greater, then, if the greater term be commensurable in length with the rational straight line set out, let the whole be called a **fourth binomial**;

5. if the lesser, a **fifth binomial**;

6. and if neither, a **sixth binomial.**

PROPOSITION 48.

To find the first binomial straight line.

Let two numbers AC, CB be set out such that the sum of them AB has to BC the ratio which a square number has to a square number, but has not to CA the ratio which a square number has to a square number; [Lemma 1 after X. 28]

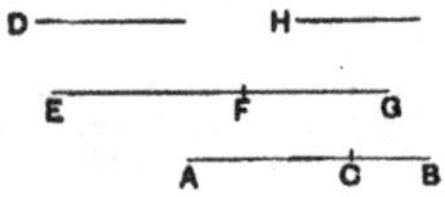

let any rational straight line D be set out, and let EF be commensurable in length with D.

Therefore EF is also rational.

Let it be contrived that,

as the number BA is to AC, so is the square on EF to the square on FG. [X. 6, Por.]

But AB has to AC the ratio which a number has to a number;

therefore the square on EF also has to the square on FG the ratio which a number has to a number,

so that the square on EF is commensurable with the square on FG. [X. 6]

And EF is rational;

therefore FG is also rational.

And, since BA has not to AC the ratio which a square number has to a square number, neither, therefore, has the square on *EF* to the square on *FG* the ratio which a square number has to a square number; therefore *EF* is incommensurable in length with *FG*. [x. 9]

Therefore *EF*, *FG* are rational straight lines commensurable in square only;
therefore *EG* is binomial. [x. 36]

I say that it is also a first binomial straight line.

For since, as the number *BA* is to *AC*, so is the square on *EF* to the square on *FG*,
while *BA* is greater than *AC*,
therefore the square on *EF* is also greater than the square on *FG*.

Let then the squares on *FG*, *H* be equal to the square on *EF*.

Now since, as *BA* is to *AC*, so is the square on *EF* to the square on *FG*,
therefore, *convertendo*,
as *AB* is to *BC*, so is the square on *EF* to the square on *H*. [v. 19, Por.]

But *AB* has to *BC* the ratio which a square number has to a square number;
therefore the square on *EF* also has to the square on *H* the ratio which a square number has to a square number.

Therefore *EF* is commensurable in length with *H*; [x. 9]
therefore the square on *EF* is greater than the square on *FG* by the square on a straight line commensurable with *EF*.

And *EF*, *FG* are rational, and *EF* is commensurable in length with *D*.

Therefore *EF* is a first binomial straight line.

Q. E. D.

Let $k\rho$ be a straight line commensurable in length with ρ, a given rational straight line.

The two numbers taken may be written $p(m^2-n^2)$, pn^2, where (m^2-n^2) is not a square.

Take x such that

$$pm^2 : p(m^2-n^2) = k^2\rho^2 : x^2 \quad \text{.........................}(1),$$

whence

$$x = k\rho\frac{\sqrt{m^2-n^2}}{m}.$$

Then $k\rho + x$, or $k\rho + k\rho\frac{\sqrt{m^2-n^2}}{m}$, is a *first binomial* straight line ......(2).

To prove this we have, from (1),

$$x^2 \frown k^2\rho^2,$$

and x is rational, but $x \smile k\rho$;
that is, x is rational and $\frown\!\!\!-\, k\rho$,
so that $k\rho + x$ is a binomial straight line.

Also, $k^2\rho^2$ being greater than x^2, suppose $k^2\rho^2 - x^2 = y^2$.

Then, from (1), $$pm^2 : pn^2 = k^2\rho^2 : y^2,$$
whence y is rational and $\frown k\rho$.

Therefore $k\rho + x$ is a *first binomial* straight line [x. Deff. II. 1].

This binomial straight line may be written thus,

$$k\rho + k\rho\sqrt{1 - \lambda^2}.$$

When we come to x. 85, we shall find that the corresponding straight line with a negative sign is the *first apotome*,

$$k\rho - k\rho\sqrt{1 - \lambda^2}.$$

Consider now the equation of which these two expressions are the roots. The equation is

$$x^2 - 2k\rho \,.\, x + \lambda^2 k^2\rho^2 = 0.$$

In other words, the first binomial and the first apotome correspond to the roots of the equation

$$x^2 - 2\alpha x + \lambda^2\alpha^2 = 0,$$

where $\alpha = k\rho$.

Proposition 49.

To find the second binomial straight line.

Let two numbers AC, CB be set out such that the sum of them AB has to BC the ratio which a square number has to a square number, but has not to AC the ratio which a square number has to a square number; let a rational straight line D be set out, and let EF be commensurable in length with D;
therefore EF is rational.

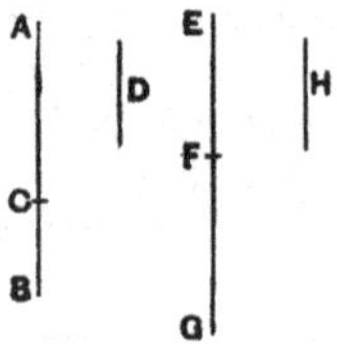

Let it be contrived then that,
as the number CA is to AB, so also is the square on EF to the square on FG; [x. 6, Por.]
therefore the square on EF is commensurable with the square on FG. [x. 6]

Therefore FG is also rational.

Now, since the number CA has not to AB the ratio which a square number has to a square number, neither has the square on EF to the square on FG the ratio which a square number has to a square number.

Therefore EF is incommensurable in length with FG; [x. 9]

therefore EF, FG are rational straight lines commensurable in square only;

therefore EG is binomial. [x. 36]

It is next to be proved that it is also a second binomial straight line.

For since, inversely, as the number BA is to AC, so is the square on GF to the square on FE,

while BA is greater than AC,

therefore the square on GF is greater than the square on FE.

Let the squares on EF, H be equal to the square on GF; therefore, *convertendo*, as AB is to BC, so is the square on FG to the square on H. [v. 19, Por.]

But AB has to BC the ratio which a square number has to a square number;

therefore the square on FG also has to the square on H the ratio which a square number has to a square number.

Therefore FG is commensurable in length with H; [x. 9]

so that the square on FG is greater than the square on FE by the square on a straight line commensurable with FG.

And FG, FE are rational straight lines commensurable in square only, and EF, the lesser term, is commensurable in length with the rational straight line D set out.

Therefore EG is a second binomial straight line.

Q. E. D.

Taking a rational straight line $k\rho$ commensurable in length with ρ, and selecting numbers of the same form as before, viz. $p(m^2 - n^2)$, pn^2, we put

$$p(m^2 - n^2) : pm^2 = k^2\rho^2 : x^2 \quad \text{.........................(1)},$$

so that

$$x = k\rho \frac{m}{\sqrt{m^2 - n^2}}$$

$$= k\rho \frac{1}{\sqrt{1 - \lambda^2}}, \text{ say} \quad \text{.........................(2)}.$$

Just as before, x is rational and $\frown k\rho$,

whence $k\rho + x$ is a *binomial* straight line.

By (1), $x^2 > k^2\rho^2$.

Let $$x^2 - k^2\rho^2 = y^2,$$
whence, from (1), $$pm^2 : pn^2 = x^2 : y^2,$$
and y is therefore rational and $\frown x$.

The greater term of the binomial straight line is x and the lesser $k\rho$, and

$$\frac{k\rho}{\sqrt{1-\lambda^2}} + k\rho$$

satisfies the definition of the ***second binomial*** straight line.

The corresponding ***second apotome*** [x. 86] is

$$\frac{k\rho}{\sqrt{1-\lambda^2}} - k\rho.$$

The equation of which the two expressions are the roots is

$$x^2 - \frac{2k\rho}{\sqrt{1-\lambda^2}} \cdot x + \frac{\lambda^2}{1-\lambda^2} k^2\rho^2 = 0,$$

or $$x^2 - 2\alpha x + \lambda^2\alpha^2 = 0,$$

where $$\alpha = \frac{k\rho}{\sqrt{1-\lambda^2}}.$$

PROPOSITION 50.

To find the third binomial straight line.

Let two numbers AC, CB be set out such that the sum of them AB has to BC the ratio which a square number has to a square number, but has not to AC the ratio which a square number has to a square number.

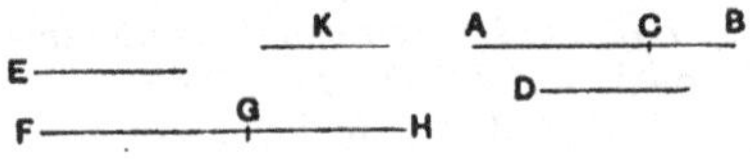

Let any other number D, not square, be set out also, and let it not have to either of the numbers BA, AC the ratio which a square number has to a square number.

Let any rational straight line E be set out,

and let it be contrived that, as D is to AB, so is the square on E to the square on FG; [x. 6, Por.]

therefore the square on E is commensurable with the square on FG. [x. 6]

And E is rational;

therefore FG is also rational.

And, since D has not to AB the ratio which a square number has to a square number,
neither has the square on E to the square on FG the ratio which a square number has to a square number;
therefore E is incommensurable in length with FG. [x. 9]

Next let it be contrived that, as the number BA is to AC, so is the square on FG to the square on GH; [x. 6, Por.]
therefore the square on FG is commensurable with the square on GH. [x. 6]

But FG is rational;
therefore GH is also rational.

And, since BA has not to AC the ratio which a square number has to a square number,
neither has the square on FG to the square on HG the ratio which a square number has to a square number;
therefore FG is incommensurable in length with GH. [x. 9]

Therefore FG, GH are rational straight lines commensurable in square only;
therefore FH is binomial. [x. 36]

I say next that it is also a third binomial straight line.

For since, as D is to AB, so is the square on E to the square on FG,
and, as BA is to AC, so is the square on FG to the square on GH,
therefore, *ex aequali*, as D is to AC, so is the square on E to the square on GH. [v. 22]

But D has not to AC the ratio which a square number has to a square number;
therefore neither has the square on E to the square on GH the ratio which a square number has to a square number;
therefore E is incommensurable in length with GH. [x. 9]

And since, as BA is to AC, so is the square on FG to the square on GH,
therefore the square on FG is greater than the square on GH.

Let then the squares on GH, K be equal to the square on FG;

therefore, *convertendo*, as *AB* is to *BC*, so is the square on *FG* to the square on *K*. [v. 19, Por.]

But *AB* has to *BC* the ratio which a square number has to a square number;
therefore the square on *FG* also has to the square on *K* the ratio which a square number has to a square number;
therefore *FG* is commensurable in length with *K*. [x. 9]

Therefore the square on *FG* is greater than the square on *GH* by the square on a straight line commensurable with *FG*.

And *FG*, *GH* are rational straight lines commensurable in square only, and neither of them is commensurable in length with *E*.

Therefore *FH* is a third binomial straight line.

Q. E. D.

Let ρ be a rational straight line.

Take the numbers $q(m^2 - n^2)$, qn^2,
and let p be a third number which is not a square and which has not to qm^2 or $q(m^2 - n^2)$ the ratio of square to square.

Take x such that $p : qm^2 = \rho^2 : x^2$ ....(1).

Thus x is rational and $\smile \rho$ ....(2).

Next suppose that $qm^2 : q(m^2 - n^2) = x^2 : y^2$ ....(3).

It follows that y is rational and $\frown x$ ....(4).

Thus $(x + y)$ is a *binomial* straight line.

Again, from (1) and (3), *ex aequali*,

$$p : q(m^2 - n^2) = \rho^2 : y^2 \quad \text{..............................(5),}$$

whence $y \smile \rho$ ....(6).

Suppose that $x^2 - y^2 = z^2$.

Then, from (3), *convertendo*,

$$qm^2 : qn^2 = x^2 : z^2,$$

whence $z \frown x$.

Thus $\sqrt{x^2 - y^2} \frown x$,

and x, y are both $\smile \rho$;
therefore $x + y$ is a *third binomial* straight line.

Now, from (1), $x = \rho \, . \, \dfrac{m\sqrt{q}}{\sqrt{p}}$,

and, by (5), $y = \rho \, . \, \dfrac{\sqrt{m^2 - n^2} \, . \, \sqrt{q}}{\sqrt{p}}$.

Thus the *third binomial* is

$$\sqrt{\frac{q}{p}} \, . \, \rho\,(m + \sqrt{m^2 - n^2}),$$

which we may write in the form

$$m\sqrt{k} \, . \, \rho + m\sqrt{k} \, . \, \rho\sqrt{1 - \lambda^2}.$$

The corresponding *third apotome* [x. 87] is

$$m\sqrt{k}\,.\,\rho - m\sqrt{k}\,.\,\rho\sqrt{1-\lambda^2}.$$

The two expressions are accordingly the roots of the equation

$$x^2 - 2m\sqrt{k}\,.\,\rho x + \lambda^2 m^2 k\rho^2 = 0,$$

or

$$x^2 - 2\alpha x + \lambda^2\alpha^2 = 0,$$

where

$$\alpha = m\sqrt{k}\,.\,\rho.$$

See also note on x. 53 (*ad fin.*).

Proposition 51.

To find the fourth binomial straight line.

Let two numbers AC, CB be set out such that AB neither has to BC, nor yet to AC, the ratio which a square number has to a square number.

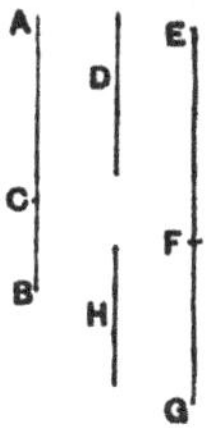

Let a rational straight line D be set out, and let EF be commensurable in length with D; therefore EF is also rational.

Let it be contrived that, as the number BA is to AC, so is the square on EF to the square on FG; [x. 6, Por.]

therefore the square on EF is commensurable with the square on FG; [x. 6]

therefore FG is also rational.

Now, since BA has not to AC the ratio which a square number has to a square number,

neither has the square on EF to the square on FG the ratio which a square number has to a square number;

therefore EF is incommensurable in length with FG. [x. 9]

Therefore EF, FG are rational straight lines commensurable in square only;

so that EG is binomial.

I say next that it is also a fourth binomial straight line.

For since, as BA is to AC, so is the square on EF to the square on FG,

therefore the square on EF is greater than the square on FG.

Let then the squares on FG, H be equal to the square on EF;

therefore, *convertendo*, as the number AB is to BC, so is the square on EF to the square on H. [v. 19, Por.]

But AB has not to BC the ratio which a square number has to a square number;
therefore neither has the square on EF to the square on H the ratio which a square number has to a square number.

Therefore EF is incommensurable in length with H; [x. 9]
therefore the square on EF is greater than the square on GF by the square on a straight line incommensurable with EF.

And EF, FG are rational straight lines commensurable in square only, and EF is commensurable in length with D.

Therefore EG is a fourth binomial straight line.

Q. E. D.

Take numbers m, n such that $(m+n)$ has not to either m or n the ratio of square to square.

Take x such that $(m+n) : m = k^2\rho^2 : x^2$,

whence
$$x = k\rho\sqrt{\frac{m}{m+n}}$$
$$= \frac{k\rho}{\sqrt{1+\lambda}}, \text{ say.}$$

Then $k\rho + x$, or $k\rho + \dfrac{k\rho}{\sqrt{1+\lambda}}$, is a *fourth binomial* straight line.

For $\sqrt{k^2\rho^2 - x^2}$ is incommensurable in length with $k\rho$, and $k\rho$ is commensurable in length with ρ.

The corresponding *fourth apotome* [x. 88] is
$$k\rho - \frac{k\rho}{\sqrt{1+\lambda}}.$$

The equation of which the two expressions are the roots is
$$x^2 - 2k\rho \cdot x + \frac{\lambda}{1+\lambda}k^2\rho^2 = 0,$$
or
$$x^2 - 2\alpha x + \frac{\lambda}{1+\lambda}\alpha^2 = 0,$$
where
$$\alpha = k\rho.$$

Proposition 52.

To find the fifth binomial straight line.

Let two numbers AC, CB be set out such that AB has not to either of them the ratio which a square number has to a square number;
let any rational straight line D be set out,

and let EF be commensurable with D;
therefore EF is rational.

Let it be contrived that, as CA is to AB, so is the square on EF to the square on FG. [x. 6, Por.]

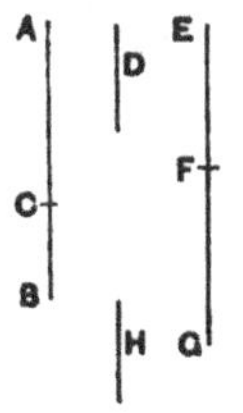

But CA has not to AB the ratio which a square number has to a square number;
therefore neither has the square on EF to the square on FG the ratio which a square number has to a square number.

Therefore EF, FG are rational straight lines commensurable in square only; [x. 9]
therefore EG is binomial. [x. 36]

I say next that it is also a fifth binomial straight line.

For since, as CA is to AB, so is the square on EF to the square on FG,
inversely, as BA is to AC, so is the square on FG to the square on FE;
therefore the square on GF is greater than the square on FE.

Let then the squares on EF, H be equal to the square on GF;
therefore, *convertendo*, as the number AB is to BC, so is the square on GF to the square on H. [v. 19, Por.]

But AB has not to BC the ratio which a square number has to a square number;
therefore neither has the square on FG to the square on H the ratio which a square number has to a square number.

Therefore FG is incommensurable in length with H; [x. 9]
so that the square on FG is greater than the square on FE by the square on a straight line incommensurable with FG.

And GF, FE are rational straight lines commensurable in square only, and the lesser term EF is commensurable in length with the rational straight line D set out.

Therefore EG is a fifth binomial straight line.

Q. E. D.

If m, n be numbers of the kind taken in the last proposition, take x such that

$$m : (m + n) = k^2\rho^2 : x^2.$$

In this case
$$x = k\rho\sqrt{\frac{m+n}{m}}$$
$$= k\rho\sqrt{1+\lambda}, \text{ say,}$$

and $x > k\rho$.

Then $k\rho\sqrt{1+\lambda} + k\rho$ is a *fifth binomial* straight line.

For $\sqrt{x^2 - k^2\rho^2}$, or $\sqrt{\lambda} \,.\, k\rho$, is incommensurable in length with $k\rho\sqrt{1+\lambda}$, or x;

and $k\rho$, but not $k\rho\sqrt{1+\lambda}$, is commensurable in length with ρ.

The corresponding *fifth apotome* [x. 89] is
$$k\rho\sqrt{1+\lambda} - k\rho.$$

The equation of which the fifth binomial and the fifth apotome are the roots is
$$x^2 - 2k\rho\sqrt{1+\lambda} \,.\, x + \lambda k^2\rho^2 = 0,$$
or
$$x^2 - 2ax + \frac{\lambda}{1+\lambda}a^2 = 0,$$
where
$$a = k\rho\sqrt{1+\lambda}.$$

Proposition 53.

To find the sixth binomial straight line.

Let two numbers AC, CB be set out such that AB has not to either of them the ratio which a square number has to a square number;

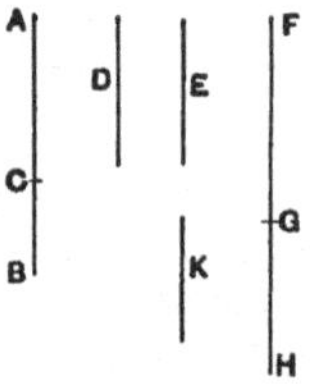

and let there also be another number D which is not square and which has not to either of the numbers BA, AC the ratio which a square number has to a square number.

Let any rational straight line E be set out,

and let it be contrived that, as D is to AB, so is the square on E to the square on FG; [x. 6, Por.]

therefore the square on E is commensurable with the square on FG. [x. 6]

And E is rational;

therefore FG is also rational.

Now, since D has not to AB the ratio which a square number has to a square number,

neither has the square on E to the square on FG the ratio which a square number has to a square number;
therefore E is incommensurable in length with FG. [x. 9]

Again, let it be contrived that, as BA is to AC, so is the square on FG to the square on GH. [x. 6, Por.]

Therefore the square on FG is commensurable with the square on HG. [x. 6]

Therefore the square on HG is rational;
therefore HG is rational.

And, since BA has not to AC the ratio which a square number has to a square number,
neither has the square on FG to the square on GH the ratio which a square number has to a square number;
therefore FG is incommensurable in length with GH. [x. 9]

Therefore FG, GH are rational straight lines commensurable in square only;
therefore FH is binomial. [x. 36]

It is next to be proved that it is also a sixth binomial straight line.

For since, as D is to AB, so is the square on E to the square on FG,
and also, as BA is to AC, so is the square on FG to the square on GH,
therefore, *ex aequali*, as D is to AC, so is the square on E to the square on GH. [v. 22]

But D has not to AC the ratio which a square number has to a square number;
therefore neither has the square on E to the square on GH the ratio which a square number has to a square number;
therefore E is incommensurable in length with GH. [x. 9]

But it was also proved incommensurable with FG;
therefore each of the straight lines FG, GH is incommensurable in length with E.

And, since, as BA is to AC, so is the square on FG to the square on GH,
therefore the square on FG is greater than the square on GH.

Let then the squares on GH, K be equal to the square on FG;

therefore, *convertendo*, as AB is to BC, so is the square on FG to the square on K. [v. 19, Por.]

But AB has not to BC the ratio which a square number has to a square number;
so that neither has the square on FG to the square on K the ratio which a square number has to a square number.

Therefore FG is incommensurable in length with K; [x. 9]
therefore the square on FG is greater than the square on GH by the square on a straight line incommensurable with FG.

And FG, GH are rational straight lines commensurable in square only, and neither of them is commensurable in length with the rational straight line E set out.

Therefore FH is a sixth binomial straight line.

Q. E. D.

Take numbers m, n such that $(m+n)$ has not to either of the numbers m, n the ratio of square to square; take also a third number p, which is not square, and which has not to either of the numbers $(m+n)$, m the ratio of square to square.

Let $$p : (m+n) = \rho^2 : x^2 \quad\text{.........(1)}$$
and $$(m+n) : m = x^2 : y^2 \quad\text{.........(2)}.$$

Then shall $(x+y)$ be a *sixth binomial* straight line.

For, by (1), x is rational and $\smile \rho$.

By (2), since x is rational,
$$y \text{ is rational and } \smile x.$$

Hence x, y are rational and commensurable in square only, so that $(x+y)$ is a binomial straight line.

Again, *ex aequali*, from (1) and (2),
$$p : m = \rho^2 : y^2 \quad\text{.........(3)},$$
whence $y \smile \rho$.

Thus x, y are both incommensurable in length with ρ.

Lastly, from (2), *convertendo*,
$$(m+n) : n = x^2 : (x^2 - y^2),$$
so that $\sqrt{x^2-y^2} \smile x$.

Therefore $(x+y)$ is a *sixth binomial* straight line.

Now, from (1) and (3),
$$x = \rho \cdot \sqrt{\frac{m+n}{p}} = \rho\sqrt{k}, \text{ say},$$
$$y = \rho \cdot \sqrt{\frac{m}{p}} = \rho\sqrt{\lambda}, \text{ say},$$
and the *sixth binomial* straight line may be written
$$\sqrt{k} \cdot \rho + \sqrt{\lambda} \cdot \rho.$$

The corresponding *sixth apotome* is [x. 90]
$$\sqrt{k} \cdot \rho - \sqrt{\lambda} \cdot \rho;$$

and the equation of which the two expressions are the roots is

$$x^2 - 2\sqrt{k} \,.\, \rho x + (k - \lambda)\,\rho^2 = 0,$$

or

$$x^2 - 2\alpha x + \frac{k - \lambda}{k}\,\alpha^2 = 0,$$

where $\alpha = \sqrt{k} \,.\, \rho$.

Tannery remarks ("De la solution géométrique des problèmes du second degré avant Euclide" in *Mémoires de la Société des sciences physiques et naturelles de Bordeaux*, 2ᵉ Série, T. IV.) that Euclid admits as binomials and apotomes the *third* and *sixth* binomials and apotomes which are the square roots of first binomials and apotomes respectively. Hence the third and sixth binomials and apotomes are the positive roots of *biquadratic* equations of the same form as the quadratics which give as roots the first and fourth binomials and apotomes. But this remark seems to be of no value because (as was pointed out a hundred years ago by Cossali, II. p. 260) the squares of *all the six* binomials and apotomes (including the first and fourth) give *first* binomials and apotomes respectively. Hence we may equally well regard them all as roots of biquadratics reducible to quadratics, or generally as roots of equations of the form

$$x^{2^n} \pm 2\alpha \,.\, x^{2^{n-1}} \pm q = 0;$$

and nothing is gained by raising the degree of the equations in this way.

It is, of course, easy to see that the most general form of binomial and apotome, viz.

$$\rho \,.\, \sqrt{k} \pm \rho \,.\, \sqrt{\lambda},$$

give *first* binomials and apotomes when squared.

For the square is $\rho\,\{(k + \lambda)\,\rho \pm 2\sqrt{k\lambda} \,.\, \rho\}$; and the expression within the bracket is a first binomial or apotome, because

(1) $k + \lambda > 2\sqrt{k\lambda}$,

(2) $\sqrt{(k + \lambda)^2 - 4k\lambda} = k - \lambda$, which is $\frown (k + \lambda)$,

(3) $(k + \lambda)\,\rho \frown \rho$.

LEMMA.

Let there be two squares *AB*, *BC*, and let them be placed so that *DB* is in a straight line with *BE*;

therefore *FB* is also in a straight line with *BG*.

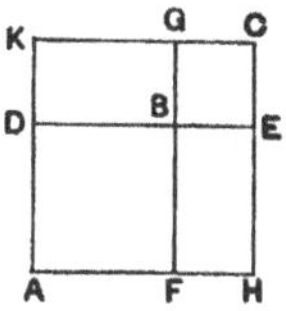

Let the parallelogram *AC* be completed; I say that *AC* is a square, that *DG* is a mean proportional between *AB*, *BC*, and further that *DC* is a mean proportional between *AC*, *CB*.

For, since *DB* is equal to *BF*, and *BE* to *BG*, therefore the whole *DE* is equal to the whole *FG*.

But *DE* is equal to each of the straight lines *AH*, *KC*, and *FG* is equal to each of the straight lines *AK*, *HC*; [I. 34] therefore each of the straight lines AH, KC is also equal to each of the straight lines AK, HC.

Therefore the parallelogram AC is equilateral.

And it is also rectangular;
therefore AC is a square.

And since, as FB is to BG, so is DB to BE,
while, as FB is to BG, so is AB to DG,
and, as DB is to BE, so is DG to BC, [VI. 1]
therefore also, as AB is to DG, so is DG to BC. [V. 11]

Therefore DG is a mean proportional between AB, BC.

I say next that DC is also a mean proportional between AC, CB.

For since, as AD is to DK, so is KG to GC—
for they are equal respectively—
and, *componendo*, as AK is to KD, so is KC to CG, [V. 18]
while, as AK is to KD, so is AC to CD,
and, as KC is to CG, so is DC to CB, [VI. 1]
therefore also, as AC is to DC, so is DC to BC. [V. 11]

Therefore DC is a mean proportional between AC, CB.

Being what it was proposed to prove.

It is here proved that

$$x^2 : xy = xy : y^2,$$

and

$$(x+y)^2 : (x+y)y = (x+y)y : y^2.$$

The first of the two results is proved in the course of x. 25 (lines 6—8 on p. 57 above). This fact may, I think, suggest doubt as to the genuineness of this Lemma.

Proposition 54.

If an area be contained by a rational straight line and the first binomial, the "side" of the area is the irrational straight line which is called binomial.

For let the area AC be contained by the rational straight line AB and the first binomial AD;
I say that the "side" of the area AC is the irrational straight line which is called binomial.

For, since AD is a first binomial straight line, let it be divided into its terms at E,
and let AE be the greater term.

It is then manifest that AE, ED are rational straight lines commensurable in square only,
the square on AE is greater than the square on ED by the square on a straight line commensurable with AE,
and AE is commensurable in length with the rational straight line AB set out. [x. Deff. II. 1]

Let ED be bisected at the point F.

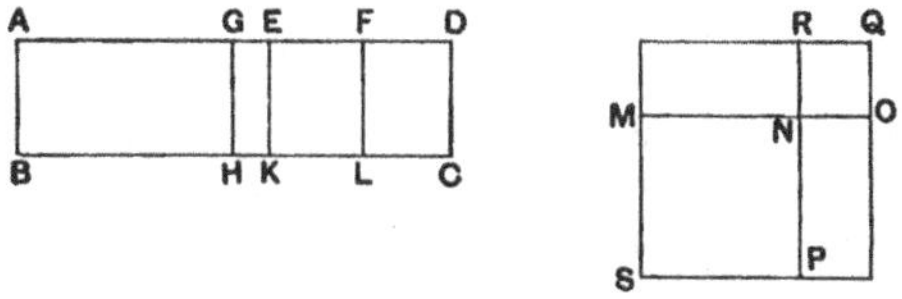

Then, since the square on AE is greater than the square on ED by the square on a straight line commensurable with AE,
therefore, if there be applied to the greater AE a parallelogram equal to the fourth part of the square on the less, that is, to the square on EF, and deficient by a square figure, it divides it into commensurable parts. [x. 17]

Let then the rectangle AG, GE equal to the square on EF be applied to AE;
therefore AG is commensurable in length with EG.

Let GH, EK, FL be drawn from G, E, F parallel to either of the straight lines AB, CD;
let the square SN be constructed equal to the parallelogram AH, and the square NQ equal to GK, [II. 14]
and let them be placed so that MN is in a straight line with NO;
therefore RN is also in a straight line with NP.

And let the parallelogram SQ be completed;
therefore SQ is a square. [Lemma]

Now, since the rectangle AG, GE is equal to the square on EF,
therefore, as AG is to EF, so is FE to EG; [VI. 17]
therefore also, as AH is to EL, so is EL to KG; [VI. 1]
therefore EL is a mean proportional between AH, GK.

But AH is equal to SN, and GK to NQ;
therefore EL is a mean proportional between SN, NQ.

But *MR* is also a mean proportional between the same *SN*, *NQ*; [Lemma]
therefore *EL* is equal to *MR*,
so that it is also equal to *PO*.

But *AH*, *GK* are also equal to *SN*, *NQ*;
therefore the whole *AC* is equal to the whole *SQ*, that is, to the square on *MO*;
therefore *MO* is the "side" of *AC*.

I say next that *MO* is binomial.

For, since *AG* is commensurable with *GE*,
therefore *AE* is also commensurable with each of the straight lines *AG*, *GE*. [X. 15]

But *AE* is also, by hypothesis, commensurable with *AB*;
therefore *AG*, *GE* are also commensurable with *AB*. [X. 12]

And *AB* is rational;
therefore each of the straight lines *AG*, *GE* is also rational;
therefore each of the rectangles *AH*, *GK* is rational, [X. 19]
and *AH* is commensurable with *GK*.

But *AH* is equal to *SN*, and *GK* to *NQ*;
therefore *SN*, *NQ*, that is, the squares on *MN*, *NO*, are rational and commensurable.

And, since *AE* is incommensurable in length with *ED*,
while *AE* is commensurable with *AG*, and *DE* is commensurable with *EF*,
therefore *AG* is also incommensurable with *EF*, [X. 13]
so that *AH* is also incommensurable with *EL*. [VI. 1, X. 11]

But *AH* is equal to *SN*, and *EL* to *MR*;
therefore *SN* is also incommensurable with *MR*.

But, as *SN* is to *MR*, so is *PN* to *NR*; [VI. 1]
therefore *PN* is incommensurable with *NR*. [X. 11]

But *PN* is equal to *MN*, and *NR* to *NO*;
therefore *MN* is incommensurable with *NO*.

And the square on *MN* is commensurable with the square on *NO*,
and each is rational;
therefore *MN*, *NO* are rational straight lines commensurable in square only.

Therefore MO is binomial [x. 36] and the "side" of AC.

Q. E. D.

2. **"side."** I use the word "side" in the sense explained in the note on x. Def. 4 (p. 13 above), i.e. as short for "side of a square equal to." The Greek is ἡ τὸ χωρίον δυναμένη.

A *first binomial* straight line being, as we have seen in x. 48, of the form

$$k\rho + k\rho\sqrt{1-\lambda^2},$$

the problem solved in this proposition is the equivalent of *finding the square root of this expression multiplied by* ρ, or of

$$\rho\,(k\rho + k\rho\sqrt{1-\lambda^2}),$$

and of proving that the said square root represents a *binomial* straight line as defined in x. 36.

The geometrical method corresponds sufficiently closely to the algebraical one which we should use.

First solve the equations

$$\left.\begin{aligned} u + v &= k\rho \\ uv &= \tfrac{1}{4}k^2\rho^2(1-\lambda^2) \end{aligned}\right\} \quad \ldots\ldots\ldots\ldots\ldots(1).$$

Then, if u, v represent the straight lines so found, put

$$\left.\begin{aligned} x^2 &= \rho u \\ y^2 &= \rho v \end{aligned}\right\} \quad \ldots\ldots\ldots\ldots\ldots(2);$$

and the straight line $(x+y)$ is the square root required.

The actual algebraical solution of (1) gives

$$u - v = k\rho\,.\,\lambda,$$

so that

$$u = \tfrac{1}{2}k\rho(1+\lambda),$$
$$v = \tfrac{1}{2}k\rho(1-\lambda),$$

and therefore

$$x = \rho\sqrt{\frac{k}{2}(1+\lambda)},$$

$$y = \rho\sqrt{\frac{k}{2}(1-\lambda)},$$

and

$$x + y = \rho\sqrt{\frac{k}{2}(1+\lambda)} + \rho\sqrt{\frac{k}{2}(1-\lambda)}.$$

This is clearly a *binomial* straight line as defined in x. 36.

Since Euclid has to express his results by straight lines in his figure, and has no symbols to make the result obvious by inspection, he is obliged to prove (1) that $(x+y)$ is the square root of $\rho\,(k\rho + k\rho\sqrt{1-\lambda^2})$, and (2) that $(x+y)$ is a binomial straight line, in the following manner.

First, he proves, by means of the preceding Lemma, that

$$xy = \frac{k}{2}\rho^2\sqrt{1-\lambda^2} \quad \ldots\ldots\ldots\ldots\ldots(3);$$

therefore

$$\begin{aligned} (x+y)^2 &= x^2 + y^2 + 2xy \\ &= \rho\,(u+v) + 2xy \\ &= k\rho^2 + k\rho^2\sqrt{1-\lambda^2}, \text{ by (1) and (3)}, \end{aligned}$$

so that

$$x + y = \sqrt{\rho\,(k\rho + k\rho\sqrt{1-\lambda^2})}.$$

Secondly, it results from (1), [by X. 17], that

$$u \frown v,$$

so that u, v are both $\frown (u+v)$, and therefore $\frown \rho$ ....(4);
thus u, v are rational,
whence ρu, ρv are both rational, and

$$\rho u \frown \rho v.$$

Therefore x^2, y^2 are rational and commensurable ....(5).

Next, $k\rho \smile k\rho\sqrt{1-\lambda^2}$,
and $k\rho \frown u$, while $k\rho\sqrt{1-\lambda^2} \frown \frac{1}{2}k\rho\sqrt{1-\lambda^2}$;

therefore $\quad u \smile \frac{1}{2}k\rho\sqrt{1-\lambda^2}$,

whence $\quad \rho u \smile \frac{1}{2}k\rho^2\sqrt{1-\lambda^2}$,

or $\quad x^2 \smile xy$,

so that $\quad x \smile y$.

By this and (5), x, y are rational and $\frown\!\!-$, so that $(x+y)$ is a binomial straight line. [X. 36]

X. 91 will prove in like manner that a like theorem holds for *apotomes*, viz. that

$$\rho\sqrt{\frac{k}{2}(1+\lambda)} - \rho\sqrt{\frac{k}{2}(1-\lambda)} = \sqrt{\rho\left(k\rho - k\rho\sqrt{1-\lambda^2}\right)}.$$

Since the *first binomial* straight line and the *first apotome* are the roots of the equation

$$x^2 - 2k\rho \,.\, x + \lambda^2k^2\rho^2 = 0,$$

this proposition and X. 91 give us the solution of the biquadratic equation

$$x^4 - 2k\rho^2 \,.\, x^2 + \lambda^2k^2\rho^4 = 0.$$

Proposition 55.

If an area be contained by a rational straight line and the second binomial, the "side" of the area is the irrational straight line which is called a first bimedial.

For let the area $ABCD$ be contained by the rational straight line AB and the second binomial AD;

I say that the "side" of the area AC is a first bimedial straight line.

For, since AD is a second binomial straight line, let it be divided into its terms at E, so that AE is the greater term; therefore AE, ED are rational straight lines commensurable in square only,

the square on AE is greater than the square on ED by the square on a straight line commensurable with AE,

and the lesser term ED is commensurable in length with AB. [X. Deff. II. 2]

Let ED be bisected at F,

and let there be applied to *AE* the rectangle *AG*, *GE* equal to the square on *EF* and deficient by a square figure;

therefore *AG* is commensurable in length with *GE*. [x. 17]

Through *G*, *E*, *F* let *GH*, *EK*, *FL* be drawn parallel to *AB*, *CD*,

let the square *SN* be constructed equal to the parallelogram *AH*, and the square *NQ* equal to *GK*,

and let them be placed so that *MN* is in a straight line with *NO*;

therefore *RN* is also in a straight line with *NP*.

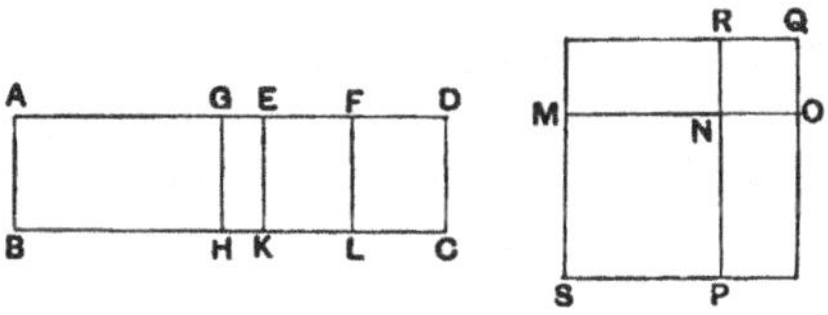

Let the square *SQ* be completed.

It is then manifest from what was proved before that *MR* is a mean proportional between *SN*, *NQ* and is equal to *EL*, and that *MO* is the "side" of the area *AC*.

It is now to be proved that *MO* is a first bimedial straight line.

Since *AE* is incommensurable in length with *ED*,

while *ED* is commensurable with *AB*,

therefore *AE* is incommensurable with *AB*. [x. 13]

And, since *AG* is commensurable with *EG*,

AE is also commensurable with each of the straight lines *AG*, *GE*. [x. 15]

But *AE* is incommensurable in length with *AB*;

therefore *AG*, *GE* are also incommensurable with *AB*. [x. 13]

Therefore *BA*, *AG* and *BA*, *GE* are pairs of rational straight lines commensurable in square only;

so that each of the rectangles *AH*, *GK* is medial. [x. 21]

Hence each of the squares *SN*, *NQ* is medial.

Therefore *MN*, *NO* are also medial.

And, since *AG* is commensurable in length with *GE*,

AH is also commensurable with *GK*, [vi. 1, x. 11]

that is, *SN* is commensurable with *NQ*,

that is, the square on *MN* with the square on *NO*.

And, since AE is incommensurable in length with ED, while AE is commensurable with AG, and ED is commensurable with EF, therefore AG is incommensurable with EF; [x. 13] so that AH is also incommensurable with EL, that is, SN is incommensurable with MR, that is, PN with NR, [vi. 1, x. 11] that is, MN is incommensurable in length with NO.

But MN, NO were proved to be both medial and commensurable in square; therefore MN, NO are medial straight lines commensurable in square only.

I say next that they also contain a rational rectangle.

For, since DE is, by hypothesis, commensurable with each of the straight lines AB, EF, therefore EF is also commensurable with EK. [x. 12]

And each of them is rational; therefore EL, that is, MR is rational, [x. 19] and MR is the rectangle MN, NO.

But, if two medial straight lines commensurable in square only and containing a rational rectangle be added together, the whole is irrational and is called a first bimedial straight line. [x. 37]

Therefore MO is a first bimedial straight line.

Q. E. D.

39. **Therefore BA, AG and BA, GE are pairs of rational straight lines commensurable in square only.** The text has "Therefore *BA*, *AG*, *GE* are rational straight lines commensurable in square only," which I have altered because it would naturally convey the impression that *any two* of the three straight lines are commensurable in square only, whereas *AG*, *GE* are commensurable in length (l. 18), and it is only the other two pairs which are commensurable in square only.

A *second binomial* straight line being [x. 49] of the form

$$\frac{k\rho}{\sqrt{1-\lambda^2}} + k\rho,$$

the present proposition is equivalent to finding the *square root of the expression*

$$\rho\left(\frac{k\rho}{\sqrt{1-\lambda^2}} + k\rho\right).$$

As in the last proposition, Euclid finds u, v from the equations

$$\left.\begin{aligned} u+v &= \frac{k\rho}{\sqrt{1-\lambda^2}} \\ uv &= \tfrac{1}{4}k^2\rho^2 \end{aligned}\right\} \quad \ldots\ldots\ldots\ldots(1),$$

then finds x, y from the equations

$$\left.\begin{aligned} x^2 &= \rho u \\ y^2 &= \rho v \end{aligned}\right\} \quad \ldots\ldots\ldots\ldots(2),$$

and then proves (α) that

$$x+y = \sqrt{\rho\left(\frac{k\rho}{\sqrt{1-\lambda^2}}+k\rho\right)},$$

and (β) that $(x+y)$ is a first bimedial straight line [X. 37].

The steps in the proof are as follows.

For (α) reference to the corresponding part of the previous proposition suffices.

(β) By (1) and X. 17,

$$u \frown v;$$

therefore u, v are both rational and $\frown (u+v)$, and therefore $\smile \rho$ [by (1)]...(3).

Hence ρu, ρv, or x^2, y^2, are *medial* areas,

so that x, y are also medial ....(4).

But, since $u \frown v$,

$$x^2 \frown y^2 \quad \ldots\ldots\ldots\ldots(5).$$

Again $(u+v)$, or $\dfrac{k\rho}{\sqrt{1-\lambda^2}}$, $\smile k\rho$,

so that $u \smile \tfrac{1}{2}k\rho$,

whence $\rho u \smile \tfrac{1}{2}k\rho^2$,

or $x^2 \smile xy$,

and $x \smile y$ ....(6).

Thus [(4), (5), (6)] x, y are medial and $\frown$.

Lastly, $xy = \tfrac{1}{2}k\rho^2$, which is rational.

Therefore $(x+y)$ is a *first bimedial* straight line.

The actual straight lines obtained from (1) are

$$\left.\begin{aligned} u &= \tfrac{1}{2}\,\frac{1+\lambda}{\sqrt{1-\lambda^2}}\,k\rho \\ v &= \tfrac{1}{2}\cdot\frac{1-\lambda}{\sqrt{1-\lambda^2}}\,k\rho \end{aligned}\right\},$$

so that

$$x+y = \rho\sqrt{\frac{k}{2}}\left(\frac{1+\lambda}{1-\lambda}\right)^{\frac{1}{4}} + \rho\sqrt{\frac{k}{2}}\left(\frac{1-\lambda}{1+\lambda}\right)^{\frac{1}{4}}.$$

The corresponding *first apotome of a medial* straight line found in X. 92 being the same thing with a *minus* sign between the terms, the two expressions are the roots of the biquadratic

$$x^4 - \frac{2k\rho^2}{\sqrt{1-\lambda^2}}\,x^2 + \frac{\lambda^2}{1-\lambda^2}\,k^2\rho^4 = 0,$$

being the equation in x^2 corresponding to that in x in X. 49.

Proposition 56.

If an area be contained by a rational straight line and the third binomial, the "side" of the area is the irrational straight line called a second bimedial.

For let the area $ABCD$ be contained by the rational straight line AB and the third binomial AD divided into its terms at E, of which terms AE is the greater;
I say that the "side" of the area AC is the irrational straight line called a second bimedial.

For let the same construction be made as before.

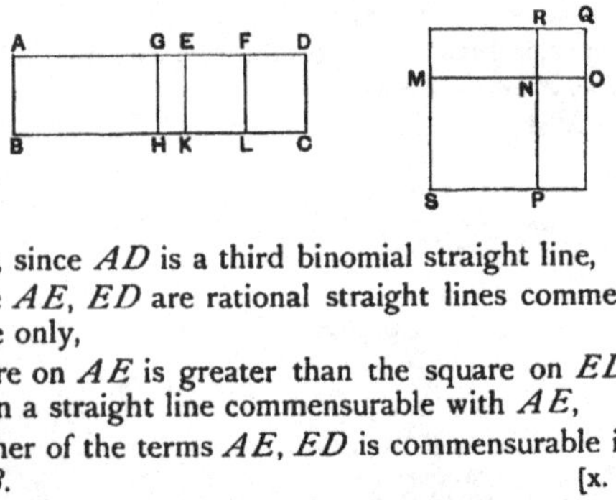

Now, since AD is a third binomial straight line,
therefore AE, ED are rational straight lines commensurable in square only,
the square on AE is greater than the square on ED by the square on a straight line commensurable with AE,
and neither of the terms AE, ED is commensurable in length with AB. [x. Deff. II. 3]

Then, in manner similar to the foregoing, we shall prove that MO is the "side" of the area AC,
and MN, NO are medial straight lines commensurable in square only;
so that MO is bimedial.

It is next to be proved that it is also a second bimedial straight line.

Since DE is incommensurable in length with AB, that is, with EK,
and DE is commensurable with EF,
therefore EF is incommensurable in length with EK. [x. 13]

And they are rational;

therefore *FE*, *EK* are rational straight lines commensurable in square only.

Therefore *EL*, that is, *MR*, is medial. [x. 21]

And it is contained by *MN*, *NO*;

therefore the rectangle *MN*, *NO* is medial.

Therefore *MO* is a second bimedial straight line. [x. 38]

Q. E. D.

This proposition in like manner is the equivalent of finding the square root of the product of ρ and the *third binomial* [x. 50], i.e. of the expression

$$\rho(\sqrt{k}.\rho + \sqrt{k}.\rho\sqrt{1-\lambda^2}).$$

As before, put

$$\left.\begin{aligned} u+v &= \sqrt{k}.\rho \\ uv &= \tfrac{1}{4}k\rho^2(1-\lambda^2) \end{aligned}\right\} \quad\ldots\ldots\ldots\ldots\ldots\ldots(1).$$

Next, u, v being found, let

$$x^2 = \rho u,$$
$$y^2 = \rho v;$$

then $(x+y)$ is the square root required and is a *second bimedial* straight line. [x. 38]

For, as in the last proposition, it is proved that $(x+y)$ is the square root, and x, y are medial and $\frown$.

Again, $xy = \frac{1}{2}\sqrt{k}.\rho^2\sqrt{1-\lambda^2}$, which is *medial*.

Hence $(x+y)$ is a *second bimedial* straight line.

By solving equations (1), we find

$$u = \tfrac{1}{2}(\sqrt{k}.\rho + \lambda\sqrt{k}.\rho),$$
$$v = \tfrac{1}{2}(\sqrt{k}.\rho - \lambda\sqrt{k}.\rho),$$

and

$$x+y = \rho\sqrt{\frac{\sqrt{k}}{2}(1+\lambda)} + \rho\sqrt{\frac{\sqrt{k}}{2}(1-\lambda)}.$$

The corresponding *second apotome of a medial* found in x. 93 is the same thing with a minus sign between the terms, and the two are the roots (cf. note on x. 50) of the biquadratic equation

$$x^4 - 2\sqrt{k}.\rho^2x^2 + \lambda^2k\rho^4 = 0.$$

Proposition 57.

If an area be contained by a rational straight line and the fourth binomial, the "side" of the area is the irrational straight line called major.

For let the area *AC* be contained by the rational straight line *AB* and the fourth binomial *AD* divided into its terms at *E*, of which terms let *AE* be the greater;

I say that the "side" of the area *AC* is the irrational straight line called major.

For, since AD is a fourth binomial straight line,
therefore AE, ED are rational straight lines commensurable in square only,
the square on AE is greater than the square on ED by the square on a straight line incommensurable with AE,
and AE is commensurable in length with AB. [x. Deff. II. 4]

Let DE be bisected at F,
and let there be applied to AE a parallelogram, the rectangle AG, GE, equal to the square on EF;
therefore AG is incommensurable in length with GE. [x. 18]

Let GH, EK, FL be drawn parallel to AB,
and let the rest of the construction be as before;
it is then manifest that MO is the "side" of the area AC.

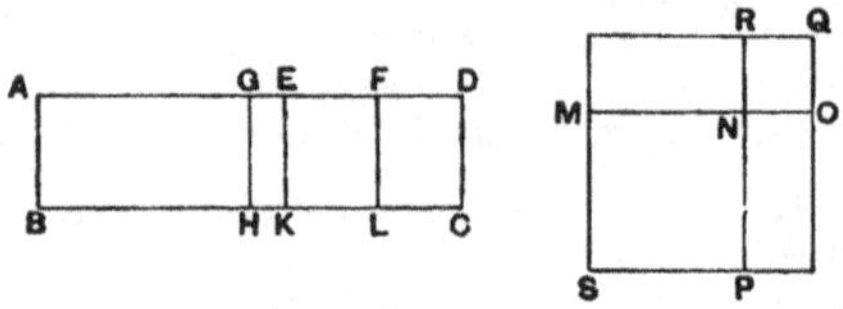

It is next to be proved that MO is the irrational straight line called major.

Since AG is incommensurable with EG,
AH is also incommensurable with GK, that is, SN with NQ; [VI. 1, X. 11]
therefore MN, NO are incommensurable in square.

And, since AE is commensurable with AB,
AK is rational; [x. 19]
and it is equal to the squares on MN, NO;
therefore the sum of the squares on MN, NO is also rational.

And, since DE is incommensurable in length with AB, that is, with EK,
while DE is commensurable with EF,
therefore EF is incommensurable in length with EK. [x. 13]

Therefore EK, EF are rational straight lines commensurable in square only;
therefore LE, that is, MR, is medial. [x. 21]

And it is contained by MN, NO;
therefore the rectangle MN, NO is medial.

And the [sum] of the squares on MN, NO is rational, and MN, NO are incommensurable in square.

But, if two straight lines incommensurable in square and making the sum of the squares on them rational, but the rectangle contained by them medial, be added together, the whole is irrational and is called major. [x. 39]

Therefore MO is the irrational straight line called major and is the "side" of the area AC. Q. E. D.

The problem here is to find the square root of the expression [cf. x. 51]

$$\rho\left(k\rho + \frac{k\rho}{\sqrt{1+\lambda}}\right).$$

The procedure is the same.

Find u, v from the equations

$$\left.\begin{aligned} u + v &= k\rho \\ uv &= \tfrac{1}{4}\,\frac{k^2\rho^2}{1+\lambda} \end{aligned}\right\} \qquad\qquad (1),$$

and, if

$$\left.\begin{aligned} x^2 &= \rho u \\ y^2 &= \rho v \end{aligned}\right\} \qquad\qquad (2),$$

$(x+y)$ is the required square root.

To prove that $(x+y)$ is the *major* irrational straight line Euclid argues thus.

By x. 18, $\qquad u \smile v$,

therefore $\qquad \rho u \smile \rho v$,

or $\qquad x^2 \smile y^2$,

so that $\qquad x \smile\!\!\!- \, y \qquad\qquad (3)$.

Now, since $(u+v) \frown \rho$,

$(u+v)\rho$, or (x^2+y^2), is a rational area $\qquad\qquad (4)$.

Lastly, $xy = \frac{1}{2}\dfrac{k\rho^2}{\sqrt{1+\lambda}}$, which is a *medial* area $\qquad\qquad (5)$.

Thus [(3), (4), (5)] $(x+y)$ is a *major* irrational straight line. [x. 39]

Actual solution gives

$$u = \tfrac{1}{2}k\rho\left(1 + \sqrt{\frac{\lambda}{1+\lambda}}\right),$$

$$v = \tfrac{1}{2}k\rho\left(1 - \sqrt{\frac{\lambda}{1+\lambda}}\right),$$

and $$x + y = \rho\,.\sqrt{\frac{k}{2}\left(1 + \sqrt{\frac{\lambda}{1+\lambda}}\right)} + \rho\,.\sqrt{\frac{k}{2}\left(1 - \sqrt{\frac{\lambda}{1+\lambda}}\right)}.$$

The corresponding square root found in x. 94 is the *minor* irrational straight line, the terms being separated by a *minus* sign, and the two straight lines are the roots (cf. note on x. 51) of the biquadratic equation

$$x^4 - 2k\rho^2\,.\,x^2 + \frac{\lambda}{1+\lambda}k^2\rho^4 = 0.$$

Proposition 58.

If an area be contained by a rational straight line and the fifth binomial, the "side" of the area is the irrational straight line called the side of a rational plus a medial area.

For let the area AC be contained by the rational straight line AB and the fifth binomial AD divided into its terms at E, so that AE is the greater term;

I say that the "side" of the area AC is the irrational straight line called the side of a rational plus a medial area.

For let the same construction be made as before shown; it is then manifest that MO is the "side" of the area AC.

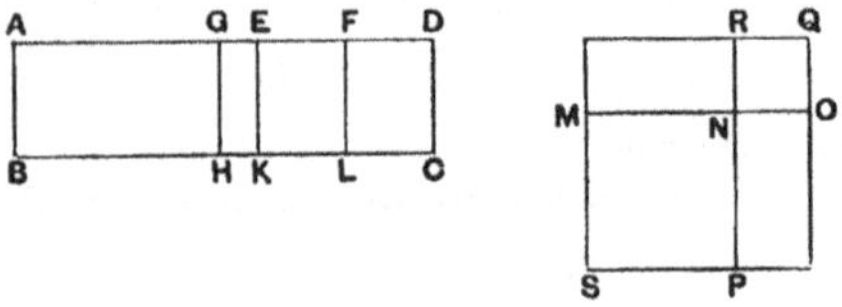

It is then to be proved that MO is the side of a rational plus a medial area.

For, since AG is incommensurable with GE, [x. 18]
therefore AH is also commensurable with HE, [vi. 1, x. 11]
that is, the square on MN with the square on NO;
therefore MN, NO are incommensurable in square.

And, since AD is a fifth binomial straight line, and ED the lesser segment,
therefore ED is commensurable in length with AB. [x. Deff. ii. 5]

But AE is incommensurable with ED;
therefore AB is also incommensurable in length with AE. [x. 13]

Therefore AK, that is, the sum of the squares on MN, NO, is medial. [x. 21]

And, since DE is commensurable in length with AB, that is, with EK,
while DE is commensurable with EF,
therefore EF is also commensurable with EK. [x. 12]

And EK is rational;
therefore EL, that is, MR, that is, the rectangle MN, NO, is also rational. [x. 19]

Therefore MN, NO are straight lines incommensurable in square which make the sum of the squares on them medial, but the rectangle contained by them rational.

Therefore MO is the side of a rational plus a medial area [x. 40] and is the "side" of the area AC.

Q. E. D.

We have here to find the square root of the expression [cf. x. 52]

$$\rho(k\rho\sqrt{1+\lambda}+k\rho).$$

As usual, we put

$$\left.\begin{aligned} u+v &= k\rho\sqrt{1+\lambda} \\ uv &= \tfrac{1}{4}k^2\rho^2 \end{aligned}\right\} \quad\text{..........................(1).}$$

Then, u, v being found, we take

$$\left.\begin{aligned} x^2 &= \rho u \\ y^2 &= \rho v \end{aligned}\right\} \quad\text{................................(2),}$$

and $(x+y)$, so found, is our required square root.

Euclid's proof of the *class* of $(x+y)$ is as follows:

By x. 18, $u \smile v$;
therefore $\rho u \smile \rho v$,
so that $x^2 \smile y^2$,
and $x \;\underset{\frown}{\smile}\; y$ ....(3).

Next $u+v \smile k\rho$
$\smile \rho$,
whence $\rho(u+v)$, or (x^2+y^2), is a *medial* area ....(4).

Lastly, $xy=\frac{1}{2}k\rho^2$, which is a *rational* area ....(5).

Hence [(3), (4), (5)] $(x+y)$ is the *side of a rational plus a medial area.* [x. 40]

If we solve algebraically, we obtain

$$u=\frac{k\rho}{2}(\sqrt{1+\lambda}+\sqrt{\lambda}),$$

$$v=\frac{k\rho}{2}(\sqrt{1+\lambda}-\sqrt{\lambda}),$$

and $$x+y=\rho\sqrt{\frac{k}{2}(\sqrt{1+\lambda}+\sqrt{\lambda})}+\rho\sqrt{\frac{k}{2}(\sqrt{1+\lambda}-\sqrt{\lambda})}.$$

The corresponding "side" found in x. 95 is a *straight line which produces with a rational area a medial whole*, being of the form $(x-y)$, where x, y have the same values as above.

The two square roots are (cf. note on x. 52) the roots of the biquadratic equation

$$x^4-2k\rho^2\sqrt{1+\lambda}\,.\,x^2+\lambda k^2\rho^4=0.$$

PROPOSITION 59.

If an area be contained by a rational straight line and the sixth binomial, the "side" of the area is the irrational straight line called the side of the sum of two medial areas.

For let the area $ABCD$ be contained by the rational straight line AB and the sixth binomial AD, divided into its terms at E, so that AE is the greater term;

I say that the "side" of AC is the side of the sum of two medial areas.

Let the same construction be made as before shown.

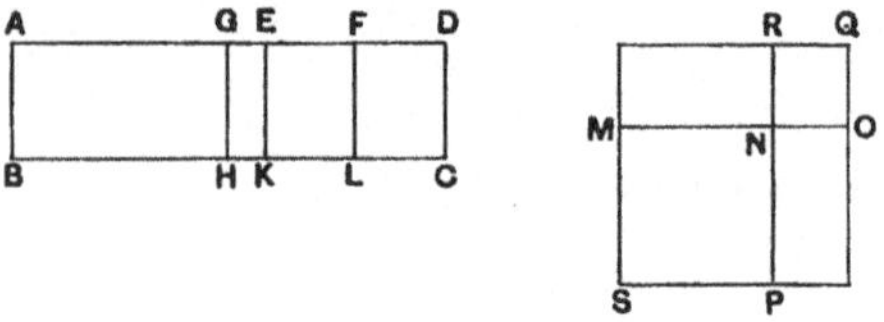

It is then manifest that MO is the "side" of AC, and that MN is incommensurable in square with NO.

Now, since EA is incommensurable in length with AB, therefore EA, AB are rational straight lines commensurable in square only;

therefore AK, that is, the sum of the squares on MN, NO, is medial. [x. 21]

Again, since ED is incommensurable in length with AB, therefore FE is also incommensurable with EK; [x. 13]

therefore FE, EK are rational straight lines commensurable in square only;

therefore EL, that is, MR, that is, the rectangle MN, NO, is medial. [x. 21]

And, since AE is incommensurable with EF,

AK is also incommensurable with EL. [vi. 1, x. 11]

But AK is the sum of the squares on MN, NO,

and EL is the rectangle MN, NO;

therefore the sum of the squares on MN, NO is incommensurable with the rectangle MN, NO.

And each of them is medial, and MN, NO are incommensurable in square.

Therefore MO is the side of the sum of two medial areas [x. 41], and is the "side" of AC.

Q. E. D.

Euclid here finds the square root of the expression [cf. x. 53]

$$\rho\,(\sqrt{k}\,.\,\rho + \sqrt{\lambda}\,.\,\rho).$$

As usual, we solve the equations

$$\left.\begin{aligned} u + v &= \sqrt{k}\,.\,\rho \\ uv &= \tfrac{1}{4}\lambda\rho^2 \end{aligned}\right\} \quad\ldots\ldots\ldots\ldots\ldots\ldots(1);$$

then, u, v being found, we put

$$\left.\begin{aligned} x^2 &= \rho u \\ y^2 &= \rho v \end{aligned}\right\} \quad\ldots\ldots\ldots\ldots\ldots\ldots(2),$$

and $(x + y)$ is the square root required.

Euclid proves that $(x + y)$ is the *side of* (*the sum of*) *two medial areas*, as follows.

As in the last two propositions, x, y are proved to be incommensurable in square.

Now $\sqrt{k}\,.\,\rho$, ρ are commensurable in square only;

therefore $\rho\,(u + v)$, or $(x^2 + y^2)$, is a *medial* area ....(3).

Next, $xy = \frac{1}{2}\sqrt{\lambda}\,.\,\rho^2$, which is again a *medial area* ....(4).

Lastly, $\sqrt{k}\,.\,\rho \smile \frac{1}{2}\sqrt{\lambda}\,.\,\rho$,

so that $\sqrt{k}\,.\,\rho^2 \smile \frac{1}{2}\sqrt{\lambda}\,.\,\rho^2$;

that is, $(x^2 + y^2) \smile xy$ ....(5).

Hence [(3), (4), (5)] $(x + y)$ is the *side of the sum of two medial areas*.

Solving the equations algebraically, we have

$$u = \frac{\rho}{2}\,(\sqrt{k} + \sqrt{k - \lambda}),$$

$$v = \frac{\rho}{2}\,(\sqrt{k} - \sqrt{k - \lambda}),$$

and $$x + y = \rho\sqrt{\tfrac{1}{2}\,(\sqrt{k} + \sqrt{k - \lambda})} + \rho\sqrt{\tfrac{1}{2}\,(\sqrt{k} - \sqrt{k - \lambda})}.$$

The corresponding square root found in x. 96 is $x - y$, where x, y are the same as here.

The two square roots are (cf. note on x. 53) the roots of the biquadratic equation

$$x^4 - 2\sqrt{k}\,.\,\rho^2 x^2 + (k - \lambda)\,\rho^4 = 0.$$

[LEMMA.

If a straight line be cut into unequal parts, the squares on the unequal parts are greater than twice the rectangle contained by the unequal parts.

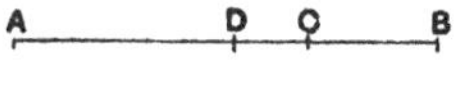

Let AB be a straight line, and let it be cut into unequal parts at C, and let AC be the greater;

I say that the squares on AC, CB are greater than twice the rectangle AC, CB.

For let AB be bisected at D.

Since then a straight line has been cut into equal parts at D, and into unequal parts at C,

therefore the rectangle AC, CB together with the square on CD is equal to the square on AD, [II. 5]

so that the rectangle AC, CB is less than the square on AD;

therefore twice the rectangle AC, CB is less than double of the square on AD.

But the squares on AC, CB are double of the squares on AD, DC; [II. 9]

therefore the squares on AC, CB are greater than twice the rectangle AC, CB.

Q. E. D.]

We have already remarked (note on x. 44) that the Lemma here proving that

$$x^2 + y^2 > 2xy$$

can hardly be genuine, since the result is used in x. 44.

Proposition 60.

The square on the binomial straight line applied to a rational straight line produces as breadth the first binomial.

Let AB be a binomial straight line divided into its terms at C, so that AC is the greater term;

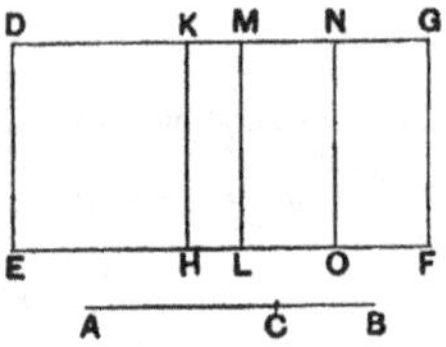

let a rational straight line DE be set out,

and let $DEFG$ equal to the square on AB be applied to DE producing DG as its breadth;

I say that DG is a first binomial straight line.

For let there be applied to DE the rectangle DH equal to the square on AC, and KL equal to the square on BC;

therefore the remainder, twice the rectangle AC, CB, is equal to MF.

Let MG be bisected at N, and let NO be drawn parallel [to ML or GF].

Therefore each of the rectangles MO, NF is equal to once the rectangle AC, CB.

Now, since AB is a binomial divided into its terms at C,

therefore AC, CB are rational straight lines commensurable in square only; [X. 36]
therefore the squares on AC, CB are rational and commensurable with one another,
so that the sum of the squares on AC, CB is also rational. [X. 15]

And it is equal to DL;
therefore DL is rational.

And it is applied to the rational straight line DE;
therefore DM is rational and commensurable in length with DE. [X. 20]

Again, since AC, CB are rational straight lines commensurable in square only,
therefore twice the rectangle AC, CB, that is MF, is medial. [X. 21]

And it is applied to the rational straight line ML;
therefore MG is also rational and incommensurable in length with ML, that is, DE. [X. 22]

But MD is also rational and is commensurable in length with DE;
therefore DM is incommensurable in length with MG. [X. 13]

And they are rational;
therefore DM, MG are rational straight lines commensurable in square only;
therefore DG is binomial. [X. 36]

It is next to be proved that it is also a first binomial straight line.

Since the rectangle AC, CB is a mean proportional between the squares on AC, CB, [cf. Lemma after X. 53]
therefore MO is also a mean proportional between DH, KL.

Therefore, as DH is to MO, so is MO to KL,
that is, as DK is to MN, so is MN to MK; [VI. 1]
therefore the rectangle DK, KM is equal to the square on MN. [VI. 17]

And, since the square on AC is commensurable with the square on CB,
DH is also commensurable with KL,
so that DK is also commensurable with KM. [VI. 1, X. 11]

And, since the squares on *AC*, *CB* are greater than twice the rectangle *AC*, *CB*, [Lemma]

therefore *DL* is also greater than *MF*,

so that *DM* is also greater than *MG*. [VI. 1]

And the rectangle *DK*, *KM* is equal to the square on *MN*, that is, to the fourth part of the square on *MG*,

and *DK* is commensurable with *KM*.

But, if there be two unequal straight lines, and to the greater there be applied a parallelogram equal to the fourth part of the square on the less and deficient by a square figure, and if it divide it into commensurable parts, the square on the greater is greater than the square on the less by the square on a straight line commensurable with the greater; [x. 17]

therefore the square on *DM* is greater than the square on *MG* by the square on a straight line commensurable with *DM*.

And *DM*, *MG* are rational,

and *DM*, which is the greater term, is commensurable in length with the rational straight line *DE* set out.

Therefore *DG* is a first binomial straight line. [x. Deff. II. 1]

Q. E. D.

In the hexad of propositions beginning with this we have the solution of the converse problem to that of x. 54—59. We find the *squares* of the irrational straight lines of x. 36—41 and prove that they are respectively equal to the rectangles contained by a rational straight line and the *first, second, third, fourth, fifth* and *sixth binomials*.

In x. 60 we prove that, $\rho + \sqrt{k}\,.\,\rho$ being a *binomial* straight line [x. 36],

$$\frac{(\rho + \sqrt{k}\,.\,\rho)^2}{\sigma}$$

is a *first binomial* straight line, and we find it geometrically.

The procedure may be represented thus.

Take x, y, z such that

$$\sigma x = \rho^2,$$
$$\sigma y = k\rho^2$$
$$\sigma\,.\,2z = 2\sqrt{k}\,.\,\rho^2,$$

ρ^2, $k\rho^2$ being of course the squares on the terms of the original binomial, and $2\sqrt{k}\,.\,\rho^2$ twice the rectangle contained by them.

Then
$$(x+y) + 2z = \frac{(\rho + \sqrt{k}\,.\,\rho)^2}{\sigma},$$

and we have to prove that $(x+y) + 2z$ is a *first binomial* straight line of which $(x+y)$, $2z$ are the terms and $(x+y)$ the greater.

Euclid divides the proof into two parts, showing first that $(x+y) + 2z$ is *some* binomial, and secondly that it is the *first* binomial.

(α) $\rho \sim \sqrt{k}\cdot\rho$, so that ρ^2, $k\rho^2$ are rational and commensurable;
therefore $\rho^2 + k\rho^2$, or $\sigma(x+y)$, is a rational area,

whence $(x+y)$ is rational and $\frown \sigma$ ....(1).

Next, $2\rho\cdot\sqrt{k}\cdot\rho$ is a medial area,
so that $\sigma\cdot 2z$ is a medial area,

whence $2z$ is rational but $\smile \sigma$ ....(2).

Hence [(1), (2)] $(x+y)$, $2z$ are rational and commensurable in square only ....(3);
thus $(x+y)+2z$ is a *binomial* straight line. [x. 36]

(β) $\rho^2 : \sqrt{k}\cdot\rho^2 = \sqrt{k}\cdot\rho^2 : k\rho^2$,

so that $\sigma x : \sigma z = \sigma z : \sigma y$,

and $x : z = z : y$,

or $xy = z^2 = \frac{1}{4}(2z)^2$...............................(4).

Now ρ^2, $k\rho^2$ are commensurable, so that σx, σy are commensurable, and therefore

$x \frown y$ ....(5).

And, since [Lemma] $\rho^2 + k\rho^2 > 2\sqrt{k}\cdot\rho^2$,

$$x + y > 2z.$$

But $(x+y)$ is given, being equal to $\dfrac{\rho^2 + k\rho^2}{\sigma}$ ....(6).

Therefore [(4), (5), (6), and x. 17] $\sqrt{(x+y)^2 - (2z)^2} \frown (x+y)$.

And $(x+y)$, $2z$ are rational and $\sim$ [(3)],
while $(x+y) \frown \sigma$ [(1)].

Hence $(x+y)+2z$ is a *first binomial.*

The actual value of $(x+y)+2z$ is, of course,

$$\frac{\rho^2}{\sigma}(\overline{1+k} + 2\sqrt{k}).$$

Proposition 61.

The square on the first bimedial straight line applied to a rational straight line produces as breadth the second binomial.

Let AB be a first bimedial straight line divided into its medials at C, of which medials AC is the greater;

let a rational straight line DE be set out,

and let there be applied to DE the parallelogram DF equal to the square on AB, producing DG as its breadth;

I say that DG is a second binomial straight line.

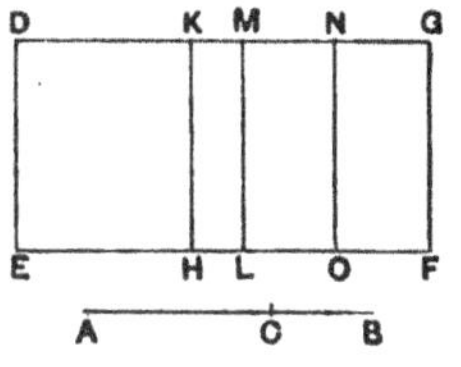

For let the same construction as before be made.

Then, since AB is a first bimedial divided at C, therefore AC, CB are medial straight lines commensurable in square only, and containing a rational rectangle, [x. 37] so that the squares on AC, CB are also medial. [x. 21]

Therefore DL is medial. [x. 15 and 23, Por.]

And it has been applied to the rational straight line DE; therefore MD is rational and incommensurable in length with DE. [x. 22]

Again, since twice the rectangle AC, CB is rational, MF is also rational.

And it is applied to the rational straight line ML; therefore MG is also rational and commensurable in length with ML, that is, DE; [x. 20] therefore DM is incommensurable in length with MG. [x. 13]

And they are rational; therefore DM, MG are rational straight lines commensurable in square only; therefore DG is binomial. [x. 36]

It is next to be proved that it is also a second binomial straight line.

For, since the squares on AC, CB are greater than twice the rectangle AC, CB, therefore DL is also greater than MF, so that DM is also greater than MG. [vi. 1]

And, since the square on AC is commensurable with the square on CB, DH is also commensurable with KL, so that DK is also commensurable with KM. [vi. 1, x. 11]

And the rectangle DK, KM is equal to the square on MN; therefore the square on DM is greater than the square on MG by the square on a straight line commensurable with DM. [x. 17]

And MG is commensurable in length with DE.

Therefore DG is a second binomial straight line. [x. Deff. ii. 2]

In this case we have to prove that, $(k^{\frac{1}{4}}\rho + k^{\frac{3}{4}}\rho)$ being a *first bimedial* straight line, as found in x. 37,

$$\frac{(k^{\frac{1}{4}}\rho + k^{\frac{3}{4}}\rho)^2}{\sigma}$$

is a *second binomial* straight line.

The form of the proposition and the figure being similar to those of x. 60, I can somewhat abbreviate the reproduction of the proof.

Take x, y, z such that

$$\sigma x = k^{\frac{1}{2}}\rho^2,$$
$$\sigma y = k^{\frac{3}{2}}\rho^2,$$
$$\sigma \,.\, 2z = 2k\rho^2.$$

Then shall $(x+y)+2z$ be a second binomial.

(α) $k^{\frac{1}{4}}\rho$, $k^{\frac{3}{4}}\rho$ are medial straight lines commensurable in square only and containing a rational rectangle. [x. 37]

The squares $k^{\frac{1}{2}}\rho^2$, $k^{\frac{3}{2}}\rho^2$ are medial;
thus the sum, or $\sigma(x+y)$, is medial. [x. 23, Por.]

Therefore $(x+y)$ is rational and $\smile \sigma$.

And $\sigma \,.\, 2z$ is rational;
therefore $2z$ is rational and $\frown \sigma$ ....(1).

Therefore $(x+y)$, $2z$ are rational and $\frown$ ....(2),
so that $(x+y)+2z$ is a *binomial*.

(β) As before, $(x+y) > 2z$.

Now, $k^{\frac{1}{2}}\rho^2$, $k^{\frac{3}{2}}\rho^2$ being commensurable,

$$x \frown y.$$

And $xy = z^2$,

while $x+y = \dfrac{k^{\frac{1}{2}}\rho^2 + k^{\frac{3}{2}}\rho^2}{\sigma}$.

Hence [x. 17] $\sqrt{(x+y)^2-(2z)^2} \frown (x+y)$ ....(3).

But $2z \frown \sigma$, by (1).

Therefore [(1), (2), (3)] $(x+y)+2z$ is a *second binomial* straight line.

Of course $(x+y)+2z = \dfrac{\rho^2}{\sigma}\{\sqrt{k}(1+k)+2k\}$.

Proposition 62.

The square on the second bimedial straight line applied to a rational straight line produces as breadth the third binomial.

Let AB be a second bimedial straight line divided into its medials at C, so that AC is the greater segment;
let DE be any rational straight line, and to DE let there be applied the parallelogram DF equal to the square on AB and producing DG as its breadth;
I say that DG is a third binomial straight line.

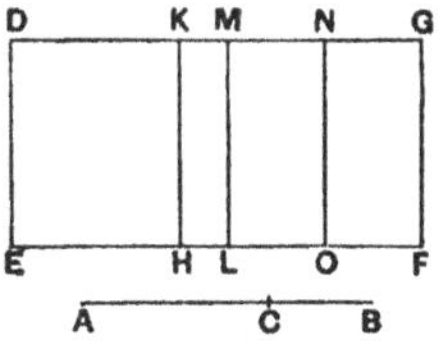

Let the same construction be made as before shown.

Then, since AB is a second bimedial divided at C, therefore AC, CB are medial straight lines commensurable in square only and containing a medial rectangle, [x. 38] so that the sum of the squares on AC, CB is also medial. [x. 15 and 23 Por.]

And it is equal to DL; therefore DL is also medial.

And it is applied to the rational straight line DE; therefore MD is also rational and incommensurable in length with DE. [x. 22]

For the same reason, MG is also rational and incommensurable in length with ML, that is, with DE; therefore each of the straight lines DM, MG is rational and incommensurable in length with DE.

And, since AC is incommensurable in length with CB, and, as AC is to CB, so is the square on AC to the rectangle AC, CB, therefore the square on AC is also incommensurable with the rectangle AC, CB. [x. 11]

Hence the sum of the squares on AC, CB is incommensurable with twice the rectangle AC, CB, [x. 12, 13] that is, DL is incommensurable with MF, so that DM is also incommensurable with MG. [vi. 1, x. 11]

And they are rational; therefore DG is binomial. [x. 36]

It is to be proved that it is also a third binomial straight line.

In manner similar to the foregoing we may conclude that DM is greater than MG, and that DK is commensurable with KM.

And the rectangle DK, KM is equal to the square on MN; therefore the square on DM is greater than the square on MG by the square on a straight line commensurable with DM.

And neither of the straight lines DM, MG is commensurable in length with DE.

Therefore DG is a third binomial straight line. [x. Deff. ii. 3]

Q. E. D.

We have to prove that [cf. x. 38]

$$\frac{1}{\sigma}\left(k^{\frac{1}{4}}\rho + \frac{\lambda^{\frac{1}{2}}\rho}{k^{\frac{1}{4}}}\right)^2$$

is a *third binomial* straight line.

Take x, y, z such that

$$\sigma x = k^{\frac{1}{2}}\rho^2,$$

$$\sigma y = \frac{\lambda\rho^2}{k^{\frac{1}{2}}},$$

$$\sigma \,.\, 2z = 2\sqrt{\lambda}\,.\,\rho^2.$$

(α) Now $k^{\frac{1}{4}}\rho$, $\dfrac{\lambda^{\frac{1}{2}}\rho}{k^{\frac{1}{4}}}$ are medial straight lines commensurable in square only and containing a *medial* rectangle. [x. 38]

The sum of the squares on them, or $\sigma(x+y)$, is *medial*;

therefore $(x+y)$ is rational and $\smile \sigma$ .... (1).

And $\sigma \,.\, 2z$ being medial also,

$2z$ is rational and $\smile \sigma$ ....(2).

Now
$$k^{\frac{1}{4}}\rho : \frac{\lambda^{\frac{1}{2}}\rho}{k^{\frac{1}{4}}} = (k^{\frac{1}{4}}\rho)^2 : k^{\frac{1}{4}}\rho \,.\, \frac{\lambda^{\frac{1}{2}}\rho}{k^{\frac{1}{4}}}$$

$$= \sigma x : \sigma z,$$

whence $\sigma x \smile \sigma z$.

But $(k^{\frac{1}{4}}\rho)^2 \frown \left\{(k^{\frac{1}{4}}\rho)^2 + \left(\dfrac{\lambda^{\frac{1}{2}}\rho}{k^{\frac{1}{4}}}\right)^2\right\}$, or $\sigma x \frown \sigma(x+y)$, and $\sigma z \frown \sigma \,.\, 2z$;

therefore $\sigma(x+y) \smile \sigma \,.\, 2z$,

or $(x+y) \smile 2z$ ....(3).

Hence [(1), (2), (3)] $(x+y)+2z$ is a binomial straight line............(4).

(β) As before, $(x+y) > 2z$,

and $x \frown y$.

Also $xy = z^2$.

Therefore [x. 17] $\sqrt{(x+y)^2-(2z)^2} \frown (x+y)$.

And [(1), (2)] neither $(x+y)$ nor $2z$ is $\frown \sigma$.

Therefore $(x+y)+2z$ is a *third binomial* straight line.

Obviously
$$(x+y)+2z = \frac{\rho^2}{\sigma}\left\{\frac{k+\lambda}{\sqrt{k}} + 2\sqrt{\lambda}\right\}.$$

PROPOSITION 63.

The square on the major straight line applied to a rational straight line produces as breadth the fourth binomial.

Let AB be a major straight line divided at C, so that AC is greater than CB;

let DE be a rational straight line,

and to DE let there be applied the parallelogram DF equal to the square on AB and producing DG as its breadth;

I say that DG is a fourth binomial straight line.

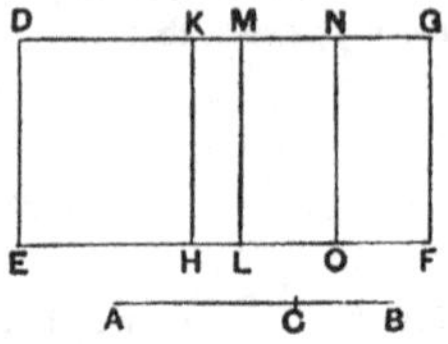

Let the same construction be made as before shown.

Then, since AB is a major straight line divided at C,

AC, CB are straight lines incommensurable in square which make the sum of the squares on them rational, but the rectangle contained by them medial. [x. 39]

Since then the sum of the squares on AC, CB is rational, therefore DL is rational;

therefore DM is also rational and commensurable in length with DE. [x. 20]

Again, since twice the rectangle AC, CB, that is, MF, is medial,

and it is applied to the rational straight line ML,

therefore MG is also rational and incommensurable in length with DE; [x. 22]

therefore DM is also incommensurable in length with MG. [x. 13]

Therefore DM, MG are rational straight lines commensurable in square only;

therefore DG is binomial. [x. 36]

It is to be proved that it is also a fourth binomial straight line.

In manner similar to the foregoing we can prove that DM is greater than MG,

and that the rectangle DK, KM is equal to the square on MN.

Since then the square on AC is incommensurable with the square on CB,

therefore DH is also incommensurable with KL,

so that DK is also incommensurable with KM. [vi. 1, x. 11]

But, if there be two unequal straight lines, and to the greater there be applied a parallelogram equal to the fourth part of the square on the less and deficient by a square figure, and if it divide it into incommensurable parts, then the square on the greater will be greater than the square on the less by the square on a straight line incommensurable in length with the greater; [x. 18]

therefore the square on *DM* is greater than the square on *MG* by the square on a straight line incommensurable with *DM*.

And *DM*, *MG* are rational straight lines commensurable in square only,

and *DM* is commensurable with the rational straight line *DE* set out.

Therefore *DG* is a fourth binomial straight line. [x. Deff. II. 4]

Q. E. D.

We have to prove that [cf. x. 39]

$$\frac{1}{\sigma}\left\{\frac{\rho}{\sqrt{2}}\sqrt{1+\frac{k}{\sqrt{1+k^2}}}+\frac{\rho}{\sqrt{2}}\sqrt{1-\frac{k}{\sqrt{1+k^2}}}\right\}^2$$

is a *fourth binomial* straight line.

For brevity we must call this expression

$$\frac{1}{\sigma}(u+v)^2.$$

Take x, y, z such that

$$\left.\begin{aligned}\sigma x &= u^2\\ \sigma y &= v^2\\ \sigma \,.\, 2z &= 2uv\end{aligned}\right\},$$

wherein it has to be remembered [x. 39] that u, v are incommensurable in square, (u^2+v^2) is rational, and uv is medial.

(α) (u^2+v^2), and therefore $\sigma(x+y)$, is rational;

therefore $(x+y)$ is rational and $\frown \sigma$ ....(1).

$2uv$, and therefore $\sigma \,.\, 2z$, is medial;

therefore $2z$ is rational and $\smile \sigma$ ....(2).

Thus $(x+y)$, $2z$ are rational and $\frown$ ....(3),

so that $(x+y)+2z$ is a binomial straight line.

(β) As before, $x+y>2z$,

and $xy=z^2$.

Now, since $u^2 \smile v^2$,

$\sigma x \smile \sigma y$, or $x \smile y$.

Hence [x. 18] $\sqrt{(x+y)^2-(2z)^2} \smile (x+y)$...................(4).

And $(x+y) \frown \sigma$, by (1).

Therefore [(3), (4)] $(x+y)+2z$ is a *fourth binomial* straight line.

It is of course $$\frac{\rho^2}{\sigma}\left\{1+\frac{1}{\sqrt{1+k^2}}\right\}.$$

Proposition 64.

The square on the side of a rational plus a medial area applied to a rational straight line produces as breadth the fifth binomial.

Let *AB* be the side of a rational plus a medial area, divided into its straight lines at *C*,

so that *AC* is the greater;

let a rational straight line *DE* be set out,

and let there be applied to *DE* the parallelogram *DF* equal to the square on *AB*, producing *DG* as its breadth;

I say that *DG* is a fifth binomial straight line.

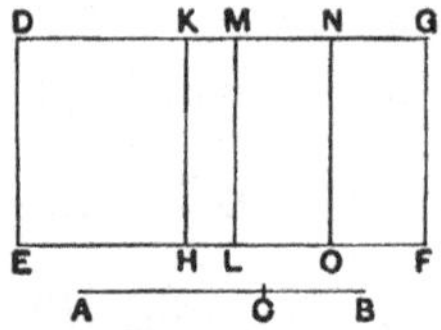

Let the same construction as before be made.

Since then *AB* is the side of a rational plus a medial area, divided at *C*,

therefore *AC*, *CB* are straight lines incommensurable in square which make the sum of the squares on them medial, but the rectangle contained by them rational. [x. 40]

Since then the sum of the squares on *AC*, *CB* is medial,

therefore *DL* is medial,

so that *DM* is rational and incommensurable in length with *DE*. [x. 22]

Again, since twice the rectangle *AC*, *CB*, that is *MF*, is rational,

therefore *MG* is rational and commensurable with *DE*. [x. 20]

Therefore *DM* is incommensurable with *MG*; [x. 13]

therefore *DM*, *MG* are rational straight lines commensurable in square only;

therefore *DG* is binomial. [x. 36]

I say next that it is also a fifth binomial straight line.

For it can be proved similarly that the rectangle *DK*, *KM* is equal to the square on *MN*,

and that *DK* is incommensurable in length with *KM*;

therefore the square on *DM* is greater than the square on *MG* by the square on a straight line incommensurable with *DM*. [x. 18]

And DM, MG are commensurable in square only, and the less, MG, is commensurable in length with DE.

Therefore DG is a fifth binomial.

Q. E. D.

To prove that [cf. x. 40]

$$\frac{1}{\sigma}\left\{\frac{\rho}{\sqrt{2(1+k^2)}}\sqrt{\sqrt{1+k^2}+k}+\frac{\rho}{\sqrt{2(1+k^2)}}\sqrt{\sqrt{1+k^2}-k}\right\}^2$$

is a *fifth binomial* straight line.

For brevity denote it by $\frac{1}{\sigma}(u+v)^2$, and put

$$\sigma x = u^2,$$
$$\sigma y = v^2,$$
$$\sigma \,.\, 2z = 2uv.$$

Remembering that [x. 40] $u^2 \smile v^2$, (u^2+v^2) is medial, and $2uv$ is rational, we proceed thus.

(α) $\sigma(x+y)$ is medial;

therefore $(x+y)$ is rational and $\smile \sigma$ ....(1).

Next, $\sigma \,.\, 2z$ is rational;

therefore $2z$ is rational and $\frown \sigma$..........(2).

Thus $(x+y)$, $2z$ are rational and $\sim$ ....(3),

so that $(x+y)+2z$ is a binomial straight line.

(β) As before, $x+y>2z$,

$$xy=z^2,$$

and $x \smile y$.

Therefore [x. 18] $\sqrt{(x+y)^2-(2z)^2} \smile (x+y)$ ....(4).

Hence [(2), (3), (4)] $(x+y)+2z$ is a *fifth binomial* straight line.

It is of course $\frac{\rho^2}{\sigma}\left\{\frac{1}{\sqrt{1+k^2}}+\frac{1}{1+k^2}\right\}$.

Proposition 65.

The square on the side of the sum of two medial areas applied to a rational straight line produces as breadth the sixth binomial.

Let AB be the side of the sum of two medial areas, divided at C,

let DE be a rational straight line,

and let there be applied to DE the parallelogram DF equal to the square on AB, producing DG as its breadth;

I say that *DG* is a sixth binomial straight line.

For let the same construction be made as before.

Then, since *AB* is the side of the sum of two medial areas, divided at *C*,

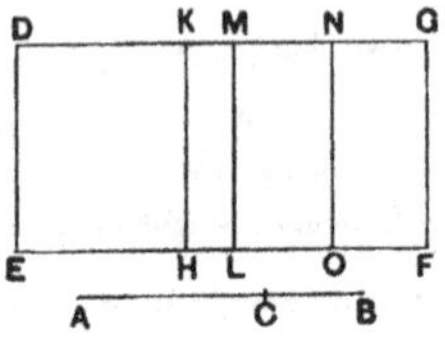

therefore *AC*, *CB* are straight lines incommensurable in square which make the sum of the squares on them medial, the rectangle contained by them medial, and moreover the sum of the squares on them incommensurable with the rectangle contained by them, [x. 41]

so that, in accordance with what was before proved, each of the rectangles *DL*, *MF* is medial.

And they are applied to the rational straight line *DE*;

therefore each of the straight lines *DM*, *MG* is rational and incommensurable in length with *DE*. [x. 22]

And, since the sum of the squares on *AC*, *CB* is incommensurable with twice the rectangle *AC*, *CB*,

therefore *DL* is incommensurable with *MF*.

Therefore *DM* is also incommensurable with *MG*; [vi. 1, x. 11]

therefore *DM*, *MG* are rational straight lines commensurable in square only;

therefore *DG* is binomial. [x. 36]

I say next that it is also a sixth binomial straight line.

Similarly again we can prove that the rectangle *DK*, *KM* is equal to the square on *MN*,

and that *DK* is incommensurable in length with *KM*;

and, for the same reason, the square on *DM* is greater than the square on *MG* by the square on a straight line incommensurable in length with *DM*.

And neither of the straight lines *DM*, *MG* is commensurable in length with the rational straight line *DE* set out.

Therefore *DG* is a sixth binomial straight line.

Q. E. D.

To prove that [cf. x. 41]

$$\frac{1}{\sigma}\left\{\frac{\rho\lambda^{\frac{1}{4}}}{\sqrt{2}}\sqrt{1+\frac{k}{\sqrt{1+k^2}}}+\frac{\rho\lambda^{\frac{1}{4}}}{\sqrt{2}}\sqrt{1-\frac{k}{\sqrt{1+k^2}}}\right\}^2$$

is a ***sixth binomial*** straight line.

Denote it by $\frac{1}{\sigma}(u+v)^2$, and put

$$\sigma x = u^2,$$
$$\sigma y = v^2,$$
$$\sigma \,.\, 2z = 2uv.$$

Now, by x. 41, $u^2 \not\smile v^2$, (u^2+v^2) is medial, $2uv$ is medial, and $(u^2+v^2) \smile 2uv$.

(α) In this case $\sigma(x+y)$ is medial;

therefore $(x+y)$ is rational and $\smile \sigma$ ....(1).

In like manner, $2z$ is rational and $\smile \sigma$ ....(2).

And, since $\sigma(x+y) \smile \sigma \,.\, 2z$,

$(x+y) \smile 2z$ ....(3).

Therefore $(x+y)+2z$ is a binomial straight line.

(β) As before, $x+y>2z$,

$$xy = z^2,$$
$$x \smile y;$$

therefore [x. 18] $\sqrt{(x+y)^2-(2z)^2} \smile (x+y)$ ....(4).

Hence [(1), (2), (3), (4)] $(x+y)+2z$ is a ***sixth binomial*** straight line.

It is obviously $\frac{\rho^2}{\sigma}\left\{\sqrt{\lambda}+\frac{\sqrt{\lambda}}{\sqrt{1+k^2}}\right\}$.

PROPOSITION 66.

A straight line commensurable in length with a binomial straight line is itself also binomial and the same in order.

Let AB be binomial, and let CD be commensurable in length with AB;

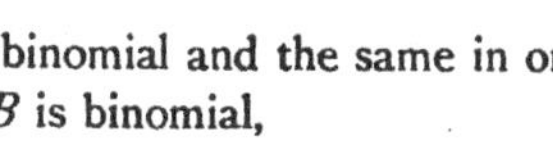

I say that CD is binomial and the same in order with AB.

For, since AB is binomial,
let it be divided into its terms at E,
and let AE be the greater term;

therefore *AE*, *EB* are rational straight lines commensurable in square only. [x. 36]

Let it be contrived that,

as *AB* is to *CD*, so is *AE* to *CF*; [vi. 12]

therefore also the remainder *EB* is to the remainder *FD* as *AB* is to *CD*. [v. 19]

But *AB* is commensurable in length with *CD*;

therefore *AE* is also commensurable with *CF*, and *EB* with *FD*. [x. 11]

And *AE*, *EB* are rational;

therefore *CF*, *FD* are also rational.

And, as *AE* is to *CF*, so is *EB* to *FD*. [v. 11]

Therefore, alternately, as *AE* is to *EB*, so is *CF* to *FD*. [v. 16]

But *AE*, *EB* are commensurable in square only;

therefore *CF*, *FD* are also commensurable in square only. [x. 11]

And they are rational;

therefore *CD* is binomial. [x. 36]

I say next that it is the same in order with *AB*.

For the square on *AE* is greater than the square on *EB* either by the square on a straight line commensurable with *AE* or by the square on a straight line incommensurable with it.

If then the square on *AE* is greater than the square on *EB* by the square on a straight line commensurable with *AE*, the square on *CF* will also be greater than the square on *FD* by the square on a straight line commensurable with *CF*. [x. 14]

And, if *AE* is commensurable with the rational straight line set out, *CF* will also be commensurable with it, [x. 12]

and for this reason each of the straight lines *AB*, *CD* is a first binomial, that is, the same in order. [x. Deff. ii. 1]

But, if *EB* is commensurable with the rational straight line set out, *FD* is also commensurable with it, [x. 12]

and for this reason again *CD* will be the same in order with *AB*,

for each of them will be a second binomial. [x. Deff. ii. 2]

But, if neither of the straight lines *AE*, *EB* is commensurable with the rational straight line set out, neither of the straight lines *CF*, *FD* will be commensurable with it, [x. 13]
and each of the straight lines *AB*, *CD* is a third binomial. [x. Deff. II. 3]

But, if the square on *AE* is greater than the square on *EB* by the square on a straight line incommensurable with *AE*,
the square on *CF* is also greater than the square on *FD* by the square on a straight line incommensurable with *CF*. [x. 14]

And, if *AE* is commensurable with the rational straight line set out, *CF* is also commensurable with it,
and each of the straight lines *AB*, *CD* is a fourth binomial. [x. Deff. II. 4]

But, if *EB* is so commensurable, so is *FD* also,
and each of the straight lines *AB*, *CD* will be a fifth binomial. [x. Deff. II. 5]

But, if neither of the straight lines *AE*, *EB* is so commensurable, neither of the straight lines *CF*, *FD* is commensurable with the rational straight line set out,
and each of the straight lines *AB*, *CD* will be a sixth binomial. [x. Deff. II. 6]

Hence a straight line commensurable in length with a binomial straight line is binomial and the same in order.

Q. E. D.

The proofs of this and the following propositions up to x. 70 inclusive are easy and require no elucidation. They are equivalent to saying that, if in each of the preceding irrational straight lines $\frac{m}{n}\rho$ is substituted for ρ, the resulting irrational is of the same kind as that from which it is altered.

Proposition 67.

A straight line commensurable in length with a bimedial straight line is itself also bimedial and the same in order.

Let *AB* be bimedial, and let *CD* be commensurable in length with *AB*;

I say that *CD* is bimedial and the same in order with *AB*.

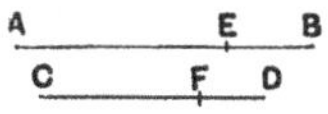

For, since *AB* is bimedial,
let it be divided into its medials at *E*;

therefore *AE*, *EB* are medial straight lines commensurable in square only. [x. 37, 38]

And let it be contrived that,

as *AB* is to *CD*, so is *AE* to *CF*;

therefore also the remainder *EB* is to the remainder *FD* as *AB* is to *CD*. [v. 19]

But *AB* is commensurable in length with *CD*;

therefore *AE*, *EB* are also commensurable with *CF*, *FD* respectively. [x. 11]

But *AE*, *EB* are medial;

therefore *CF*, *FD* are also medial. [x. 23]

And since, as *AE* is to *EB*, so is *CF* to *FD*, [v. 11]

and *AE*, *EB* are commensurable in square only,

CF, *FD* are also commensurable in square only. [x. 11]

But they were also proved medial;

therefore *CD* is bimedial.

I say next that it is also the same in order with *AB*.

For since, as *AE* is to *EB*, so is *CF* to *FD*,

therefore also, as the square on *AE* is to the rectangle *AE*, *EB*, so is the square on *CF* to the rectangle *CF*, *FD*;

therefore, alternately,

as the square on *AE* is to the square on *CF*, so is the rectangle *AE*, *EB* to the rectangle *CF*, *FD*. [v. 16]

But the square on *AE* is commensurable with the square on *CF*;

therefore the rectangle *AE*, *EB* is also commensurable with the rectangle *CF*, *FD*.

If therefore the rectangle *AE*, *EB* is rational,

the rectangle *CF*, *FD* is also rational,

[and for this reason *CD* is a first bimedial]; [x. 37]

but if medial, medial, [x. 23, Por.]

and each of the straight lines *AB*, *CD* is a second bimedial. [x. 38]

And for this reason *CD* will be the same in order with *AB*.

Q. E. D.

PROPOSITION 68.

A straight line commensurable with a major straight line is itself also major.

Let *AB* be major, and let *CD* be commensurable with *AB*; I say that *CD* is major.

Let *AB* be divided at *E*;
therefore *AE*, *EB* are straight lines incommensurable in square which make the sum of the squares on them rational, but the rectangle contained by them medial. [x. 39]

Let the same construction be made as before.

Then since, as *AB* is to *CD*, so is *AE* to *CF*, and *EB* to *FD*,
therefore also, as *AE* is to *CF*, so is *EB* to *FD*. [v. 11]

But *AB* is commensurable with *CD*;
therefore *AE*, *EB* are also commensurable with *CF*, *FD* respectively. [x. 11]

And since, as *AE* is to *CF*, so is *EB* to *FD*,
alternately also,
as *AE* is to *EB*, so is *CF* to *FD*; [v. 16]
therefore also, *componendo*,
as *AB* is to *BE*, so is *CD* to *DF*; [v. 18]
therefore also, as the square on *AB* is to the square on *BE*, so is the square on *CD* to the square on *DF*. [vi. 20]

Similarly we can prove that, as the square on *AB* is to the square on *AE*, so also is the square on *CD* to the square on *CF*.

Therefore also, as the square on *AB* is to the squares on *AE*, *EB*, so is the square on *CD* to the squares on *CF*, *FD*;
therefore also, alternately,
as the square on *AB* is to the square on *CD*, so are the squares on *AE*, *EB* to the squares on *CF*, *FD*. [v. 16]

But the square on *AB* is commensurable with the square on *CD*;
therefore the squares on *AE*, *EB* are also commensurable with the squares on *CF*, *FD*.

And the squares on AE, EB together are rational;
therefore the squares on CF, FD together are rational.

Similarly also twice the rectangle AE, EB is commensurable with twice the rectangle CF, FD.

And twice the rectangle AE, EB is medial;
therefore twice the rectangle CF, FD is also medial. [x. 23, Por.]

Therefore CF, FD are straight lines incommensurable in square which make, at the same time, the sum of the squares on them rational, but the rectangle contained by them medial; therefore the whole CD is the irrational straight line called major. [x. 39]

Therefore a straight line commensurable with the major straight line is major.

Q. E. D.

Proposition 69.

A straight line commensurable with the side of a rational plus a medial area is itself also the side of a rational plus a medial area.

Let AB be the side of a rational plus a medial area, and let CD be commensurable with AB;
it is to be proved that CD is also the side of a rational plus a medial area.

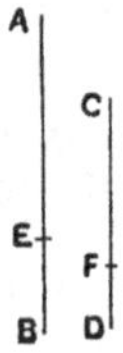

Let AB be divided into its straight lines at E;
therefore AE, EB are straight lines incommensurable in square which make the sum of the squares on them medial, but the rectangle contained by them rational. [x. 40]

Let the same construction be made as before.

We can then prove similarly that
CF, FD are incommensurable in square,
and the sum of the squares on AE, EB is commensurable with the sum of the squares on CF, FD,
and the rectangle AE, EB with the rectangle CF, FD;
so that the sum of the squares on CF, FD is also medial, and the rectangle CF, FD rational.

Therefore CD is the side of a rational plus a medial area.

Q. E. D.

PROPOSITION 70.

A straight line commensurable with the side of the sum of two medial areas is the side of the sum of two medial areas.

Let *AB* be the side of the sum of two medial areas, and *CD* commensurable with *AB*;

it is to be proved that *CD* is also the side of the sum of two medial areas.

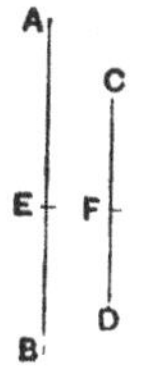

For, since *AB* is the side of the sum of two medial areas,

let it be divided into its straight lines at *E*;

therefore *AE*, *EB* are straight lines incommensurable in square which make the sum of the squares on them medial, the rectangle contained by them medial, and furthermore the sum of the squares on *AE*, *EB* incommensurable with the rectangle *AE*, *EB*. [x. 41]

Let the same construction be made as before.

We can then prove similarly that

CF, *FD* are also incommensurable in square,

the sum of the squares on *AE*, *EB* is commensurable with the sum of the squares on *CF*, *FD*,

and the rectangle *AE*, *EB* with the rectangle *CF*, *FD*;

so that the sum of the squares on *CF*, *FD* is also medial,

the rectangle *CF*, *FD* is medial,

and moreover the sum of the squares on *CF*, *FD* is incommensurable with the rectangle *CF*, *FD*.

Therefore *CD* is the side of the sum of two medial areas.

Q. E. D.

PROPOSITION 71.

If a rational and a medial area be added together, four irrational straight lines arise, namely a binomial or a first bimedial or a major or a side of a rational plus a medial area.

Let *AB* be rational, and *CD* medial;

I say that the "side" of the area *AD* is a binomial or a first bimedial or a major or a side of a rational plus a medial area.

For AB is either greater or less than CD.

First, let it be greater;
let a rational straight line EF be set out,
let there be applied to EF the rectangle EG equal to AB, producing EH as breadth,
and let HI, equal to DC, be applied to EF, producing HK as breadth.

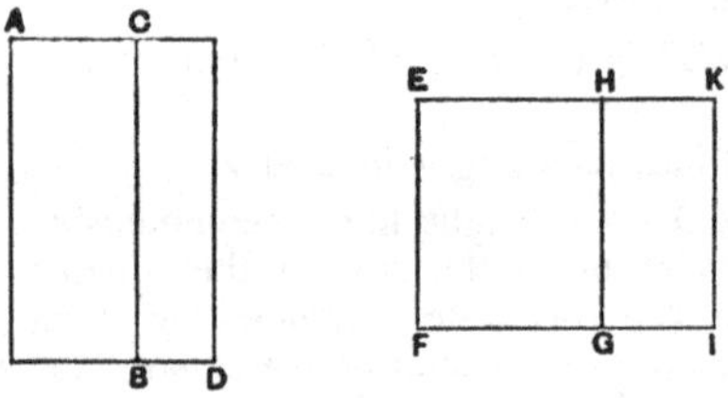

Then, since AB is rational and is equal to EG,
therefore EG is also rational.

And it has been applied to EF, producing EH as breadth;
therefore EH is rational and commensurable in length with EF. [X. 20]

Again, since CD is medial and is equal to HI,
therefore HI is also medial.

And it is applied to the rational straight line EF, producing HK as breadth;
therefore HK is rational and incommensurable in length with EF. [X. 22]

And, since CD is medial,
while AB is rational,
therefore AB is incommensurable with CD,
so that EG is also incommensurable with HI.

But, as EG is to HI, so is EH to HK; [VI. 1]
therefore EH is also incommensurable in length with HK. [X. 11]

And both are rational;
therefore EH, HK are rational straight lines commensurable in square only;
therefore EK is a binomial straight line, divided at H. [X. 36]

And, since AB is greater than CD,
while AB is equal to EG and CD to HI,
therefore EG is also greater than HI;
therefore EH is also greater than HK.

The square, then, on EH is greater than the square on HK either by the square on a straight line commensurable in length with EH or by the square on a straight line incommensurable with it.

First, let the square on it be greater by the square on a straight line commensurable with itself.

Now the greater straight line HE is commensurable in length with the rational straight line EF set out;
therefore EK is a first binomial. [x. Deff. II. 1]

But EF is rational;
and, if an area be contained by a rational straight line and the first binomial, the side of the square equal to the area is binomial. [x. 54]

Therefore the "side" of EI is binomial;
so that the "side" of AD is also binomial.

Next, let the square on EH be greater than the square on HK by the square on a straight line incommensurable with EH.

Now the greater straight line EH is commensurable in length with the rational straight line EF set out;
therefore EK is a fourth binomial. [x. Deff. II. 4]

But EF is rational;
and, if an area be contained by a rational straight line and the fourth binomial, the "side" of the area is the irrational straight line called major. [x. 57]

Therefore the "side" of the area EI is major;
so that the "side" of the area AD is also major.

Next, let AB be less than CD;
therefore EG is also less than HI,
so that EH is also less than HK.

Now the square on HK is greater than the square on EH either by the square on a straight line commensurable with HK or by the square on a straight line incommensurable with it.

First, let the square on it be greater by the square on a straight line commensurable in length with itself.

Now the lesser straight line *EH* is commensurable in length with the rational straight line *EF* set out;

therefore *EK* is a second binomial. [x. Deff. II. 2]

But *EF* is rational;

and, if an area be contained by a rational straight line and the second binomial, the side of the square equal to it is a first bimedial; [x. 55]

therefore the "side" of the area *EI* is a first bimedial,

so that the "side" of *AD* is also a first bimedial.

Next, let the square on *HK* be greater than the square on *HE* by the square on a straight line incommensurable with *HK*.

Now the lesser straight line *EH* is commensurable with the rational straight line *EF* set out;

therefore *EK* is a fifth binomial. [x. Deff. II. 5]

But *EF* is rational;

and, if an area be contained by a rational straight line and the fifth binomial, the side of the square equal to the area is a side of a rational plus a medial area. [x. 58]

Therefore the "side" of the area *EI* is a side of a rational plus a medial area,

so that the "side" of the area *AD* is also a side of a rational plus a medial area.

Therefore etc. Q. E. D.

A *rational* area being of the form $k\rho^2$, and a *medial* area of the form $\sqrt{\lambda}\,.\,\rho^2$, the problem is to classify

$$\sqrt{k\rho^2 + \sqrt{\lambda}\,.\,\rho^2}$$

according to the different possible relations between k, λ.

Put $$\sigma u = k\rho^2,$$
$$\sigma v = \sqrt{\lambda}\,.\,\rho^2.$$

Then, since the former rectangle is rational, the latter medial,

u is rational and $\frown \sigma$,

v is rational and $\smile \sigma$.

Also the rectangles are incommensurable;

so that $$u \smile v.$$

Hence u, v are rational and $\frown$;

whence $(u + v)$ is a binomial straight line.

The possibilities now are as follows:

I. $u > v$.

Then either

(1) $\sqrt{u^2 - v^2} \frown u$,

or (2) $\sqrt{u^2 - v^2} \smile u$,

while in both cases $u \frown \sigma$.

In case (1) $(u + v)$ is a *first binomial* straight line,
and in case (2) $(u + v)$ is a *fourth binomial* straight line.

Thus $\sqrt{\sigma(u + v)}$ is either (1) a *binomial* straight line [x. 54] or (2) a *major* irrational straight line [x. 57].

II. $v > u$.

Then either

(1) $\sqrt{v^2 - u^2} \frown v$,

or (2) $\sqrt{v^2 - u^2} \smile v$,

while in both cases $v \smile \sigma$, but $u \frown \sigma$.

Hence, in case (1), $(v + u)$ is a *second binomial* straight line,
and, in case (2), $(v + u)$ is a *fifth binomial* straight line.

Thus $\sqrt{\sigma(v + u)}$ is either (1) a *first bimedial* straight line [x. 55], or (2) a *side of a rational plus a medial area* [x. 58].

Proposition 72.

If two medial areas incommensurable with one another be added together, the remaining two irrational straight lines arise, namely either a second bimedial or a side of the sum of two medial areas.

For let two medial areas *AB*, *CD* incommensurable with one another be added together;
I say that the "side" of the area *AD* is either a second bimedial or a side of the sum of two medial areas.

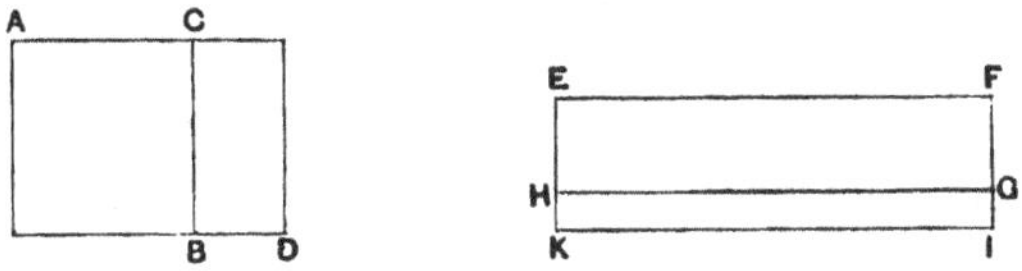

For *AB* is either greater or less than *CD*.

First, if it so chance, let *AB* be greater than *CD*.

Let the rational straight line *EF* be set out,
and to *EF* let there be applied the rectangle *EG* equal to

AB and producing EH as breadth, and the rectangle HI equal to CD and producing HK as breadth.

Now, since each of the areas AB, CD is medial,

therefore each of the areas EG, HI is also medial.

And they are applied to the rational straight line FE, producing EH, HK as breadth;

therefore each of the straight lines EH, HK is rational and incommensurable in length with EF. [x. 22]

And, since AB is incommensurable with CD,

and AB is equal to EG, and CD to HI,

therefore EG is also incommensurable with HI.

But, as EG is to HI, so is EH to HK; [vi. 1]

therefore EH is incommensurable in length with HK. [x. 11]

Therefore EH, HK are rational straight lines commensurable in square only;

therefore EK is binomial. [x. 36]

But the square on EH is greater than the square on HK either by the square on a straight line commensurable with EH or by the square on a straight line incommensurable with it.

First, let the square on it be greater by the square on a straight line commensurable in length with itself.

Now neither of the straight lines EH, HK is commensurable in length with the rational straight line EF set out;

therefore EK is a third binomial. [x. Deff. ii. 3]

But EF is rational;

and, if an area be contained by a rational straight line and the third binomial, the "side" of the area is a second bimedial; [x. 56]

therefore the "side" of EI, that is, of AD, is a second bimedial.

Next, let the square on EH be greater than the square on HK by the square on a straight line incommensurable in length with EH.

Now each of the straight lines EH, HK is incommensurable in length with EF;

therefore EK is a sixth binomial. [x. Deff. ii. 6]

But, if an area be contained by a rational straight line and

the sixth binomial, the "side" of the area is the side of the sum of two medial areas; [x. 59]
so that the "side" of the area AD is also the side of the sum of two medial areas.

Therefore etc.

Q. E. D.

We have to classify, according to the different possible relations between k, λ, the straight line

$$\sqrt{\sqrt{k}\cdot\rho^2+\sqrt{\lambda}\cdot\rho^2},$$

where $\sqrt{k}\cdot\rho^2$ and $\sqrt{\lambda}\cdot\rho^2$ are incommensurable.

Suppose that
$$\sigma u=\sqrt{k}\cdot\rho^2,$$
$$\sigma v=\sqrt{\lambda}\cdot\rho^2.$$

It is immaterial whether $\sqrt{k}\cdot\rho^2$ or $\sqrt{\lambda}\cdot\rho^2$ is the greater. Suppose, e.g., that the former is.

Now, $\sqrt{k}\cdot\rho^2$, $\sqrt{\lambda}\cdot\rho^2$ being both *medial* areas, and σ rational,

u, v are both rational and $\smile \sigma$ ....(1).

Again, by hypothesis, $\sigma u \smile \sigma v$,

or $u \smile v$ ....(2).

Hence [(1), (2)] $(u+v)$ is a binomial straight line.

Next, $\sqrt{u^2-v^2}$ is either commensurable or incommensurable in length with u.

(a) Suppose $\sqrt{u^2-v^2} \frown u$.

In this case $(u+v)$ is a *third binomial* straight line,
and therefore [x. 56]

$\sqrt{\sigma(u+v)}$ is a *second bimedial* straight line.

(β) If $\sqrt{u^2-v^2} \smile u$,
$(u+v)$ is a *sixth binomial* straight line,
and therefore [x. 59]

$\sqrt{\sigma(u+v)}$ is a *side of the sum of two medial areas.*

The binomial straight line and the irrational straight lines after it are neither the same with the medial nor with one another.

For the square on a medial, if applied to a rational straight line, produces as breadth a straight line rational and incommensurable in length with that to which it is applied. [x. 22]

But the square on the binomial, if applied to a rational straight line, produces as breadth the first binomial. [x. 60]

The square on the first bimedial, if applied to a rational straight line, produces as breadth the second binomial. [x. 61]

The square on the second bimedial, if applied to a rational straight line, produces as breadth the third binomial. [x. 62]

The square on the major, if applied to a rational straight line, produces as breadth the fourth binomial. [x. 63]

The square on the side of a rational plus a medial area, if applied to a rational straight line, produces as breadth the fifth binomial. [x. 64]

The square on the side of the sum of two medial areas, if applied to a rational straight line, produces as breadth the sixth binomial. [x. 65]

And the said breadths differ both from the first and from one another: from the first because it is rational, and from one another because they are not the same in order;

so that the irrational straight lines themselves also differ from one another.

The explanation after x. 72 is for the purpose of showing that all the irrational straight lines treated hitherto are different from one another, viz. the medial, the six irrational straight lines beginning with the binomial, and the six consisting of the first, second, third, fourth, fifth and sixth binomials.

Proposition 73.

If from a rational straight line there be subtracted a rational straight line commensurable with the whole in square only, the remainder is irrational; and let it be called an **apotome.**

For from the rational straight line AB let the rational straight line BC, commensurable with the whole in square only, be subtracted;

I say that the remainder AC is the irrational straight line called **apotome.**

For, since AB is incommensurable in length with BC,

and, as AB is to BC, so is the square on AB to the rectangle AB, BC,

therefore the square on AB is incommensurable with the rectangle AB, BC. [x. 11]

But the squares on AB, BC are commensurable with the square on AB, [x. 15]

and twice the rectangle AB, BC is commensurable with the rectangle AB, BC. [x. 6]

And, inasmuch as the squares on AB, BC are equal to twice the rectangle AB, BC together with the square on CA, [II. 7]

therefore the squares on AB, BC are also incommensurable with the remainder, the square on AC. [x. 13, 16]

But the squares on AB, BC are rational;

therefore AC is irrational. [x. Def. 4]

And let it be called an **apotome**.

Q. E. D.

Euclid now passes to the irrational straight lines which are the *difference* and not, as before, the sum of two straight lines. *Apotome* ("portion cut off") accordingly takes the place of *binomial* and the other terms follow *mutatis mutandis*. The first hexad of propositions (73 to 78) exhibit the six irrational straight lines which are really the result of extracting the *square root* of the six irrationals in the later propositions 85 to 90 (or, strictly speaking, of finding the sides of squares equal to the rectangles formed by each of those six irrational straight lines respectively with a rational straight line). Thus, just as in the corresponding propositions about the irrational straight lines formed by addition, the further removed irrationals, so to speak, come first.

We shall denote the *apotome* etc. by $(x-y)$, which is formed by subtracting a certain lesser straight line y from a greater x. In x. 79 and later propositions y is called by Euclid the *annex* (ἡ προσαρμόζουσα), being the straight line which, when added to the apotome or other irrational formed by subtraction, makes up the greater x.

The methods of proof are exactly the same as in the preceding propositions about the irrational straight lines formed by *addition*.

In this proposition x, y are rational straight lines commensurable in square only, and we have to prove that $(x-y)$, the *apotome*, is irrational.

$x \frown\!\!\!\!- y$, so that $x \smile y$:

therefore, since $x : y = x^2 : xy$,

$$x^2 \smile xy.$$

But $x^2 \frown (x^2+y^2)$, and $xy \frown 2xy$;

therefore $x^2+y^2 \smile 2xy$,

whence $(x-y)^2 \smile (x^2+y^2)$.

But (x^2+y^2) is rational;

therefore $(x-y)^2$, and consequently $(x-y)$, is irrational.

The *apotome* $(x-y)$ is of the form $\rho \sim \sqrt{k}\,.\,\rho$, just as the binomial straight line is of the form $\rho + \sqrt{k}\,.\,\rho$.

Proposition 74.

If from a medial straight line there be subtracted a medial straight line which is commensurable with the whole in square only, and which contains with the whole a rational rectangle, the remainder is irrational. And let it be called a **first apotome of a medial** *straight line.*

For from the medial straight line AB let there be subtracted the medial straight line BC which is commensurable with AB in square only and with AB makes the rectangle AB, BC rational;

I say that the remainder AC is irrational; and let it be called a **first apotome of a medial** straight line.

For, since AB, BC are medial,

the squares on AB, BC are also medial.

But twice the rectangle AB, BC is rational;

therefore the squares on AB, BC are incommensurable with twice the rectangle AB, BC;

therefore twice the rectangle AB, BC is also incommensurable with the remainder, the square on AC, [cf. II. 7]

since, if the whole is incommensurable with one of the magnitudes, the original magnitudes will also be incommensurable. [x. 16]

But twice the rectangle AB, BC is rational;

therefore the square on AC is irrational;

therefore AC is irrational. [x. Def. 4]

And let it be called a **first apotome of a medial** straight line.

The *first apotome of a medial* straight line is the difference between straight lines of the form $k^{\frac{1}{4}}\rho$, $k^{\frac{3}{4}}\rho$, which are medial straight lines commensurable in square only and forming a rational rectangle.

By hypothesis, x^2, y^2 are *medial* areas.

And, since xy is rational, $(x^2 + y^2) \smile xy$

$\smile 2xy$,

whence $(x - y)^2 \smile 2xy$.

But $2xy$ is rational;

therefore $(x - y)^2$, and consequently $(x - y)$, is irrational.

This irrational, which is of the form $(k^{\frac{1}{4}}\rho \sim k^{\frac{3}{4}}\rho)$, is the *first apotome of a medial* straight line; the term corresponding of course to *first bimedial*, which applies where the sign is positive.

PROPOSITION 75.

If from a medial straight line there be subtracted a medial straight line which is commensurable with the whole in square only, and which contains with the whole a medial rectangle, the remainder is irrational; and let it be called a **second apotome of a medial** *straight line.*

For from the medial straight line *AB* let there be subtracted the medial straight line *CB* which is commensurable with the whole *AB* in square only and such that the rectangle *AB*, *BC*, which it contains with the whole *AB*, is medial; [x. 28]
I say that the remainder *AC* is irrational; and let it be called a **second apotome of a medial** straight line.

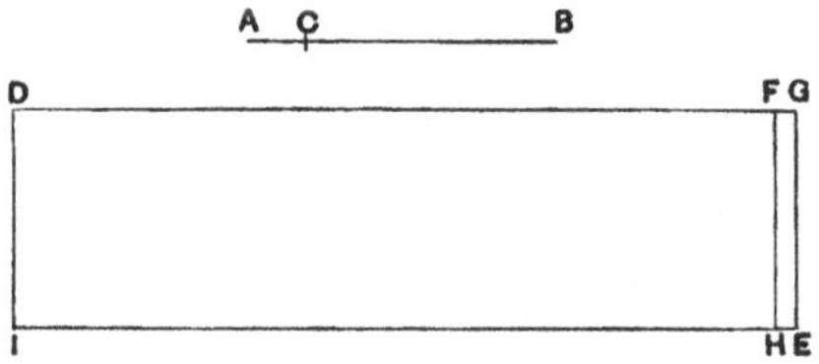

For let a rational straight line *DI* be set out,
let *DE* equal to the squares on *AB*, *BC* be applied to *DI*, producing *DG* as breadth,
and let *DH* equal to twice the rectangle *AB*, *BC* be applied to *DI*, producing *DF* as breadth;
therefore the remainder *FE* is equal to the square on *AC*. [II. 7]

Now, since the squares on *AB*, *BC* are medial and commensurable,
therefore *DE* is also medial. [x. 15 and 23, Por.]

And it is applied to the rational straight line *DI*, producing *DG* as breadth;
therefore *DG* is rational and incommensurable in length with *DI*. [x. 22]

Again, since the rectangle *AB*, *BC* is medial,
therefore twice the rectangle *AB*, *BC* is also medial. [x. 23, Por.]

And it is equal to DH;
therefore DH is also medial.

And it has been applied to the rational straight line DI, producing DF as breadth;
therefore DF is rational and incommensurable in length with DI. [X. 22]

And, since AB, BC are commensurable in square only,
therefore AB is incommensurable in length with BC;
therefore the square on AB is also incommensurable with the rectangle AB, BC. [X. 11]

But the squares on AB, BC are commensurable with the square on AB, [X. 15]
and twice the rectangle AB, BC is commensurable with the rectangle AB, BC; [X. 6]
therefore twice the rectangle AB, BC is incommensurable with the squares on AB, BC. [X. 13]

But DE is equal to the squares on AB, BC,
and DH to twice the rectangle AB, BC;
therefore DE is incommensurable with DH.

But, as DE is to DH, so is GD to DF; [VI. 1]
therefore GD is incommensurable with DF. [X. 11]

And both are rational;
therefore GD, DF are rational straight lines commensurable in square only;
therefore FG is an apotome. [X. 73]

But DI is rational,
and the rectangle contained by a rational and an irrational straight line is irrational, [deduction from X. 20]
and its "side" is irrational.

And AC is the "side" of FE;
therefore AC is irrational.

And let it be called a **second apotome of a medial straight line.**

Q. E. D.

We have here the difference between $k^{\frac{1}{4}}\rho$, $\sqrt{\lambda}\,.\,\rho/k^{\frac{1}{4}}$, two medial straight lines commensurable in square only and containing a medial rectangle.

Apply each of the areas (x^2+y^2), $2xy$ to a rational straight line σ, i.e. suppose that

$$x^2+y^2=\sigma u,$$
$$2xy=\sigma v.$$

Then σu, σv are medial areas,
so that u, v are both rational and $\smile \sigma$ ....(1).

Again, $x \smile y$;
therefore $x^2 \smile xy$,
and consequently $x^2 + y^2 \smile 2xy$,
or $\sigma u \smile \sigma v$,
and $u \smile v$ ....(2).

Thus [(1), (2)] u, v are rational and $\frown$;
therefore [x. 73] $(u - v)$ is an ***apotome***,
and, $(u - v)$ being thus irrational,

$$(u - v)\,\sigma \text{ is an irrational area.}$$

Hence $(x - y)^2$, and consequently $(x - y)$, is irrational.

The irrational straight line $k^{\frac{1}{4}}\rho \sim \dfrac{\sqrt{\lambda}\,.\,\rho}{k^{\frac{1}{4}}}$ is called a ***second apotome of a medial*** straight line.

Proposition 76.

If from a straight line there be subtracted a straight line which is incommensurable in square with the whole and which with the whole makes the squares on them added together rational, but the rectangle contained by them medial, the remainder is irrational; and let it be called **minor**.

For from the straight line AB let there be subtracted the straight line BC which is incommensurable in square with the whole and fulfils the given conditions. [x. 33]

I say that the remainder AC is the irrational straight line called **minor**.

For, since the sum of the squares on AB, BC is rational, while twice the rectangle AB, BC is medial,
therefore the squares on AB, BC are incommensurable with twice the rectangle AB, BC;
and, *convertendo*, the squares on AB, BC are incommensurable with the remainder, the square on AC. [II. 7, x. 16]

But the squares on AB, BC are rational;
therefore the square on AC is irrational;
therefore AC is irrational.

And let it be called **minor**.

Q. E. D.

x, y are here of the form found in x. 33, viz.

$$\frac{\rho}{\sqrt{2}}\sqrt{1+\frac{k}{\sqrt{1+k^2}}},\quad \frac{\rho}{\sqrt{2}}\sqrt{1-\frac{k}{\sqrt{1+k^2}}}.$$

By hypothesis (x^2+y^2) is a *rational*, xy a *medial*, area.

Therefore $(x^2+y^2) \smile 2xy$,

whence $(x-y)^2 \smile (x^2+y^2)$.

Therefore $(x-y)^2$, and consequently $(x-y)$, is irrational.

The *minor* (irrational) straight line is thus of the form

$$\frac{\rho}{\sqrt{2}}\sqrt{1+\frac{k}{\sqrt{1+k^2}}}-\frac{\rho}{\sqrt{2}}\sqrt{1-\frac{k}{\sqrt{1+k^2}}}.$$

Observe the use of *convertendo* (ἀναστρέψαντι) for the inference that, since $(x^2+y^2) \smile 2xy$, $(x^2+y^2) \smile (x-y)^2$. The use of the word corresponds exactly to its use in proportions.

PROPOSITION 77.

If from a straight line there be subtracted a straight line which is incommensurable in square with the whole, and which with the whole makes the sum of the squares on them medial, but twice the rectangle contained by them rational, the remainder is irrational: and let it be called **that which produces with a rational area a medial whole.**

For from the straight line *AB* let there be subtracted the straight line *BC* which is incommensurable in square with *AB* and fulfils the given conditions; [x. 34]

I say that the remainder *AC* is the irrational straight line aforesaid.

For, since the sum of the squares on *AB*, *BC* is medial,

while twice the rectangle *AB*, *BC* is rational,

therefore the squares on *AB*, *BC* are incommensurable with twice the rectangle *AB*, *BC*;

therefore the remainder also, the square on *AC*, is incommensurable with twice the rectangle *AB*, *BC*. [II. 7, x. 16]

And twice the rectangle *AB*, *BC* is rational;

therefore the square on *AC* is irrational;

therefore *AC* is irrational.

And let it be called **that which produces with a rational area a medial whole.**

Q. E. D.

Here x, y are of the form [cf. x. 34]

$$\frac{\rho}{\sqrt{2(1+k^2)}}\sqrt{\sqrt{1+k^2}+k},\quad \frac{\rho}{\sqrt{2(1+k^2)}}\sqrt{\sqrt{1+k^2}-k}.$$

By hypothesis, (x^2+y^2) is a *medial*, xy a *rational*, area;

thus $$(x^2+y^2) \smile 2xy,$$

and therefore $$(x-y)^2 \smile 2xy,$$

whence $(x-y)^2$, and consequently $(x-y)$, is irrational.

The irrational straight line

$$\frac{\rho}{\sqrt{2(1+k^2)}}\sqrt{\sqrt{1+k^2}+k}-\frac{\rho}{\sqrt{2(1+k^2)}}\sqrt{\sqrt{1+k^2}-k}$$

is called *that which produces with a rational area a medial whole* or more literally *that which with a rational area makes the whole medial* (ἡ μετὰ ῥητοῦ μέσον τὸ ὅλον ποιοῦσα). Here "produces" means "produces when a square is described on it." A clearer way of expressing the meaning would be to call this straight line the "*side*" *of a medial minus a rational area* corresponding to the "*side*" *of a rational plus a medial area* [x. 40].

Proposition 78.

If from a straight line there be subtracted a straight line which is incommensurable in square with the whole and which with the whole makes the sum of the squares on them medial, twice the rectangle contained by them medial, and further the squares on them incommensurable with twice the rectangle contained by them, the remainder is irrational; and let it be called **that which produces with a medial area a medial whole.**

For from the straight line AB let there be subtracted the straight line BC incommensurable in square with AB and fulfilling the given conditions; [x. 35]

I say that the remainder AC is the irrational straight line called **that which produces with a medial area a medial whole.**

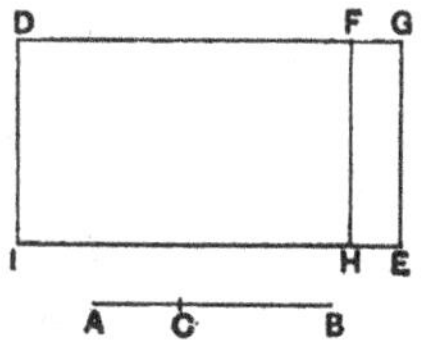

For let a rational straight line DI be set out,

to DI let there be applied DE equal to the squares on AB, BC, producing DG as breadth,

and let DH equal to twice the rectangle AB, BC be subtracted.

Therefore the remainder FE is equal to the square on AC, [II. 7]

so that AC is the "side" of FE.

Now, since the sum of the squares on AB, BC is medial and is equal to DE,

therefore DE is medial.

And it is applied to the rational straight line DI, producing DG as breadth;

therefore DG is rational and incommensurable in length with DI. [X. 22]

Again, since twice the rectangle AB, BC is medial and is equal to DH,

therefore DH is medial.

And it is applied to the rational straight line DI, producing DF as breadth;

therefore DF is also rational and incommensurable in length with DI. [X. 22]

And, since the squares on AB, BC are incommensurable with twice the rectangle AB, BC,

therefore DE is also incommensurable with DH.

But, as DE is to DH, so also is DG to DF; [VI. 1]

therefore DG is incommensurable with DF. [X. 11]

And both are rational;

therefore GD, DF are rational straight lines commensurable in square only.

Therefore FG is an apotome. [X. 73]

And FH is rational;

but the rectangle contained by a rational straight line and an apotome is irrational, [deduction from X. 20]

and its "side" is irrational.

And AC is the "side" of FE;

therefore AC is irrational.

And let it be called **that which produces with a medial area a medial whole.**

Q. E. D.

In this case x, y have respectively the forms [cf. x. 35]

$$\frac{\rho\lambda^{\frac{1}{4}}}{\sqrt{2}}\sqrt{1+\frac{k}{\sqrt{1+k^2}}},\quad \frac{\rho\lambda^{\frac{1}{4}}}{\sqrt{2}}\sqrt{1-\frac{k}{\sqrt{1+k^2}}}.$$

Suppose that $$x^2+y^2=\sigma u,$$
$$2xy=\sigma v.$$

By hypothesis, the areas σu, σv are medial;

therefore u, v are both rational and $\smile \sigma$ ....(1).

Further $$\sigma u \smile \sigma v,$$

so that $$u \smile v \quad \text{..................................} (2).$$

Hence [(1), (2)] u, v are rational and $\frown$,

so that $(u-v)$ is the irrational straight line called *apotome* [x. 73].

Thus $\sigma(u-v)$ is an irrational area,

so that $(x-y)^2$, and consequently $(x-y)$, is irrational.

The irrational straight line

$$\frac{\rho\lambda^{\frac{1}{4}}}{\sqrt{2}}\sqrt{1+\frac{k}{\sqrt{1+k^2}}}-\frac{\rho\lambda^{\frac{1}{4}}}{\sqrt{2}}\sqrt{1-\frac{k}{\sqrt{1+k^2}}}$$

is called *that which produces* [i.e. when a square is described on it] *with a medial area a medial whole*, more literally *that which with a medial area makes the whole medial* (ἡ μετὰ μέσου μέσον τὸ ὅλον ποιοῦσα). A clearer phrase (to us) would be the "*side*" *of the difference between two medial areas*, corresponding to the "*side*" *of* (*the sum of*) *two medial areas* [x. 41].

Proposition 79.

To an apotome only one rational straight line can be annexed which is commensurable with the whole in square only.

Let AB be an apotome, and BC an annex to it;

therefore AC, CB are rational straight lines commensurable in square only. [x. 73]

I say that no other rational straight line can be annexed to AB which is commensurable with the whole in square only.

For, if possible, let BD be so annexed;

therefore AD, DB are also rational straight lines commensurable in square only. [x. 73]

Now, since the excess of the squares on AD, DB over twice the rectangle AD, DB is also the excess of the squares on AC, CB over twice the rectangle AC, CB,

for both exceed by the same, the square on AB, [II. 7]

therefore, alternately, the excess of the squares on AD, DB over the squares on AC, CB is the excess of twice the rectangle AD, DB over twice the rectangle AC, CB.

But the squares on AD, DB exceed the squares on AC, CB by a rational area,
for both are rational;
therefore twice the rectangle AD, DB also exceeds twice the rectangle AC, CB by a rational area:
which is impossible,
for both are medial [X. 21], and a medial area does not exceed a medial by a rational area. [X. 26]

Therefore no other rational straight line can be annexed to AB which is commensurable with the whole in square only.

Therefore only one rational straight line can be annexed to an apotome which is commensurable with the whole in square only.

Q. E. D.

This proposition proves the equivalent of the well-known theorem of surds that,
if $a - \sqrt{b} = x - \sqrt{y}$, then $a = x$, $b = y$;
and, if $\sqrt{a} - \sqrt{b} = \sqrt{x} - \sqrt{y}$, then $a = x$, $b = y$.

The method of proof corresponds to that of X. 42 for positive signs.

Suppose, if possible, that an *apotome* can be expressed as $(x - y)$ and also as $(x' - y')$, where x, y are rational straight lines commensurable in square only, and x', y' are so also.

Of x, x', let x be the greater.

Now, since $$x - y = x' - y',$$
$$x^2 + y^2 - (x'^2 + y'^2) = 2xy - 2x'y'.$$

But $(x^2 + y^2)$, $(x'^2 + y'^2)$ are both rational, so that their difference is a rational area.

On the other hand, $2xy$, $2x'y'$ are both *medial* areas, being of the form $\sqrt{k} \cdot \rho^2$;
therefore the difference between two medial areas is rational:
which is impossible [X. 26].

Therefore etc.

PROPOSITION 80.

To a first apotome of a medial straight line only one medial straight line can be annexed which is commensurable with the whole in square only and which contains with the whole a rational rectangle.

For let AB be a first apotome of a medial straight line, and let BC be an annex to AB;

therefore AC, CB are medial straight lines commensurable in square only and such that the rectangle AC, CB which they contain is rational; [x. 74]

I say that no other medial straight line can be annexed to AB which is commensurable with the whole in square only and which contains with the whole a rational area.

For, if possible, let DB also be so annexed;

therefore AD, DB are medial straight lines commensurable in square only and such that the rectangle AD, DB which they contain is rational. [x. 74]

Now, since the excess of the squares on AD, DB over twice the rectangle AD, DB is also the excess of the squares on AC, CB over twice the rectangle AC, CB,

for they exceed by the same, the square on AB, [II. 7]

therefore, alternately, the excess of the squares on AD, DB over the squares on AC, CB is also the excess of twice the rectangle AD, DB over twice the rectangle AC, CB.

But twice the rectangle AD, DB exceeds twice the rectangle AC, CB by a rational area,

for both are rational.

Therefore the squares on AD, DB also exceed the squares on AC, CB by a rational area:

which is impossible,

for both are medial [x. 15 and 23, Por.], and a medial area does not exceed a medial by a rational area. [x. 26]

Therefore etc.

Q. E. D.

Suppose, if possible, that the same *first apotome of a medial* straight line can be expressed in terms of the required character in two ways, so that

$$x - y = x' - y',$$

and suppose that $x > x'$.

In this case $x^2 + y^2$, $(x'^2 + y'^2)$ are both *medial* areas, and $2xy$, $2x'y'$ are both *rational* areas;

and $$x^2 + y^2 - (x'^2 + y'^2) = 2xy - 2x'y'.$$

Hence x. 26 is contradicted again;

therefore etc.

Proposition 81.

To a second apotome of a medial straight line only one medial straight line can be annexed which is commensurable with the whole in square only and which contains with the whole a medial rectangle.

Let AB be a second apotome of a medial straight line and BC an annex to AB;

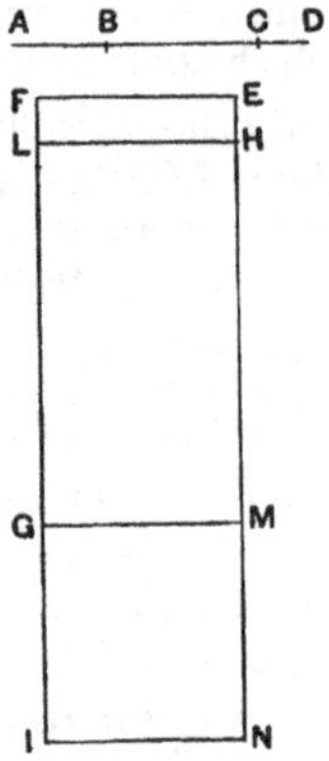

therefore AC, CB are medial straight lines commensurable in square only and such that the rectangle AC, CB which they contain is medial. [X. 75]

I say that no other medial straight line can be annexed to AB which is commensurable with the whole in square only and which contains with the whole a medial rectangle.

For, if possible, let BD also be so annexed;

therefore AD, DB are also medial straight lines commensurable in square only and such that the rectangle AD, DB which they contain is medial. [X. 75]

Let a rational straight line EF be set out,

let EG equal to the squares on AC, CB be applied to EF, producing EM as breadth,

and let HG equal to twice the rectangle AC, CB be subtracted, producing HM as breadth;

therefore the remainder EL is equal to the square on AB, [II. 7]

so that AB is the "side" of EL.

Again, let EI equal to the squares on AD, DB be applied to EF, producing EN as breadth.

But EL is also equal to the square on AB;

therefore the remainder HI is equal to twice the rectangle AD, DB. [II. 7]

Now, since AC, CB are medial straight lines, therefore the squares on AC, CB are also medial.

And they are equal to EG;
therefore EG is also medial. [x. 15 and 23, Por.]

And it is applied to the rational straight line EF, producing EM as breadth;
therefore EM is rational and incommensurable in length with EF. [x. 22]

Again, since the rectangle AC, CB is medial,
twice the rectangle AC, CB is also medial. [x. 23, Por.]

And it is equal to HG;
therefore HG is also medial.

And it is applied to the rational straight line EF, producing HM as breadth;
therefore HM is also rational and incommensurable in length with EF. [x. 22]

And, since AC, CB are commensurable in square only,
therefore AC is incommensurable in length with CB.

But, as AC is to CB, so is the square on AC to the rectangle AC, CB;
therefore the square on AC is incommensurable with the rectangle AC, CB. [x. 11]

But the squares on AC, CB are commensurable with the square on AC,
while twice the rectangle AC, CB is commensurable with the rectangle AC, CB; [x. 6]
therefore the squares on AC, CB are incommensurable with twice the rectangle AC, CB. [x. 13]

And EG is equal to the squares on AC, CB,
while GH is equal to twice the rectangle AC, CB;
therefore EG is incommensurable with HG.

But, as EG is to HG, so is EM to HM; [vi. 1]
therefore EM is incommensurable in length with MH. [x. 11]

And both are rational;
therefore EM, MH are rational straight lines commensurable in square only;
therefore EH is an apotome, and HM an annex to it. [x. 73]

Similarly we can prove that HN is also an annex to it; therefore to an apotome different straight lines are annexed which are commensurable with the wholes in square only: which is impossible. [X. 79]

Therefore etc.

Q. E. D.

As the irrationality of the *second apotome of a medial* straight line was deduced [X. 75] from the irrationality of an apotome, so the present theorem is reduced to X. 79.

Suppose, if possible, that $(x-y)$, $(x'-y')$ are the *same* second apotome of a medial straight line;
and let (say) x be greater than x'.

Apply (x^2+y^2), $2xy$ and also $(x'^2+y'^2)$, $2x'y'$ to a rational straight line σ, i.e. put

$$\left.\begin{aligned} x^2+y^2 &= \sigma u \\ 2xy &= \sigma v \end{aligned}\right\} \text{ and } \left.\begin{aligned} x'^2+y'^2 &= \sigma u' \\ 2x'y' &= \sigma v' \end{aligned}\right\}.$$

Dealing with $(x-y)$ first, we have:
(x^2+y^2) is a medial area, and $2xy$ is also a medial area.

Therefore u, v are both rational and $\smile \sigma$ ....(1).

Also, since $x \frown y$, $x \smile y$,
so that $x^2 \smile xy$,
whence, as usual, $x^2+y^2 \smile 2xy$,
that is, $\sigma u \smile \sigma v$,
and therefore $u \smile v$ ....(2).

Thus [(1) and (2)] u, v are rational and $\frown$,
so that $(u-v)$ is an *apotome*.

Similarly $(u'-v')$ is proved to be the *same* apotome.

Hence this apotome is formed in two ways:
which contradicts X. 79.

Therefore the original hypothesis is false, and a *second apotome of a medial* straight line is uniquely formed.

Proposition 82.

To a minor straight line only one straight line can be annexed which is incommensurable in square with the whole and which makes, with the whole, the sum of the squares on them rational but twice the rectangle contained by them medial.

Let AB be the minor straight line, and let BC be an annex to AB;

therefore AC, CB are straight lines incommensurable in square which make the sum of the squares on them rational, but twice the rectangle contained by them medial. [x. 76]

I say that no other straight line can be annexed to *AB* fulfilling the same conditions.

For, if possible, let *BD* be so annexed;

therefore *AD*, *DB* are also straight lines incommensurable in square which fulfil the aforesaid conditions. [x. 76]

Now, since the excess of the squares on *AD*, *DB* over the squares on *AC*, *CB* is also the excess of twice the rectangle *AD*, *DB* over twice the rectangle *AC*, *CB*,

while the squares on *AD*, *DB* exceed the squares on *AC*, *CB* by a rational area,

for both are rational,

therefore twice the rectangle *AD*, *DB* also exceeds twice the rectangle *AC*, *CB* by a rational area:

which is impossible, for both are medial. [x. 26]

Therefore to a minor straight line only one straight line can be annexed which is incommensurable in square with the whole and which makes the squares on them added together rational, but twice the rectangle contained by them medial.

Q. E. D.

Suppose, if possible, that, with the usual notation,

$$x - y = x' - y';$$

and let x (say) be greater than x'.

In this case $(x^2 + y^2)$, $(x'^2 + y'^2)$ are both *rational* areas, and $2xy$, $2x'y'$ are both *medial* areas.

But, as before, $(x^2 + y^2) - (x'^2 + y'^2) = 2xy - 2x'y'$,

so that the difference between two medial areas is rational: which is impossible [x. 26].

Therefore etc.

Proposition 83.

To a straight line which produces with a rational area a medial whole only one straight line can be annexed which is incommensurable in square with the whole straight line and which with the whole straight line makes the sum of the squares on them medial, but twice the rectangle contained by them rational.

Let AB be the straight line which produces with a rational area a medial whole,

and let BC be an annex to AB;

therefore AC, CB are straight lines incommensurable in square which fulfil the given conditions. [x. 77]

I say that no other straight line can be annexed to AB which fulfils the same conditions.

For, if possible, let BD be so annexed;

therefore AD, DB are also straight lines incommensurable in square which fulfil the given conditions. [x. 77]

Since then, as in the preceding cases,

the excess of the squares on AD, DB over the squares on AC, CB is also the excess of twice the rectangle AD, DB over twice the rectangle AC, CB,

while twice the rectangle AD, DB exceeds twice the rectangle AC, CB by a rational area,

for both are rational,

therefore the squares on AD, DB also exceed the squares on AC, CB by a rational area:

which is impossible, for both are medial. [x. 26]

Therefore no other straight line can be annexed to AB which is incommensurable in square with the whole and which with the whole fulfils the aforesaid conditions;

therefore only one straight line can be so annexed.

Q. E. D.

Suppose, with the same notation, that

$$x - y = x' - y'. \qquad (x > x')$$

Here, $(x^2 + y^2)$, $(x'^2 + y'^2)$ being both medial areas, and $2xy$, $2x'y'$ both rational areas,

while $$(x^2 + y^2) - (x'^2 + y'^2) = 2xy - 2x'y',$$

x. 26 is contradicted again.

Therefore etc.

Proposition 84.

To a straight line which produces with a medial area a medial whole only one straight line can be annexed which is incommensurable in square with the whole straight line and which with the whole straight line makes the sum of the squares on them medial and twice the rectangle contained by them both medial and also incommensurable with the sum of the squares on them.

Let *AB* be the straight line which produces with a medial area a medial whole,

and *BC* an annex to it;

therefore *AC*, *CB* are straight lines incommensurable in square which fulfil the aforesaid conditions. [x. 78]

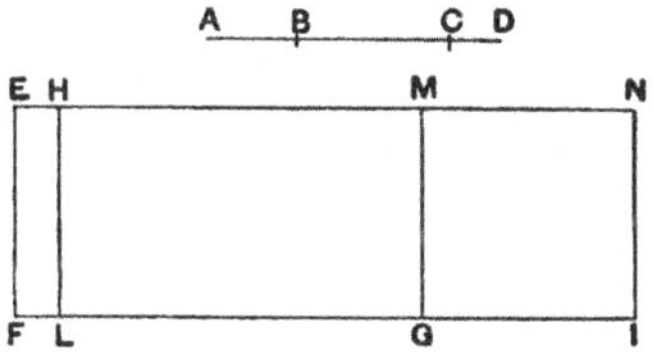

I say that no other straight line can be annexed to *AB* which fulfils the aforesaid conditions.

For, if possible, let *BD* be so annexed,

so that *AD*, *DB* are also straight lines incommensurable in square which make the squares on *AD*, *DB* added together medial, twice the rectangle *AD*, *DB* medial, and also the squares on *AD*, *DB* incommensurable with twice the rectangle *AD*, *DB*. [x. 78]

Let a rational straight line *EF* be set out,

let *EG* equal to the squares on *AC*, *CB* be applied to *EF*, producing *EM* as breadth,

and let *HG* equal to twice the rectangle *AC*, *CB* be applied to *EF*, producing *HM* as breadth;

therefore the remainder, the square on *AB* [II. 7], is equal to *EL*;

therefore *AB* is the "side" of *EL*.

Again, let *EI* equal to the squares on *AD*, *DB* be applied to *EF*, producing *EN* as breadth.

But the square on *AB* is also equal to *EL*;

therefore the remainder, twice the rectangle *AD*, *DB* [II. 7], is equal to *HI*.

Now, since the sum of the squares on *AC*, *CB* is medial and is equal to *EG*,

therefore *EG* is also medial.

And it is applied to the rational straight line *EF*, producing *EM* as breadth;

therefore *EM* is rational and incommensurable in length with *EF*. [x. 22]

Again, since twice the rectangle *AC*, *CB* is medial and is equal to *HG*,

therefore *HG* is also medial.

And it is applied to the rational straight line *EF*, producing *HM* as breadth;

therefore *HM* is rational and incommensurable in length with *EF*. [x. 22]

And, since the squares on *AC*, *CB* are incommensurable with twice the rectangle *AC*, *CB*,

EG is also incommensurable with *HG*;

therefore *EM* is also incommensurable in length with *MH*. [vi. 1, x. 11]

And both are rational;

therefore *EM*, *MH* are rational straight lines commensurable in square only;

therefore *EH* is an apotome, and *HM* an annex to it. [x. 73]

Similarly we can prove that *EH* is again an apotome and *HN* an annex to it.

Therefore to an apotome different rational straight lines are annexed which are commensurable with the wholes in square only:

which was proved impossible. [x. 79]

Therefore no other straight line can be so annexed to *AB*.

Therefore to *AB* only one straight line can be annexed which is incommensurable in square with the whole and which with the whole makes the squares on them added together medial, twice the rectangle contained by them medial, and also the squares on them incommensurable with twice the rectangle contained by them.

Q. E. D.

With the usual notation, suppose that

$$x - y = x' - y'. \qquad (x > x')$$

Let $\left.\begin{aligned} x^2 + y^2 &= \sigma u \\ 2xy &= \sigma v \end{aligned}\right\}$ and $\left.\begin{aligned} x'^2 + y'^2 &= \sigma u' \\ 2x'y' &= \sigma v' \end{aligned}\right\}$.

Consider $(x - y)$ first;
it follows, since $(x^2 + y^2)$, $2xy$ are both medial areas, that
u, v are both rational and $\smile \sigma$ ....(1).

But $x^2 + y^2 \smile 2xy$,

that is, $\sigma u \smile \sigma v$,

and therefore $u \smile v$ ....(2).

Therefore [(1) and (2)] u, v are rational and $\frown$;
hence $(u - v)$ is an *apotome*.

Similarly $(u' - v')$ is proved to be the *same* apotome.

Thus the same apotome is formed as such in two ways:
which is impossible [x. 79].

Therefore, etc.

DEFINITIONS III.

1. Given a rational straight line and an apotome, if the square on the whole be greater than the square on the annex by the square on a straight line commensurable in length with the whole, and the whole be commensurable in length with the rational straight line set out, let the apotome be called a **first apotome.**

2. But if the annex be commensurable in length with the rational straight line set out, and the square on the whole be greater than that on the annex by the square on a straight line commensurable with the whole, let the apotome be called a **second apotome.**

3. But if neither be commensurable in length with the rational straight line set out, and the square on the whole be greater than the square on the annex by the square on a straight line commensurable with the whole, let the apotome be called a **third apotome.**

4. Again, if the square on the whole be greater than the square on the annex by the square on a straight line incommensurable with the whole, then, if the whole be commensurable in length with the rational straight line set out, let the apotome be called a **fourth apotome**;

5. if the annex be so commensurable, a **fifth**;

6. and, if neither, a **sixth.**

PROPOSITION 85.

To find the first apotome.

Let a rational straight line A be set out, and let BG be commensurable in length with A; therefore BG is also rational.

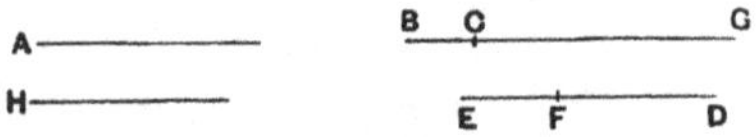

Let two square numbers DE, EF be set out, and let their difference FD not be square; therefore neither has ED to DF the ratio which a square number has to a square number.

Let it be contrived that, as ED is to DF, so is the square on BG to the square on GC; [x. 6, Por.]

therefore the square on BG is commensurable with the square on GC. [x. 6]

But the square on BG is rational; therefore the square on GC is also rational; therefore GC is also rational.

And, since ED has not to DF the ratio which a square number has to a square number, therefore neither has the square on BG to the square on GC the ratio which a square number has to a square number; therefore BG is incommensurable in length with GC. [x. 9]

And both are rational; therefore BG, GC are rational straight lines commensurable in square only; therefore BC is an apotome. [x. 73]

I say next that it is also a first apotome.

For let the square on H be that by which the square on BG is greater than the square on GC.

Now since, as ED is to FD, so is the square on BG to the square on GC, therefore also, *convertendo*, [v. 19, Por.]

as DE is to EF, so is the square on GB to the square on H.

But DE has to EF the ratio which a square number has to a square number,

for each is square;

therefore the square on GB also has to the square on H the ratio which a square number has to a square number;

therefore BG is commensurable in length with H. [x. 9]

And the square on BG is greater than the square on GC by the square on H;

therefore the square on BG is greater than the square on GC by the square on a straight line commensurable in length with BG.

And the whole BG is commensurable in length with the rational straight line A set out.

Therefore BC is a first apotome. [x. Deff. III. 1]

Therefore the first apotome BC has been found.

(Being) that which it was required to find.

Take $k\rho$ commensurable in length with ρ, the given rational straight line.

Let m^2, n^2 be square numbers such that $(m^2 - n^2)$ is not square.

Take x such that $m^2 : (m^2 - n^2) = k^2\rho^2 : x^2$ ....(1),

so that
$$x = k\rho \frac{\sqrt{m^2 - n^2}}{m}$$
$$= k\rho \sqrt{1 - \lambda^2}, \text{ say.}$$

Then shall $k\rho - x$, or $k\rho - k\rho\sqrt{1 - \lambda^2}$, be a ***first apotome***.

For (α) it follows from (1) that x is rational but incommensurable with $k\rho$, whence $k\rho$, x are rational and $\smile$,

so that $(k\rho - x)$ is an apotome.

(β) If $y^2 = k^2\rho^2 - x^2$, then, by (1), ***convertendo***,
$$m^2 : n^2 = k^2\rho^2 : y^2,$$
whence y, that is, $\sqrt{k^2\rho^2 - x^2}$, is commensurable in length with $k\rho$.

And $k\rho \frown \rho$;

therefore $k\rho - x$ is a ***first apotome***.

As explained in the note to x. 48, the first apotome
$$k\rho - k\rho\sqrt{1 - \lambda^2}$$
is one of the roots of the equation
$$x^2 - 2k\rho \, . \, x + \lambda^2 k^2\rho^2 = 0.$$

PROPOSITION 86.

To find the second apotome.

Let a rational straight line *A* be set out, and *GC* commensurable in length with *A*;

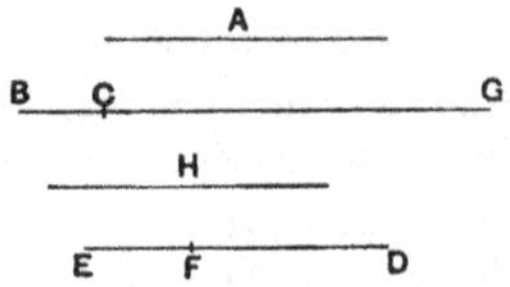

therefore *GC* is rational.

Let two square numbers *DE*, *EF* be set out, and let their difference *DF* not be square.

Now let it be contrived that, as *FD* is to *DE*, so is the square on *CG* to the square on *GB*. [x. 6, Por.]

Therefore the square on *CG* is commensurable with the square on *GB*. [x. 6]

But the square on *CG* is rational;
therefore the square on *GB* is also rational;
therefore *BG* is rational.

And, since the square on *GC* has not to the square on *GB* the ratio which a square number has to a square number,
CG is incommensurable in length with *GB*. [x. 9]

And both are rational;
therefore *CG*, *GB* are rational straight lines commensurable in square only;
therefore *BC* is an apotome. [x. 73]

I say next that it is also a second apotome.

For let the square on *H* be that by which the square on *BG* is greater than the square on *GC*.

Since then, as the square on *BG* is to the square on *GC*, so is the number *ED* to the number *DF*,
therefore, *convertendo*,
as the square on *BG* is to the square on *H*, so is *DE* to *EF*. [v. 19, Por.]

And each of the numbers *DE*, *EF* is square;
therefore the square on *BG* has to the square on *H* the ratio which a square number has to a square number;
therefore *BG* is commensurable in length with *H*. [x. 9]

And the square on *BG* is greater than the square on *GC* by the square on *H*;
therefore the square on *BG* is greater than the square on *GC*

by the square on a straight line commensurable in length with *BG*.

And *CG*, the annex, is commensurable with the rational straight line *A* set out.

Therefore *BC* is a second apotome. [x. Deff. III. 2]

Therefore the second apotome *BC* has been found.

Q. E. D.

Take, as before, $k\rho$ commensurable in length with ρ.

Let m^2, n^2 be again square numbers, but $(m^2 - n^2)$ not square.

Take x such that $(m^2 - n^2) : m^2 = k^2\rho^2 : x^2$ ....(1),

whence
$$x = k\rho \frac{m}{\sqrt{m^2 - n^2}}$$
$$= \frac{k\rho}{\sqrt{1 - \lambda^2}}, \text{ say.}$$

Thus x is greater than $k\rho$.

Then $x - k\rho$, or $\dfrac{k\rho}{\sqrt{1 - \lambda^2}} - k\rho$, is a *second apotome.*

For (α), as before, x is rational and $\sim k\rho$.

(β) If $x^2 - k^2\rho^2 = y^2$, we have, from (1),
$$m^2 : n^2 = x^2 : y^2.$$

Thus y, or $\sqrt{x^2 - k^2\rho^2}$, is commensurable in length with x.

And $k\rho$ is $\frown \rho$.

Therefore $x - k\rho$ is a *second apotome.*

As explained in the note on x. 49, the *second apotome*
$$\frac{k\rho}{\sqrt{1 - \lambda^2}} - k\rho$$
is the lesser root of the equation
$$x^2 - \frac{2k\rho}{\sqrt{1 - \lambda^2}} \cdot x + \frac{\lambda^2}{1 - \lambda^2} k^2\rho^2 = 0.$$

PROPOSITION 87.

To find the third apotome.

Let a rational straight line *A* be set out, let three numbers *E*, *BC*, *CD* be set out which have not to one another the ratio which a square number has to a square number, but let *CB* have to *BD* the ratio which a square number has to a square number.

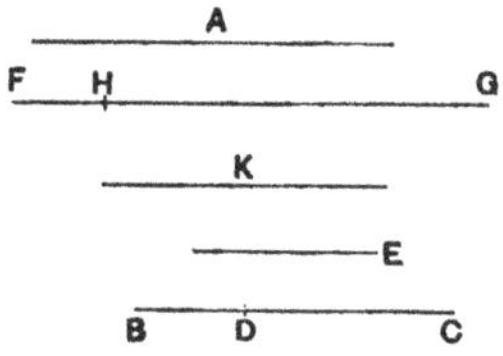

Let it be contrived that, as *E* is to *BC*, so is the square on *A* to the square on *FG*, and, as BC is to CD, so is the square on FG to the square on GH. [x. 6, Por.]

Since then, as E is to BC, so is the square on A to the square on FG,
therefore the square on A is commensurable with the square on FG. [x. 6]

But the square on A is rational;
therefore the square on FG is also rational;
therefore FG is rational.

And, since E has not to BC the ratio which a square number has to a square number,
therefore neither has the square on A to the square on FG the ratio which a square number has to a square number;
therefore A is incommensurable in length with FG. [x. 9]

Again, since, as BC is to CD, so is the square on FG to the square on GH,
therefore the square on FG is commensurable with the square on GH. [x. 6]

But the square on FG is rational;
therefore the square on GH is also rational;
therefore GH is rational.

And, since BC has not to CD the ratio which a square number has to a square number,
therefore neither has the square on FG to the square on GH the ratio which a square number has to a square number;
therefore FG is incommensurable in length with GH. [x. 9]

And both are rational;
therefore FG, GH are rational straight lines commensurable in square only;
therefore FH is an apotome. [x. 73]

I say next that it is also a third apotome.

For since, as E is to BC, so is the square on A to the square on FG,
and, as BC is to CD, so is the square on FG to the square on HG,
therefore, *ex aequali*, as E is to CD, so is the square on A to the square on HG. [v. 22]

But E has not to CD the ratio which a square number has to a square number;

therefore neither has the square on A to the square on GH the ratio which a square number has to a square number;

therefore A is incommensurable in length with GH. [x. 9]

Therefore neither of the straight lines FG, GH is commensurable in length with the rational straight line A set out.

Now let the square on K be that by which the square on FG is greater than the square on GH.

Since then, as BC is to CD, so is the square on FG to the square on GH,

therefore, *convertendo*, as BC is to BD, so is the square on FG to the square on K. [v. 19, Por.]

But BC has to BD the ratio which a square number has to a square number;

therefore the square on FG also has to the square on K the ratio which a square number has to a square number.

Therefore FG is commensurable in length with K, [x. 9] and the square on FG is greater than the square on GH by the square on a straight line commensurable with FG.

And neither of the straight lines FG, GH is commensurable in length with the rational straight line A set out;

therefore FH is a third apotome. [x. Deff. III. 3]

Therefore the third apotome FH has been found.

Q. E. D.

Let ρ be a rational straight line.

Take numbers p, qm^2, $q(m^2 - n^2)$ which have not to one another the ratio of square to square.

Now let x, y be such that

$$p : qm^2 = \rho^2 : x^2 \quad \text{.........................}(1)$$

and

$$qm^2 : q(m^2 - n^2) = x^2 : y^2 \quad \text{.........................}(2).$$

Then shall $(x - y)$ be a *third apotome*.

For (a), from (1),

x is rational but $\smile \rho$ ....(3).

And, from (2), y is rational but $\smile x$.

Therefore x, y are rational and $\frown$,

so that $(x - y)$ is an apotome.

(β) By (1), (2), *ex aequali*,

$$p : q(m^2 - n^2) = \rho^2 : y^2,$$

whence $y \smile \rho$.

Thus, by this and (3), x, y are both $\smile \rho$ ....(4).

Lastly, let $z^2 = x^2 - y^2$, so that, from (2), *convertendo*,

$$qm^2 : qn^2 = x^2 : z^2;$$

therefore z, or $\sqrt{x^2 - y^2}$, $\frown x$ ....(5).

Thus [(4) and (5)] $(x - y)$ is a *third apotome*.

To find its form, we have, from (1) and (2),

$$x = \rho \cdot \frac{m\sqrt{q}}{\sqrt{p}},$$

$$y = \rho \cdot \frac{\sqrt{m^2 - n^2} \cdot \sqrt{q}}{\sqrt{p}},$$

so that $$x - y = \sqrt{\frac{q}{p}} \cdot \rho\,(m - \sqrt{m^2 - n^2}).$$

This may be written in the form

$$m\sqrt{k} \cdot \rho - m\sqrt{k} \cdot \rho\sqrt{1 - \lambda^2}.$$

As explained in the note on x. 50, this is the lesser root of the equation

$$x^2 - 2m\sqrt{k} \cdot \rho x + \lambda^2 m^2 k \rho^2 = 0.$$

Proposition 88.

To find the fourth apotome.

Let a rational straight line A be set out, and BG commensurable in length with it;
therefore BG is also rational.

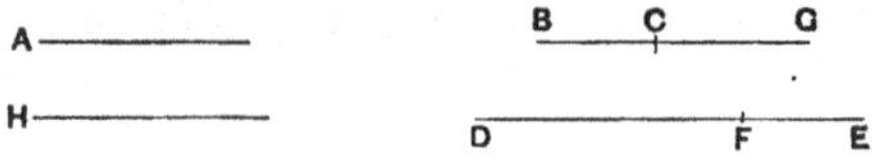

Let two numbers DF, FE be set out such that the whole DE has not to either of the numbers DF, EF the ratio which a square number has to a square number.

Let it be contrived that, as DE is to EF, so is the square on BG to the square on GC; [x. 6, Por.]
therefore the square on BG is commensurable with the square on GC. [x. 6]

But the square on BG is rational;
therefore the square on GC is also rational;
therefore GC is rational.

Now, since DE has not to EF the ratio which a square number has to a square number,
therefore neither has the square on BG to the square on GC the ratio which a square number has to a square number;
therefore BG is incommensurable in length with GC. [x. 9]

And both are rational;
therefore BG, GC are rational straight lines commensurable in square only;
therefore BC is an apotome. [x. 73]

Now let the square on H be that by which the square on BG is greater than the square on GC.

Since then, as DE is to EF, so is the square on BG to the square on GC,
therefore also, *convertendo*, as ED is to DF, so is the square on GB to the square on H. [v. 19, Por.]

But ED has not to DF the ratio which a square number has to a square number;
therefore neither has the square on GB to the square on H the ratio which a square number has to a square number;
therefore BG is incommensurable in length with H. [x. 9]

And the square on BG is greater than the square on GC by the square on H;
therefore the square on BG is greater than the square on GC by the square on a straight line incommensurable with BG.

And the whole BG is commensurable in length with the rational straight line A set out.

Therefore BC is a fourth apotome. [x. Deff. III. 4]

Therefore the fourth apotome has been found.

Q. E. D.

Beginning with ρ, $k\rho$, as in x. 85, 86, we take numbers m, n such that $(m+n)$ has not to either of the numbers m, n the ratio of a square number to a square number.

Take x such that $(m+n) : n = k^2\rho^2 : x^2$ ....(1),

whence
$$x = k\rho\sqrt{\frac{n}{m+n}}$$
$$= \frac{k\rho}{\sqrt{1+\lambda}}, \text{ say.}$$

Then shall $(k\rho - x)$, or $\left(k\rho - \frac{k\rho}{\sqrt{1+\lambda}}\right)$, be a *fourth apotome*.

For, by (1), x is rational and $\sim kp$.

Also $\sqrt{k^2\rho^2 - x^2}$ is incommensurable with $k\rho$, since

$$(m+n) : m = k^2\rho^2 : (k^2\rho^2 - x^2),$$

and the ratio $(m+n) : m$ is not that of a square number to a square number.

And $k\rho \frown \rho$.

As explained in the note on x. 51, the *fourth apotome*

$$k\rho - \frac{k\rho}{\sqrt{1+\lambda}}$$

is the lesser root of the quadratic equation

$$x^2 - 2k\rho \, . \, x + \frac{\lambda}{1+\lambda} k^2\rho^2 = 0.$$

Proposition 89.

To find the fifth apotome.

Let a rational straight line A be set out, and let CG be commensurable in length with A;

therefore CG is rational.

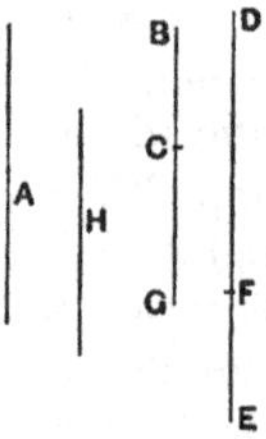

Let two numbers DF, FE be set out such that DE again has not to either of the numbers DF, FE the ratio which a square number has to a square number;

and let it be contrived that, as FE is to ED, so is the square on CG to the square on GB.

Therefore the square on GB is also rational; [x. 6]

therefore BG is also rational.

Now since, as DE is to EF, so is the square on BG to the square on GC,

while DE has not to EF the ratio which a square number has to a square number,

therefore neither has the square on BG to the square on GC the ratio which a square number has to a square number;

therefore BG is incommensurable in length with GC. [x. 9]

And both are rational;

therefore BG, GC are rational straight lines commensurable in square only;

therefore BC is an apotome. [x. 73]

I say next that it is also a fifth apotome.

For let the square on H be that by which the square on BG is greater than the square on GC.

Since then, as the square on BG is to the square on GC, so is DE to EF,

therefore, *convertendo*, as ED is to DF, so is the square on BG to the square on H. [v. 19, Por.]

But ED has not to DF the ratio which a square number has to a square number;

therefore neither has the square on BG to the square on H the ratio which a square number has to a square number;

therefore BG is incommensurable in length with H. [x. 9]

And the square on BG is greater than the square on GC by the square on H;

therefore the square on GB is greater than the square on GC by the square on a straight line incommensurable in length with GB.

And the annex CG is commensurable in length with the rational straight line A set out;

therefore BC is a fifth apotome. [x. Deff. III. 5]

Therefore the fifth apotome BC has been found.

Q. E. D.

Let ρ, $k\rho$ and the numbers m, n of the last proposition be taken.

Take x such that $n : (m+n) = k^2\rho^2 : x^2$ ....(1).

In this case $x > k\rho$, and $x = k\rho\sqrt{\frac{m+n}{n}}$

$= k\rho\sqrt{1+\lambda}$, say.

Then shall $(x - k\rho)$, or $(k\rho\sqrt{1+\lambda} - k\rho)$, be a *fifth apotome*.

For, by (1), x is rational and $\frown k\rho$.

And since, by (1), $(m+n) : m = x^2 : (x^2 - k^2\rho^2)$,

$\sqrt{x^2 - k^2\rho^2}$ is incommensurable with x.

Also $k\rho \frown \rho$.

As explained in the note on x. 52, the *fifth apotome*

$$k\rho\sqrt{1+\lambda} - k\rho$$

is the lesser root of the quadratic

$$x^2 - 2k\rho\sqrt{1+\lambda}\,.\,x + \lambda k^2\rho^2 = 0.$$

PROPOSITION 90.

To find the sixth apotome.

Let a rational straight line A be set out, and three numbers E, BC, CD not having to one another the ratio which a square number has to a square number;

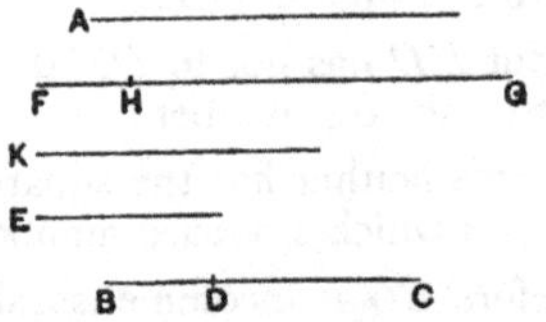

and further let CB also not have to BD the ratio which a square number has to a square number.

Let it be contrived that, as E is to BC, so is the square on A to the square on FG, and, as BC is to CD, so is the square on FG to the square on GH. [x. 6, Por.]

Now since, as E is to BC, so is the square on A to the square on FG,

therefore the square on A is commensurable with the square on FG. [x. 6]

But the square on A is rational;

therefore the square on FG is also rational;

therefore FG is also rational.

And, since E has not to BC the ratio which a square number has to a square number,

therefore neither has the square on A to the square on FG the ratio which a square number has to a square number;

therefore A is incommensurable in length with FG. [x. 9]

Again, since, as BC is to CD, so is the square on FG to the square on GH,

therefore the square on FG is commensurable with the square on GH. [x. 6]

But the square on FG is rational;

therefore the square on GH is also rational;

therefore GH is also rational.

And, since BC has not to CD the ratio which a square number has to a square number,

therefore neither has the square on *FG* to the square on *GH* the ratio which a square number has to a square number;
therefore *FG* is incommensurable in length with *GH*. [x. 9]

And both are rational;
therefore *FG*, *GH* are rational straight lines commensurable in square only;
therefore *FH* is an apotome. [x. 73]

I say next that it is also a sixth apotome.

For since, as *E* is to *BC*, so is the square on *A* to the square on *FG*,
and, as *BC* is to *CD*, so is the square on *FG* to the square on *GH*,
therefore, *ex aequali*, as *E* is to *CD*, so is the square on *A* to the square on *GH*. [v. 22]

But *E* has not to *CD* the ratio which a square number has to a square number;
therefore neither has the square on *A* to the square on *GH* the ratio which a square number has to a square number;
therefore *A* is incommensurable in length with *GH*; [x. 9]
therefore neither of the straight lines *FG*, *GH* is commensurable in length with the rational straight line *A*.

Now let the square on *K* be that by which the square on *FG* is greater than the square on *GH*.

Since then, as *BC* is to *CD*, so is the square on *FG* to the square on *GH*,
therefore, *convertendo*, as *CB* is to *BD*, so is the square on *FG* to the square on *K*. [v. 19, Por.]

But *CB* has not to *BD* the ratio which a square number has to a square number;
therefore neither has the square on *FG* to the square on *K* the ratio which a square number has to a square number;
therefore *FG* is incommensurable in length with *K*. [x. 9]

And the square on *FG* is greater than the square on *GH* by the square on *K*;
therefore the square on *FG* is greater than the square on *GH* by the square on a straight line incommensurable in length with *FG*.

And neither of the straight lines *FG*, *GH* is commensurable with the rational straight line *A* set out.

Therefore *FH* is a sixth apotome. [x. Deff. III. 6]

Therefore the sixth apotome *FH* has been found.

Q. E. D.

Let ρ be the given rational straight line.

Take numbers p, $(m+n)$, n which have not to one another the ratio of a square number to a square number, m, n being also chosen such that the ratio $(m+n):m$ is not that of square to square.

Take x, y such that $p:(m+n)=\rho^2:x^2$ ....(1),

$(m+n):n=x^2:y^2$ ....(2).

Then shall $(x-y)$ be a *sixth apotome*.

For, by (1), x is rational and $\smile \rho$ ....(3).

By (2), since x is rational,

y is rational and $\smile x$ ....(4).

Thus [(3), (4)] $(x-y)$ is an apotome.

Again, *ex aequali*, $p:n=\rho^2:y^2$,

whence $y \smile \rho$.

Thus x, y are both $\smile \rho$.

Lastly, *convertendo* from (2),

$$(m+n):m=x^2:(x^2-y^2),$$

whence $\sqrt{x^2-y^2} \smile x$.

Therefore $(x-y)$ is a *sixth apotome*.

From (1) and (2) we have

$$x=\rho\sqrt{\frac{m+n}{p}},$$

$$y=\rho\sqrt{\frac{n}{p}},$$

so that the *sixth apotome* may be written

$$\rho\sqrt{\frac{m+n}{p}}-\rho\sqrt{\frac{n}{p}},$$

or, more simply, $\sqrt{k}\cdot\rho-\sqrt{\lambda}\cdot\rho$.

As explained in the note on x. 53, the *sixth apotome* is the lesser root of the equation

$$x^2-2\sqrt{k}\cdot\rho x+(k-\lambda)\rho^2=0.$$

Proposition 91.

If an area be contained by a rational straight line and a first apotome, the "side" of the area is an apotome.

For let the area *AB* be contained by the rational straight line *AC* and the first apotome *AD*;

I say that the "side" of the area *AB* is an apotome.

For, since AD is a first apotome, let DG be its annex;
therefore AG, GD are rational straight lines commensurable in square only. [x. 73]

And the whole AG is commensurable with the rational straight line AC set out,
and the square on AG is greater than the square on GD by the square on a straight line commensurable in length with AG; [x. Deff. III. 1]
if therefore there be applied to AG a parallelogram equal to the fourth part of the square on DG and deficient by a square figure, it divides it into commensurable parts. [x. 17]

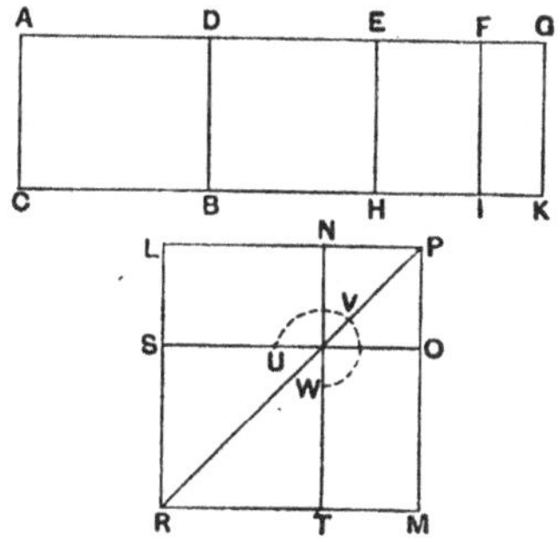

Let DG be bisected at E,
let there be applied to AG a parallelogram equal to the square on EG and deficient by a square figure,
and let it be the rectangle AF, FG;
therefore AF is commensurable with FG.

And through the points E, F, G let EH, FI, GK be drawn parallel to AC.

Now, since AF is commensurable in length with FG,
therefore AG is also commensurable in length with each of the straight lines AF, FG. [x. 15]

But AG is commensurable with AC;
therefore each of the straight lines AF, FG is commensurable in length with AC. [x. 12]

And AC is rational;
therefore each of the straight lines AF, FG is also rational,
so that each of the rectangles AI, FK is also rational. [X. 19]

Now, since DE is commensurable in length with EG,
therefore DG is also commensurable in length with each of the straight lines DE, EG. [X. 15]

But DG is rational and incommensurable in length with AC;
therefore each of the straight lines DE, EG is also rational and incommensurable in length with AC; [X. 13]
therefore each of the rectangles DH, EK is medial. [X. 21]

Now let the square LM be made equal to AI, and let there be subtracted the square NO having a common angle with it, the angle LPM, and equal to FK;
therefore the squares LM, NO are about the same diameter. [VI. 26]

Let PR be their diameter, and let the figure be drawn.

Since then the rectangle contained by AF, FG is equal to the square on EG,
therefore, as AF is to EG, so is EG to FG. [VI. 17]

But, as AF is to EG, so is AI to EK,
and, as EG is to FG, so is EK to KF; [VI. 1]
therefore EK is a mean proportional between AI, KF. [V. 11]

But MN is also a mean proportional between LM, NO, as was before proved, [Lemma after X. 53]
and AI is equal to the square LM, and KF to NO;
therefore MN is also equal to EK.

But EK is equal to DH, and MN to LO;
therefore DK is equal to the gnomon UVW and NO.

But AK is also equal to the squares LM, NO;
therefore the remainder AB is equal to ST.

But ST is the square on LN;
therefore the square on LN is equal to AB;
therefore LN is the "side" of AB.

I say next that LN is an apotome.

For, since each of the rectangles AI, FK is rational, and they are equal to LM, NO,
therefore each of the squares LM, NO, that is, the squares on LP, PN respectively, is also rational;
therefore each of the straight lines LP, PN is also rational.

Again, since DH is medial and is equal to LO,
therefore LO is also medial.

Since then LO is medial,
while NO is rational,
therefore LO is incommensurable with NO.

But, as LO is to NO, so is LP to PN; [VI. 1]
therefore LP is incommensurable in length with PN. [X. 11]

And both are rational;
therefore LP, PN are rational straight lines commensurable in square only;
therefore LN is an apotome. [X. 73]

And it is the "side" of the area AB;
therefore the "side" of the area AB is an apotome.

Therefore etc.

This proposition corresponds to X. 54, and the problem solved in it is to find and to classify the *side of a square equal to the rectangle contained by a first apotome and* ρ, or (algebraically) to find

$$\sqrt{\rho(k\rho - k\rho\sqrt{1-\lambda^2})}.$$

First find u, v from the equations

$$\left.\begin{aligned} u + v &= k\rho \\ uv &= \tfrac{1}{4}k^2\rho^2(1-\lambda^2) \end{aligned}\right\} \quad\text{.............................(1).}$$

If u, v represent the values so found, put

$$\left.\begin{aligned} x^2 &= \rho u \\ y^2 &= \rho v \end{aligned}\right\} \quad\text{......................................(2),}$$

and $(x - y)$ shall be the square root required.

To prove this Euclid argues thus.

By (1), $\quad u : \tfrac{1}{2}k\rho\sqrt{1-\lambda^2} = \tfrac{1}{2}k\rho\sqrt{1-\lambda^2} : v,$

whence $\quad \rho u : \tfrac{1}{2}k\rho^2\sqrt{1-\lambda^2} = \tfrac{1}{2}k\rho^2\sqrt{1-\lambda^2} : \rho v,$

or $\quad x^2 : \tfrac{1}{2}k\rho^2\sqrt{1-\lambda^2} = \tfrac{1}{2}k\rho^2\sqrt{1-\lambda^2} : y^2.$

But [Lemma after X. 53]

$$x^2 : xy = xy : y^2,$$

so that $\quad xy = \tfrac{1}{2}k\rho^2\sqrt{1-\lambda^2}$ ....(3).

Therefore $$(x-y)^2 = x^2 + y^2 - 2xy$$
$$= \rho(u+v) - k\rho^2\sqrt{1-\lambda^2}$$
$$= k\rho^2 - k\rho^2\sqrt{1-\lambda^2}.$$

Thus $(x-y)$ is equal to $\sqrt{\rho(k\rho - k\rho\sqrt{1-\lambda^2})}$.

It has next to be proved that $(x-y)$ is an *apotome*.

From (1) it follows, by X. 17, that
$$u \frown v;$$
thus u, v are both commensurable with $(u+v)$ and therefore with ρ......(4).

Hence u, v are both rational,
so that ρu, ρv are rational areas;
therefore, by (2), x^2, y^2 are rational and commensurable ....(5),
whence also x, y are rational straight lines ....(6).

Next, $k\rho\sqrt{1-\lambda^2}$ is rational and $\smile \rho$;
therefore $\frac{1}{2}k\rho^2\sqrt{1-\lambda^2}$ is a *medial* area.

That is, by (3), xy is a medial area.

But [(5)] y^2 is a *rational* area;
therefore $$xy \smile y^2,$$
or $$x \smile y.$$

But [(6)] x, y are both rational.

Therefore x, y are rational and $\frown$;
so that $(x-y)$ is an *apotome*.

To find the form of $(x-y)$ algebraically, we have, by solving (1),
$$u = \tfrac{1}{2}k\rho(1+\lambda),$$
$$v = \tfrac{1}{2}k\rho(1-\lambda),$$
whence, from (2), $$x = \rho\sqrt{\frac{k}{2}(1+\lambda)},$$
$$y = \rho\sqrt{\frac{k}{2}(1-\lambda)},$$
and $$x - y = \rho\sqrt{\frac{k}{2}(1+\lambda)} - \rho\sqrt{\frac{k}{2}(1-\lambda)}.$$

As explained in the note on X. 54, $(x-y)$ is the lesser positive root of the biquadratic equation
$$x^4 - 2k\rho^2 . x^2 + \lambda^2 k^2\rho^4 = 0.$$

PROPOSITION 92.

If an area be contained by a rational straight line and a second apotome, the "side" of the area is a first apotome of a medial straight line.

For let the area AB be contained by the rational straight line AC and the second apotome AD;

I say that the "side" of the area AB is a first apotome of a medial straight line.

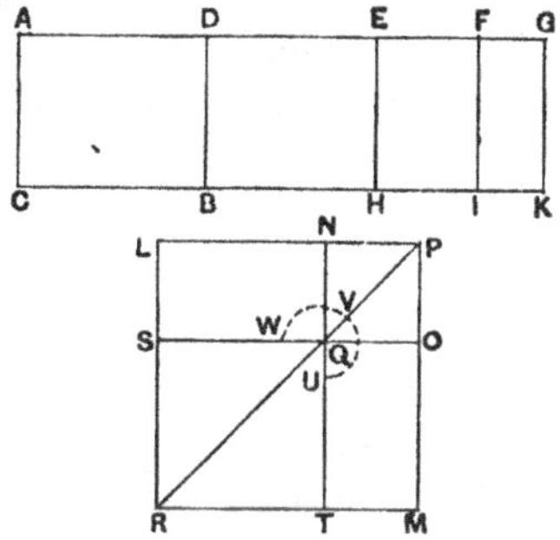

For let DG be the annex to AD;

therefore AG, GD are rational straight lines commensurable in square only, [x. 73]

and the annex DG is commensurable with the rational straight line AC set out,

while the square on the whole AG is greater than the square on the annex GD by the square on a straight line commensurable in length with AG. [x. Deff. III. 2]

Since then the square on AG is greater than the square on GD by the square on a straight line commensurable with AG,

therefore, if there be applied to AG a parallelogram equal to the fourth part of the square on GD and deficient by a square figure, it divides it into commensurable parts. [x. 17]

Let then DG be bisected at E,

let there be applied to AG a parallelogram equal to the square on EG and deficient by a square figure,

and let it be the rectangle AF, FG;

therefore AF is commensurable in length with FG.

Therefore AG is also commensurable in length with each of the straight lines AF, FG. [x. 15]

But AG is rational and incommensurable in length with AC;

therefore each of the straight lines AF, FG is also rational and incommensurable in length with AC; [x. 13]

therefore each of the rectangles AI, FK is medial. [x. 21]

Again, since DE is commensurable with EG,

therefore DG is also commensurable with each of the straight lines DE, EG. [x. 15]

But DG is commensurable in length with AC.

Therefore each of the rectangles DH, EK is rational. [x. 19]

Let then the square LM be constructed equal to AI, and let there be subtracted NO equal to FK and being about the same angle with LM, namely the angle LPM;

therefore the squares LM, NO are about the same diameter. [vi. 26]

Let PR be their diameter, and let the figure be drawn.

Since then AI, FK are medial and are equal to the squares on LP, PN,

the squares on LP, PN are also medial;

therefore LP, PN are also medial straight lines commensurable in square only.

And, since the rectangle AF, FG is equal to the square on EG,

therefore, as AF is to EG, so is EG to FG, [vi. 17]

while, as AF is to EG, so is AI to EK,

and, as EG is to FG, so is EK to FK; [vi. 1]

therefore EK is a mean proportional between AI, FK. [v. 11]

But MN is also a mean proportional between the squares LM, NO,

and AI is equal to LM, and FK to NO;

therefore MN is also equal to EK.

But DH is equal to EK, and LO equal to MN;

therefore the whole DK is equal to the gnomon UVW and NO.

Since then the whole AK is equal to LM, NO,

and, in these, DK is equal to the gnomon UVW and NO,

therefore the remainder AB is equal to TS.

But TS is the square on LN;
therefore the square on LN is equal to the area AB;
therefore LN is the "side" of the area AB.

I say that LN is a first apotome of a medial straight line.

For, since EK is rational and is equal to LO,
therefore LO, that is, the rectangle LP, PN, is rational.

But NO was proved medial;
therefore LO is incommensurable with NO.

But, as LO is to NO, so is LP to PN; [VI. 1]
therefore LP, PN are incommensurable in length. [X. 11]

Therefore LP, PN are medial straight lines commensurable in square only which contain a rational rectangle;
therefore LN is a first apotome of a medial straight line. [X. 74]

And it is the "side" of the area AB.

Therefore the "side" of the area AB is a first apotome of a medial straight line.

Q. E. D.

There is an evident flaw in the text in the place (Heiberg, p. 282, ll. 17—20: translation p. 196 above) where it is said that "since then AI, FK are medial and are equal to the squares on LP, PN, the squares on LP, PN are also medial; *therefore* LP, PN *are also medial straight lines commensurable in square only*." It is not till the last lines of the proposition (Heiberg, p. 284, ll. 17, 18) that it is proved that LP, PN are *incommensurable in length*. What should have been proved in the former passage is that the *squares* on LP, PN are commensurable, so that LP, PN are commensurable in square (not commensurable in square *only*). I have supplied the step in the note below: "Also $x^2 \frown y^2$, since $u \frown v$." Theon seems to have observed the omission and to have put "and commensurable with one another" after "medial" in the passage quoted, though even this does not show *why* the squares on LP, PN are commensurable. One MS. (V) also has "only" (μόνον) erased after "commensurable in square."

This proposition amounts to finding and classifying

$$\sqrt{\rho\left(\frac{k\rho}{\sqrt{1-\lambda^2}} - k\rho\right)}.$$

The method is that of the last proposition. Euclid solves, first, the equations

$$\left.\begin{aligned} u + v &= \frac{k\rho}{\sqrt{1-\lambda^2}} \\ uv &= \tfrac{1}{4}k^2\rho^2 \end{aligned}\right\} \quad \ldots\ldots\ldots\ldots\ldots\ldots(1).$$

Then, using the values of u, v so found, he puts

$$\left.\begin{aligned} x^2 &= \rho u \\ y^2 &= \rho v \end{aligned}\right\} \qquad \ldots\ldots\ldots\ldots\ldots\ldots\ldots\ldots(2),$$

and $(x - y)$ is the square root required.

That $$(x-y) = \sqrt{\rho\left(\frac{k\rho}{\sqrt{1-\lambda^2}} - k\rho\right)}$$

is proved in the same way as is the corresponding fact in X. 91.

From (1) $$u : \tfrac{1}{2}k\rho = \tfrac{1}{2}k\rho : v,$$

so that $$\rho u : \tfrac{1}{2}k\rho^2 = \tfrac{1}{2}k\rho^2 : \rho v.$$

But $$x^2 : xy = xy : y^2,$$

whence, by (2), $$xy = \tfrac{1}{2}k\rho^2 \qquad \ldots\ldots\ldots\ldots\ldots\ldots\ldots\ldots(3).$$

Therefore $$\begin{aligned} (x-y)^2 &= x^2 + y^2 - 2xy \\ &= \rho(u+v) - k\rho^2 \\ &= \rho\left(\frac{k\rho}{\sqrt{1-\lambda^2}} - k\rho\right). \end{aligned}$$

Next, we have to prove that $(x - y)$ is a ***first apotome of a medial*** straight line.

From (1) it follows, by X. 17, that

$$u \frown v \qquad \ldots\ldots\ldots\ldots\ldots\ldots\ldots\ldots(4),$$

therefore u, v are both $\frown (u + v)$.

But [(1)] $(u + v)$ is rational and $\smile \rho$;

therefore u, v are both rational and $\smile \rho$ ....(5).

Therefore ρu, ρv, or x^2, y^2, are both *medial* areas, and x, y are medial straight lines ....(6).

Also $x^2 \frown y^2$, since $u \frown v$ [(4)]..(7).

Now xy, or $\frac{1}{2}k\rho^2$, is a *rational* area;

therefore $$xy \smile y^2,$$

and $$x \smile y.$$

Hence [(6), (7), (3)] x, y are medial straight lines commensurable in square only and containing a rational rectangle;

therefore $(x - y)$ is a ***first apotome of a medial*** straight line.

Algebraical solution of the equations gives

$$u = \tfrac{1}{2}\frac{1+\lambda}{\sqrt{1-\lambda^2}}k\rho,$$

$$v = \tfrac{1}{2}\frac{1-\lambda}{\sqrt{1-\lambda^2}}k\rho,$$

and $$x - y = \rho\sqrt{\frac{k}{2}\left(\frac{1+\lambda}{1-\lambda}\right)^{\frac{1}{2}}} - \rho\sqrt{\frac{k}{2}\left(\frac{1-\lambda}{1+\lambda}\right)^{\frac{1}{2}}}.$$

As explained in the note on X. 55, this is the lesser positive root of the equation

$$x^4 - \frac{2k\rho^2}{\sqrt{1-\lambda^2}}x^2 + \frac{\lambda^2}{1-\lambda^2}k^2\rho^4 = 0.$$

Proposition 93.

If an area be contained by a rational straight line and a third apotome, the "side" of the area is a second apotome of a medial straight line.

For let the area AB be contained by the rational straight line AC and the third apotome AD;

I say that the "side" of the area AB is a second apotome of a medial straight line.

For let DG be the annex to AD;

therefore AG, GD are rational straight lines commensurable in square only,

and neither of the straight lines AG, GD is commensurable in length with the rational straight line AC set out,

while the square on the whole AG is greater than the square on the annex DG by the square on a straight line commensurable with AG. [x. Deff. III. 3]

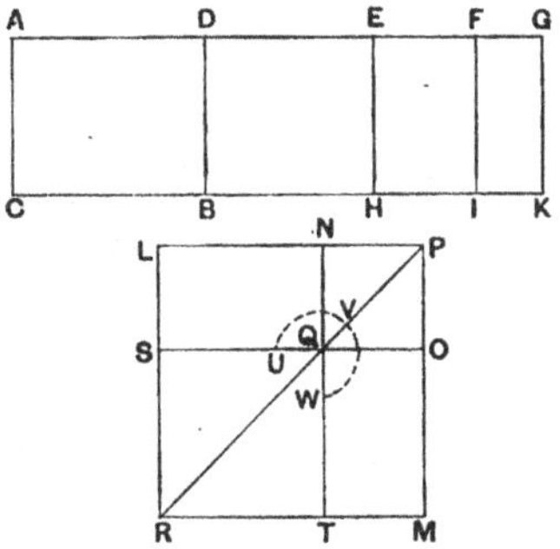

Since then the square on AG is greater than the square on GD by the square on a straight line commensurable with AG,

therefore, if there be applied to AG a parallelogram equal to the fourth part of the square on DG and deficient by a square figure, it will divide it into commensurable parts. [x. 17]

Let then DG be bisected at E,

let there be applied to AG a parallelogram equal to the square on EG and deficient by a square figure,

and let it be the rectangle AF, FG.

Let EH, FI, GK be drawn through the points E, F, G parallel to AC.

Therefore AF, FG are commensurable;

therefore AI is also commensurable with FK. [vi. 1, x. 11]

And, since AF, FG are commensurable in length,

therefore AG is also commensurable in length with each of the straight lines AF, FG. [x. 15]

But AG is rational and incommensurable in length with AC;

so that AF, FG are so also. [x. 13]

Therefore each of the rectangles AI, FK is medial. [x. 21]

Again, since DE is commensurable in length with EG,

therefore DG is also commensurable in length with each of the straight lines DE, EG. [x. 15]

But GD is rational and incommensurable in length with AC;

therefore each of the straight lines DE, EG is also rational and incommensurable in length with AC; [x. 13]

therefore each of the rectangles DH, EK is medial. [x. 21]

And, since AG, GD are commensurable in square only,

therefore AG is incommensurable in length with GD.

But AG is commensurable in length with AF, and DG with EG;

therefore AF is incommensurable in length with EG. [x. 13]

But, as AF is to EG, so is AI to EK; [vi. 1]

therefore AI is incommensurable with EK. [x. 11]

Now let the square LM be constructed equal to AI,

and let there be subtracted NO equal to FK and being about the same angle with LM;

therefore LM, NO are about the same diameter. [vi. 26]

Let PR be their diameter, and let the figure be drawn.

Now, since the rectangle AF, FG is equal to the square on EG,

therefore, as AF is to EG, so is EG to FG. [vi. 17]

But, as *AF* is to *EG*, so is *AI* to *EK*,
and, as *EG* is to *FG*, so is *EK* to *FK*; [vi. 1]
therefore also, as *AI* is to *EK*, so is *EK* to *FK*; [v. 11]
therefore *EK* is a mean proportional between *AI*, *FK*.

But *MN* is also a mean proportional between the squares *LM*, *NO*,
and *AI* is equal to *LM*, and *FK* to *NO*;
therefore *EK* is also equal to *MN*.

But *MN* is equal to *LO*, and *EK* equal to *DH*;
therefore the whole *DK* is also equal to the gnomon *UVW* and *NO*.

But *AK* is also equal to *LM*, *NO*;
therefore the remainder *AB* is equal to *ST*, that is, to the square on *LN*;
therefore *LN* is the "side" of the area *AB*.

I say that *LN* is a second apotome of a medial straight line.

For, since *AI*, *FK* were proved medial, and are equal to the squares on *LP*, *PN*,
therefore each of the squares on *LP*, *PN* is also medial;
therefore each of the straight lines *LP*, *PN* is medial.

And, since *AI* is commensurable with *FK*, [vi. 1, x. 11]
therefore the square on *LP* is also commensurable with the square on *PN*.

Again, since *AI* was proved incommensurable with *EK*,
therefore *LM* is also incommensurable with *MN*,
that is, the square on *LP* with the rectangle *LP*, *PN*;
so that *LP* is also incommensurable in length with *PN*; [vi. 1, x. 11]
therefore *LP*, *PN* are medial straight lines commensurable in square only.

I say next that they also contain a medial rectangle.

For, since *EK* was proved medial, and is equal to the rectangle *LP*, *PN*,
therefore the rectangle *LP*, *PN* is also medial,
so that *LP*, *PN* are medial straight lines commensurable in square only which contain a medial rectangle.

Therefore LN is a second apotome of a medial straight line; [x. 75]

and it is the "side" of the area AB.

Therefore the "side" of the area AB is a second apotome of a medial straight line.

Q. E. D.

Here we are to find and classify the irrational straight line

$$\sqrt{\rho(\sqrt{k}.\rho-\sqrt{k}.\rho\sqrt{1-\lambda^2})}.$$

Following the same method, we put

$$\left.\begin{aligned} u+v &= \sqrt{k}.\rho \\ uv &= \tfrac{1}{4}k\rho^2(1-\lambda^2) \end{aligned}\right\} \quad\ldots\ldots\ldots\ldots(1).$$

Next, u, v being found, let

$$\left.\begin{aligned} x^2 &= \rho u \\ y^2 &= \rho v \end{aligned}\right\} \quad\ldots\ldots\ldots\ldots(2);$$

then $(x-y)$ is the square root required and is a *second apotome of a medial* straight line.

That $(x-y)$ is the square root required and that x^2, y^2 are medial areas, so that x, y are medial straight lines, is proved exactly as in the last proposition.

The rectangle xy, being equal to $\frac{1}{2}\sqrt{k}.\rho^2\sqrt{1-\lambda^2}$, is also medial.

Now, from (1), by x. 17, $u \frown v$,

whence $u+v \frown u$.

But $(u+v)$, or $\sqrt{k}.\rho$, $\smile \frac{1}{2}\sqrt{k}.\rho\sqrt{1-\lambda^2}$;

therefore $u \smile \frac{1}{2}\sqrt{k}.\rho\sqrt{1-\lambda^2}$,

and consequently $\rho u \smile \frac{1}{2}\sqrt{k}.\rho^2\sqrt{1-\lambda^2}$,

or $x^2 \smile xy$,

whence $x \smile y$.

And, since $u \frown v$, $\rho u \frown \rho v$,

or $x^2 \frown y^2$.

Thus x, y are medial straight lines commensurable in square only.

And xy is a medial area.

Therefore $(x-y)$ is a *second apotome of a medial* straight line.

Its actual form is found by solving equations (1), (2);

thus $u = \frac{1}{2}(\sqrt{k}.\rho+\lambda\sqrt{k}.\rho)$,

$v = \frac{1}{2}(\sqrt{k}.\rho-\lambda\sqrt{k}.\rho)$,

and $$x-y=\rho\sqrt{\frac{\sqrt{k}}{2}(1+\lambda)}-\rho\sqrt{\frac{\sqrt{k}}{2}(1-\lambda)}.$$

As explained in the note on x. 56, this is the lesser positive root of the equation

$$x^4-2\sqrt{k}.\rho^2x^2+\lambda^2k\rho^4=0.$$

PROPOSITION 94.

If an area be contained by a rational straight line and a fourth apotome, the "side" of the area is minor.

For let the area AB be contained by the rational straight line AC and the fourth apotome AD;

I say that the "side" of the area AB is minor.

For let DG be the annex to AD;

therefore AG, GD are rational straight lines commensurable in square only,

AG is commensurable in length with the rational straight line AC set out,

and the square on the whole AG is greater than the square on the annex DG by the square on a straight line incommensurable in length with AG, [x. Deff. III. 4]

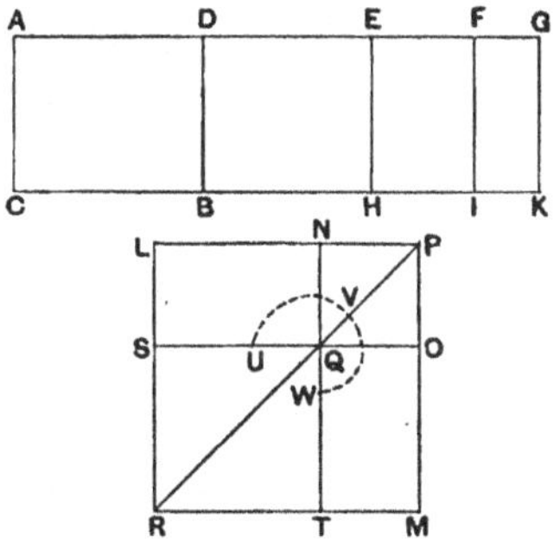

Since then the square on AG is greater than the square on GD by the square on a straight line incommensurable in length with AG,

therefore, if there be applied to AG a parallelogram equal to the fourth part of the square on DG and deficient by a square figure, it will divide it into incommensurable parts. [x. 18]

Let then DG be bisected at E,

let there be applied to AG a parallelogram equal to the square on EG and deficient by a square figure,

and let it be the rectangle AF, FG;

therefore AF is incommensurable in length with FG.

Let *EH*, *FI*, *GK* be drawn through *E*, *F*, *G* parallel to *AC*, *BD*.

Since then *AG* is rational and commensurable in length with *AC*,
therefore the whole *AK* is rational. [X. 19]

Again, since *DG* is incommensurable in length with *AC*, and both are rational,
therefore *DK* is medial. [X. 21]

Again, since *AF* is incommensurable in length with *FG*, therefore *AI* is also incommensurable with *FK*. [VI. 1, X. 11]

Now let the square *LM* be constructed equal to *AI*,
and let there be subtracted *NO* equal to *FK* and about the same angle, the angle *LPM*.

Therefore the squares *LM*, *NO* are about the same diameter. [VI. 26]

Let *PR* be their diameter, and let the figure be drawn.

Since then the rectangle *AF*, *FG* is equal to the square on *EG*,
therefore, proportionally, as *AF* is to *EG*, so is *EG* to *FG*. [VI. 17]

But, as *AF* is to *EG*, so is *AI* to *EK*,
and, as *EG* is to *FG*, so is *EK* to *FK*; [VI. 1]
therefore *EK* is a mean proportional between *AI*, *FK*. [V. 11]

But *MN* is also a mean proportional between the squares *LM*, *NO*,
and *AI* is equal to *LM*, and *FK* to *NO*;
therefore *EK* is also equal to *MN*.

But *DH* is equal to *EK*, and *LO* is equal to *MN*;
therefore the whole *DK* is equal to the gnomon *UVW* and *NO*.

Since, then, the whole *AK* is equal to the squares *LM*, *NO*,
and, in these, *DK* is equal to the gnomon *UVW* and the square *NO*,
therefore the remainder *AB* is equal to *ST*, that is, to the square on *LN*;
therefore *LN* is the "side" of the area *AB*.

I say that LN is the irrational straight line called minor.

For, since AK is rational and is equal to the squares on LP, PN,

therefore the sum of the squares on LP, PN is rational.

Again, since DK is medial,

and DK is equal to twice the rectangle LP, PN,

therefore twice the rectangle LP, PN is medial.

And, since AI was proved incommensurable with FK,

therefore the square on LP is also incommensurable with the square on PN.

Therefore LP, PN are straight lines incommensurable in square which make the sum of the squares on them rational, but twice the rectangle contained by them medial.

Therefore LN is the irrational straight line called minor; [x. 76]

and it is the "side" of the area AB.

Therefore the "side" of the area AB is minor.

Q. E. D.

We have here to find and classify the straight line

$$\sqrt{\rho\left(k\rho - \frac{k\rho}{\sqrt{1+\lambda}}\right)}.$$

As usual, we find u, v from the equations

$$\left.\begin{aligned} u + v &= k\rho \\ uv &= \tfrac{1}{4}\frac{k^2\rho^2}{1+\lambda} \end{aligned}\right\} \qquad\qquad (1),$$

and then, giving u, v their values, we put

$$\left.\begin{aligned} x^2 &= \rho u \\ y^2 &= \rho v \end{aligned}\right\} \qquad\qquad (2).$$

Then $(x-y)$ is the required square root.

This is proved in the same way as before, and, as before, it is proved that

$$xy = \tfrac{1}{2}\frac{k\rho^2}{\sqrt{1+\lambda}}.$$

Now, from (1), by x. 18, $\quad u \smile v$;

therefore $\quad \rho u \smile \rho v$,

or $\quad x^2 \smile y^2$,

so that x, y are incommensurable in square.

And $x^2 + y^2$, or $\rho(u+v)$, is a *rational* area $(k\rho^2)$.

But $2xy = \dfrac{k\rho^2}{\sqrt{1+\lambda}}$, which is a *medial* area.

Hence [x. 76] $(x-y)$ is the irrational straight line called *minor*.

Algebraical solution gives

$$u = \tfrac{1}{2}k\rho\left(1 + \sqrt{\frac{\lambda}{1+\lambda}}\right),$$

$$v = \tfrac{1}{2}k\rho\left(1 - \sqrt{\frac{\lambda}{1+\lambda}}\right),$$

whence $$x - y = \rho\sqrt{\frac{k}{2}\left(1 + \sqrt{\frac{\lambda}{1+\lambda}}\right)} - \rho\sqrt{\frac{k}{2}\left(1 - \sqrt{\frac{\lambda}{1+\lambda}}\right)}.$$

As explained in the note on x. 57, this is the lesser positive root of the equation

$$x^4 - 2k\rho^2 . x^2 + \frac{\lambda}{1+\lambda}k^2\rho^4 = 0.$$

PROPOSITION 95.

If an area be contained by a rational straight line and a fifth apotome, the "side" of the area is a straight line which produces with a rational area a medial whole.

For let the area AB be contained by the rational straight line AC and the fifth apotome AD;

I say that the "side" of the area AB is a straight line which produces with a rational area a medial whole.

For let DG be the annex to AD;

therefore AG, GD are rational straight lines commensurable in square only,

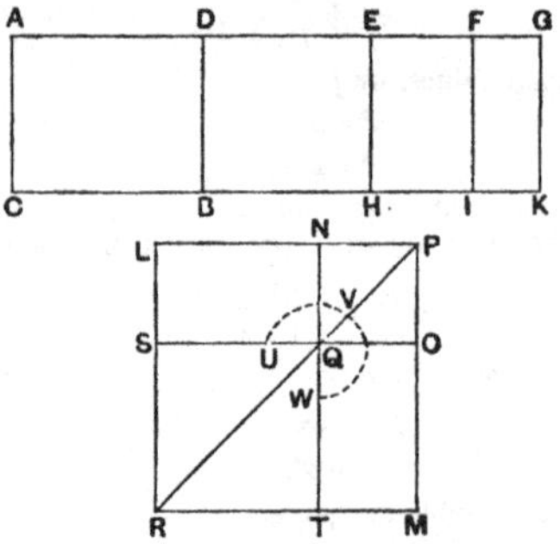

the annex GD is commensurable in length with the rational straight line AC set out,

and the square on the whole AG is greater than the square

on the annex *DG* by the square on a straight line incommensurable with *AG*. [x. Deff. iii. 5]

Therefore, if there be applied to *AG* a parallelogram equal to the fourth part of the square on *DG* and deficient by a square figure, it will divide it into incommensurable parts. [x. 18]

Let then *DG* be bisected at the point *E*,

let there be applied to *AG* a parallelogram equal to the square on *EG* and deficient by a square figure, and let it be the rectangle *AF*, *FG*;

therefore *AF* is incommensurable in length with *FG*.

Now, since *AG* is incommensurable in length with *CA*, and both are rational,

therefore *AK* is medial. [x. 21]

Again, since *DG* is rational and commensurable in length with *AC*,

DK is rational. [x. 19]

Now let the square *LM* be constructed equal to *AI*, and let the square *NO* equal to *FK* and about the same angle, the angle *LPM*, be subtracted;

therefore the squares *LM*, *NO* are about the same diameter. [vi. 26]

Let *PR* be their diameter, and let the figure be drawn.

Similarly then we can prove that *LN* is the "side" of the area *AB*.

I say that *LN* is the straight line which produces with a rational area a medial whole.

For, since *AK* was proved medial and is equal to the squares on *LP*, *PN*,

therefore the sum of the squares on *LP*, *PN* is medial.

Again, since *DK* is rational and is equal to twice the rectangle *LP*, *PN*,

the latter is itself also rational.

And, since *AI* is incommensurable with *FK*,

therefore the square on *LP* is also incommensurable with the square on *PN*;

therefore *LP*, *PN* are straight lines incommensurable in

square which make the sum of the squares on them medial but twice the rectangle contained by them rational.

Therefore the remainder *LN* is the irrational straight line called that which produces with a rational area a medial whole; [x. 77]

and it is the "side" of the area *AB*.

Therefore the "side" of the area *AB* is a straight line which produces with a rational area a medial whole.

Q. E. D.

Here the problem is to find and classify

$$\sqrt{\rho\left(k\rho\sqrt{1+\lambda}-k\rho\right)}.$$

As usual, we put

$$\left.\begin{aligned} u+v &= k\rho\sqrt{1+\lambda} \\ uv &= \tfrac{1}{4}k^2\rho^2 \end{aligned}\right\} \qquad\ldots\ldots\ldots\ldots\ldots\ldots\ldots\ldots(1),$$

and, u, v being found, we take

$$\left.\begin{aligned} x^2 &= \rho u \\ y^2 &= \rho v \end{aligned}\right\} \qquad\ldots\ldots\ldots\ldots\ldots\ldots\ldots\ldots(2).$$

Then $(x-y)$ so found is our required square root.

This fact is proved as before, and, as before, we see that

$$xy = \tfrac{1}{2}k\rho^2.$$

Now from (1), by x. 18, $\qquad u \smile v$,

whence $\qquad \rho u \smile \rho v$,

or $\qquad x^2 \smile y^2$,

and x, y are incommensurable in square.

Next $(x^2+y^2) = \rho(u+v) = k\rho^2\sqrt{1+\lambda}$, which is a *medial* area.

And $2xy = k\rho^2$, which is a *rational* area.

Hence $(x-y)$ is the "*side*" *of a medial, minus a rational, area.* [x. 77]

Algebraical solution gives

$$u = \frac{k\rho}{2}\left(\sqrt{1+\lambda}+\sqrt{\lambda}\right),$$

$$v = \frac{k\rho}{2}\left(\sqrt{1+\lambda}-\sqrt{\lambda}\right),$$

and therefore

$$x-y = \rho\sqrt{\frac{k}{2}\left(\sqrt{1+\lambda}+\sqrt{\lambda}\right)} - \rho\sqrt{\frac{k}{2}\left(\sqrt{1+\lambda}-\sqrt{\lambda}\right)},$$

which is, as explained in the note to x. 58, the lesser positive root of the equation

$$x^4 - 2k\rho^2\sqrt{1+\lambda}\,.\,x^2 + \lambda k^2\rho^4 = 0.$$

PROPOSITION 96.

If an area be contained by a rational straight line and a sixth apotome, the "side" of the area is a straight line which produces with a medial area a medial whole.

For let the area AB be contained by the rational straight line AC and the sixth apotome AD;

I say that the "side" of the area AB is a straight line which produces with a medial area a medial whole.

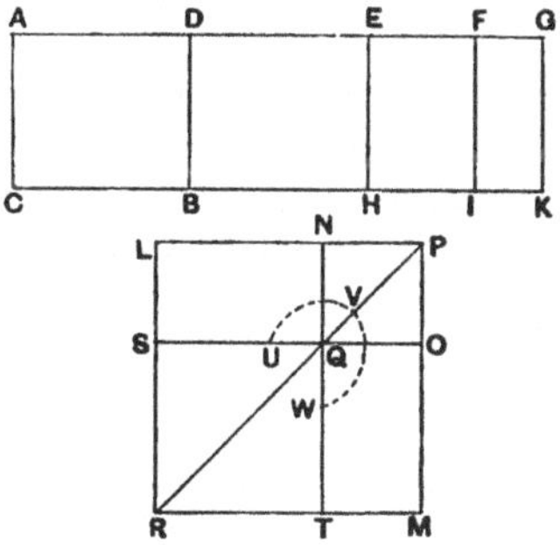

For let DG be the annex to AD;

therefore AG, GD are rational straight lines commensurable in square only,

neither of them is commensurable in length with the rational straight line AC set out,

and the square on the whole AG is greater than the square on the annex DG by the square on a straight line incommensurable in length with AG. [x. Deff. III. 6]

Since then the square on AG is greater than the square on GD by the square on a straight line incommensurable in length with AG,

therefore, if there be applied to AG a parallelogram equal to the fourth part of the square on DG and deficient by a square figure, it will divide it into incommensurable parts. [x. 18]

Let then DG be bisected at E,

let there be applied to AG a parallelogram equal to the square

on EG and deficient by a square figure, and let it be the rectangle AF, FG;
therefore AF is incommensurable in length with FG.

But, as AF is to FG, so is AI to FK; [vi. 1]
therefore AI is incommensurable with FK. [x. 11]

And, since AG, AC are rational straight lines commensurable in square only,
AK is medial. [x. 21]

Again, since AC, DG are rational straight lines and incommensurable in length,
DK is also medial. [x. 21]

Now, since AG, GD are commensurable in square only, therefore AG is incommensurable in length with GD.

But, as AG is to GD, so is AK to KD; [vi. 1]
therefore AK is incommensurable with KD. [x. 11]

Now let the square LM be constructed equal to AI,
and let NO equal to FK, and about the same angle, be subtracted;
therefore the squares LM, NO are about the same diameter. [vi. 26]

Let PR be their diameter, and let the figure be drawn.

Then in manner similar to the above we can prove that LN is the "side" of the area AB.

I say that LN is a straight line which produces with a medial area a medial whole.

For, since AK was proved medial and is equal to the squares on LP, PN,
therefore the sum of the squares on LP, PN is medial.

Again, since DK was proved medial and is equal to twice the rectangle LP, PN,
twice the rectangle LP, PN is also medial.

And, since AK was proved incommensurable with DK,
the squares on LP, PN are also incommensurable with twice the rectangle LP, PN.

And, since AI is incommensurable with FK,
therefore the square on LP is also incommensurable with the square on PN;

therefore LP, PN are straight lines incommensurable in square which make the sum of the squares on them medial, twice the rectangle contained by them medial, and further the squares on them incommensurable with twice the rectangle contained by them.

Therefore LN is the irrational straight line called that which produces with a medial area a medial whole; [x. 78] and it is the "side" of the area AB.

Therefore the "side" of the area is a straight line which produces with a medial area a medial whole.

Q. E. D.

We have to find and classify

$$\sqrt{\rho(\sqrt{k}\cdot\rho - \sqrt{\lambda}\cdot\rho)}.$$

Put, as usual,

$$\left.\begin{aligned} u+v &= \sqrt{k}\cdot\rho \\ uv &= \tfrac{1}{4}\lambda\rho^2 \end{aligned}\right\} \quad\ldots\ldots\ldots\ldots(1),$$

and, u, v being thus found, let

$$\left.\begin{aligned} x^2 &= \rho u \\ y^2 &= \rho v \end{aligned}\right\} \quad\ldots\ldots\ldots\ldots(2).$$

Then, as before, $(x-y)$ is the square root required.

For, from (1), by x. 18, $u \smile v$,

whence $\rho u \smile \rho v$,

or $x^2 \smile y^2$,

and x, y are incommensurable in square.

Next, $x^2+y^2 = \rho(u+v) = \sqrt{k}\cdot\rho^2$, which is a *medial* area.

Also $2xy = \sqrt{\lambda}\cdot\rho^2$, which is again a *medial* area.

Lastly, $\sqrt{k}\cdot\rho$, $\sqrt{\lambda}\cdot\rho$ are by hypothesis $\frown$, so that

$$\sqrt{k}\cdot\rho \smile \sqrt{\lambda}\cdot\rho,$$

whence $\sqrt{k}\cdot\rho^2 \smile \sqrt{\lambda}\cdot\rho^2$,

or $(x^2+y^2) \smile 2xy$.

Thus $(x-y)$ is the "*side*" *of a medial, minus a medial, area* [x. 78].

Algebraical solution gives

$$u = \frac{\rho}{2}(\sqrt{k}+\sqrt{k-\lambda}),$$

$$v = \frac{\rho}{2}(\sqrt{k}-\sqrt{k-\lambda}),$$

whence $x - y = \rho\sqrt{\tfrac{1}{2}(\sqrt{k}+\sqrt{k-\lambda})} - \rho\sqrt{\tfrac{1}{2}(\sqrt{k}-\sqrt{k-\lambda})}$.

This, as explained in the note on x. 59, is the lesser positive root of the equation

$$x^4 - 2\sqrt{k}\cdot\rho^2x^2 + (k-\lambda)\rho^4 = 0.$$

PROPOSITION 97.

The square on an apotome applied to a rational straight line produces as breadth a first apotome.

Let *AB* be an apotome, and *CD* rational, and to *CD* let there be applied *CE* equal to the square on *AB* and producing *CF* as breadth;
I say that *CF* is a first apotome.

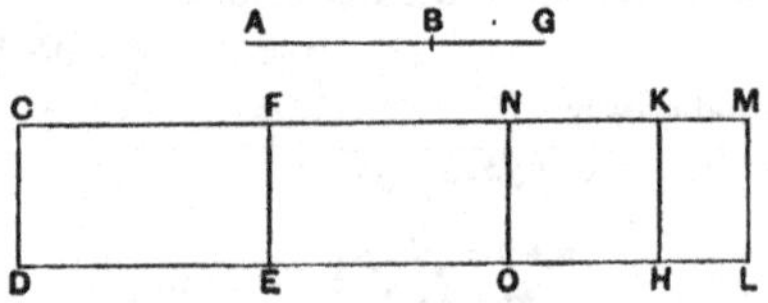

For let *BG* be the annex to *AB*;
therefore *AG*, *GB* are rational straight lines commensurable in square only. [x. 73]

To *CD* let there be applied *CH* equal to the square on *AG*, and *KL* equal to the square on *BG*.

Therefore the whole *CL* is equal to the squares on *AG*, *GB*, and, in these, *CE* is equal to the square on *AB*;
therefore the remainder *FL* is equal to twice the rectangle *AG*, *GB*. [II. 7]

Let *FM* be bisected at the point *N*,
and let *NO* be drawn through *N* parallel to *CD*;
therefore each of the rectangles *FO*, *LN* is equal to the rectangle *AG*, *GB*.

Now, since the squares on *AG*, *GB* are rational,
and *DM* is equal to the squares on *AG*, *GB*,
therefore *DM* is rational.

And it has been applied to the rational straight line *CD*, producing *CM* as breadth;
therefore *CM* is rational and commensurable in length with *CD*. [x. 20]

Again, since twice the rectangle *AG*, *GB* is medial, and *FL* is equal to twice the rectangle *AG*, *GB*,
therefore *FL* is medial.

And it is applied to the rational straight line *CD*, producing *FM* as breadth;

therefore *FM* is rational and incommensurable in length with *CD*. [x. 22]

And, since the squares on *AG*, *GB* are rational,

while twice the rectangle *AG*, *GB* is medial,

therefore the squares on *AG*, *GB* are incommensurable with twice the rectangle *AG*, *GB*.

And *CL* is equal to the squares on *AG*, *GB*,

and *FL* to twice the rectangle *AG*, *GB*;

therefore *DM* is incommensurable with *FL*.

But, as *DM* is to *FL*, so is *CM* to *FM*; [vi. 1]

therefore *CM* is incommensurable in length with *FM*. [x. 11]

And both are rational;

therefore *CM*, *MF* are rational straight lines commensurable in square only;

therefore *CF* is an apotome. [x. 73]

I say next that it is also a first apotome.

For, since the rectangle *AG*, *GB* is a mean proportional between the squares on *AG*, *GB*,

and *CH* is equal to the square on *AG*,

KL equal to the square on *BG*,

and *NL* equal to the rectangle *AG*, *GB*,

therefore *NL* is also a mean proportional between *CH*, *KL*;

therefore, as *CH* is to *NL*, so is *NL* to *KL*.

But, as *CH* is to *NL*, so is *CK* to *NM*,

and, as *NL* is to *KL*, so is *NM* to *KM*; [vi. 1]

therefore the rectangle *CK*, *KM* is equal to the square on *NM* [vi. 17], that is, to the fourth part of the square on *FM*.

And, since the square on *AG* is commensurable with the square on *GB*,

CH is also commensurable with *KL*.

But, as *CH* is to *KL*, so is *CK* to *KM*; [vi. 1]

therefore *CK* is commensurable with *KM*. [x. 11]

Since then *CM*, *MF* are two unequal straight lines,

and to CM there has been applied the rectangle CK, KM equal to the fourth part of the square on FM and deficient by a square figure,

while CK is commensurable with KM,

therefore the square on CM is greater than the square on MF by the square on a straight line commensurable in length with CM. [x. 17]

And CM is commensurable in length with the rational straight line CD set out;

therefore CF is a first apotome. [x. Deff. III. 1]

Therefore etc.

Q. E. D.

Here begins the hexad of propositions solving the problems which are the converse of those in the hexad just concluded. Props. 97 to 102 correspond of course to Props. 60 to 65 relating to the binomials etc.

We have in x. 97 to prove that, $(\rho - \sqrt{k}\cdot\rho)$ being an *apotome*,

$$\frac{(\rho \sim \sqrt{k}\cdot\rho)^2}{\sigma}$$

is a *first apotome*, and we have to find it geometrically.

Euclid's procedure may be represented thus.

Take x, y, z such that

$$\left.\begin{aligned}\sigma x &= \rho^2\\ \sigma y &= k\rho^2\\ \sigma\cdot 2z &= 2\sqrt{k}\cdot\rho^2\end{aligned}\right\}\ldots\ldots\ldots\ldots(1).$$

Thus $$(x+y) - 2z = \frac{(\rho - \sqrt{k}\cdot\rho)^2}{\sigma},$$

and we have to prove that $(x+y) - 2z$ is a *first apotome*.

(α) Now $\rho^2 + k\rho^2$, or $\sigma(x+y)$, is rational;

therefore $(x+y)$ is rational and $\frown \sigma$ ....(2).

And $2\sqrt{k}\cdot\rho^2$, or $\sigma\cdot 2z$, is medial:

therefore $2z$ is rational and $\smile \sigma$............(3).

But, $\sigma(x+y)$ being *rational*, and $\sigma\cdot 2z$ *medial*,

$$\sigma(x+y) \smile \sigma\cdot 2z,$$

whence $$(x+y) \smile 2z.$$

Therefore, since $(x+y)$, $2z$ are both rational [(2), (3)],

$(x+y)$, $2z$ are rational and $\frown$ ....(4).

Hence $(x+y) - 2z$ is an *apotome*.

(β) Since $\sqrt{k}\cdot\rho^2$ is a mean proportional between ρ^2, $k\rho^2$,

σz is a mean proportional between σx, σy [by (1)].

That is, $$\sigma x : \sigma z = \sigma z : \sigma y,$$

or $$x : z = z : y,$$

and $$xy = z^2, \text{ or } \tfrac{1}{4}(2z)^2 \ldots\ldots\ldots\ldots(5).$$

And, since $\rho^2 \frown k\rho^2$, $\sigma x \frown \sigma y$,

or $x \frown y$ ....(6).

Hence [(5), (6)], by x. 17,

$$\sqrt{(x+y)^2 - (2z)^2} \frown (x+y).$$

And [(4)] $(x+y)$, $2z$ are rational and $\smile$,

while [(2)] $(x+y) \frown \sigma$;

therefore $(x+y) - 2z$ is a *first apotome.*

The actual value of $(x+y) - 2z$ is of course

$$\frac{\rho^2}{\sigma}\{(1+k) - 2\sqrt{k}\}.$$

Proposition 98.

The square on a first apotome of a medial straight line applied to a rational straight line produces as breadth a second apotome.

Let AB be a first apotome of a medial straight line and CD a rational straight line,

and to CD let there be applied CE equal to the square on AB, producing CF as breadth;

I say that CF is a second apotome.

For let BG be the annex to AB;

therefore AG, GB are medial straight lines commensurable in square only which contain a rational rectangle. [x. 74]

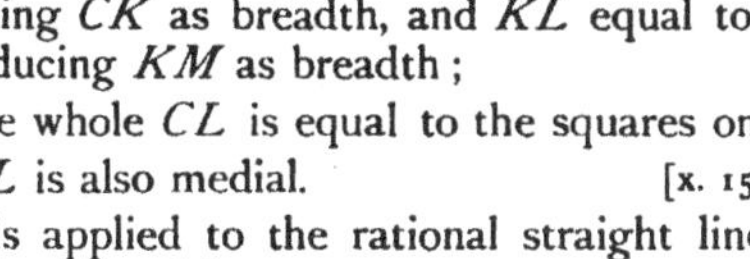

To CD let there be applied CH equal to the square on AG, producing CK as breadth, and KL equal to the square on GB, producing KM as breadth;

therefore the whole CL is equal to the squares on AG, GB;

therefore CL is also medial. [x. 15 and 23, Por.]

And it is applied to the rational straight line CD, producing CM as breadth;

therefore CM is rational and incommensurable in length with CD. [x. 22]

Now, since CL is equal to the squares on AG, GB,
and, in these, the square on AB is equal to CE,
therefore the remainder, twice the rectangle AG, GB, is equal to FL. [II. 7]

But twice the rectangle AG, GB is rational;
therefore FL is rational.

And it is applied to the rational straight line FE, producing FM as breadth;
therefore FM is also rational and commensurable in length with CD. [X. 20]

Now, since the sum of the squares on AG, GB, that is, CL, is medial, while twice the rectangle AG, GB, that is, FL, is rational,
therefore CL is incommensurable with FL.

But, as CL is to FL, so is CM to FM; [VI. 1]
therefore CM is incommensurable in length with FM. [X. 11]

And both are rational;
therefore CM, MF are rational straight lines commensurable in square only;
therefore CF is an apotome. [X. 73]

I say next that it is also a second apotome.

For let FM be bisected at N,
and let NO be drawn through N parallel to CD;
therefore each of the rectangles FO, NL is equal to the rectangle AG, GB.

Now, since the rectangle AG, GB is a mean proportional between the squares on AG, GB,
and the square on AG is equal to CH,
the rectangle AG, GB to NL,
and the square on BG to KL,
therefore NL is also a mean proportional between CH, KL;
therefore, as CH is to NL, so is NL to KL.

But, as CH is to NL, so is CK to NM,
and, as NL is to KL, so is NM to MK; [VI. 1]
therefore, as CK is to NM, so is NM to KM; [V. 11]
therefore the rectangle CK, KM is equal to the square on NM [VI. 17], that is, to the fourth part of the square on FM.

Since then CM, MF are two unequal straight lines, and the rectangle CK, KM equal to the fourth part of the square on MF and deficient by a square figure has been applied to the greater, CM, and divides it into commensurable parts, therefore the square on CM is greater than the square on MF by the square on a straight line commensurable in length with CM. [x. 17]

And the annex FM is commensurable in length with the rational straight line CD set out;
therefore CF is a second apotome. [x. Deff. III. 2]

Therefore etc.

Q. E. D.

In this case we have to find and classify

$$\frac{(k^{\frac{1}{4}}\rho \sim k^{\frac{3}{4}}\rho)^2}{\sigma}.$$

Take x, y, z such that

$$\left.\begin{aligned} \sigma x &= k^{\frac{1}{2}}\rho^2 \\ \sigma y &= k^{\frac{3}{2}}\rho^2 \\ \sigma \,.\, 2z &= 2k\rho^2 \end{aligned}\right\} \quad \ldots\ldots\ldots\ldots (1).$$

(α) Now $k^{\frac{1}{2}}\rho^2$, $k^{\frac{3}{2}}\rho^2$ are medial areas;
therefore $\sigma(x+y)$ is medial,
whence $(x+y)$ is rational and $\smile \sigma$ .... (2).

But $2k\rho^2$, and therefore $\sigma \,.\, 2z$, is rational,
whence $2z$ is rational and $\frown \sigma$ .... (3).

And, $\sigma(x+y)$ being medial, and $\sigma \,.\, 2z$ rational,

$$\sigma(x+y) \smile \sigma \,.\, 2z,$$

or

$$(x+y) \smile 2z.$$

Hence $(x+y)$, $2z$ are rational straight lines commensurable in square only, and therefore $(x+y) - 2z$ is an *apotome*.

(β) We prove, as before, that

$$xy = \tfrac{1}{4}(2z)^2 \quad \ldots\ldots\ldots\ldots (4).$$

Also $k^{\frac{1}{2}}\rho^2 \frown k^{\frac{3}{2}}\rho^2$, or $\sigma x \frown \sigma y$,
so that

$$x \frown y \quad \ldots\ldots\ldots\ldots (5).$$

[This step is omitted in P, and Heiberg accordingly brackets it. The result is, however, assumed.]

Therefore [(4), (5)], by x. 17,

$$\sqrt{(x+y)^2 - (2z)^2} \frown (x+y).$$

And $2z \frown \sigma$.

Therefore $(x+y) - 2z$ is a *second apotome*.

Obviously $$(x+y) - 2z = \frac{\rho^2}{\sigma}\{\sqrt{k}(1+k) - 2k\}.$$

PROPOSITION 99.

The square on a second apotome of a medial straight line applied to a rational straight line produces as breadth a third apotome.

Let *AB* be a second apotome of a medial straight line, and *CD* rational,

and to *CD* let there be applied *CE* equal to the square on *AB*, producing *CF* as breadth;

I say that *CF* is a third apotome.

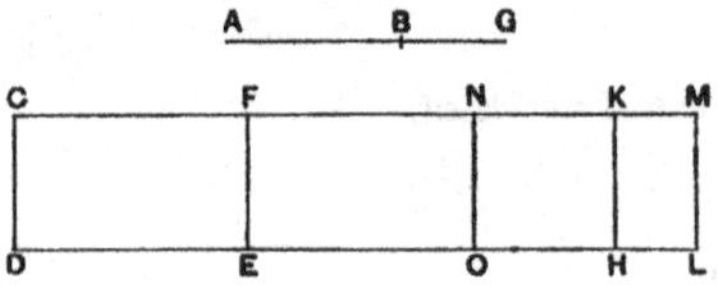

For let *BG* be the annex to *AB*;

therefore *AG*, *GB* are medial straight lines commensurable in square only which contain a medial rectangle. [x. 75]

Let *CH* equal to the square on *AG* be applied to *CD*, producing *CK* as breadth,

and let *KL* equal to the square on *BG* be applied to *KH*, producing *KM* as breadth;

therefore the whole *CL* is equal to the squares on *AG*, *GB*;

therefore *CL* is also medial. [x. 15 and 23, Por.]

And it is applied to the rational straight line *CD*, producing *CM* as breadth;

therefore *CM* is rational and incommensurable in length with *CD*. [x. 22]

Now, since the whole *CL* is equal to the squares on *AG*, *GB*, and, in these, *CE* is equal to the square on *AB*,

therefore the remainder *LF* is equal to twice the rectangle *AG*, *GB*. [II. 7]

Let then *FM* be bisected at the point *N*,

and let *NO* be drawn parallel to *CD*;

therefore each of the rectangles *FO*, *NL* is equal to the rectangle *AG*, *GB*.

But the rectangle AG, GB is medial;
therefore FL is also medial.

And it is applied to the rational straight line EF, producing FM as breadth;
therefore FM is also rational and incommensurable in length with CD. [x. 22]

And, since AG, GB are commensurable in square only,
therefore AG is incommensurable in length with GB;
therefore the square on AG is also incommensurable with the rectangle AG, GB. [vi. 1, x. 11]

But the squares on AG, GB are commensurable with the square on AG,
and twice the rectangle AG, GB with the rectangle AG, GB;
therefore the squares on AG, GB are incommensurable with twice the rectangle AG, GB. [x. 13]

But CL is equal to the squares on AG, GB,
and FL is equal to twice the rectangle AG, GB;
therefore CL is also incommensurable with FL.

But, as CL is to FL, so is CM to FM; [vi. 1]
therefore CM is incommensurable in length with FM. [x. 11]

And both are rational;
therefore CM, MF are rational straight lines commensurable in square only;
therefore CF is an apotome. [x. 73]

I say next that it is also a third apotome.

For, since the square on AG is commensurable with the square on GB,
therefore CH is also commensurable with KL,
so that CK is also commensurable with KM. [vi. 1, x. 11]

And, since the rectangle AG, GB is a mean proportional between the squares on AG, GB,
and CH is equal to the square on AG,
KL equal to the square on GB,
and NL equal to the rectangle AG, GB,
therefore NL is also a mean proportional between CH, KL;
therefore, as CH is to NL, so is NL to KL.

But, as CH is to NL, so is CK to NM,
and, as NL is to KL, so is NM to KM; [VI. 1]
therefore, as CK is to MN, so is MN to KM; [V. 11]
therefore the rectangle CK, KM is equal to [the square on MN, that is, to] the fourth part of the square on FM.

Since then CM, MF are two unequal straight lines, and a parallelogram equal to the fourth part of the square on FM and deficient by a square figure has been applied to CM, and divides it into commensurable parts,
therefore the square on CM is greater than the square on MF by the square on a straight line commensurable with CM. [X. 17]

And neither of the straight lines CM, MF is commensurable in length with the rational straight line CD set out;
therefore CF is a third apotome. [X. Deff. III. 3]

Therefore etc.

Q. E. D.

We have to find and classify

$$\frac{1}{\sigma}\left(k^{\frac{1}{4}}\rho \sim \frac{\sqrt{\lambda}\,.\,\rho}{k^{\frac{1}{4}}}\right)^2.$$

Take x, y, z such that

$$\left.\begin{aligned}\sigma x &= \sqrt{k}\,.\,\rho^2\\ \sigma y &= \frac{\lambda}{\sqrt{k}}\cdot\rho^2\\ \sigma\,.\,2z &= 2\sqrt{\lambda}\,.\,\rho^2\end{aligned}\right\}.$$

(α) Then $\sigma(x+y)$ is a medial area,
whence $(x+y)$ is rational and $\smile \sigma$ ....(1).

Also $\sigma\,.\,2z$ is medial,
whence $2z$ is rational and $\smile \sigma$ ....(2).

Again $$k^{\frac{1}{4}}\rho \smile \frac{\sqrt{\lambda}\,.\,\rho}{k^{\frac{1}{4}}},$$

whence $$\sqrt{k}\,.\,\rho^2 \smile \sqrt{\lambda}\,.\,\rho^2.$$

And $$\sqrt{k}\,.\,\rho^2 \frown \left(\sqrt{k}\,.\,\rho^2 + \frac{\lambda}{\sqrt{k}}\rho^2\right),$$

while $$\sqrt{\lambda}\,.\,\rho^2 \frown 2\sqrt{\lambda}\,.\,\rho^2;$$

therefore $$\left(\sqrt{k}\,.\,\rho^2 + \frac{\lambda}{\sqrt{k}}\rho^2\right) \smile 2\sqrt{\lambda}\,.\,\rho^2,$$

or $$\sigma(x+y) \smile \sigma\,.\,2z,$$

and $$(x+y) \smile 2z \quad\text{....................(3).}$$

Thus [(1), (2), (3)] $(x+y)$, $2z$ are rational and $\smile$,
so that $(x+y)-2z$ is an *apotome*.

(β) $\sigma x \frown \sigma y$, so that $x \frown y$.

And, as before, $$xy = \tfrac{1}{4}(2z)^2.$$

Therefore [x. 17] $$\sqrt{(x+y)^2-(2z)^2} \frown (x+y).$$

And neither $(x+y)$ nor $2z$ is $\frown \sigma$.

Therefore $(x+y)-2z$ is a *third apotome*.

It is of course equal to

$$\frac{\rho^2}{\sigma}\left\{\frac{k+\lambda}{\sqrt{k}} - 2\sqrt{\lambda}\right\}.$$

PROPOSITION 100.

The square on a minor straight line applied to a rational straight line produces as breadth a fourth apotome.

Let *AB* be a minor and *CD* a rational straight line, and to the rational straight line *CD* let *CE* be applied equal to the square on *AB* and producing *CF* as breadth;

I say that *CF* is a fourth apotome.

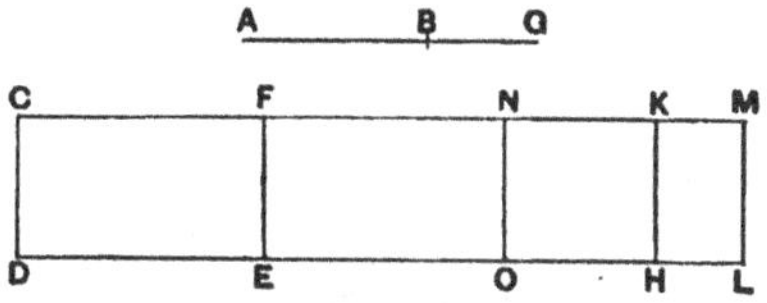

For let *BG* be the annex to *AB*;
therefore *AG*, *GB* are straight lines incommensurable in square which make the sum of the squares on *AG*, *GB* rational, but twice the rectangle *AG*, *GB* medial. [x. 76]

To *CD* let there be applied *CH* equal to the square on *AG* and producing *CK* as breadth,
and *KL* equal to the square on *BG*, producing *KM* as breadth;
therefore the whole *CL* is equal to the squares on *AG*, *GB*.

And the sum of the squares on *AG*, *GB* is rational;
therefore *CL* is also rational.

And it is applied to the rational straight line *CD*, producing *CM* as breadth;
therefore *CM* is also rational and commensurable in length with *CD*. [x. 20]

And, since the whole CL is equal to the squares on AG, GB, and, in these, CE is equal to the square on AB,

therefore the remainder FL is equal to twice the rectangle AG, GB. [II. 7]

Let then FM be bisected at the point N,

and let NO be drawn through N parallel to either of the straight lines CD, ML;

therefore each of the rectangles FO, NL is equal to the rectangle AG, GB.

And, since twice the rectangle AG, GB is medial and is equal to FL,

therefore FL is also medial.

And it is applied to the rational straight line FE, producing FM as breadth;

therefore FM is rational and incommensurable in length with CD. [x. 22]

And, since the sum of the squares on AG, GB is rational, while twice the rectangle AG, GB is medial,

the squares on AG, GB are incommensurable with twice the rectangle AG, GB.

But CL is equal to the squares on AG, GB,

and FL equal to twice the rectangle AG, GB;

therefore CL is incommensurable with FL.

But, as CL is to FL, so is CM to MF; [VI. 1]

therefore CM is incommensurable in length with MF. [x. 11]

And both are rational;

therefore CM, MF are rational straight lines commensurable in square only;

therefore CF is an apotome. [x. 73]

I say that it is also a fourth apotome.

For, since AG, GB are incommensurable in square,

therefore the square on AG is also incommensurable with the square on GB.

And CH is equal to the square on AG,

and KL equal to the square on GB;

therefore CH is incommensurable with KL.

But, as CH is to KL, so is CK to KM; [VI. 1]
therefore CK is incommensurable in length with KM. [X. 11]

And, since the rectangle AG, GB is a mean proportional between the squares on AG, GB,
and the square on AG is equal to CH,
the square on GB to KL,
and the rectangle AG, GB to NL,
therefore NL is a mean proportional between CH, KL;
therefore, as CH is to NL, so is NL to KL.

But, as CH is to NL, so is CK to NM,
and, as NL is to KL, so is NM to KM; [VI. 1]
therefore, as CK is to MN, so is MN to KM; [V. 11]
therefore the rectangle CK, KM is equal to the square on MN [VI. 17], that is, to the fourth part of the square on FM.

Since then CM, MF are two unequal straight lines, and the rectangle CK, KM equal to the fourth part of the square on MF and deficient by a square figure has been applied to CM and divides it into incommensurable parts,
therefore the square on CM is greater than the square on MF by the square on a straight line incommensurable with CM. [X. 18]

And the whole CM is commensurable in length with the rational straight line CD set out;
therefore CF is a fourth apotome. [X. Deff. III. 4]

Therefore etc.

Q. E. D.

We have to find and classify

$$\frac{1}{\sigma}\left\{\frac{\rho}{\sqrt{2}}\sqrt{1+\frac{k}{\sqrt{1+k^2}}}-\frac{\rho}{\sqrt{2}}\sqrt{1-\frac{k}{\sqrt{1+k^2}}}\right\}^2.$$

We will call this, for brevity,

$$\frac{1}{\sigma}(u-v)^2.$$

Take x, y, z such that

$$\left.\begin{aligned}\sigma x &= u^2\\ \sigma y &= v^2\\ \sigma\,.\,2z &= 2uv\end{aligned}\right\},$$

where it has to be remembered that u^2, v^2 are incommensurable, (u^2+v^2) is rational, and $2uv$ medial.

It follows that $\sigma(x+y)$ is rational and $\sigma \, . \, 2z$ medial,
so that $(x+y)$ is rational and $\frown \sigma$ ....(1),
while $2z$ is rational and $\smile \sigma$ ....(2),
and $$\sigma(x+y) \smile \sigma \, . \, 2z,$$
so that $$(x+y) \smile 2z \quad \text{....................(3).}$$

Thus [(1), (2), (3)] $(x+y)$, $2z$ are rational and $\frown\!-$,
so that $(x+y) - 2z$ is an *apotome*.

Next, since $$u^2 \smile v^2,$$
$$\sigma x \smile \sigma y,$$
or $$x \smile y.$$

And it is proved, as usual, that
$$xy = z^2 = \tfrac{1}{4}(2z)^2.$$

Therefore [x. 18] $$\sqrt{(x+y)^2 - (2z)^2} \smile (x+y).$$

But $(x+y) \frown \sigma$,
therefore $x + y - 2z$ is a *fourth apotome*.

Its value is of course $$\frac{\rho^2}{\sigma}\left(1 - \frac{1}{\sqrt{1+k^2}}\right).$$

PROPOSITION 101.

The square on the straight line which produces with a rational area a medial whole, if applied to a rational straight line, produces as breadth a fifth apotome.

Let AB be the straight line which produces with a rational area a medial whole, and CD a rational straight line, and to CD let CE be applied equal to the square on AB and producing CF as breadth;
I say that CF is a fifth apotome.

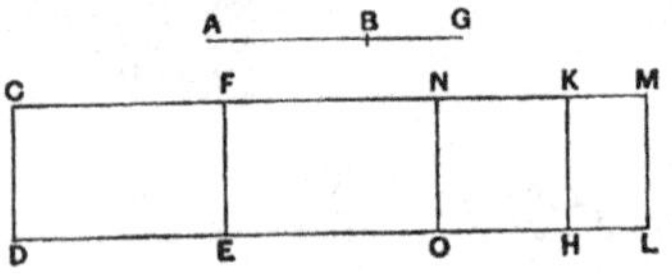

For let BG be the annex to AB;
therefore AG, GB are straight lines incommensurable in square which make the sum of the squares on them medial but twice the rectangle contained by them rational. [x. 77]

To CD let there be applied CH equal to the square on AG, and KL equal to the square on GB;
therefore the whole CL is equal to the squares on AG, GB.

But the sum of the squares on AG, GB together is medial;

therefore CL is medial.

And it is applied to the rational straight line CD, producing CM as breadth;

therefore CM is rational and incommensurable with CD. [X. 22]

And, since the whole CL is equal to the squares on AG, GB,

and, in these, CE is equal to the square on AB,

therefore the remainder FL is equal to twice the rectangle AG, GB. [II. 7]

Let then FM be bisected at N,

and through N let NO be drawn parallel to either of the straight lines CD, ML;

therefore each of the rectangles FO, NL is equal to the rectangle AG, GB.

And, since twice the rectangle AG, GB is rational and equal to FL,

therefore FL is rational.

And it is applied to the rational straight line EF, producing FM as breadth;

therefore FM is rational and commensurable in length with CD. [X. 20]

Now, since CL is medial, and FL rational,

therefore CL is incommensurable with FL.

But, as CL is to FL, so is CM to MF; [VI. 1]

therefore CM is incommensurable in length with MF. [X. 11]

And both are rational;

therefore CM, MF are rational straight lines commensurable in square only;

therefore CF is an apotome. [X. 73]

I say next that it is also a fifth apotome.

For we can prove similarly that the rectangle CK, KM is equal to the square on NM, that is, to the fourth part of the square on FM.

And, since the square on AG is incommensurable with the square on GB,

while the square on AG is equal to CH,
and the square on GB to KL,
therefore CH is incommensurable with KL.

But, as CH is to KL, so is CK to KM; [VI. 1]
therefore CK is incommensurable in length with KM. [X. 11]

Since then CM, MF are two unequal straight lines,
and a parallelogram equal to the fourth part of the square on FM and deficient by a square figure has been applied to CM, and divides it into incommensurable parts,
therefore the square on CM is greater than the square on MF by the square on a straight line incommensurable with CM. [X. 18]

And the annex FM is commensurable with the rational straight line CD set out;
therefore CF is a fifth apotome. [X. Deff. III. 5]

Q. E. D.

We have to find and classify

$$\frac{1}{\sigma}\left\{\frac{\rho}{\sqrt{2(1+k^2)}}\sqrt{\sqrt{1+k^2}+k}-\frac{\rho}{\sqrt{2(1+k^2)}}\sqrt{\sqrt{1+k^2}-k}\right\}^2.$$

Call this $\frac{1}{\sigma}(u-v)^2$, and take x, y, z such that

$$\left.\begin{array}{r}\sigma x = u^2\\ \sigma y = v^2\\ \sigma\,.\,2z = 2uv\end{array}\right\}.$$

In this case u^2, v^2 are incommensurable, (u^2+v^2) is a medial area and $2uv$ a rational area.

Since $\sigma(x+y)$ is medial and $\sigma\,.\,2z$ rational,
$(x+y)$ is rational and $\smile \sigma$,
$2z$ is rational and $\frown \sigma$,
while $(x+y) \smile 2z$.

It follows that $(x+y)$, $2z$ are rational and $\frown$,
so that $(x+y)-2z$ is an apotome.

Again, as before, $xy = z^2 = \frac{1}{4}(2z)^2$,
and, since $u^2 \smile v^2$, $\sigma x \smile \sigma y$,
or $x \smile y$.

Hence [X. 18] $\sqrt{(x+y)^2-(2z)^2} \smile (x+y)$.

And $2z \frown \sigma$.

Therefore $(x+y)-2z$ is a *fifth apotome*.

It is of course equal to

$$\frac{\rho^2}{\sigma}\left(\frac{1}{\sqrt{1+k^2}}-\frac{1}{1+k^2}\right).$$

Proposition 102.

The square on the straight line which produces with a medial area a medial whole, if applied to a rational straight line, produces as breadth a sixth apotome.

Let AB be the straight line which produces with a medial area a medial whole, and CD a rational straight line,

and to CD let CE be applied equal to the square on AB and producing CF as breadth;

I say that CF is a sixth apotome.

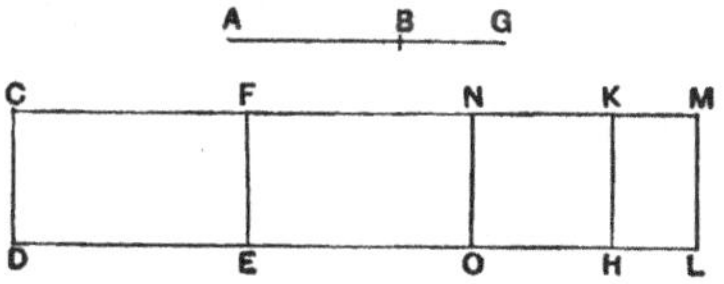

For let BG be the annex to AB;

therefore AG, GB are straight lines incommensurable in square which make the sum of the squares on them medial, twice the rectangle AG, GB medial, and the squares on AG, GB incommensurable with twice the rectangle AG, GB. [X. 78]

Now to CD let there be applied CH equal to the square on AG and producing CK as breadth,

and KL equal to the square on BG;

therefore the whole CL is equal to the squares on AG, GB;

therefore CL is also medial.

And it is applied to the rational straight line CD, producing CM as breadth;

therefore CM is rational and incommensurable in length with CD. [X. 22]

Since now CL is equal to the squares on AG, GB,

and, in these, CE is equal to the square on AB,

therefore the remainder FL is equal to twice the rectangle AG, GB. [II. 7]

And twice the rectangle AG, GB is medial;

therefore FL is also medial.

And it is applied to the rational straight line *FE*, producing *FM* as breadth;
therefore *FM* is rational and incommensurable in length with *CD*. [x. 22]

And, since the squares on *AG*, *GB* are incommensurable with twice the rectangle *AG*, *GB*,
and *CL* is equal to the squares on *AG*, *GB*,
and *FL* equal to twice the rectangle *AG*, *GB*,
therefore *CL* is incommensurable with *FL*.

But, as *CL* is to *FL*, so is *CM* to *MF*; [vi. 1]
therefore *CM* is incommensurable in length with *MF*. [x. 11]

And both are rational.

Therefore *CM*, *MF* are rational straight lines commensurable in square only;
therefore *CF* is an apotome. [x. 73]

I say next that it is also a sixth apotome.

For, since *FL* is equal to twice the rectangle *AG*, *GB*,
let *FM* be bisected at *N*,
and let *NO* be drawn through *N* parallel to *CD*;
therefore each of the rectangles *FO*, *NL* is equal to the rectangle *AG*, *GB*.

And, since *AG*, *GB* are incommensurable in square,
therefore the square on *AG* is incommensurable with the square on *GB*.

But *CH* is equal to the square on *AG*,
and *KL* is equal to the square on *GB*;
therefore *CH* is incommensurable with *KL*.

But, as *CH* is to *KL*, so is *CK* to *KM*; [vi. 1]
therefore *CK* is incommensurable with *KM*. [x. 11]

And, since the rectangle *AG*, *GB* is a mean proportional between the squares on *AG*, *GB*,
and *CH* is equal to the square on *AG*,
KL equal to the square on *GB*,
and *NL* equal to the rectangle *AG*, *GB*,
therefore *NL* is also a mean proportional between *CH*, *KL*;
therefore, as *CH* is to *NL*, so is *NL* to *KL*.

And for the same reason as before the square on *CM* is greater than the square on *MF* by the square on a straight line incommensurable with *CM*. [x. 18]

And neither of them is commensurable with the rational straight line *CD* set out;

therefore *CF* is a sixth apotome. [x. Deff. III. 6]

Q. E. D.

We have to find and classify

$$\frac{1}{\sigma}\left\{\frac{\rho\lambda^{\frac{1}{4}}}{\sqrt{2}}\sqrt{1+\frac{k}{\sqrt{1+k^2}}}-\frac{\rho\lambda^{\frac{1}{4}}}{\sqrt{2}}\sqrt{1-\frac{k}{\sqrt{1+k^2}}}\right\}^2.$$

Call this $\frac{1}{\sigma}(u-v)^2$, and put

$$\sigma x = u^2,$$
$$\sigma y = v^2,$$
$$\sigma \,.\, 2z = 2uv.$$

Here u^2, v^2 are incommensurable,

(u^2+v^2), $2uv$ are both medial areas,

and $$(u^2+v^2) \smile 2uv.$$

Since $\sigma(x+y)$, $\sigma \,.\, 2z$ are medial and incommensurable,

$(x+y)$ is rational and $\smile \sigma$,

$2z$ is rational and $\smile \sigma$,

and $$(x+y) \smile 2z.$$

Hence $(x+y)$, $2z$ are rational and $\frown$,

so that $(x+y)-2z$ is an apotome.

Again, since u^2, v^2, or σx, σy, are incommensurable,

$$x \smile y.$$

And, as before, $$xy = z^2 = \tfrac{1}{4}(2z)^2.$$

Therefore [x. 18] $$\sqrt{(x+y)^2-(2z)^2} \smile (x+y).$$

And neither $(x+y)$ nor $2z$ is $\frown z$;

therefore $(x+y)-2z$ is a *sixth apotome*.

It is of course $$\frac{\rho^2}{\sigma}\left(\sqrt{\lambda}-\frac{\sqrt{\lambda}}{\sqrt{1+k^2}}\right).$$

PROPOSITION 103.

A straight line commensurable in length with an apotome is an apotome and the same in order.

Let *AB* be an apotome,

and let *CD* be commensurable in length with *AB*;

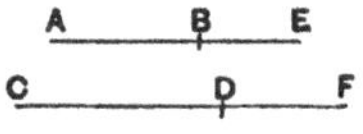

I say that *CD* is also an apotome and the same in order with *AB*.

For, since AB is an apotome, let BE be the annex to it; therefore AE, EB are rational straight lines commensurable in square only. [x. 73]

Let it be contrived that the ratio of BE to DF is the same as the ratio of AB to CD; [vi. 12]

therefore also, as one is to one, so are all to all; [v. 12]

therefore also, as the whole AE is to the whole CF, so is AB to CD.

But AB is commensurable in length with CD.

Therefore AE is also commensurable with CF, and BE with DF. [x. 11]

And AE, EB are rational straight lines commensurable in square only;

therefore CF, FD are also rational straight lines commensurable in square only. [x. 13]

Now since, as AE is to CF, so is BE to DF,

alternately therefore, as AE is to EB, so is CF to FD. [v. 16]

And the square on AE is greater than the square on EB either by the square on a straight line commensurable with AE or by the square on a straight line incommensurable with it.

If then the square on AE is greater than the square on EB by the square on a straight line commensurable with AE, the square on CF will also be greater than the square on FD by the square on a straight line commensurable with CF. [x. 14]

And, if AE is commensurable in length with the rational straight line set out,

CF is so also, [x. 12]

if BE, then DF also, [*id.*]

and, if neither of the straight lines AE, EB, then neither of the straight lines CF, FD. [x. 13]

But, if the square on AE is greater than the square on EB by the square on a straight line incommensurable with AE,

the square on CF will also be greater than the square on FD by the square on a straight line incommensurable with CF. [x. 14]

And, if *AE* is commensurable in length with the rational straight line set out,
CF is so also,
if *BE*, then *DF* also, [X. 12]
and, if neither of the straight lines *AE*, *EB*, then neither of the straight lines *CF*, *FD*. [X. 13]

Therefore *CD* is an apotome and the same in order with *AB*.

Q. E. D.

This and the following propositions to 107 inclusive (like the corresponding theorems X. 66 to 70) are easy and require no elucidation. They are equivalent to saying that, if in any of the preceding irrational straight lines $\frac{m}{n}\rho$ is substituted for ρ, the resulting irrational is of the same kind and order as that from which it is altered.

Proposition 104.

A straight line commensurable with an apotome of a medial straight line is an apotome of a medial straight line and the same in order.

Let *AB* be an apotome of a medial straight line,
and let *CD* be commensurable in length with *AB*;
I say that *CD* is also an apotome of a medial straight line and the same in order with *AB*.

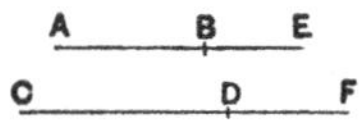

For, since *AB* is an apotome of a medial straight line, let *EB* be the annex to it.

Therefore *AE*, *EB* are medial straight lines commensurable in square only. [X. 74, 75]

Let it be contrived that, as *AB* is to *CD*, so is *BE* to *DF*; [VI. 12]
therefore *AE* is also commensurable with *CF*, and *BE* with *DF*. [V. 12, X. 11]

But *AE*, *EB* are medial straight lines commensurable in square only;
therefore *CF*, *FD* are also medial straight lines [X. 23] commensurable in square only; [X. 13]
therefore *CD* is an apotome of a medial straight line. [X. 74, 75]

I say next that it is also the same in order with AB.

Since, as AE is to EB, so is CF to FD,
therefore also, as the square on AE is to the rectangle AE, EB, so is the square on CF to the rectangle CF, FD.

But the square on AE is commensurable with the square on CF;
therefore the rectangle AE, EB is also commensurable with the rectangle CF, FD. [V. 16, X. 11]

Therefore, if the rectangle AE, EB is rational, the rectangle CF, FD will also be rational, [X. Def. 4]
and if the rectangle AE, EB is medial, the rectangle CF, FD is also medial. [X. 23, Por.]

Therefore CD is an apotome of a medial straight line and the same in order with AB. [X. 74, 75]

Q. E. D.

Proposition 105.

A straight line commensurable with a minor straight line is minor.

Let AB be a minor straight line, and CD commensurable with AB;

I say that CD is also minor.

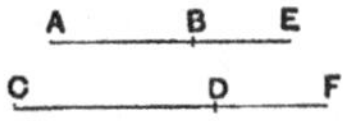

Let the same construction be made as before;
then, since AE, EB are incommensurable in square, [X. 76]
therefore CF, FD are also incommensurable in square. [X. 13]

Now since, as AE is to EB, so is CF to FD, [V. 12, V. 16]
therefore also, as the square on AE is to the square on EB, so is the square on CF to the square on FD. [VI. 22]

Therefore, *componendo*, as the squares on AE, EB are to the square on EB, so are the squares on CF, FD to the square on FD. [V. 18]

But the square on BE is commensurable with the square on DF;
therefore the sum of the squares on AE, EB is also commensurable with the sum of the squares on CF, FD. [V. 16, X. 11]

But the sum of the squares on AE, EB is rational; [X. 76]
therefore the sum of the squares on CF, FD is also rational. [X. Def. 4]

Again, since, as the square on *AE* is to the rectangle *AE*, *EB*, so is the square on *CF* to the rectangle *CF*, *FD*,

while the square on *AE* is commensurable with the square on *CF*,

therefore the rectangle *AE*, *EB* is also commensurable with the rectangle *CF*, *FD*.

But the rectangle *AE*, *EB* is medial; [x. 76]

therefore the rectangle *CF*, *FD* is also medial; [x. 23, Por.]

therefore *CF*, *FD* are straight lines incommensurable in square which make the sum of the squares on them rational, but the rectangle contained by them medial.

Therefore *CD* is minor. [x. 76]

Q. E. D.

PROPOSITION 106.

A straight line commensurable with that which produces with a rational area a medial whole is a straight line which produces with a rational area a medial whole.

Let *AB* be a straight line which produces with a rational area a medial whole,

and *CD* commensurable with *AB*;

I say that *CD* is also a straight line which produces with a rational area a medial whole.

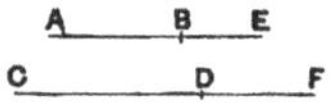

For let *BE* be the annex to *AB*;

therefore *AE*, *EB* are straight lines incommensurable in square which make the sum of the squares on *AE*, *EB* medial, but the rectangle contained by them rational. [x. 77]

Let the same construction be made.

Then we can prove, in manner similar to the foregoing, that *CF*, *FD* are in the same ratio as *AE*, *EB*,

the sum of the squares on *AE*, *EB* is commensurable with the sum of the squares on *CF*, *FD*,

and the rectangle *AE*, *EB* with the rectangle *CF*, *FD*;

so that *CF*, *FD* are also straight lines incommensurable in square which make the sum of the squares on *CF*, *FD* medial, but the rectangle contained by them rational.

Therefore CD is a straight line which produces with a rational area a medial whole. [x. 77]

Q. E. D.

Proposition 107.

A straight line commensurable with that which produces with a medial area a medial whole is itself also a straight line which produces with a medial area a medial whole.

Let AB be a straight line which produces with a medial area a medial whole,

and let CD be commensurable with AB;

I say that CD is also a straight line which produces with a medial area a medial whole.

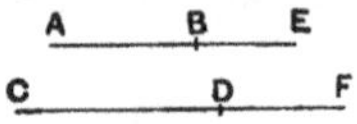

For let BE be the annex to AB,

and let the same construction be made;

therefore AE, EB are straight lines incommensurable in square which make the sum of the squares on them medial, the rectangle contained by them medial, and further the sum of the squares on them incommensurable with the rectangle contained by them. [x. 78]

Now, as was proved, AE, EB are commensurable with CF, FD,

the sum of the squares on AE, EB with the sum of the squares on CF, FD,

and the rectangle AE, EB with the rectangle CF, FD;

therefore CF, FD are also straight lines incommensurable in square which make the sum of the squares on them medial, the rectangle contained by them medial, and further the sum of the squares on them incommensurable with the rectangle contained by them.

Therefore CD is a straight line which produces with a medial area a medial whole. [x. 78]

Q. E. D.

Proposition 108.

If from a rational area a medial area be subtracted, the "side" of the remaining area becomes one of two irrational straight lines, either an apotome or a minor straight line.

For from the rational area *BC* let the medial area *BD* be subtracted;

I say that the "side" of the remainder *EC* becomes one of two irrational straight lines, either an apotome or a minor straight line.

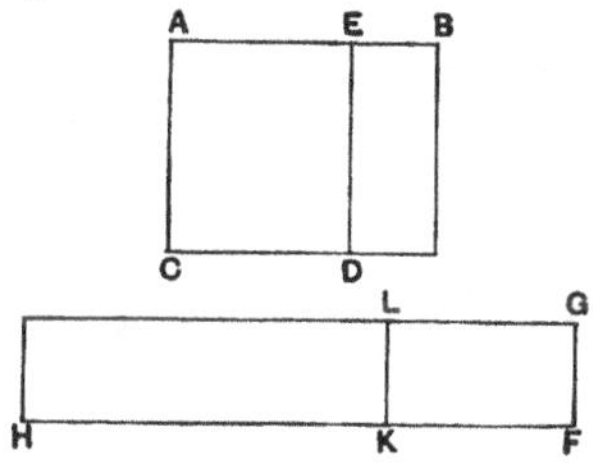

For let a rational straight line *FG* be set out,

to *FG* let there be applied the rectangular parallelogram *GH* equal to *BC*,

and let *GK* equal to *DB* be subtracted;

therefore the remainder *EC* is equal to *LH*.

Since then *BC* is rational, and *BD* medial,

while *BC* is equal to *GH*, and *BD* to *GK*,

therefore *GH* is rational, and *GK* medial.

And they are applied to the rational straight line *FG*;

therefore *FH* is rational and commensurable in length with *FG*, [x. 20]

while *FK* is rational and incommensurable in length with *FG*; [x. 22]

therefore *FH* is incommensurable in length with *FK*. [x. 13]

Therefore *FH*, *FK* are rational straight lines commensurable in square only;

therefore *KH* is an apotome [x. 73], and *KF* the annex to it.

Now the square on *HF* is greater than the square on *FK* by the square on a straight line either commensurable with *HF* or not commensurable.

First, let the square on it be greater by the square on a straight line commensurable with it.

Now the whole *HF* is commensurable in length with the rational straight line *FG* set out;

therefore *KH* is a first apotome. [x. Deff. iii. 1]

But the "side" of the rectangle contained by a rational straight line and a first apotome is an apotome. [x. 91]

Therefore the "side" of LH, that is, of EC, is an apotome.

But, if the square on HF is greater than the square on FK by the square on a straight line incommensurable with HF,

while the whole FH is commensurable in length with the rational straight line FG set out,

KH is a fourth apotome. [x. Deff. III. 4]

But the "side" of the rectangle contained by a rational straight line and a fourth apotome is minor. [x. 94]

Q. E. D.

A rational area being of the form $k\rho^2$, and a medial area of the form $\sqrt{\lambda} \cdot \rho^2$, the problem is to classify

$$\sqrt{k\rho^2 - \sqrt{\lambda} \cdot \rho^2}$$

according to the different possible relations between k, λ.

Suppose that $$\sigma u = k\rho^2,$$ $$\sigma v = \sqrt{\lambda} \cdot \rho^2.$$

Since σu is rational and σv medial,

u is rational and $\frown \sigma$,

while v is rational and $\smile \sigma$.

Therefore $$u \smile v;$$

thus u, v are rational and $\sim$,

whence $(u - v)$ is an apotome.

The possibilities are now as follows.

(1) $\sqrt{u^2 - v^2} \frown u$,

(2) $\sqrt{u^2 - v^2} \smile u$.

In both cases $u \frown \sigma$,

so that $(u - v)$ is either (1) a *first apotome*,

or (2) a *fourth apotome*.

In case (1) $\sqrt{\sigma(u - v)}$ is an *apotome* [x. 91],

but in case (2) $\sqrt{\sigma(u - v)}$ is a *minor* irrational straight line [x. 94].

Proposition 109.

If from a medial area a rational area be subtracted, there arise two other irrational straight lines, either a first apotome of a medial straight line or a straight line which produces with a rational area a medial whole.

For from the medial area BC let the rational area BD be subtracted.

I say that the "side" of the remainder EC becomes one of two irrational straight lines, either a first apotome of a medial straight line or a straight line which produces with a rational area a medial whole.

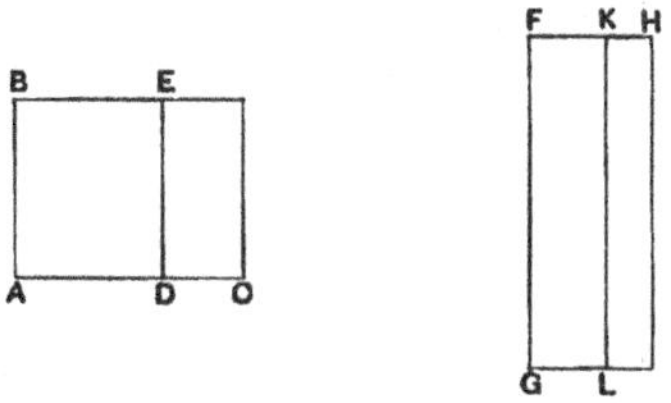

For let a rational straight line FG be set out, and let the areas be similarly applied.

It follows then that FH is rational and incommensurable in length with FG,
while KF is rational and commensurable in length with FG;
therefore FH, FK are rational straight lines commensurable in square only; [x. 13]
therefore KH is an apotome, and FK the annex to it. [x. 73]

Now the square on HF is greater than the square on FK either by the square on a straight line commensurable with HF or by the square on a straight line incommensurable with it.

If then the square on HF is greater than the square on FK by the square on a straight line commensurable with HF, while the annex FK is commensurable in length with the rational straight line FG set out,
KH is a second apotome. [x. Deff. III. 2]

But FG is rational;
so that the "side" of LH, that is, of EC, is a first apotome of a medial straight line. [x. 92]

But, if the square on HF is greater than the square on FK by the square on a straight line incommensurable with HF, while the annex FK is commensurable in length with the rational straight line FG set out,
KH is a fifth apotome; [x. Deff. III. 5]
so that the "side" of EC is a straight line which produces with a rational area a medial whole. [x. 95]

Q. E. D.

In this case we have to classify

$$\sqrt{\sqrt{k}\,.\,\rho^2 - \lambda\rho^2}.$$

Suppose that $\qquad \sigma u = \sqrt{k}\,.\,\rho^2,$

$$\sigma v = \lambda\rho^2.$$

Thus, σu being medial and σv rational,
u is rational and $\smile \sigma$,
while v is rational and $\frown \sigma$.

Thus, as before, u, v are rational and $\underset{-}{\frown}$,
so that $(u - v)$ is an apotome.

Now either

(1) $\sqrt{u^2 - v^2} \frown u$,

or (2) $\sqrt{u^2 - v^2} \smile u$,

while in both cases v is commensurable with σ.

Therefore $(u - v)$ is either (1) a second apotome,
or (2) a fifth apotome,

and hence in case (1) $\sqrt{\sigma(u - v)}$ is the ***first apotome of a medial*** straight line, [X. 92]

and in case (2) $\sqrt{\sigma(u - v)}$ is the "***side*" *of a medial, minus a rational, area.*** [X. 95]

PROPOSITION 110.

If from a medial area there be subtracted a medial area incommensurable with the whole, the two remaining irrational straight lines arise, either a second apotome of a medial straight line or a straight line which produces with a medial area a medial whole.

For, as in the foregoing figures, let there be subtracted from the medial area *BC* the medial area *BD* incommensurable with the whole;

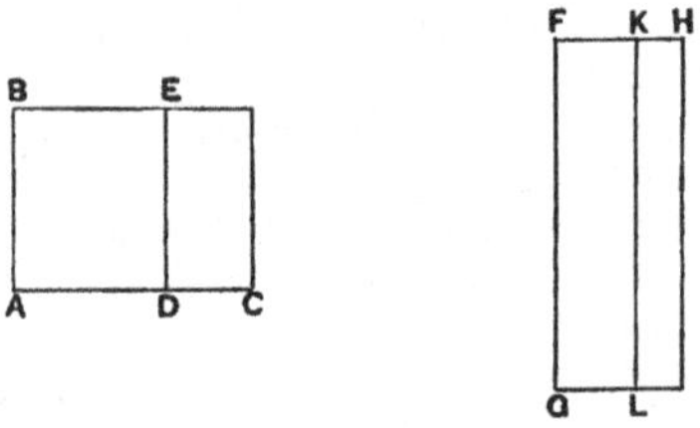

I say that the "side" of *EC* is one of two irrational straight lines, either a second apotome of a medial straight line or a straight line which produces with a medial area a medial whole.

For, since each of the rectangles *BC*, *BD* is medial,
and *BC* is incommensurable with *BD*,
it follows that each of the straight lines *FH*, *FK* will be rational and incommensurable in length with *FG*. [x. 22]

And, since *BC* is incommensurable with *BD*,
that is, *GH* with *GK*,
HF is also incommensurable with *FK*; [vi. 1, x. 11]
therefore *FH*, *FK* are rational straight lines commensurable in square only;
therefore *KH* is an apotome. [x. 73]

If then the square on *FH* is greater than the square on *FK* by the square on a straight line commensurable with *FH*,
while neither of the straight lines *FH*, *FK* is commensurable in length with the rational straight line *FG* set out,
KH is a third apotome. [x. Deff. iii. 3]

But *KL* is rational,
and the rectangle contained by a rational straight line and a third apotome is irrational,
and the "side" of it is irrational, and is called a second apotome of a medial straight line; [x. 93]
so that the "side" of *LH*, that is, of *EC*, is a second apotome of a medial straight line.

But, if the square on *FH* is greater than the square on *FK* by the square on a straight line incommensurable with *FH*,
while neither of the straight lines *HF*, *FK* is commensurable in length with *FG*,
KH is a sixth apotome. [x. Deff. iii. 6]

But the "side" of the rectangle contained by a rational straight line and a sixth apotome is a straight line which produces with a medial area a medial whole. [x. 96]

Therefore the "side" of *LH*, that is, of *EC*, is a straight line which produces with a medial area a medial whole.

Q. E. D.

We have to classify $\sqrt{\sqrt{k}\cdot\rho^2 - \sqrt{\lambda}\cdot\rho^2}$,
where $\sqrt{k}\cdot\rho^2$ is incommensurable with $\sqrt{\lambda}\cdot\rho^2$.

Put $$\sigma u = \sqrt{k}\cdot\rho^2,$$
$$\sigma v = \sqrt{\lambda}\cdot\rho^2.$$

Then u is rational and $\smile \sigma$,
v is rational and $\smile \sigma$,
and $u \smile v$.

Therefore u, v are rational and $\frown\!\!-$,
so that $(u - v)$ is an apotome.

Now either

(1) $\sqrt{u^2 - v^2} \frown u$,

or (2) $\sqrt{u^2 - v^2} \smile u$,

while in both cases both u and v are $\smile \sigma$.

In case (1) $(u - v)$ is a *third apotome*,
and in case (2) $(u - v)$ is a *sixth apotome*,
so that $\sqrt{\sigma (u - v)}$ is either (1) a *second apotome of a medial* straight line [x. 93], or (2) a "*side*" *of the difference between two medial areas* [x. 96].

PROPOSITION 111.

The apotome is not the same with the binomial straight line.

Let AB be an apotome;
I say that AB is not the same with the binomial straight line.

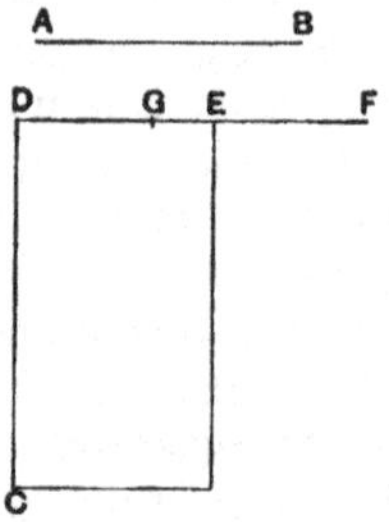

For, if possible, let it be so;
let a rational straight line DC be set out, and to CD let there be applied the rectangle CE equal to the square on AB and producing DE as breadth.

Then, since AB is an apotome,
DE is a first apotome. [x. 97]

Let EF be the annex to it;
therefore DF, FE are rational straight lines commensurable in square only,
the square on DF is greater than the square on FE by the square on a straight line commensurable with DF,
and DF is commensurable in length with the rational straight line DC set out. [x. Deff. III. 1]

Again, since AB is binomial,
therefore DE is a first binomial straight line. [x. 60]

Let it be divided into its terms at G,
and let DG be the greater term;
therefore DG, GE are rational straight lines commensurable in square only,

the square on *DG* is greater than the square on *GE* by the square on a straight line commensurable with *DG*, and the greater term *DG* is commensurable in length with the rational straight line *DC* set out. [x. Deff. ii. 1]

Therefore *DF* is also commensurable in length with *DG*; [x. 12]

therefore the remainder *GF* is also commensurable in length with *DF*. [x. 15]

But *DF* is incommensurable in length with *EF*;

therefore *FG* is also incommensurable in length with *EF*. [x. 13]

Therefore *GF*, *FE* are rational straight lines commensurable in square only;

therefore *EG* is an apotome. [x. 73]

But it is also rational:

which is impossible.

Therefore the apotome is not the same with the binomial straight line.

Q. E. D.

This proposition proves the equivalent of the fact that

$\sqrt{x} + \sqrt{y}$ cannot be equal to $\sqrt{x'} - \sqrt{y'}$, and

$x + \sqrt{y}$ cannot be equal to $x' - \sqrt{y'}$.

We should prove these results by squaring the respective expressions; and Euclid's procedure corresponds to this exactly.

He has to prove that

$$\rho + \sqrt{k} \cdot \rho \text{ cannot be equal to } \rho' - \sqrt{\lambda} \cdot \rho'.$$

For, if possible, let this be so.

Take the straight lines $\dfrac{(\rho + \sqrt{k} \cdot \rho)^2}{\sigma}$, $\dfrac{(\rho' - \sqrt{\lambda} \cdot \rho')^2}{\sigma}$;

these must be equal, and therefore

$$\frac{\rho^2}{\sigma}(1 + k + 2\sqrt{k}) = \frac{\rho'^2}{\sigma}(1 + \lambda - 2\sqrt{\lambda}) \quad \ldots\ldots\ldots\ldots(1).$$

Now $\dfrac{\rho^2}{\sigma}(1 + k)$, $\dfrac{\rho'^2}{\sigma}(1 + \lambda)$ are rational and $\frown$;

therefore
$$\left\{\frac{\rho'^2}{\sigma}(1 + \lambda) - \frac{\rho^2}{\sigma}(1 + k)\right\} \frown \frac{\rho'^2}{\sigma}(1 + \lambda)$$
$$\smile \frac{\rho'^2}{\sigma} \cdot 2\sqrt{\lambda}.$$

And, since both sides are rational, it follows that

$$\left\{\frac{\rho'^2}{\sigma}(1 + \lambda) - \frac{\rho^2}{\sigma}(1 + k)\right\} - \frac{\rho'^2}{\sigma} \cdot 2\sqrt{\lambda} \text{ is an } \textit{apotome}.$$

But, by (1), this expression is equal to $\frac{\rho^2}{\sigma} \cdot 2\sqrt{k}$, which is *rational.*

Hence an *apotome*, which is *irrational*, is also *rational*:
which is impossible.

This proposition is the connecting link which enables Euclid to prove that *all* the compound irrationals with positive signs above discussed are different from *all* the corresponding compound irrationals with negative signs, while the two sets are all different from one another and from the *medial* straight line. The recapitulation following makes this clear.

The apotome and the irrational straight lines following it are neither the same with the medial straight line nor with one another.

For the square on a medial straight line, if applied to a rational straight line, produces as breadth a straight line rational and incommensurable in length with that to which it is applied, [x. 22]

while the square on an apotome, if applied to a rational straight line, produces as breadth a first apotome, [x. 97]

the square on a first apotome of a medial straight line, if applied to a rational straight line, produces as breadth a second apotome, [x. 98]

the square on a second apotome of a medial straight line, if applied to a rational straight line, produces as breadth a third apotome, [x. 99]

the square on a minor straight line, if applied to a rational straight line, produces as breadth a fourth apotome, [x. 100]

the square on the straight line which produces with a rational area a medial whole, if applied to a rational straight line, produces as breadth a fifth apotome, [x. 101]

and the square on the straight line which produces with a medial area a medial whole, if applied to a rational straight line, produces as breadth a sixth apotome. [x. 102]

Since then the said breadths differ from the first and from one another, from the first because it is rational, and from one another since they are not the same in order,

it is clear that the irrational straight lines themselves also differ from one another.

And, since the apotome has been proved not to be the same as the binomial straight line, [x. 111]

but, if applied to a rational straight line, the straight lines following the apotome produce, as breadths, each according to its own order, apotomes, and those following the binomial straight line themselves also, according to their order, produce the binomials as breadths,

therefore those following the apotome are different, and those following the binomial straight line are different, so that there are, in order, thirteen irrational straight lines in all,

Medial,
Binomial,
First bimedial,
Second bimedial,
Major,
"Side" of a rational plus a medial area,
"Side" of the sum of two medial areas,
Apotome,
First apotome of a medial straight line,
Second apotome of a medial straight line,
Minor,
Producing with a rational area a medial whole,
Producing with a medial area a medial whole.

Proposition 112.

The square on a rational straight line applied to the binomial straight line produces as breadth an apotome the terms of which are commensurable with the terms of the binomial and moreover in the same ratio; and further the apotome so arising will have the same order as the binomial straight line.

Let A be a rational straight line,
let BC be a binomial, and let DC be its greater term;
let the rectangle BC, EF be equal to the square on A;

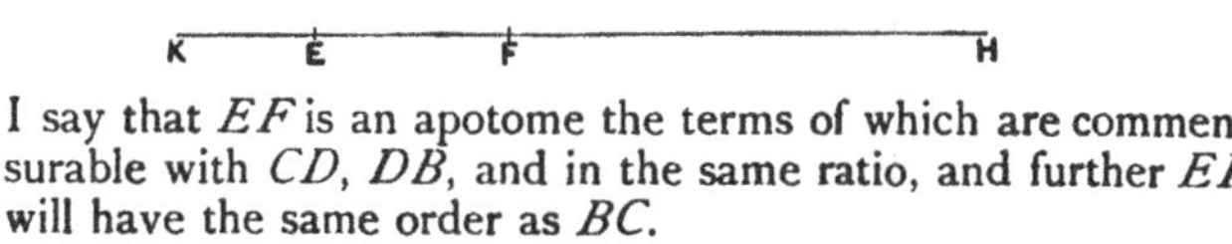

I say that EF is an apotome the terms of which are commensurable with CD, DB, and in the same ratio, and further EF will have the same order as BC.

For again let the rectangle *BD*, *G* be equal to the square on *A*.

Since then the rectangle *BC*, *EF* is equal to the rectangle *BD*, *G*,

therefore, as *CB* is to *BD*, so is *G* to *EF*. [VI. 16]

But *CB* is greater than *BD*;

therefore *G* is also greater than *EF*. [V. 16, V. 14]

Let *EH* be equal to *G*;

therefore, as *CB* is to *BD*, so is *HE* to *EF*;

therefore, *separando*, as *CD* is to *BD*, so is *HF* to *FE*. [V. 17]

Let it be contrived that, as *HF* is to *FE*, so is *FK* to *KE*;

therefore also the whole *HK* is to the whole *KF* as *FK* is to *KE*;

for, as one of the antecedents is to one of the consequents, so are all the antecedents to all the consequents. [V. 12]

But, as *FK* is to *KE*, so is *CD* to *DB*; [V. 11]

therefore also, as *HK* is to *KF*, so is *CD* to *DB*. [*id.*]

But the square on *CD* is commensurable with the square on *DB*; [X. 36]

therefore the square on *HK* is also commensurable with the square on *KF*. [VI. 22, X. 11]

And, as the square on *HK* is to the square on *KF*, so is *HK* to *KE*, since the three straight lines *HK*, *KF*, *KE* are proportional. [V. Def. 9]

Therefore *HK* is commensurable in length with *KE*,

so that *HE* is also commensurable in length with *EK*. [X. 15]

Now, since the square on *A* is equal to the rectangle *EH*, *BD*,

while the square on *A* is rational,

therefore the rectangle *EH*, *BD* is also rational.

And it is applied to the rational straight line *BD*;

therefore *EH* is rational and commensurable in length with *BD*; [X. 20]

so that *EK*, being commensurable with it, is also rational and commensurable in length with *BD*.

Since, then, as CD is to DB, so is FK to KE,

while CD, DB are straight lines commensurable in square only,

therefore FK, KE are also commensurable in square only. [x. 11]

But KE is rational;

therefore FK is also rational.

Therefore FK, KE are rational straight lines commensurable in square only;

therefore EF is an apotome. [x. 73]

Now the square on CD is greater than the square on DB either by the square on a straight line commensurable with CD or by the square on a straight line incommensurable with it.

If then the square on CD is greater than the square on DB by the square on a straight line commensurable with CD, the square on FK is also greater than the square on KE by the square on a straight line commensurable with FK. [x. 14]

And, if CD is commensurable in length with the rational straight line set out,

so also is FK; [x. 11, 12]

if BD is so commensurable,

so also is KE; [x. 12]

but, if neither of the straight lines CD, DB is so commensurable,

neither of the straight lines FK, KE is so.

But, if the square on CD is greater than the square on DB by the square on a straight line incommensurable with CD,

the square on FK is also greater than the square on KE by the square on a straight line incommensurable with FK. [x. 14]

And, if CD is commensurable with the rational straight line set out,

so also is FK;

if BD is so commensurable,

so also is KE;

but, if neither of the straight lines *CD*, *DB* is so commensurable,

neither of the straight lines *FK*, *KE* is so;

so that *FE* is an apotome, the terms of which *FK*, *KE* are commensurable with the terms *CD*, *DB* of the binomial straight line and in the same ratio, and it has the same order as *BC*.

Q. E. D.

Heiberg considers that this proposition and the succeeding ones are interpolated, though the interpolation must have taken place before Theon's time. His argument is that X. 112—115 are nowhere used, but that X. 111 rounds off the complete discussion of the 13 irrationals (as indicated in the recapitulation), thereby giving what was necessary for use in connexion with the investigation of the five regular solids. For besides X. 73 (used in XIII. 6, 11) X. 94 and 97 are used in XIII. 11, 6 respectively; and Euclid could not have stopped at X. 97 without leaving the discussion of irrationals imperfect, for X. 98—102 are closely connected with X. 97, and X. 103—111 add, as it were, the coping-stone to the whole doctrine. On the other hand, X. 112—115 are not connected with the rest of the treatise on the 13 irrationals and are not used in the stereometric books. They are rather the germ of a new study and a more abstruse investigation of irrationals *in themselves*. Prop. 115 in particular extends the number of the different kinds of irrationals. As however X. 112—115 are old and serviceable theorems, Heiberg thinks that, though Euclid did not give them, they may have been taken from Apollonius.

I will only point out what seems to me open to doubt in the above, namely that X. 112—114 (excluding 115) are not connected with the rest of the exposition of the 13 irrationals. It seems to me that they *are* so connected. X. 111 has shown us that a *binomial* straight line cannot also be an *apotome*. But X. 112—114 show us *how either of them can be used to rationalise the other*, thus giving what is surely an important relation between them.

X. 112 is the equivalent of rationalising the denominators of the fractions

$$\frac{c^2}{\sqrt{A}+\sqrt{B}},\quad \frac{c^2}{a+\sqrt{B}},$$

by multiplying numerator and denominator by $\sqrt{A}-\sqrt{B}$ and $a-\sqrt{B}$ respectively.

Euclid proves that $\dfrac{\sigma^2}{\rho+\sqrt{k}\,.\,\rho}=\lambda\rho-\sqrt{k}\,.\,\lambda\rho\ (k<1)$, and his method enables us to see that $\lambda=\sigma^2/(\rho^2-k\rho^2)$.

The proof is a remarkable instance of the dexterity of the Greeks in using geometry as the equivalent of our algebra. Like so many proofs in Archimedes and Apollonius, it leaves us completely in the dark as to how it was evolved. That the Greeks must have had some analytical method which suggested the steps of such proofs seems certain; but *what* it was must remain apparently an insoluble mystery.

I will reproduce by means of algebraical symbols the exact course of Euclid's proof.

He has to prove that $\dfrac{\sigma^2}{\rho+\sqrt{k}\,.\,\rho}$ is an apotome related in a certain way to the binomial straight line $\rho + \sqrt{k}\,.\,\rho$. If u be the straight line required, $(u + w) - w$ is shown to be an apotome of the kind described, where w is determined in the following manner.

We have $$(\rho + \sqrt{k}\,.\,\rho)\,u = \sigma^2 = \sqrt{k}\,.\,\rho\,.\,x, \text{ say,}$$
whence $$x > u.$$
Let $$x = u + v.$$ ....(1).

Then $$(\rho + \sqrt{k}\,.\,\rho) : \sqrt{k}\,.\,\rho = (u + v) : u,$$
and hence $$\rho : \sqrt{k}\,.\,\rho = v : u \text{ ....(2).}$$

Let w be taken such that
$$v : u = (u + w) : w \text{ ....(3).}$$

Thus $$v : u = (u + v + w) : (u + w) \text{ ....(4),}$$
and therefore $$\rho : \sqrt{k}\,.\,\rho = (u + v + w) : (u + w).$$

From the last proportion,
$$(u + v + w)^2 \frown (u + w)^2,$$
and, from the two preceding, $(u + w)$ is a mean proportional between $(u + v + w)$, w, so that
$$(u + v + w)^2 : (u + w)^2 = (u + v + w) : w.$$

Therefore $$(u + v + w) \frown w,$$
whence $$(u + v) \frown w.$$

Now $$\sqrt{k}\,.\,\rho\,(u + v) = \sigma^2, \text{ which is rational;}$$
therefore $$(u + v) \text{ is rational and } \frown \sqrt{k}\,.\,\rho;$$
hence $$w \text{ is also rational and } \frown \sqrt{k}\,.\,\rho \text{ ....(5).}$$

Next, by (2), (3), since ρ, $\sqrt{k}\,.\,\rho$ are $\smile\!\!-$,
$$(u + w) \smile\!\!- w,$$
and w is rational;
therefore $$(u + w) \text{ is rational,}$$
and $$(u + w),\ w \text{ are rational and } \smile\!\!-.$$

Hence $$(u + w) - w \text{ is an } \textit{apotome}.$$

Now either $$\text{(I)} \quad \sqrt{\rho^2 - k\rho^2} \frown \rho,$$
or $$\text{(II)} \quad \sqrt{\rho^2 - k\rho^2} \smile \rho.$$

In case (I) $$\sqrt{(u + w)^2 - w^2} \frown (u + w),$$ [(2), (3) and X. 14]
and in case (II) $$\sqrt{(u + w)^2 - w^2} \smile (u + w).$$ [*id.*]

Then, since [(5)] $$w \frown \sqrt{k}\,.\,\rho,$$
by X. 11 and (2), (3), $$(u + w) \frown \rho \text{ ....(6).}$$

[This step is omitted in Euclid, but the result is assumed.]

If therefore $\rho \frown \sigma$, $(u + w) \frown \sigma$;
if $\sqrt{k}\,.\,\rho \frown \sigma$, $w \frown \sigma$; [(5)]
and, if neither ρ nor $\sqrt{k}\,.\,\rho$ is $\frown \sigma$, neither $(u + w)$ nor w will be $\frown \sigma$.

Thus the order of the apotome $(u + w) - w$ is the same as that of the binomial straight line $\rho + \sqrt{k}\,.\,\rho$; while [(2), (3)] the terms are proportional and [(5), (6)] commensurable respectively.

We find $(u+w)$, w algebraically thus.

By (1), $$u = \frac{\sigma^2}{\rho + \sqrt{k} \cdot \rho};$$

and, by (2), (3), $$\frac{u+w}{w} = \frac{\rho}{\sqrt{k} \cdot \rho},$$

whence $$w = \frac{u \cdot \sqrt{k} \cdot \rho}{\rho - \sqrt{k} \cdot \rho}$$
$$= \frac{\sigma^2 \cdot \sqrt{k} \cdot \rho}{\rho^2 - k\rho^2}.$$

Thus $$u + w = w \cdot \frac{1}{\sqrt{k}} = \frac{\sigma^2 \cdot \rho}{\rho^2 - k\rho^2}.$$

Therefore $$(u+w) - w = \sigma^2 \cdot \frac{\rho - \sqrt{k} \cdot \rho}{\rho^2 - k\rho^2}.$$

PROPOSITION 113.

The square on a rational straight line, if applied to an apotome, produces as breadth the binomial straight line the terms of which are commensurable with the terms of the apotome and in the same ratio; and further the binomial so arising has the same order as the apotome.

Let A be a rational straight line and BD an apotome, and let the rectangle BD, KH be equal to the square on A, so that the square on the rational straight line A when applied to the apotome BD produces KH as breadth;

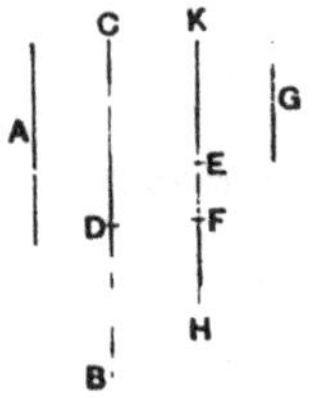

I say that KH is a binomial straight line the terms of which are commensurable with the terms of BD and in the same ratio: and further KH has the same order as BD.

For let DC be the annex to BD;

therefore BC, CD are rational straight lines commensurable in square only. [X. 73]

Let the rectangle BC, G be also equal to the square on A.

But the square on A is rational:

therefore the rectangle BC, G is also rational.

And it has been applied to the rational straight line BC;

therefore G is rational and commensurable in length with BC. [X. 20]

Since now the rectangle *BC*, *G* is equal to the rectangle *BD*, *KH*,

therefore, proportionally, as *CB* is to *BD*, so is *KH* to *G*. [VI. 16]

But *BC* is greater than *BD*;

therefore *KH* is also greater than *G*. [v. 16, v. 14]

Let *KE* be made equal to *G*;

therefore *KE* is commensurable in length with *BC*.

And since, as *CB* is to *BD*, so is *HK* to *KE*,

therefore, *convertendo*, as *BC* is to *CD*, so is *KH* to *HE*. [v. 19, Por.]

Let it be contrived that, as *KH* is to *HE*, so is *HF* to *FE*;

therefore also the remainder *KF* is to *FH* as *KH* is to *HE*, that is, as *BC* is to *CD*. [v. 19]

But *BC*, *CD* are commensurable in square only;

therefore *KF*, *FH* are also commensurable in square only. [x. 11]

And since, as *KH* is to *HE*, so is *KF* to *FH*,

while, as *KH* is to *HE*, so is *HF* to *FE*,

therefore also, as *KF* is to *FH*, so is *HF* to *FE*, [v. 11]

so that also, as the first is to the third, so is the square on the first to the square on the second; [v. Def. 9]

therefore also, as *KF* is to *FE*, so is the square on *KF* to the square on *FH*.

But the square on *KF* is commensurable with the square on *FH*,

for *KF*, *FH* are commensurable in square;

therefore *KF* is also commensurable in length with *FE*, [x. 11]

so that *KF* is also commensurable in length with *KE*. [x. 15]

But *KE* is rational and commensurable in length with *BC*;

therefore *KF* is also rational and commensurable in length with *BC*. [x. 12]

And, since, as *BC* is to *CD*, so is *KF* to *FH*,

alternately, as *BC* is to *KF*, so is *DC* to *FH*. [v. 16]

But *BC* is commensurable with *KF*;

therefore *FH* is also commensurable in length with *CD*. [x. 11]

But BC, CD are rational straight lines commensurable in square only;
therefore KF, FH are also rational straight lines [x. Def. 3] commensurable in square only;
therefore KH is binomial. [x. 36]

If now the square on BC is greater than the square on CD by the square on a straight line commensurable with BC,
the square on KF will also be greater than the square on FH by the square on a straight line commensurable with KF. [x. 14]

And, if BC is commensurable in length with the rational straight line set out,
so also is KF;
if CD is commensurable in length with the rational straight line set out,
so also is FH,
but, if neither of the straight lines BC, CD,
then neither of the straight lines KF, FH.

But, if the square on BC is greater than the square on CD by the square on a straight line incommensurable with BC,
the square on KF is also greater than the square on FH by the square on a straight line incommensurable with KF. [x. 14]

And, if BC is commensurable with the rational straight line set out,
so also is KF;
if CD is so commensurable,
so also is FH;
but, if neither of the straight lines BC, CD,
then neither of the straight lines KF, FH.

Therefore KH is a binomial straight line, the terms of which KF, FH are commensurable with the terms BC, CD of the apotome and in the same ratio,
and further KH has the same order as BD.

Q. E. D.

This proposition, which is companion to the preceding, gives us the equivalent of the rationalisation of the denominator of

$$\frac{c^2}{\sqrt{A}-\sqrt{B}} \text{ or } \frac{c^2}{a \sim \sqrt{B}}.$$

Euclid (or the writer) proves that

$$\frac{\sigma^2}{\rho - \sqrt{k}\,.\,\rho} = \lambda\rho + \lambda\sqrt{k}\,.\,\rho, \qquad (k < 1)$$

and his method enables us to see that $\lambda = \sigma^2/(\rho^2 - k\rho^2)$.

Let $$\frac{\sigma^2}{\rho - \sqrt{k}\,.\,\rho} = u;$$

and it is proved that u is the binomial straight line $(u - w) + w$, where w is determined as shown below.

$$u(\rho - \sqrt{k}\,.\,\rho) = \sigma^2 = \rho x, \text{ say,}$$

whence $\rho : (\rho - \sqrt{k}\,.\,\rho) = u : x$ ....(1),

so that $x < u$.

Let then $x = u - v$.

Since $(u - v)\rho = \sigma^2$, a rational area,

$(u - v)$ is rational and $\frown \rho$ ....(2).

And [(1)] $\rho : (\rho - \sqrt{k}\,.\,\rho) = u : (u - v)$,

so that, *convertendo*, $\rho : \sqrt{k}\,.\,\rho = u : v$.

Suppose that $u : v = w : (v - w)$,

so that [V. 19] $(u - w) : w = u : v = w : (v - w)$.

Thus, w being a mean proportional between $(u - w)$, $(v - w)$,

$$(u - w)^2 : w^2 = (u - w) : (v - w).$$

But $(u - w)^2 : w^2 = u^2 : v^2$

$= \rho^2 : k\rho^2$ ....(3),

so that $(u - w)^2 \frown w^2$.

Therefore $(u - w) \frown (v - w)$

$\frown \{(u - w) - (v - w)\}$

$\frown (u - v)$.

Therefore [(2)] $(u - w)$ is rational and $\frown \rho$ ....(4).

And, since $\rho : \sqrt{k}\,.\,\rho = (u - w) : w$,

w is rational and $\frown \sqrt{k}\,.\,\rho$ ....(5).

Hence [(4), (5)] $(u - w)$, w are rational and $\frown\!\!\!-$,

so that $(u - w) + w$ is a *binomial* straight line.

Now either (I) $\sqrt{\rho^2 - k\rho^2} \frown \rho$,

or (II) $\sqrt{\rho^2 - k\rho^2} \smile \rho$.

In case (I) $\sqrt{(u - w)^2 - w^2} \frown (u - w)$,

and in case (II) $\sqrt{(u - w)^2 - w^2} \smile (u - w)$. [(3) and X. 14]

And, if $\rho \frown \sigma$, $(u - w) \frown \sigma$; [(4)]

if $\sqrt{k}\,.\,\rho \frown \sigma$, $w \frown \sigma$; [5]

while, if neither ρ nor $\sqrt{k}\,.\,\rho$ is $\frown \sigma$, neither $(u - w)$ nor w is $\frown \sigma$.

Hence $(u - w) + w$ is a binomial straight line of the same order as the apotome $\rho - \sqrt{k}\,.\,\rho$, its terms are proportional to those of the apotome [(3)], and commensurable with them respectively [(4), (5)].

To find $(u - w)$, w algebraically we have

$$u = \frac{\sigma^2}{\rho - \sqrt{k}\,.\,\rho},$$

$$\frac{u - w}{w} = \frac{\rho}{\sqrt{k}\,.\,\rho}.$$

From the latter
$$w = \frac{u\,.\,\sqrt{k}\,.\,\rho}{\rho + \sqrt{k}\,.\,\rho}$$
$$= \frac{\sigma^2\,.\,\sqrt{k}\,.\,\rho}{\rho^2 - k\rho^2}.$$

Thus
$$u - w = w\,.\,\frac{1}{\sqrt{k}} = \frac{\sigma^2\rho}{\rho^2 - k\rho^2}.$$

Therefore
$$(u - w) + w = \sigma^2\,.\,\frac{\rho + \sqrt{k}\,.\,\rho}{\rho^2 - k\rho^2}.$$

PROPOSITION 114.

If an area be contained by an apotome and the binomial straight line the terms of which are commensurable with the terms of the apotome and in the same ratio, the "side" of the area is rational.

For let an area, the rectangle *AB*, *CD*, be contained by the apotome *AB* and the binomial straight line *CD*,

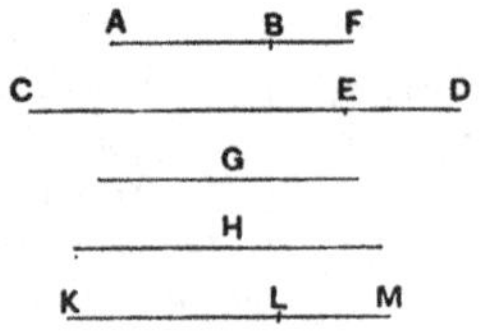

and let *CE* be the greater term of the latter;

let the terms *CE*, *ED* of the binomial straight line be commensurable with the terms *AF*, *FB* of the apotome and in the same ratio;

and let the "side" of the rectangle *AB*, *CD* be *G*;

I say that *G* is rational.

For let a rational straight line *H* be set out,

and to *CD* let there be applied a rectangle equal to the square on *H* and producing *KL* as breadth.

Therefore *KL* is an apotome.

Let its terms be *KM*, *ML* commensurable with the terms *CE*, *ED* of the binomial straight line and in the same ratio.

[X. 112]

But CE, ED are also commensurable with AF, FB and in the same ratio;
therefore, as AF is to FB, so is KM to ML.

Therefore, alternately, as AF is to KM, so is BF to LM; therefore also the remainder AB is to the remainder KL as AF is to KM. [v. 19]

But AF is commensurable with KM; [x. 12]
therefore AB is also commensurable with KL. [x. 11]

And, as AB is to KL, so is the rectangle CD, AB to the rectangle CD, KL; [vi. 1]
therefore the rectangle CD, AB is also commensurable with the rectangle CD, KL. [x. 11]

But the rectangle CD, KL is equal to the square on H;
therefore the rectangle CD, AB is commensurable with the square on H.

But the square on G is equal to the rectangle CD, AB;
therefore the square on G is commensurable with the square on H.

But the square on H is rational;
therefore the square on G is also rational;
therefore G is rational.

And it is the "side" of the rectangle CD, AB.

Therefore etc.

Porism. And it is made manifest to us by this also that it is possible for a rational area to be contained by irrational straight lines.

Q. E. D.

This theorem is equivalent to the proof of the fact that

$$\sqrt{(\sqrt{A} - \sqrt{B})(\lambda\sqrt{A} + \lambda\sqrt{B})} = \sqrt{\lambda(A - B)},$$

and

$$\sqrt{(a \sim \sqrt{B})(\lambda a + \lambda\sqrt{B})} = \sqrt{\lambda(a^2 \sim B)}.$$

The result of the theorem x. 112 is used for the purpose thus.

We have to prove that

$$\sqrt{(\rho - \sqrt{k}\,.\,\rho)(\lambda\rho + \lambda\sqrt{k}\,.\,\rho)}$$

is *rational*.

By x. 112 we have, if σ is a rational straight line,

$$\frac{\sigma^2}{\lambda\rho + \lambda\sqrt{k}\,.\,\rho} = \lambda'\rho - \lambda'\sqrt{k}\,.\,\rho \quad \text{.....................(1).}$$

Now $\rho : \lambda'\rho = \sqrt{k} . \rho : \lambda' \sqrt{k} . \rho = (\rho - \sqrt{k} . \rho) : (\lambda'\rho - \lambda' \sqrt{k} . \rho)$,

so that $(\rho - \sqrt{k} . \rho) \frown (\lambda'\rho - \lambda' \sqrt{k} . \rho)$.

Multiplying each by $(\lambda\rho + \lambda \sqrt{k} . \rho)$, we have

$$(\rho - \sqrt{k} . \rho)(\lambda\rho + \lambda \sqrt{k} . \rho) \frown (\lambda\rho + \lambda \sqrt{k} . \rho)(\lambda'\rho - \lambda' \sqrt{k} . \rho)$$

$$\frown \sigma^2, \quad \text{by (1).}$$

That is, $(\rho - \sqrt{k} . \rho)(\lambda\rho + \lambda \sqrt{k} . \rho)$ is a rational area,

and therefore $\sqrt{(\rho - \sqrt{k} . \rho)(\lambda\rho + \lambda \sqrt{k} . \rho)}$ is rational.

PROPOSITION 115.

From a medial straight line there arise irrational straight lines infinite in number, and none of them is the same as any of the preceding.

Let A be a medial straight line;

I say that from A there arise irrational straight lines infinite in number, and none of them is the same as any of the preceding.

Let a rational straight line B be set out,

and let the square on C be equal to the rectangle B, A;

therefore C is irrational; [x. Def. 4]

for that which is contained by an irrational and a rational straight line is irrational. [deduction from x. 20]

And it is not the same with any of the preceding;

for the square on none of the preceding, if applied to a rational straight line produces as breadth a medial straight line.

Again, let the square on D be equal to the rectangle B, C;

therefore the square on D is irrational. [deduction from x. 20]

Therefore D is irrational; [x. Def. 4]

and it is not the same with any of the preceding, for the square on none of the preceding, if applied to a rational straight line, produces C as breadth.

Similarly, if this arrangement proceeds *ad infinitum*, it is manifest that from the medial straight line there arise irrational straight lines infinite in number, and none is the same with any of the preceding.

Q. E. D.

Heiberg is clearly right in holding that this proposition, at all events, is alien to the general scope of Book x, and is therefore probably an interpolation, made however before Theon's time. It is of the same character as a *scholium* at the end of the Book, which is (along with the interpolated proposition proving, in two ways, the incommensurability of the diagonal of a square with its side) relegated by August as well as Heiberg to an Appendix.

The proposition amounts to this.

The straight line $k^{\frac{1}{4}}\rho$ being medial, if σ be a rational straight line, $\sqrt{k^{\frac{1}{4}}\rho\sigma}$ is a new irrational straight line. So is the mean proportional between this and another rational straight line σ', and so on indefinitely.

ANCIENT EXTENSIONS OF THE THEORY OF BOOK X.

From the hints given by the author of the commentary found in Arabic by Woepcke (cf. pp. 3—4 above) it would seem probable that Apollonius' extensions of the theory of irrationals took two directions: (1) generalising the *medial* straight line of Euclid, and (2) forming compound irrationals by the addition and subtraction of more than two terms of the sort composing the binomials, apotomes, etc. The commentator writes (Woepcke's article, pp. 694 sqq.):

"It is also necessary that we should know that, not only when we join together two straight lines rational and commensurable in square do we obtain the binomial straight line, but three or four lines produce in an analogous manner the same thing. In the first case, we obtain the trinomial straight line, since the whole line is irrational; and in the second case we obtain the quadrinomial, and so on *ad infinitum*. The proof of the (irrationality of the) line composed of three lines rational and commensurable in square is exactly the same as the proof relating to the combination of two lines.

"But we must start afresh and remark that not only can we take one sole medial line between two lines commensurable in square, but we can take three or four of them and so on *ad infinitum*, since we can take, between any two given straight lines, as many lines as we wish in continued proportion.

"Likewise, in the lines formed by addition not only can we construct the binomial straight line, but we can also construct the trinomial, as well as the first and second trimedial; and, further, the line composed of three straight lines incommensurable in square and such that the one of them gives with each of the two others a sum of squares (which is) rational, while the rectangle contained by the two lines is medial, so that there results a *major* (irrational) composed of three lines.

"And, in an analogous manner, we obtain the straight line which is the 'side' of a rational plus a medial area, composed of three straight lines, and, likewise, that which is the 'side' of (the sum of) two medials."

The generalisation of the medial is apparently after the following manner. Let x, y be two straight lines rational and commensurable in square only and suppose that m means are interposed, so that

$$x : x_1 = x_1 : x_2 = x_2 : x_3 = \ldots = x_{m-1} : x_m = x_m : y.$$

We easily derive herefrom
$$\frac{x}{x_r} = \left(\frac{x}{x_1}\right)^r,$$
$$\frac{x}{y} = \left(\frac{x}{x_1}\right)^{m+1},$$

and hence $$x_1^r = x_r \,.\, x^{r-1},$$
$$x_1^{m+1} = y \,.\, x^m,$$
so that $$(x_r \,.\, x^{r-1})^{m+1} = (y \,.\, x^m)^r,$$
and therefore $$x_r^{m+1} = x^{m-r+1} \,.\, y^r,$$
or $$x_r = (x^{m-r+1} \,.\, y^r)^{\frac{1}{m+1}},$$
which is the generalised medial.

We now pass to the trinomial etc., with the commentator's further remarks about them.

(1) *The trinomial.* "Suppose three rational straight lines commensurable in square only. The line composed of two of these lines, that is, the binomial straight line, is irrational, and, in consequence, the area contained by this line and the remaining line is irrational, and, likewise, the double of the area contained by these two lines will be irrational. Thus the square on the whole line composed of three lines is irrational and consequently the line is irrational, and it is called a trinomial straight line."

It is easy to see that this "proof" is not conclusive as stated. Nor does Woepcke seem to show how the proposition can be proved on Euclidean lines. But I think it would be somewhat as follows.

Suppose x, y, z to be rational and $\frown$.

Then x^2, y^2, z^2 are rational, and $2yz$, $2zx$, $2xy$ are all medial.

First, $(2yz + 2zx + 2xy)$ cannot be rational.

For suppose this sum equal to a rational area, say σ^2.

Since $$2yz + 2zx + 2xy = \sigma^2,$$
$$2zx + 2xy = \sigma^2 - 2yz,$$
or the sum of two medial areas incommensurable with one another is equal to the difference between a rational area and a medial area.

But the "side" of the sum of the two medial areas must [x. 72] be one of two irrationals with a positive sign; and the "side" of the difference between a rational area and a medial area must [x. 108] be one of two irrationals with a negative sign.

And the first "side" cannot be the same as the second [x. 111 and explanation following].

Therefore $$2zx + 2xy \neq \sigma^2 - 2yz,$$
and $2yz + 2zx + 2xy$ is consequently *irrational.*

Therefore $$(x^2 + y^2 + z^2) \smile (2yz + 2zx + 2xy),$$
whence $$(x + y + z)^2 \smile (x^2 + y^2 + z^2),$$
so that $(x + y + z)^2$, and therefore also $(x + y + z)$, is irrational.

The commentator goes on:

"And, if we have four lines commensurable in square, as we have said, the procedure will be exactly the same; and we shall treat the succeeding lines in an analogous manner."

Without speculating further as to how the extension was made to the *quadrinomial* etc., we may suppose with Woepcke that Apollonius probably investigated the multinomial
$$\rho + \sqrt{\kappa} \,.\, \rho + \sqrt{\lambda} \,.\, \rho + \sqrt{\mu} \,.\, \rho + \ldots$$

(2) The *first trimedial* straight line.

The commentator here says: "Suppose we have three medial lines commensurable in square [only], one of which contains with each of the two others a rational rectangle; then the straight line composed of the two lines is irrational and is called the first bimedial; the remaining line is medial, and the area contained by these two lines is irrational. Consequently the square on the whole line is irrational."

To begin with, the conditions here given are incompatible. If x, y, z be medial straight lines such that xy, xz are both rational,

$$y : z = xy : xz = m : n,$$

and y, z are commensurable *in length* and not *in square only*.

Hence it seems that we must, with Woepcke, understand "three medial straight lines such that *one is commensurable with each of the other two* in square only and makes with it a rational rectangle."

If x, y, z be the three medial straight lines,

$$(x^2 + y^2 + z^2) \frown x^2,$$

so that $(x^2 + y^2 + z^2)$ is medial.

Also we have $2xy$, $2xz$ both rational and $2yz$ medial.

Now $(x^2 + y^2 + z^2) + 2yz + 2xy + 2xz$ cannot be rational, for, if it were, the sum of two medial areas, $(x^2 + y^2 + z^2)$, $2yz$, would be rational: which is impossible. [Cf. x. 72.]

Hence $(x + y + z)$ is irrational.

(3) The *second trimedial* straight line.

Suppose x, y, z to be medial straight lines commensurable in square only and containing with each other medial rectangles.

Then $(x^2 + y^2 + z^2) \frown x^2$, and is medial.

Also $2yz$, $2zx$, $2xy$ are all medial areas.

To prove the irrationality in this case I presume that the method would be like that of x. 38 about the *second bimedial.*

Suppose σ to be a rational straight line and let

$$\left.\begin{aligned}(x^2 + y^2 + z^2) &= \sigma t\\ 2yz &= \sigma u\\ 2zx &= \sigma v\\ 2xy &= \sigma w\end{aligned}\right\}.$$

Here, since, e.g., $xz : xy = v : w$,

or $z : y = v : w$,

and similarly $x : z = w : u$,

u, v, w are *commensurable in square only.*

Also, since $(x^2 + y^2 + z^2) \frown x^2$
$\smile xy,$

t is incommensurable with w.

Similarly t is incommensurable with u, v.

But t, u, v, w are all rational and $\frown \sigma$.

Therefore $(t+u+v+w)$ is a quadrinomial and therefore irrational.

Therefore $\sigma(t+u+v+w)$, or $(x+y+z)^2$, is irrational,

whence $(x+y+z)$ is irrational.

(4) The *major* made up of three straight lines.

The commentator describes this as "the line composed of three straight lines incommensurable in square and such that one of them gives with each of the other two a sum of squares (which is) rational, while the rectangle contained by the two lines is medial."

If x, y, z are the three straight lines, this would indicate

$$(x^2+y^2) \text{ rational},$$
$$(x^2+z^2) \text{ rational},$$
$$2yz \text{ medial}.$$

Woepcke points out (pp. 696—8, note) the difficulties connected with this supposition or the supposition of

$$(x^2+y^2) \text{ rational},$$
$$(x^2+z^2) \text{ rational},$$
$$2xy \text{ (or } 2xz) \text{ medial},$$

and concludes that what is meant is the supposition

$$\left.\begin{array}{r}(x^2+y^2) \text{ rational}\\ xy \text{ medial}\\ xz \text{ medial}\end{array}\right\}$$

(though the text is against this).

The assumption of (x^2+y^2) and (x^2+z^2) being concurrently rational is certainly further removed from Euclid, for x. 33 only enables us to find *one pair* of lines having the property, as x, y.

But we will not pursue these speculations further.

As regards further irrationals formed by *subtraction* the commentator writes as follows.

"Again, it is not necessary that, in the irrational straight lines formed by means of subtraction, we should confine ourselves to making one subtraction only, so as to obtain the apotome, or the first apotome of the medial, or the second apotome of the medial, or the minor, or the straight line which produces with a rational area a medial whole, or that which produces with a medial area a medial whole; but we shall be able here to make two or three or four subtractions.

"When we do that, we show in manner analogous to the foregoing that the lines which remain are irrational and that each of them is one of the lines formed by subtraction. That is to say that, if from a rational line we cut off another rational line commensurable with the whole line in square, we obtain, for remainder, an apotome; and, if we subtract from this line (which is) cut off and rational—that which Euclid calls the *annex* (προσαρμόζουσα)—another rational line which is commensurable with it in square, we obtain, as the remainder, an apotome; likewise, if we cut off from the rational line cut off from this line (i.e. the annex of the apotome last arrived at) another line which is commensurable with it in square, the remainder is an apotome. The same thing occurs in the subtraction of the other lines."

As Woepcke remarks, the idea is the formation of the successive apotomes $\sqrt{a} - \sqrt{b}$, $\sqrt{b} - \sqrt{c}$, $\sqrt{c} - \sqrt{d}$, etc. We should naturally have expected to see the writer form and discuss the following expressions

$$(\sqrt{a} - \sqrt{b}) - \sqrt{c},$$
$$\{(\sqrt{a} - \sqrt{b}) - \sqrt{c}\} - \sqrt{d}, \text{ etc.}$$

BOOK XI.

DEFINITIONS.

1. A **solid** is that which has length, breadth, and depth.

2. An extremity of a solid is a surface.

3. A **straight line** is **at right angles to a plane,** when it makes right angles with all the straight lines which meet it and are in the plane.

4. A **plane** is **at right angles to a plane** when the straight lines drawn, in one of the planes, at right angles to the common section of the planes are at right angles to the remaining plane.

5. The **inclination of a straight line to a plane** is, assuming a perpendicular drawn from the extremity of the straight line which is elevated above the plane to the plane, and a straight line joined from the point thus arising to the extremity of the straight line which is in the plane, the angle contained by the straight line so drawn and the straight line standing up.

6. The **inclination of a plane to a plane** is the acute angle contained by the straight lines drawn at right angles to the common section at the same point, one in each of the planes.

7. A plane is said to be **similarly inclined** to a plane as another is to another when the said angles of the inclinations are equal to one another.

8. **Parallel planes** are those which do not meet.

9. **Similar solid figures** are those contained by similar planes equal in multitude.

10. **Equal and similar solid figures** are those contained by similar planes equal in multitude and in magnitude.

11. A **solid angle** is the inclination constituted by more than two lines which meet one another and are not in the same surface, towards all the lines.

Otherwise: A **solid angle** is that which is contained by more than two plane angles which are not in the same plane and are constructed to one point.

12. A **pyramid** is a solid figure, contained by planes, which is constructed from one plane to one point.

13. A **prism** is a solid figure contained by planes two of which, namely those which are opposite, are equal, similar and parallel, while the rest are parallelograms.

14. When, the diameter of a semicircle remaining fixed, the semicircle is carried round and restored again to the same position from which it began to be moved, the figure so comprehended is a **sphere**.

15. The **axis of the sphere** is the straight line which remains fixed and about which the semicircle is turned.

16. The **centre of the sphere** is the same as that of the semicircle.

17. A **diameter of the sphere** is any straight line drawn through the centre and terminated in both directions by the surface of the sphere.

18. When, one side of those about the right angle in a right-angled triangle remaining fixed, the triangle is carried round and restored again to the same position from which it began to be moved, the figure so comprehended is a **cone**.

And, if the straight line which remains fixed be equal to the remaining side about the right angle which is carried round, the cone will be **right-angled**; if less, **obtuse-angled**; and if greater, **acute-angled**.

19. The **axis of the cone** is the straight line which remains fixed and about which the triangle is turned.

20. And the **base** is the circle described by the straight line which is carried round.

21. When, one side of those about the right angle in a rectangular parallelogram remaining fixed, the parallelogram is carried round and restored again to the same position from which it began to be moved, the figure so comprehended is a **cylinder**.

22. The **axis of the cylinder** is the straight line which remains fixed and about which the parallelogram is turned.

23. And the **bases** are the circles described by the two sides opposite to one another which are carried round.

24. **Similar cones and cylinders** are those in which the axes and the diameters of the bases are proportional.

25. A **cube** is a solid figure contained by six equal squares.

26. An **octahedron** is a solid figure contained by eight equal and equilateral triangles.

27. An **icosahedron** is a solid figure contained by twenty equal and equilateral triangles.

28. A **dodecahedron** is a solid figure contained by twelve equal, equilateral, and equiangular pentagons.

DEFINITION I.

Στερεόν ἐστι τὸ μῆκος καὶ πλάτος καὶ βάθος ἔχον.

This definition was evidently traditional, as may be inferred from a number of passages in Plato and Aristotle. Thus Plato speaks (*Sophist*, 235 D) of making an imitation of a model (παράδειγμα) "in length and breadth and depth" and (*Laws*, 817 E) of "the art of measuring length, surface and depth" as one of three μαθήματα. *Depth*, the third dimension, is used alone as a description of "body" by Aristotle, the term being regarded as connoting the other two dimensions; thus (*Metaph.* 1020 a 13, 11) "length is a line, breadth a surface, and depth body"; "that which is continuous in one direction is length, in two directions breadth, and in three depth." Similarly Plato (*Rep.* 528 B, D), when reconsidering his classification of astronomy as next to (plane) geometry: "although the science dealing with the additional dimension of depth is next in order, yet, owing to the fact that it is studied absurdly, I passed it over and put next to geometry astronomy, the *motion* of (bodies having) depth." In Aristotle (*Topics* VI. 5, 142 b 24) we find "the definition of body, that which has three dimensions (διαστάσεις)"; elsewhere he speaks of it as "that which has all the dimensions" (*De caelo* I. 1, 268 b 6), "that which has dimension every way" (τὸ πάντῃ διάστασιν ἔχον, *Metaph.* 1066 b 32) etc. In the *Physics*

(IV. 1, 208 b 13 sqq.) he speaks of the "dimensions" as *six*, dividing each of the three into two opposites, "up and down, before and behind, right and left," though of course, as he explains, these terms are relative.

Heron, as might be expected, combines the two forms of the definition. "A solid body is that which has length, breadth, and depth: or that which possesses the three dimensions." (Def. 13.)

Similarly Theon of Smyrna (p. 111, 19, ed. Hiller): "that which is extended (διαστατόν) and divisible in three directions is solid, having length, breadth and depth."

DEFINITION 2.

Στερεοῦ δὲ πέρας ἐπιφάνεια.

In like manner Aristotle says (*Metaph.* 1066 b 23) that the notion (λόγος) of body is "that which is bounded by surfaces" (ἐπιπέδοις in this case) and (*Metaph.* 1060 b 15) "surfaces (ἐπιφάνειαι) are divisions of bodies."

So Heron (Def. 13): "Every solid is bounded (περατοῦται) by surfaces, and is produced when a surface is moved from a forward position in a backward direction."

DEFINITION 3.

Εὐθεῖα πρὸς ἐπίπεδον ὀρθή ἐστιν, ὅταν πρὸς πάσας τὰς ἁπτομένας αὐτῆς εὐθείας καὶ οὔσας ἐν τῷ ἐπιπέδῳ ὀρθὰς ποιῇ γωνίας.

This definition and the next are given almost word for word by Heron (Def. 115).

That a straight line *can* be so related to a plane as described in Def. 3 is established in XI. 4. The fact has been made the basis of a definition of a *plane* which is attributed by Crelle to Fourier, and is as follows. "A plane is formed by the totality of all the straight lines which, passing through one and the same point of a straight line in space, stand perpendicular to it." Stated in this form, the definition is open to the objection that the conception of a right angle, involving the measurement of angles, presupposes a plane, inasmuch as the measurement of angles depends ultimately upon the superposition of two planes and their coincidence throughout when two lines in one coincide with two lines in the other respectively. Cf. my note on I. Def. 7, Vol. I. pp. 173—5.

DEFINITION 4.

Ἐπίπεδον πρὸς ἐπίπεδον ὀρθόν ἐστιν, ὅταν αἱ τῇ κοινῇ τομῇ τῶν ἐπιπέδων πρὸς ὀρθὰς ἀγόμεναι εὐθεῖαι ἐν ἑνὶ τῶν ἐπιπέδων τῷ λοιπῷ ἐπιπέδῳ πρὸς ὀρθὰς ὦσιν.

Both this definition and Def. 6 use the *common section* of two planes, though it is not till XI. 3 that this common section is proved to be a straight line. The definition however, just like Def. 3, is legitimate, because the object is to explain the meaning of terms, not to prove anything.

The definition of perpendicular planes is made by Legendre a particular case of Def. 6, the limiting case, namely, where the angle representing the "inclination of a plane to a plane" is a right angle.

DEFINITION 5.

Εὐθείας πρὸς ἐπίπεδον κλίσις ἐστίν, ὅταν ἀπὸ τοῦ μετεώρου πέρατος τῆς εὐθείας ἐπὶ τὸ ἐπίπεδον κάθετος ἀχθῇ, καὶ ἀπὸ τοῦ γενομένου σημείου ἐπὶ τὸ ἐν τῷ ἐπιπέδῳ πέρας τῆς εὐθείας εὐθεῖα ἐπιζευχθῇ, ἡ περιεχομένη γωνία ὑπὸ τῆς ἀχθείσης καὶ τῆς ἐφεστώσης.

In other words, the inclination of a straight line to a plane is the angle between the straight line and its *projection* on the plane. This angle is of course less than the angle between the straight line and any other straight line in the plane through the intersection of the straight line and plane; and the fact is sometimes made the subject of a proposition in modern text-books. It is easily proved by means of the propositions XI. 4, I. 19 and 18.

DEFINITION 6.

Ἐπιπέδου πρὸς ἐπίπεδον κλίσις ἐστὶν ἡ περιεχομένη ὀξεῖα γωνία ὑπὸ τῶν πρὸς ὀρθὰς τῇ κοινῇ τομῇ ἀγομένων πρὸς τῷ αὐτῷ σημείῳ ἐν ἑκατέρῳ τῶν ἐπιπέδων.

When two planes meet in a straight line, they form what is called in modern text-books a *dihedral angle*, which is defined as the *opening* or *angular opening* between the two planes. This *dihedral angle* is an "angle" altogether different in kind from a plane angle, as again it is different from a *solid angle* as defined by Euclid (i.e. a trihedral, tetrahedral, etc. angle). Adopting for the moment Apollonius' conception of an angle as the "bringing together of a surface or solid towards one point under a broken line or surface" (Proclus, p. 123, 16), we may regard a dihedral angle as the bringing together of the broken surface formed by two intersecting planes not to a *point* but to a *straight line*, namely the intersection of the planes. Legendre, in a proposition on the subject, applied provisionally the term *corner* to describe the dihedral angle between two planes; and this would be a better word, I think, than *opening* to use in the definition.

The distinct species of "angle" which we call dihedral is, however, *measured* by a certain plane angle, namely that which Euclid describes in the present definition and calls the *inclination of a plane to a plane*, and which in some modern text-books is called the *plane angle of the dihedral angle*.

It is necessary to show that this plane angle is a proper measure of the dihedral angle, and accordingly Legendre has a proposition to this effect. In order to prove it, it is necessary to show that, given two planes meeting in a straight line,

(1) the plane angle in question is the same at all points of the straight line forming the common section;

(2) if the dihedral angle between two planes increases or diminishes in a certain ratio, the plane angle in question will increase or diminish in the same ratio.

(1) If MAN, MAP be two planes intersecting in MA, and if AN, AP be drawn in the planes respectively and at right angles to MA, the angle NAP is the *inclination of the plane to the plane* or the *plane angle of the dihedral angle*.

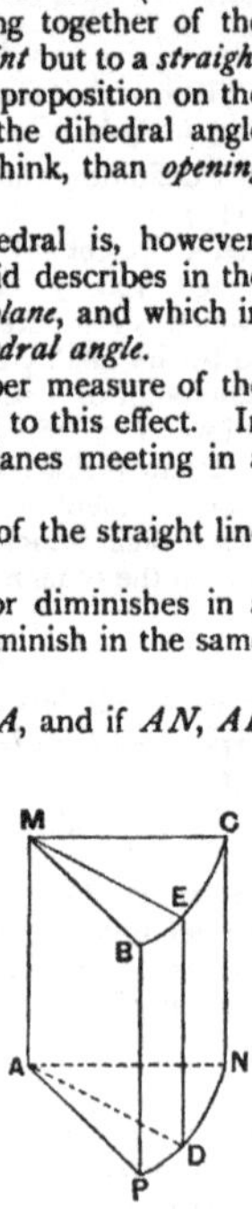

Let MC, MB be also drawn in the respective planes at right angles to MA.

Then since, in the plane MAN, MC and AN are drawn at right angles to the same straight line MA,

MC, AN are parallel.

For the same reason, MB, AP are parallel.

Therefore [XI. 10] the angle BMC is equal to the angle PAN.

And M may be any point on MA. Therefore the plane angle described in the definition is the same at all points of AM.

(2) In the plane NAP draw the arc NDP of any circle with centre A, and draw the radius AD.

Now the planes NAP, CMB, being both at right angles to the straight line MA, are parallel; [XI. 14]

therefore the intersections AD, ME of these planes with the plane MAD are parallel, [XI. 16]

and consequently the angles BME, PAD are equal. [XI. 10]

If now the plane angle NAD were equal to the plane angle DAP, the dihedral angle $NAMD$ would be equal to the dihedral angle $DAMP$;

for, if the angle PAD were applied to the angle DAN, AM remaining the same, the corresponding dihedral angles would coincide.

Successive applications of this result show that, if the angles NAD, DAP each contain a certain angle a certain number of times, the dihedral angles $NAMD$, $DAMP$ will contain the corresponding dihedral angle the same number of times respectively.

Hence, where the angles NAD, DAP are *commensurable*, the dihedral angles corresponding to them are in the same ratio.

Legendre then extends the proof to the case where the plane angles are incommensurable by reference to an exactly similar extension in his proposition corresponding to Euclid VI. 1, for which see the note on that proposition.

Modern text-books make the extension by an appeal to *limits*.

DEFINITION 7.

Ἐπίπεδον πρὸς ἐπίπεδον ὁμοίως κεκλίσθαι λέγεται καὶ ἕτερον πρὸς ἕτερον, ὅταν αἱ εἰρημέναι τῶν κλίσεων γωνίαι ἴσαι ἀλλήλαις ὦσιν.

DEFINITION 8.

Παράλληλα ἐπίπεδά ἐστι τὰ ἀσύμπτωτα.

Heron has the same definition of parallel planes (Def. 115). The Greek word which is translated "which do not meet" is ἀσύμπτωτα, the term which has been adopted for the *asymptotes* of a curve.

DEFINITION 9.

Ὅμοια στερεὰ σχήματά ἐστι τὰ ὑπὸ ὁμοίων ἐπιπέδων περιεχόμενα ἴσων τὸ πλῆθος.

DEFINITION 10.

Ἴσα δὲ καὶ ὅμοιά στερεὰ σχήματά ἐστι τὰ ὑπὸ ὁμοίων ἐπιπέδων περιεχόμενα ἴσων τῷ πλήθει καὶ τῷ μεγέθει.

These definitions, the second of which practically only substitutes the words "equal and similar" for the word "similar" in the first, have been the mark of much criticism.

Simson holds that the equality of solid figures is a thing which ought to be *proved*, by the method of superposition, or otherwise, and hence that Def. 10 is not a definition but a *theorem* which ought not to have been placed among the definitions. Secondly, he gives an example to show that the definition or theorem is not universally true. He takes a pyramid and then erects on the base, on opposite sides of it, two equal pyramids smaller than the first. The addition and subtraction of these pyramids respectively from the first give two solid figures which satisfy the definition but are clearly not equal (the smaller having a re-entrant angle); whence it also appears that two unequal solid angles may be contained by the same number of equal plane angles.

Maintaining then that Def. 10 is an interpolation by "an unskilful hand," Simson transfers to a place before Def. 9 the definition of a solid angle, and then defines similar solid figures as follows:

Similar solid figures are such as have all their solid angles equal, each to each, and which are contained by the same number of similar planes.

Legendre has an invaluable discussion of the whole subject of these definitions (Note XII., pp. 323—336, of the 14th edition of his *Éléments de Géométrie*). He remarks in the first place that, as Simson said, Def. 10 is not properly a definition, but a theorem which it is necessary to prove; for it is not evident that two solids are equal for the sole reason that they have an equal number of equal faces, and, if true, the fact should be proved by superposition or otherwise. The fault of Def. 10 is also common to Def. 9. For, if Def. 10 is not proved, one might suppose that there exist two unequal and dissimilar solids with equal faces; but, in that case, according to Definition 9, a solid having faces similar to those of the two first would be similar to both of them, i.e. to two solids of different form: a conclusion implying a contradiction or at least not according with the natural meaning of the word "similar."

What then is to be said in defence of the two definitions as given by Euclid? It is to be observed that the figures which Euclid actually proves equal or similar by reference to Deff. 9, 10 are such that their solid angles do not consist of more than *three* plane angles; and he proves sufficiently clearly that, if three plane angles forming one solid angle be respectively equal to three plane angles forming another solid angle, the two solid angles are equal. If now two polyhedra have their faces equal respectively, the corresponding solid angles will be made up of the same number of plane angles, and the plane angles forming each solid angle in one polyhedron will be respectively equal to the plane angles forming the corresponding solid angle in the other. Therefore, if the plane angles in each solid angle are not more than three in number, the corresponding solid angles will be equal. But if the corresponding faces are equal, and the corresponding solid angles equal, the solids must be equal; for they can be superposed, or at least they will be symmetrical with one another. Hence the statement of Deff. 9, 10 is true and admissible at all events in the case of figures with trihedral angles, which is the only case taken by Euclid.

Again, the example given by Simson to prove the incorrectness of Def. 10 introduces a solid with a re-entrant angle. But it is more than probable that Euclid deliberately intended to exclude such solids and to take cognizance of *convex* polyhedra only; hence Simson's example is not conclusive against the definition.

Legendre observes that Simson's own definition, though true, has the disadvantage that it contains a number of superfluous conditions. To get over the difficulties, Legendre himself divides the definition of similar solids into two, the first of which defines similar *triangular pyramids* only, and the second (which defines similar polyhedra in general) is based on the first.

Two triangular pyramids are similar when they have pairs of faces respectively similar, similarly placed and equally inclined to one another.

Then, having formed a triangle with the vertices of three angles taken on the same face or base of a polyhedron, we may imagine the vertices of the different solid angles of the polyhedron situated outside of the plane of this base to be the vertices of as many triangular pyramids which have the triangle for common base, and each of these pyramids will determine the position of one solid angle of the polyhedron. This being so,

Two polyhedra are similar when they have similar bases, and the vertices of their corresponding solid angles outside the bases are determined by triangular pyramids similar each to each.

As a matter of fact, Cauchy proved that two *convex* solid figures are equal if they are contained by equal plane figures similarly arranged. Legendre gives a proof which, he says, is nearly the same as Cauchy's, depending on two lemmas which lead to the theorem that, *Given a convex polyhedron in which all the solid angles are made up of more than three plane angles, it is impossible to vary the inclinations of the planes of this solid so as to produce a second polyhedron formed by the same planes arranged in the same manner as in the given polyhedron.* The convex polyhedron *in which all the solid angles are made up of more than three plane angles* is obtained by cutting off from *any* given polyhedron all the triangular pyramids forming trihedral angles (if one and the same edge is common to *two* trihedral angles, only one of these angles is suppressed in the first operation). This is legitimate because trihedral angles are invariable from their nature.

Hence it would appear that Heron's definition of equal solid figures, which adds "similarly situated" to Euclid's "similar" is correct, if it be understood to apply to *convex* polyhedra only: *Equal solid figures are those which are contained by equal and similarly situated planes, equal in number and magnitude:* where, however, the words "equal and" before "similarly situated" might be dispensed with.

Heron (Def. 118) defines *similar solid figures* as *those which are contained by planes similar and similarly situated.* If understood of *convex* polyhedra, there would not appear to be any objection to this, in view of the truth of Cauchy's proposition about equal solid figures.

DEFINITION 11.

Στερεὰ γωνία ἐστὶν ἡ ὑπὸ πλειόνων ἢ δύο γραμμῶν ἁπτομένων ἀλλήλων καὶ μὴ ἐν τῇ αὐτῇ ἐπιφανείᾳ οὐσῶν πρὸς πάσαις ταῖς γραμμαῖς κλίσις. Ἄλλως· στερεὰ γωνία ἐστὶν ἡ ὑπὸ πλειόνων ἢ δύο γωνιῶν ἐπιπέδων περιεχομένη μὴ οὐσῶν ἐν τῷ αὐτῷ ἐπιπέδῳ πρὸς ἑνὶ σημείῳ συνισταμένων.

Heiberg conjectures that the first of these two definitions, which is not in Euclid's manner, was perhaps taken by him from some earlier *Elements*.

The phraseology of the second definition is exactly that of Plato when he is speaking of solid angles in the *Timaeus* (p. 55). Thus he speaks (1) of four equilateral triangles so put together (*ξυνιστάμενα*) that each set of three plane angles makes one solid angle, (2) of eight equilateral triangles put together so that each set of four plane angles makes one solid angle, and (3) of six squares making eight solid angles, each composed of three plane right angles.

As we know, Apollonius defined an angle as the "bringing together of a surface or solid to one point under a broken line or surface." Heron (Def. 24) even omits the word "broken" and says that *A solid angle is in general* (*κοινῶς*) *the bringing together of a surface which has its concavity in one and the same direction to one point.* It is clear from an allusion in Proclus (p. 123, 1—6) to the half of a cone cut off by a triangle through the axis, and from a scholium to this definition, that there was controversy as to the correctness of describing as a solid angle the "angle" enclosed by fewer than three surfaces (including curved surfaces). Thus the scholiast says that Euclid's definition of a solid angle as made up of three or more plane angles is deficient because it does not e.g. cover the case of the angle of a "fourth part of a sphere," which is contained by more than two surfaces, though not all plane. But he declines to admit that the half-cone forms a solid angle at the vertex, for in that case the vertex of the cone would itself be an angle, and a solid angle would then be formed both by two surfaces and by one surface: "which is not true." Heron on the other hand (Def. 24) distinctly speaks of solid angles which are not contained by plane rectilineal angles, "e.g. the angles of cones." The conception of the latter "angles" as the *limit* of solid angles with an infinite number of infinitely small constituent plane angles does not appear in the Greek geometers so far as I know.

In modern text-books a polyhedral angle is usually spoken of as *formed* (or *bounded*) *by three or more planes meeting at a point*, or it is *the angular opening between such planes at the point where they meet.*

DEFINITION 12.

Πυραμίς ἐστι σχῆμα στερεὸν ἐπιπέδοις περιεχόμενον ἀπὸ ἑνὸς ἐπιπέδου πρὸς ἑνὶ σημείῳ συνεστώς.

This definition is by no means too clear, nor is the slightly amplified definition added to it by Heron (Def. 100). *A pyramid is the figure brought together to one point, by putting together triangles, from a triangular, quadrilateral or polygonal, that is, any rectilineal, base.*

As we might expect, there is great variety in the definitions given in modern text-books. Legendre says *a pyramid is the solid formed when several triangular planes start from one point and are terminated at the different sides of one polygonal plane.*

Mr H. M. Taylor and Smith and Bryant call it *a polyhedron all but one of whose faces meet in a point.*

Mehler reverses Legendre's form and gives the content of Euclid's in clearer language. "*An* n-*sided pyramid is bounded by an* n-*sided polygon as base and* n *triangles which connect its sides with one and the same point outside it.*"

Rausenberger points out that a pyramid is the figure cut off from a solid angle formed of any number of plane angles by a plane which intersects the solid angle.

DEFINITION 13.

Πρίσμα ἐστὶ σχῆμα στερεὸν ἐπιπέδοις περιεχόμενον, ὧν δύο τὰ ἀπεναντίον ἴσα τε καὶ ὅμοιά ἐστι καὶ παράλληλα, τὰ δὲ λοιπὰ παραλληλόγραμμα.

Mr H. M. Taylor, followed by Smith and Bryant, defines a prism as *a polyhedron all but two of the faces of which are parallel to one straight line.*

Mehler calls an *n*-sided prism *a body contained between two parallel planes and enclosed by* n *other planes with parallel lines of intersection.*

Heron's definition of a prism is much wider (Def. 105). *Prisms are those figures which are connected* (συνάπτοντα) *from a rectilineal base to a rectilineal area by rectilineal collocation* (κατ' εὐθύγραμμον σύνθεσιν). By this Heron must apparently mean any convex solid formed by connecting the sides and angles of two polygons in different planes, and each having any number of sides, by straight lines forming triangular faces (where of course two adjacent triangles may be in one plane and so form one quadrilateral face) in the manner shown in the annexed figure, where $ABCD$, EFG represent the base and its opposite.

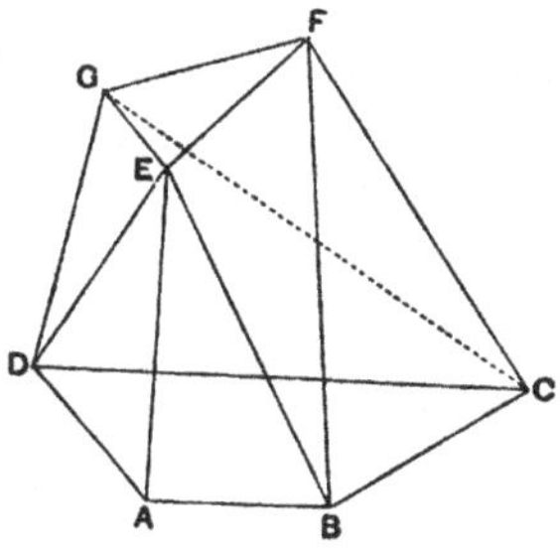

Heron goes on to explain that, if the face opposite to the base reduces to a straight line, and a solid is formed by connecting the base to its extremities by straight lines, as in the other case, the resulting figure is neither a pyramid nor a prism.

Further, he defines *parallelogrammic* (in the body of the definition *parallel-sided*) *prisms* as being those prisms which have six faces and have their opposite planes parallel.

DEFINITION 14.

Σφαῖρά ἐστιν, ὅταν ἡμικυκλίου μενούσης τῆς διαμέτρου περιενεχθὲν τὸ ἡμικύκλιον εἰς τὸ αὐτὸ πάλιν ἀποκατασταθῇ, ὅθεν ἤρξατο φέρεσθαι, τὸ περιληφθὲν σχῆμα.

The scholiast observes that this definition is not properly a definition of a sphere but a description of the mode of generating it. But it will be seen, in the last propositions of Book XIII., why Euclid put the definition in this form. It is because it is this particular view of a sphere which he uses to prove that the vertices of the regular solids which he wishes to "comprehend" in certain spheres do lie on the surfaces of those spheres. He proves in fact that the said vertices lie on *semicircles described on certain diameters of the spheres*. For the real definition the scholiast refers to Theodosius' *Sphaerica*. But of course the proper definition was given much earlier. In Aristotle the characteristic of a sphere is that *its extremity is equally distant from its centre* (τὸ ἴσον ἀπέχειν τοῦ μέσου τὸ ἔσχατον, *De caelo* II. 14, 297 a 24). Heron (Def. 77) uses the same form as that in which Euclid defines the circle: *A sphere is a solid figure bounded by one surface, such that all the straight lines falling on it from one point of those which lie within the figure are equal to one another.* So the usual definition in the text-books: *A sphere is a closed surface such that all points of it are equidistant from a fixed point within it.*

DEFINITION 15.

Ἄξων δὲ τῆς σφαίρας ἐστὶν ἡ μένουσα εὐθεῖα, περὶ ἣν τὸ ἡμικύκλιον στρέφεται.

That *any* diameter of a sphere may be called an axis is made clear by Heron (Def. 79). *The diameter of the sphere is called an axis, and is any straight line drawn through the centre and bounded in both directions by the sphere, immovable, about which the sphere is moved and turned.* Cf. Euclid's Def. 17.

DEFINITION 16.

Κέντρον δὲ τῆς σφαίρας ἐστὶ τὸ αὐτό, ὃ καὶ τοῦ ἡμικυκλίου.

Heron, Def. 78. *The middle (point) of the sphere is called its centre; and this same point is also the centre of the hemisphere.*

DEFINITION 17.

Διάμετρος δὲ τῆς σφαίρας ἐστὶν εὐθεῖά τις διὰ τοῦ κέντρου ἠγμένη καὶ περατουμενη ἐφ᾽ ἑκάτερα τὰ μέρη ὑπὸ τῆς ἐπιφανείας τῆς σφαίρας.

DEFINITION 18.

Κῶνός ἐστιν, ὅταν ὀρθογωνίου τριγώνου μενούσης μιᾶς πλευρᾶς τῶν περὶ τὴν ὀρθὴν γωνίαν περιενεχθὲν τὸ τρίγωνον εἰς τὸ αὐτὸ πάλιν ἀποκατασταθῇ, ὅθεν ἤρξατο φέρεσθαι, τὸ περιληφθὲν σχῆμα. κἂν μὲν ἡ μένουσα εὐθεῖα ἴση ᾖ τῇ λοιπῇ [τῇ] περὶ τὴν ὀρθὴν περιφερομένῃ, ὀρθογώνιος ἔσται ὁ κῶνος, ἐὰν δὲ ἐλάττων, ἀμβλυγώνιος, ἐὰν δὲ μείζων, ὀξυγώνιος.

This definition, or rather description of the genesis, of a (right) cone is interesting on account of the second sentence distinguishing between *right-angled*, *obtuse-angled* and *acute-angled* cones. This distinction is quite unnecessary for Euclid's purpose and is not used by him in Book XII.; it is no doubt a relic of the method, still in use in Euclid's time, by which the earlier Greek geometers produced conic sections, namely, by cutting right cones only by sections always perpendicular to an edge. With this system the parabola was a *section of a right-angled cone*, the hyperbola a *section of an obtuse-angled cone*, and the ellipse a *section of an acute-angled cone*. The conic sections were so called by Archimedes, and generally until Apollonius, who was the first to give the complete theory of their generation by means of sections not perpendicular to an edge, and from cones which are in general *oblique circular* cones. Thus Apollonius begins his *Conics* with the more scientific definition of a cone. If, he says, a straight line infinite in length, and passing always through a fixed point, be made to move round the circumference of a circle which is not in the same plane with the point, so as to pass successively through every point of that circumference, the moving straight line will trace out the surface of a *double cone*, or two similar cones lying in opposite directions and meeting in the fixed point, which is the *apex* of each cone. The circle about which the straight line moves is called the *base* of the cone lying between the said circle and the fixed point, and the *axis* is defined as the straight line drawn from the fixed point, or the apex, to the centre of the circle forming the base. Apollonius goes on to say that the cone is a *scalene* or *oblique* cone except in the particular case where the axis is perpendicular to the base. In this latter case it is a *right* cone.

Archimedes called the right cone an *isosceles* cone. This fact, coupled with the appearance in his treatise *On Conoids and Spheroids* (7, 8, 9) of *sections of acute-angled cones* (ellipses) as sections of conical surfaces which are proved to be oblique circular cones by finding their circular sections, makes it sufficiently clear that Archimedes, if he had defined a cone, would have defined it in the same way as Apollonius does.

DEFINITION 19.

Ἄξων δὲ τοῦ κώνου ἐστὶν ἡ μένουσα εὐθεῖα, περὶ ἣν τὸ τρίγωνον στρέφεται.

DEFINITION 20.

Βάσις δὲ ὁ κύκλος ὁ ὑπὸ τῆς περιφερομένης εὐθείας γραφόμενος.

DEFINITION 21.

Κύλινδρός ἐστιν, ὅταν ὀρθογωνίου παραλληλογράμμου μενούσης μιᾶς πλευρᾶς τῶν περὶ τὴν ὀρθὴν γωνίαν περιενεχθὲν τὸ παραλληλόγραμμον εἰς τὸ αὐτὸ πάλιν ἀποκατασταθῇ, ὅθεν ἤρξατο φέρεσθαι, τὸ περιληφθὲν σχῆμα.

DEFINITION 22.

Ἄξων δὲ τοῦ κυλίνδρου ἐστὶν ἡ μένουσα εὐθεῖα, περὶ ἣν τὸ παραλληλόγραμμον στρέφεται.

DEFINITION 23.

Βάσεις δὲ οἱ κύκλοι οἱ ὑπὸ τῶν ἀπεναντίον περιαγομένων δύο πλευρῶν γραφόμενοι.

DEFINITION 24.

Ὅμοιοι κῶνοι καὶ κύλινδροί εἰσιν, ὧν οἵ τε ἄξονες καὶ αἱ διάμετροι τῶν βάσεων ἀνάλογόν εἰσιν.

DEFINITION 25.

Κύβος ἐστὶ σχῆμα στερεὸν ὑπὸ ἓξ τετραγώνων ἴσων περιεχόμενον.

DEFINITION 26.

Ὀκτάεδρόν ἐστι σχῆμα στερεὸν ὑπὸ ὀκτὼ τριγώνων ἴσων καὶ ἰσοπλεύρων περιεχόμενον.

DEFINITION 27.

Εἰκοσάεδρόν ἐστι σχῆμα στερεὸν ὑπὸ εἴκοσι τριγώνων ἴσων καὶ ἰσοπλεύρων περιεχόμενον.

DEFINITION 28.

Δωδεκάεδρόν ἐστι σχῆμα στερεὸν ὑπὸ δώδεκα πενταγώνων ἴσων καὶ ἰσοπλεύρων καὶ ἰσογωνίων περιεχόμενον.

BOOK XI. PROPOSITIONS.

Proposition 1.

A part of a straight line cannot be in the plane of reference and a part in a plane more elevated.

For, if possible, let a part AB of the straight line ABC be in the plane of reference, and a part BC in a plane more elevated.

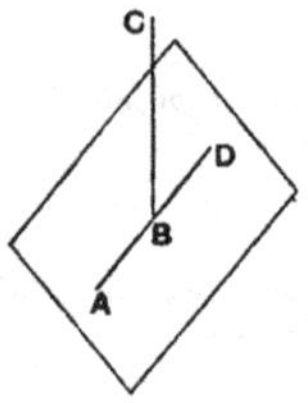

There will then be in the plane of reference some straight line continuous with AB in a straight line.

Let it be BD;

therefore AB is a common segment of the two straight lines ABC, ABD:

which is impossible, inasmuch as, if we describe a circle with centre B and distance AB, the diameters will cut off unequal circumferences of the circle.

Therefore a part of a straight line cannot be in the plane of reference, and a part in a plane more elevated.

Q. E. D.

1. **the plane of reference,** τὸ ὑποκείμενον ἐπίπεδον, the plane laid down or assumed.
2. **more elevated,** μετεωροτέρῳ.

There is no doubt that the proofs of the first three propositions are unsatisfactory owing to the fact that Euclid is not able to make any use of his definition of a plane for the purpose of these proofs, and they really depend upon truths which can only be assumed as axiomatic. The definition of a plane as *that surface which lies evenly with the straight lines on itself*, whatever its exact meaning may be, is nowhere appealed to as a criterion to show whether a particular surface is or is not a plane. If the meaning of it is what I conjecture in the note on Book I., Def. 7 (Vol. I. p. 171), if, namely, it only tries to express without an appeal to sight what Plato meant by the "middle covering the extremities" (i.e. apparently, in the case of a plane, the fact that a plane looked at edgewise takes the form of a straight line), then it is perhaps possible to connect the definition with a method of generating a plane which has commended itself to many writers as giving a better definition. Thus, if we conceive a straight line in space and a point outside it placed so that, in Plato's words, the line "covers" the point as we look at them, the line will also "cover" every straight line which passes through the given point and some one point on the given straight line. Hence, if a straight line passing always through a fixed point moves in such a way as to pass successively through every point of a given straight line which does not contain the given point, the moving straight line describes a surface which satisfies the Euclidean definition of a plane as I have interpreted it. But if we adopt the definition of a plane as *the surface described by a straight line which, passing through a given point, turns about it in such a way as always to intersect a given straight line not passing through the given point*, this definition, though it would help us to prove Eucl. XI. 2, does not give us the fundamental properties of a plane; some postulate is necessary in addition. The same is true even if we take a definition which gives *more* than is required to determine a plane, the definition known as Simson's, though it is at least as early as the time of Theon of Smyrna, who says (p. 112, 5) that *a plane is a surface such that, if a straight line meet it in two points, the straight line lies wholly in it* (ὅλη αὐτῷ ἐφαρμόζεται). This is also called the *axiom of the plane*. (For some attempts to *prove* this on the basis of other definitions of a plane see my note on the definition of a *plane surface*, I. Def. 7.) If this definition or axiom be assumed, Prop. 1 becomes evident, for, as Legendre says, "In accordance with the definition of the plane, when a straight line has two points common with a plane, it lies wholly in the plane."

Euclid practically assumes the axiom when he says in this proposition "there will be in the plane of reference some straight line continuous with *AB*." Clavius tries, unsuccessfully, to deduce this from Euclid's own definition of a plane; and he seems to admit his failure, because he proceeds to try another tack. Draw, he says, in the plane *DE*, the straight line *CG* at right angles to *AC*, and, again in the plane *DE*, *CF* at right angles to *CG* [I. 11]. Then *AC*, *CF* make right angles with *CG* in the same plane; therefore (I. 14) *ACF* is a straight line. But this does not really help, because Euclid assumes tacitly, in Book I. as well as Book XI., that a straight line joining two points in a plane lies wholly in that plane.

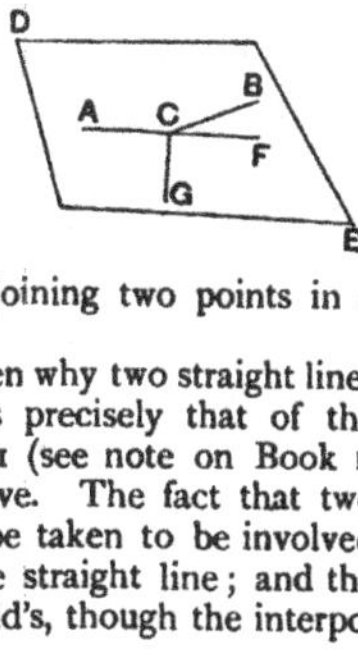

A curious point in Euclid's proof is the reason given why two straight lines cannot have a common segment. The argument is precisely that of the "proof" of the same thing given by Proclus on I. 1 (see note on Book I. Post. 2, Vol. I. p. 197) and is of course inconclusive. The fact that two straight lines cannot have a common segment must be taken to be involved in the definition of, and the postulates relating to, the straight line; and the "proof" given here can hardly, I should say, be Euclid's, though the interpolation, if it be such, must have been made very early.

The proof assumes too that a circle *can* be described so as to cut *BA*, *BC* and *BD*, or, in other words, it assumes that *AD*, *BC* are in one plane; that is, Prop. 1 as we have it really assumes the result of Prop. 2. There is therefore ground for Simson's alteration of the proof (after the point where *BD* has been taken in the given plane in a straight line with *AB*) to the following:

"Let any plane pass through the straight line *AD* and be turned about it until it pass through the point *C*.

And, because the points B, C are in this plane, the straight line BC is in it. [Simson's def.]

Therefore there are two straight lines ABC, ABD in the same plane that have a common segment AB:
which is impossible."

Simson, of course, justifies the last inference by reference to his Corollary to I. 11, which, however, as we have seen, is not a valid proof of the assumption, which is really implied in I. Post. 2.

An alternative reading, perhaps due to Theon, says, after the words "which is impossible" in the Greek text, "for a straight line does not meet a straight line in more points than one; otherwise the straight lines will coincide." Simson (who however does not seem to have had the second clause beginning "otherwise" in the text which he used) attacks this alternative reading in a rather confused note chiefly directed against a criticism by Thomas Simpson, without (as it seems to me) sufficient reason. It contains surely a legitimate argument. The supposed straight lines ABC, ABD meet in more than two points, namely in all the points between A and B. But two straight lines cannot have two points common without coinciding altogether; therefore ABC must coincide with ABD.

Proposition 2.

If two straight lines cut one another, they are in one plane, and every triangle is in one plane.

For let the two straight lines AB, CD cut one another at the point E;
I say that AB, CD are in one plane, and every triangle is in one plane.

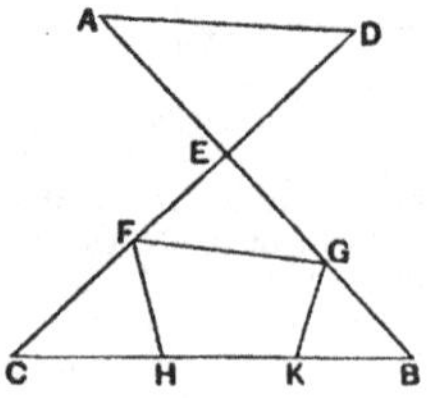

For let points F, G be taken at random on EC, EB,
let CB, FG be joined,
and let FH, GK be drawn across;
I say first that the triangle ECB is in one plane.

For, if part of the triangle ECB, either FHC or GBK, is in the plane of reference, and the rest in another,
a part also of one of the straight lines EC, EB will be in the plane of reference, and a part in another.

But, if the part $FCBG$ of the triangle ECB be in the plane of reference, and the rest in another,
a part also of both the straight lines EC, EB will be in the plane of reference and a part in another:
which was proved absurd. [XI. 1]

Therefore the triangle ECB is in one plane.

But, in whatever plane the triangle ECB is, in that plane also is each of the straight lines EC, EB,

and, in whatever plane each of the straight lines EC, EB is, in that plane are AB, CD also. [XI. 1]

Therefore the straight lines AB, CD are in one plane,

and every triangle is in one plane.

Q. E. D.

It must be admitted that the "proof" of this proposition is not of any value. For one thing, Euclid only takes certain triangles and a certain quadrilateral respectively forming part of the original triangle, and argues about these. But, for anything we are supposed to know, there may be some part of the triangle bounded (let us say) by some curve which is not in the same plane with the triangle.

We may agree with Simson that it would be preferable to enunciate the proposition as follows.

Two straight lines which intersect are in one plane, and three straight lines which intersect two and two are in one plane.

Adopting Smith and Bryant's figure in preference to Simson's, we suppose three straight lines PQ, RS, XY to intersect two and two in A, B, C.

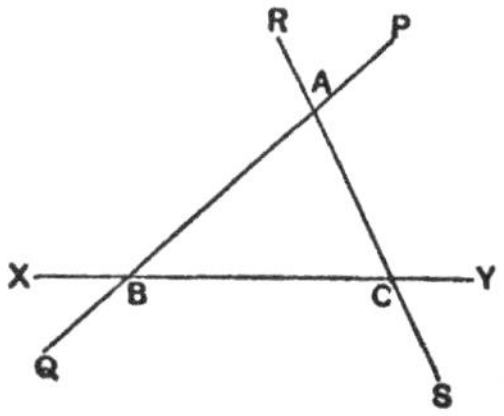

Then Simson's proof (adopted by Legendre also) proceeds thus.

Let any plane pass through the straight line PQ, and let this plane be turned about PQ (produced indefinitely) as axis until it passes through the point C.

Then, since the points A, C are in this plane, the straight line AC (and therefore the straight line RS produced indefinitely) lies wholly in the plane. [Simson's def.]

For the same reason, since the points B, C are in the plane, the straight line XY lies wholly in the plane.

Hence all three straight lines PQ, RS, XY (and of course any pair of them) lie in one plane.

But it has still to be proved that there is *only* one plane passing through the three straight lines.

This may be done, as in Mr Taylor's Euclid, thus.

Suppose, if possible, that there are *two* different planes through A, B, C.

The straight lines BC, CA, AB then lie wholly in each of the two planes.

Now any straight line in one of the two planes must intersect at least two of the straight lines (produced if necessary);

let it intersect two of them in K, L.

Then, since K, L are also in the second plane, the line KL lies wholly in that plane.

Hence every straight line in either of the planes lies wholly in the other also; and therefore the planes are coincident throughout their whole surface.

It follows from the above that

A plane is determined (i.e. uniquely *determined*) *by any of the following data:*

(1) *by three straight lines meeting one another two and two,*

(2) *by three points not in a straight line,*

(3) *by two straight lines meeting one another,*

(4) *by a straight line and a point without it.*

Proposition 3.

If two planes cut one another, their common section is a straight line.

For let the two planes AB, BC cut one another, and let the line DB be their common section;

I say that the line DB is a straight line.

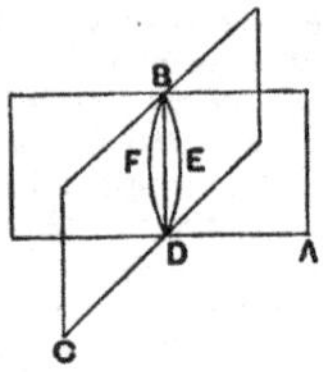

For, if not, from D to B let the straight line DEB be joined in the plane AB, and in the plane BC the straight line DFB.

Then the two straight lines DEB, DFB will have the same extremities, and will clearly enclose an area:

which is absurd.

Therefore DEB, DFB are not straight lines.

Similarly we can prove that neither will there be any other straight line joined from D to B except DB the common section of the planes AB, BC.

Therefore etc.

Q. E. D.

I think Simson is right in objecting to the words after "which is absurd," to the effect that DEB, DFB are not straight lines, and that neither can there be any other straight line joined from D to B except DB, as being unnecessary. It is right to conclude at once from the absurdity that BD cannot *but* be a straight line.

Legendre makes his proof depend on Prop. 2. "For, if, among the points common to the two planes, three should be found which are not in a straight line, the two planes in question, each passing through three points, would only amount to one and the same plane." [This of course assumes that three points determine one and *only* one plane, which, strictly speaking, involves more than Prop. 2 itself, as shown in the last note.]

A favourite proposition in modern text-books is the following. The proof seems to be due to von Staudt (Killing, *Grundlagen der Geometrie*, Vol. II. p. 43).

If two planes meet in a point, they meet in a straight line.

Let ABC, ADE be two given planes meeting at A.

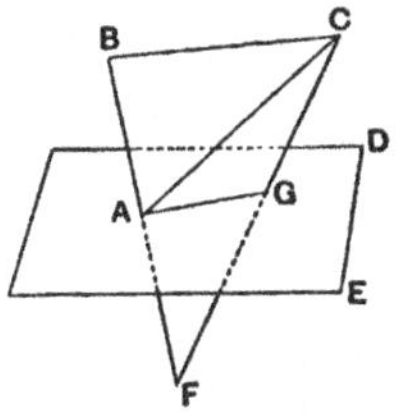

Take any points B, C lying on the plane ABC, and not on the plane ADE but on the same side of it.

Join AB, AC, and produce BA to F.

Join CF.

Then, since B, F are on opposite sides of the plane ADE,

C, F are also on opposite sides of it.

Therefore CF must meet the plane ADE in some point, say G.

Then, since A, G are both in each of the planes ABC, ADE, the straight line AG is in both planes. [Simson's def.]

This is also the place to insert the proposition that, *If three planes intersect two and two, their lines of intersection either meet in a point or are parallel two and two.*

Let there be three planes intersecting in the straight lines AB, CD, EF.

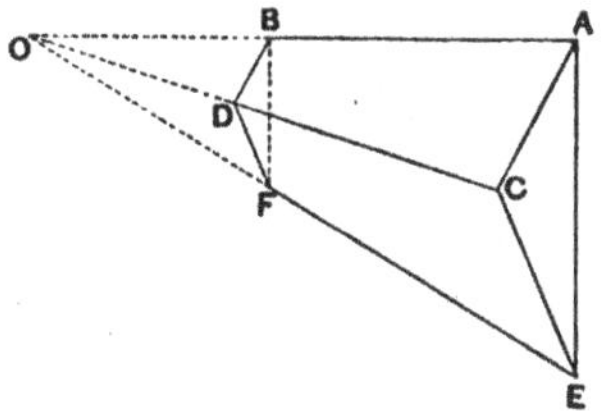

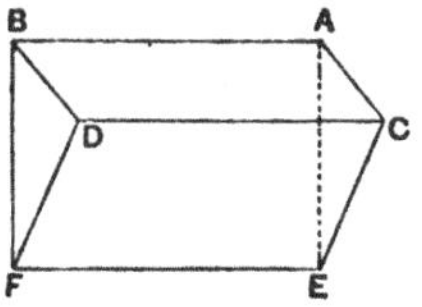

Now AB, EF are in a plane; therefore they either meet in a point or are parallel.

(1) Let them meet in O.

Then O, being a point in AB, lies in the plane AD, and, being also a point in EF, lies also in the plane ED.

Therefore O, being common to the planes AD, DE, must lie on CD, the line of their intersection;

i.e. CD, if produced, passes through O.

(2) Let AB, EF not meet, but let them be parallel.

Then CD cannot meet AB; for, if it did, it must necessarily meet EF, by the first case.

Therefore CD, AB, being in one plane, are parallel.

Similarly CD, EF are parallel.

PROPOSITION 4.

If a straight line be set up at right angles to two straight lines which cut one another, at their common point of section, it will also be at right angles to the plane through them.

For let a straight line *EF* be set up at right angles to the two straight lines *AB*, *CD*, which cut one another at the point *E*, from *E*;

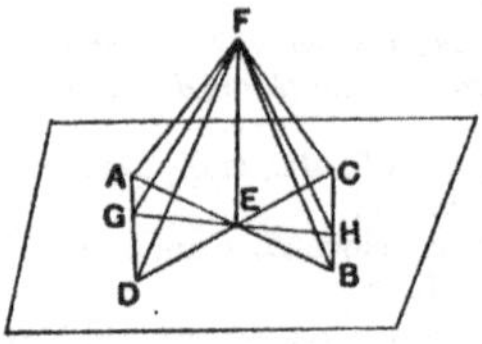

I say that *EF* is also at right angles to the plane through *AB*, *CD*.

For let *AE*, *EB*, *CE*, *ED* be cut off equal to one another,
and let any straight line *GEH* be drawn across through *E*, at random;
let *AD*, *CB* be joined,
and further let *FA*, *FG*, *FD*, *FC*, *FH*, *FB* be joined from the point *F* taken at random <on *EF*>.

Now, since the two straight lines *AE*, *ED* are equal to the two straight lines *CE*, *EB*, and contain equal angles, [I. 15]
therefore the base *AD* is equal to the base *CB*,
and the triangle *AED* will be equal to the triangle *CEB*; [I. 4]
so that the angle *DAE* is also equal to the angle *EBC*.

But the angle *AEG* is also equal to the angle *BEH*; [I. 15]
therefore *AGE*, *BEH* are two triangles which have two angles equal to two angles respectively, and one side equal to one side, namely that adjacent to the equal angles, that is to say, *AE* to *EB*;
therefore they will also have the remaining sides equal to the remaining sides. [I. 26]

Therefore *GE* is equal to *EH*, and *AG* to *BH*.

And, since *AE* is equal to *EB*,
while *FE* is common and at right angles,
therefore the base *FA* is equal to the base *FB*. [I. 4]

For the same reason
FC is also equal to *FD*.

And, since *AD* is equal to *CB*,
and *FA* is also equal to *FB*,
the two sides *FA*, *AD* are equal to the two sides *FB*, *BC* respectively;
and the base *FD* was proved equal to the base *FC*;
therefore the angle *FAD* is also equal to the angle *FBC*. [I. 8]

And since, again, *AG* was proved equal to *BH*, and further *FA* also equal to *FB*, the two sides *FA*, *AG* are equal to the two sides *FB*, *BH*.

And the angle *FAG* was proved equal to the angle *FBH*; therefore the base *FG* is equal to the base *FH*. [I. 4]

Now since, again, *GE* was proved equal to *EH*, and *EF* is common, the two sides *GE*, *EF* are equal to the two sides *HE*, *EF*; and the base *FG* is equal to the base *FH*; therefore the angle *GEF* is equal to the angle *HEF*. [I. 8]

Therefore each of the angles *GEF*, *HEF* is right.

Therefore *FE* is at right angles to *GH* drawn at random through *E*.

Similarly we can prove that *FE* will also make right angles with all the straight lines which meet it and are in the plane of reference.

But a straight line is at right angles to a plane when it makes right angles with all the straight lines which meet it and are in that same plane; [XI. Def. 3] therefore *FE* is at right angles to the plane of reference.

But the plane of reference is the plane through the straight lines *AB*, *CD*.

Therefore *FE* is at right angles to the plane through *AB*, *CD*.

Therefore etc.

Q. E. D.

The steps to be successively proved in order to establish this proposition by Euclid's method are

(1) triangles *AED*, *BEC* equal in all respects, [by I. 4]

(2) triangles *AEG*, *BEH* equal in all respects, [by I. 26] so that *AG* is equal to *BH*, and *GE* to *EH*,

(3) triangles *AEF*, *BEF* equal in all respects, [I. 4] so that *AF* is equal to *BF*,

(4) likewise triangles *CEF*, *DEF*, so that *CF* is equal to *DF*,

(5) triangles *FAD*, *FBC* equal in all respects, [I. 8] so that the angles *FAG*, *FBH* are equal,

(6) triangles *FAG*, *FBH* equal in all respects, [by (2), (3), (5) and I. 4] so that *FG* is equal to *FH*,

(7) triangles FEG, FEH equal in all respects, [by (2), (6) and I. 8]
so that the angles FEG, FEH are equal,
and therefore FE is at right angles to GH.

In consequence of the length of the above proof others have been suggested, and the proof which now finds most general acceptance is that of Cauchy, which is as follows.

Let AB be perpendicular to two straight lines BC, BD in the plane MN at their point of intersection B.

In the plane MN draw BE, *any* straight line through B.

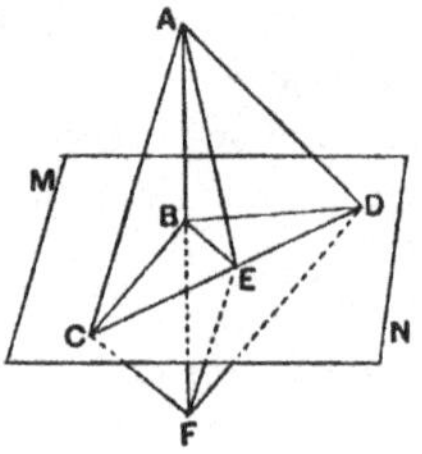

Join CD, and let CD meet BE in E.

Produce AB to F so that BF is equal to AB.

Join AC, AE, AD, CF, EF, DF.

Since BC is perpendicular to AF at its middle point B,
AC is equal to CF.

Similarly AD is equal to DF.

Since in the triangles ACD, FCD the two sides AC, CD are respectively equal to the two sides FC, CD, and the third sides AD, FD are also equal,

the angles ACD, FCD are equal. [I. 8]

The triangles ACE, FCE thus have two sides and the included angle equal, whence

EA is equal to EF. [I. 4]

The triangles ABE, FBE have now all their sides equal respectively; therefore the angles ABE, FBE are equal, [I. 8]

and AB is perpendicular to BE.

And BE is in *any* straight line through B in the plane MN.

Legendre's proof is not so easy, but it is interesting. We are first required to draw through any point E within the angle CBD a straight line CD bisected at E.

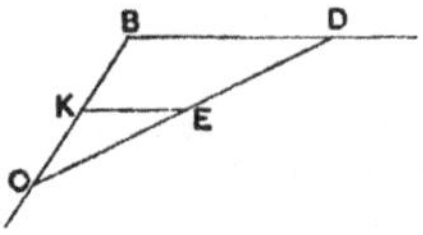

To do this we draw EK parallel to DB meeting BC in K, and then mark off KC equal to BK.

CE is then joined and produced to D; and CD is the straight line required.

Now, joining AC, AE, AD in the figure above, we have, since CD is bisected at E,

(1) in the triangle ACD,

$$AC^2 + AD^2 = 2AE^2 + 2ED^2,$$

and also (2) in the triangle BCD,

$$BC^2 + BD^2 = 2BE^2 + 2ED^2.$$

Subtracting, and remembering that the triangles ABC, ABD are right-angled, so that

$$AC^2 - BC^2 = AB^2,$$

and $$AD^2 - BD^2 = AB^2,$$

we have $$2AB^2 = 2AE^2 - 2BE^2,$$

or $$AE^2 = AB^2 + BE^2,$$

whence [I. 48] the angle ABE is a right angle, and AB is perpendicular to BE.

It follows of course from this proposition that the perpendicular AB is the *shortest distance from* A *to the plane* MN.

And it can readily be proved that,

If from a point without a plane oblique straight lines be drawn to the plane, (1) *those meeting the plane at equal distances from the foot of the perpendicular are equal, and*

(2) *of two straight lines meeting the plane at unequal distances from the foot of the perpendicular, the more remote is the greater.*

Lastly, it is easily seen that

From a point outside a plane only one perpendicular can be drawn to that plane.

For, if possible, let there be two perpendiculars. Then a plane can be drawn through them, and this will cut the original plane in a straight line.

This straight line and the two perpendiculars will form a plane triangle which has two right angles: which is impossible.

PROPOSITION 5.

If a straight line be set up at right angles to three straight lines which meet one another, at their common point of section, the three straight lines are in one plane.

For let a straight line AB be set up at right angles to the three straight lines BC, BD, BE, at their point of meeting at B;

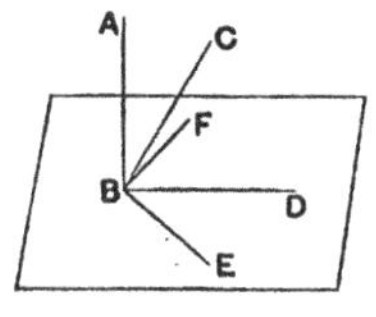

I say that BC, BD, BE are in one plane.

For suppose they are not, but, if possible, let BD, BE be in the plane of reference and BC in one more elevated;

let the plane through AB, BC be produced;

it will thus make, as common section in the plane of reference, a straight line. [XI. 3]

Let it make BF.

Therefore the three straight lines AB, BC, BF are in one plane, namely that drawn through AB, BC.

Now, since AB is at right angles to each of the straight lines BD, BE,

therefore AB is also at right angles to the plane through BD, BE. [XI. 4]

But the plane through BD, BE is the plane of reference; therefore AB is at right angles to the plane of reference.

Thus AB will also make right angles with all the straight lines which meet it and are in the plane of reference. [XI. Def. 3]

But BF which is in the plane of reference meets it; therefore the angle ABF is right.

But, by hypothesis, the angle ABC is also right; therefore the angle ABF is equal to the angle ABC.

And they are in one plane:
which is impossible.

Therefore the straight line BC is not in a more elevated plane;
therefore the three straight lines BC, BD, BE are in one plane.

Therefore, if a straight line be set up at right angles to three straight lines, at their point of meeting, the three straight lines are in one plane. Q. E. D.

It follows that, *if a right angle be turned about one of the straight lines containing it the other will describe a plane.*

At any point in a straight line it is possible to draw *only one* plane which is at right angles to the straight line.

One such plane can be found by taking any two planes through the given straight line, drawing perpendiculars to the straight line in the respective planes, e.g. BO, CO in the planes AOB, AOC, each perpendicular to AO, and then drawing a plane (BOC) through the perpendiculars.

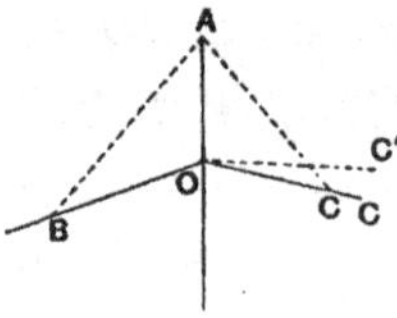

If there were another plane through O perpendicular to AO, it must meet the plane through AO and some perpendicular to it as OC in a straight line OC' different from OC.

Then, by XI. 4, AOC' is a right angle, and in the same plane with the right angle AOC: which is impossible.

Next, *one plane and only one can be drawn through a point outside a straight line at right angles to that line.*

Let P be the given point, AB the given straight line.

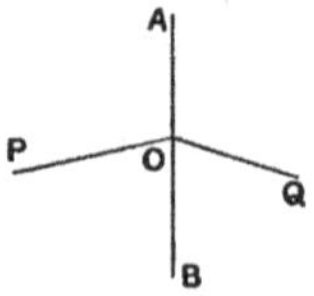

In the plane through P and AB, draw PO perpendicular to AB, and through O draw another straight line OQ at right angles to AB.

Then the plane through OP, OQ is perpendicular to AB.

If there were another plane through P perpendicular to AB, either

(1) it would intersect AB at O but not pass through OQ, or

(2) it would intersect AB at a point different from O.

In either case, an absurdity would result.

PROPOSITION 6.

If two straight lines be at right angles to the same plane, the straight lines will be parallel.

For let the two straight lines AB, CD be at right angles to the plane of reference;

I say that AB is parallel to CD.

For let them meet the plane of reference at the points B, D,

let the straight line BD be joined,

let DE be drawn, in the plane of reference, at right angles to BD,

let DE be made equal to AB,

and let BE, AE, AD be joined.

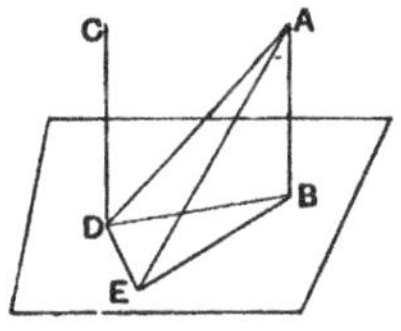

Now, since AB is at right angles to the plane of reference, it will also make right angles with all the straight lines which meet it and are in the plane of reference. [XI. Def. 3]

But each of the straight lines BD, BE is in the plane of reference and meets AB;

therefore each of the angles ABD, ABE is right.

For the same reason

each of the angles CDB, CDE is also right.

And, since AB is equal to DE,

and BD is common,

the two sides AB, BD are equal to the two sides ED, DB;

and they include right angles;

therefore the base AD is equal to the base BE. [I. 4]

And, since AB is equal to DE,

while AD is also equal to BE,

the two sides AB, BE are equal to the two sides ED, DA;

and AE is their common base;

therefore the angle ABE is equal to the angle EDA. [I. 8]

But the angle ABE is right;
therefore the angle EDA is also right;
therefore ED is at right angles to DA.

But it is also at right angles to each of the straight lines BD, DC;
therefore ED is set up at right angles to the three straight lines BD, DA, DC at their point of meeting;
therefore the three straight lines BD, DA, DC are in one plane. [XI. 5]

But, in whatever plane DB, DA are, in that plane is AB also,
for every triangle is in one plane; [XI. 2]
therefore the straight lines AB, BD, DC are in one plane.

And each of the angles ABD, BDC is right;
therefore AB is parallel to CD. [I. 28]

Therefore etc. Q. E. D.

If anyone wishes to convince himself of the real necessity for some general agreement as to the order in which propositions in elementary geometry should be taken, let him contemplate the hopeless result of too much independence on the part of editors in the matter of this proposition and its converse, XI. 8.

Legendre adopts a different, and elegant, method of proof; but he applies it to XI. 8, which he gives first, and then deduces XI. 6 from it by *reductio ad absurdum*. Dr Mehler uses Legendre's method of proof but applies it to XI. 6, and then gives XI. 8 as a deduction from it. Lardner follows Legendre. Holgate, the editor of a recent American book, gives Euclid's proof of XI. 6 and deduces XI. 8 by *reductio ad absurdum*. His countrymen, Schultze and Sevenoak, give XI. 8 first, but put it after, and deduce it from, Eucl. XI. 10; they then give XI. 6, practically as a deduction from XI. 8 by *reductio ad absurdum*, after a proposition corresponding to Eucl. XI. 11 and 12, and a corollary to the effect that through a given point one and only one perpendicular can be drawn to a given plane.

We will now give the proof of XI. 6 by Legendre's method (adopted by Smith and Bryant as well as by Mehler).

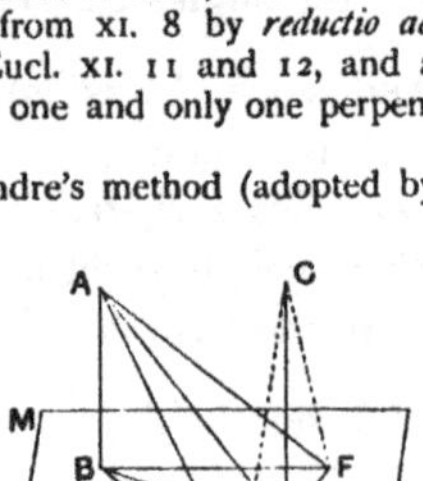

Let AB, CD be both perpendicular to the same plane MN.

Join BD.

Now, since BD meets AB, CD, both of which are perpendicular to the plane MN in which BD is,
the angles ABD, CDB are right angles.

AB, CD will therefore be parallel *provided that they are in the same plane.*

Through D draw EDF, in the plane MN, at right angles to BD, and make ED equal to DF.

Join *BE*, *BF*, *AE*, *AD*, *AF*.

Then the triangles *BDE*, *BDF* are equal in all respects (by I. 4), so that

BE is equal to *BF*.

It follows, since the angles *ABE*, *ABF* are right, that the triangles *ABE*, *ABF* are equal in all respects, and

AE is equal to *AF*.

[Mehler now argues elegantly thus. If *CE*, *CF* be also joined, it is clear that

CE is equal to *CF*.

Hence each of the four points *A*, *B*, *C*, *D* is equidistant from the two points *E*, *F*.

Therefore the points A, B, C, D *are in one plane*, so that *AB*, *CD* are parallel.

If, however, we do not use the locus of points equidistant from two fixed points, we proceed as follows.]

The triangles *AED*, *AFD* have their sides equal respectively;

hence [I. 8] the angles *ADE*, *ADF* are equal,

so that *ED* is at right angles to *AD*.

Thus *ED* is at right angles to *BD*, *AD*, *CD*;

therefore *CD* is in the plane through *AD*, *BD*. [XI. 5]

But *AB* is in that same plane; [XI. 2]

therefore *AB*, *CD* are in the same plane.

And the angles *ABD*, *CDB* are right;

therefore *AB*, *CD* are parallel.

PROPOSITION 7.

If two straight lines be parallel and points be taken at random on each of them, the straight line joining the points is in the same plane with the parallel straight lines.

Let *AB*, *CD* be two parallel straight lines,

and let points *E*, *F* be taken at random on them respectively;

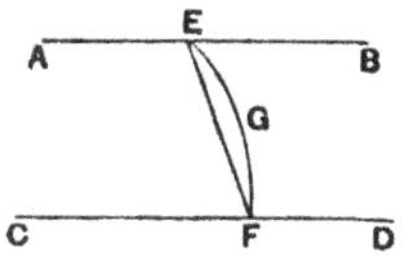

I say that the straight line joining the points *E*, *F* is in the same plane with the parallel straight lines.

For suppose it is not, but, if possible, let it be in a more elevated plane as *EGF*,

and let a plane be drawn through *EGF*;

it will then make, as section in the plane of reference, a straight line. [XI. 3]

Let it make it, as EF;
therefore the two straight lines EGF, EF will enclose an area:
which is impossible.

Therefore the straight line joined from E to F is not in a plane more elevated;
therefore the straight line joined from E to F is in the plane through the parallel straight lines AB, CD.

Therefore etc.

Q. E. D.

It is true that this proposition, in the form in which Euclid enunciates it, is hardly necessary if the plane is defined as a surface such that, if any two points be taken in it, the straight line joining them lies wholly in the surface. But Euclid did not give this definition; and, moreover, Prop. 2 would be usefully supplemented by a proposition which should prove that *two parallel straight lines determine a plane* (i.e. one plane and one *only*) *which also contains all the straight lines which join a point on one of the parallels to a point on the other.* That there cannot be *two* planes through a pair of parallels would be proved in the same way as we prove that two or three intersecting straight lines cannot be in two different planes, inasmuch as each transversal lying in one of the two supposed planes through the parallels would lie wholly in the other also, so that the two supposed planes must coincide throughout (cf. note on Prop. 2 above).

But, whatever be the value of the proposition as it is, Simson seems to have spoilt it completely. He leaves out the construction of a plane through EGF, which, as Euclid says, must cut the plane containing the parallels in a straight line; and, instead, he says, "In the plane $ABCD$ in which the parallels are draw the straight line EHF from E to F." Now, although we can easily draw a straight line from E to F, to claim that we can draw it *in the plane in which the parallels are* is surely to assume the very result which is to be proved. All that we could properly say is that the straight line joining E to F is in *some* plane which contains the parallels; we do not know that there is no more than *one* such plane, or that the parallels determine a plane *uniquely*, without some such argument as that which Euclid gives.

Nor can I subscribe to the remarks in Simson's note on the proposition. He says (1) "This proposition has been put into this book by some unskilful editor, as is evident from this, that straight lines which are drawn from one point to another in a plane are, in the preceding books, supposed to be in that plane; and if they were not, some demonstrations in which one straight line is supposed to meet another would not be conclusive. For instance, in Prop. 30, Book I, the straight line GK would not meet EF, if GK were not in the plane in which are the parallels AB, CD, and in which, by hypothesis, the straight line EF is." But the subject-matter of Book I. and Book XI. is quite different; in Book I. everything is in one plane, and when Euclid, in defining parallels, says they are straight lines *in the same plane* etc., he only does so because he must, in order to exclude non-intersecting straight lines which are *not* parallel. Thus in I. 30 there is nothing wrong in assuming that there may be three parallels in one plane, and that the straight line GHK cuts all three.

But in Book XI. it becomes a question whether there can be *more* than one plane through parallel straight lines.

Simson goes on to say (2) "Besides, this 7th Proposition is demonstrated by the preceding 3rd; in which the very same thing which is proposed to be demonstrated in the 7th is twice assumed, viz., that the straight line drawn from one point to another in a plane is in that plane." But there is nothing in Prop. 3 about a plane in which two parallel straight lines are; therefore there is no assumption of the result of Prop. 7. What is assumed is that, given two points in *a plane*, they can be joined by a straight line in the plane: a legitimate assumption.

Lastly, says Simson, "And the same thing is assumed in the preceding 6th Prop. in which the straight line which joins the points *B*, *D* that are in the plane to which *AB* and *CD* are at right angles is supposed to be in that plane." Here again there is no question of a *plane in which two parallels are*; so that the criticism here, as with reference to Prop. 3, appears to rest on a misapprehension.

PROPOSITION 8.

If two straight lines be parallel, and one of them be at right angles to any plane, the remaining one will also be at right angles to the same plane.

Let *AB*, *CD* be two parallel straight lines,

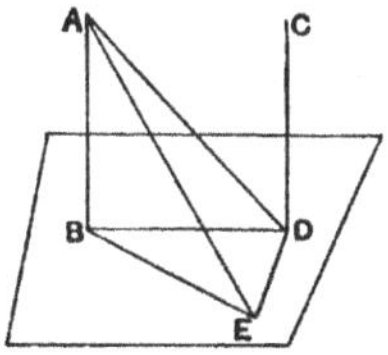

and let one of them, *AB*, be at right angles to the plane of reference;

I say that the remaining one, *CD*, will also be at right angles to the same plane.

For let *AB*, *CD* meet the plane of reference at the points *B*, *D*,

and let *BD* be joined;

therefore *AB*, *CD*, *BD* are in one plane. [XI. 7]

Let *DE* be drawn, in the plane of reference, at right angles to *BD*,

let *DE* be made equal to *AB*,

and let *BE*, *AE*, *AD* be joined.

Now, since *AB* is at right angles to the plane of reference, therefore *AB* is also at right angles to all the straight lines which meet it and are in the plane of reference; [XI. Def. 3]

therefore each of the angles *ABD*, *ABE* is right.

And, since the straight line *BD* has fallen on the parallels *AB*, *CD*, therefore the angles *ABD*, *CDB* are equal to two right angles. [I. 29]

But the angle *ABD* is right;
therefore the angle *CDB* is also right;
therefore *CD* is at right angles to *BD*.

And, since *AB* is equal to *DE*,
and *BD* is common,
the two sides *AB*, *BD* are equal to the two sides *ED*, *DB*;
and the angle *ABD* is equal to the angle *EDB*,
for each is right;
therefore the base *AD* is equal to the base *BE*.

And, since *AB* is equal to *DE*,
and *BE* to *AD*,
the two sides *AB*, *BE* are equal to the two sides *ED*, *DA* respectively,
and *AE* is their common base;
therefore the angle *ABE* is equal to the angle *EDA*.

But the angle *ABE* is right;
therefore the angle *EDA* is also right;
therefore *ED* is at right angles to *AD*.

But it is also at right angles to *DB*;
therefore *ED* is also at right angles to the plane through *BD*, *DA*. [XI. 4]

Therefore *ED* will also make right angles with all the straight lines which meet it and are in the plane through *BD*, *DA*.

But *DC* is in the plane through *BD*, *DA*, inasmuch as *AB*, *BD* are in the plane through *BD*, *DA*, [XI. 2]
and *DC* is also in the plane in which *AB*, *BD* are.

Therefore *ED* is at right angles to *DC*,
so that *CD* is also at right angles to *DE*.

But *CD* is also at right angles to *BD*.

Therefore *CD* is set up at right angles to the two straight lines *DE*, *DB* which cut one another, from the point of section at *D*;

so that CD is also at right angles to the plane through DE, DB. [XI. 4]

But the plane through DE, DB is the plane of reference; therefore CD is at right angles to the plane of reference.

Therefore etc.

Q. E. D.

Simson objects to the words which explain why DC is in the plane through BD, DA, viz. "inasmuch as AB, BD are in the plane through BD, DA, and DC is also in the plane in which AB, BD are," as being too roundabout. He concludes that they are corrupt or interpolated, and that we ought only to have the words "because all three are in the plane in which are the parallels AB, CD" (by Prop. 7 preceding). But I think Euclid's words can be defended. Prop. 7 says nothing of a plane determined by *two* transversals as BD, DA are. Hence it is natural to say that DC is in the same plane in which AB, BD are [Prop. 7], and AB, BD are in the same plane as BD, DA [Prop. 2], so that DC is in the plane through BD, DA.

Legendre's alternative proof is split by him into two propositions.

(1) *Let* AB *be a perpendicular to the plane* MN *and* EF *a line situated in that plane; if from* B, *the foot of the perpendicular*, BD *be drawn perpendicular to* EF, *and* AD *be joined, I say that* AD *will be perpendicular to* EF.

(2) *If* AB *is perpendicular to the plane* MN, *every straight line* CD *parallel to* AB *will be perpendicular to the same plane.*

To prove both propositions together we suppose CD given, join BD, and draw EF perpendicular to BD in the plane MN.

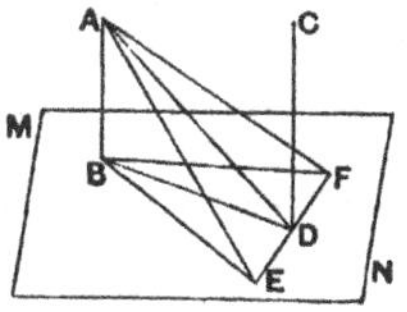

(1) As before, we make DE equal to DF and join BE, BF, AE, AF.

Then, since the angles BDE, BDF are right, and DE, DF equal,

BE is equal to BF. [I. 4]

And, since AB is perpendicular to the plane,

the angles ABE, ABF are both right.

Therefore, in the triangles ABE, ABF,

AE is equal to AF. [I. 4]

Lastly, in the triangles ADE, ADF, since AE is equal to AF, and DE to DF, while AD is common,

the angle ADE is equal to the angle ADF, [I. 8]

so that AD is perpendicular to EF.

(2) ED being thus perpendicular to DA, and also (by construction) perpendicular to DB,

ED is perpendicular to the plane ADB. [XI. 4]

But CD, being parallel to AB, is in the plane ABD;

therefore ED is perpendicular to CD. [XI. Def. 3]

Also, since AB, CD are parallel,
and ABD is a right angle,
CDB is also a right angle.

Thus CD is perpendicular to both DE and DB, and therefore to the plane MN through DE, DB.

PROPOSITION 9.

Straight lines which are parallel to the same straight line and are not in the same plane with it are also parallel to one another.

For let each of the straight lines AB, CD be parallel to EF, not being in the same plane with it;

I say that AB is parallel to CD.

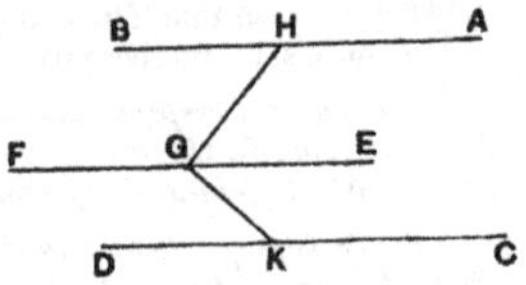

For let a point G be taken at random on EF, and from it let there be drawn GH, in the plane through EF, AB, at right angles to EF, and GK in the plane through FE, CD again at right angles to EF.

Now, since EF is at right angles to each of the straight lines GH, GK, therefore EF is also at right angles to the plane through GH, GK. [XI. 4]

And EF is parallel to AB;
therefore AB is also at right angles to the plane through HG, GK. [XI. 8]

For the same reason
CD is also at right angles to the plane through HG, GK;
therefore each of the straight lines AB, CD is at right angles to the plane through HG, GK.

But, if two straight lines be at right angles to the same plane, the straight lines are parallel; [XI. 6]
therefore AB is parallel to CD.

Q. E. D.

Proposition 10.

If two straight lines meeting one another be parallel to two straight lines meeting one another not in the same plane, they will contain equal angles.

For let the two straight lines *AB*, *BC* meeting one another be parallel to the two straight lines *DE*, *EF* meeting one another, not in the same plane;

I say that the angle *ABC* is equal to the angle *DEF*.

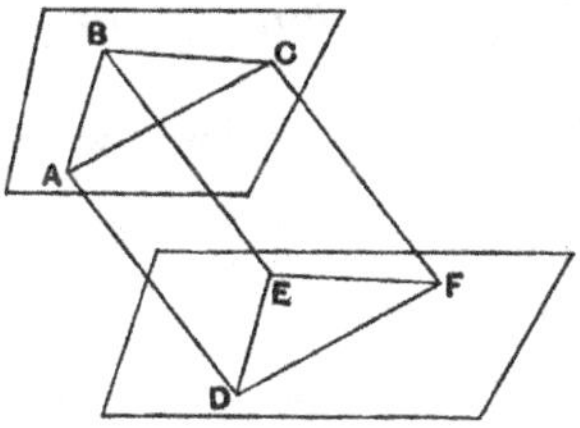

For let *BA*, *BC*, *ED*, *EF* be cut off equal to one another, and let *AD*, *CF*, *BE*, *AC*, *DF* be joined.

Now, since *BA* is equal and parallel to *ED*, therefore *AD* is also equal and parallel to *BE*. [I. 33]

For the same reason

CF is also equal and parallel to *BE*.

Therefore each of the straight lines *AD*, *CF* is equal and parallel to *BE*.

But straight lines which are parallel to the same straight line and are not in the same plane with it are parallel to one another; [XI. 9]

therefore *AD* is parallel and equal to *CF*.

And *AC*, *DF* join them;

therefore *AC* is also equal and parallel to *DF*. [I. 33]

Now, since the two sides *AB*, *BC* are equal to the two sides *DE*, *EF*,

and the base *AC* is equal to the base *DF*,

therefore the angle *ABC* is equal to the angle *DEF*. [I. 8]

Therefore etc.

Q. E. D.

The result of this proposition does not appear to be quoted in Euclid until XII. 3; but Euclid no doubt inserted it here advisedly, because it has the effect of incidentally proving that the "inclination of two planes to one another," as defined in XI. Def. 6, is one and the same angle at whatever point of the common section the plane angle measuring it is drawn.

PROPOSITION 11.

From a given elevated point to draw a straight line perpendicular to a given plane.

Let *A* be the given elevated point, and the plane of reference the given plane;

thus it is required to draw from the point *A* a straight line perpendicular to the plane of reference.

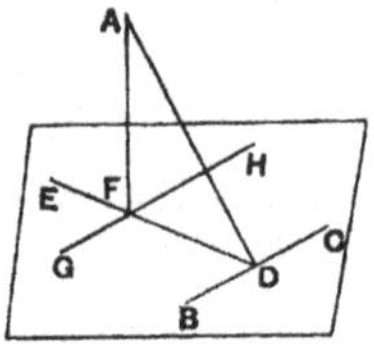

Let any straight line *BC* be drawn, at random, in the plane of reference, and let *AD* be drawn from the point *A* perpendicular to *BC*. [I. 12]

If then *AD* is also perpendicular to the plane of reference, that which was enjoined will have been done.

But, if not, let *DE* be drawn from the point *D* at right angles to *BC* and in the plane of reference, [I. 11]

let *AF* be drawn from *A* perpendicular to *DE*, [I. 12]

and let *GH* be drawn through the point *F* parallel to *BC*. [I. 31]

Now, since *BC* is at right angles to each of the straight lines *DA*, *DE*,

therefore *BC* is also at right angles to the plane through *ED*, *DA*. [XI. 4]

And *GH* is parallel to it;

but, if two straight lines be parallel, and one of them be at right angles to any plane, the remaining one will also be at right angles to the same plane; [XI. 8]

therefore *GH* is also at right angles to the plane through *ED*, *DA*.

Therefore *GH* is also at right angles to all the straight lines which meet it and are in the plane through *ED*, *DA*. [XI. Def. 3]

But *AF* meets it and is in the plane through *ED*, *DA*; therefore *GH* is at right angles to *FA*, so that *FA* is also at right angles to *GH*.

But *AF* is also at right angles to *DE*; therefore *AF* is at right angles to each of the straight lines *GH*, *DE*.

But, if a straight line be set up at right angles to two straight lines which cut one another, at the point of section, it will also be at right angles to the plane through them; [XI. 4] therefore *FA* is at right angles to the plane through *ED*, *GH*.

But the plane through *ED*, *GH* is the plane of reference; therefore *AF* is at right angles to the plane of reference.

Therefore from the given elevated point *A* the straight line *AF* has been drawn perpendicular to the plane of reference.

Q. E. F.

The text-books differ in the *form* which they give to this proposition rather than in substance. They commonly assume the construction of a *plane* through the point *A* at right angles to any straight line *BC* in the given plane (the construction being effected in the manner shown at the end of the note on XI. 5 above). The advantage of this method is that it enables a perpendicular to be drawn from a point *in* the plane also, by the same construction. (Where the letters for the two figures differ, those referring to the second figure are put in brackets.)

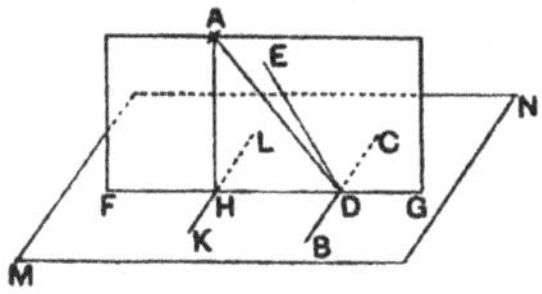

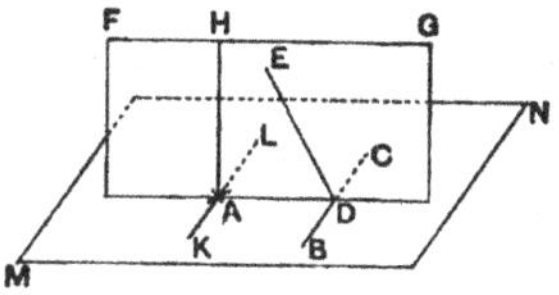

We can include the construction of the plane through *A* perpendicular to *BC*, and make the whole into one proposition, thus.

BC being any straight line in the given plane *MN*, draw *AD* perpendicular to *BC*.

In any plane passing through *BC* but not through *A* draw *DE* at right angles to *BC*.

Through *DA*, *DE* draw a plane; this will intersect the given plane *MN* in a straight line, as *FD* (*AD*).

In the plane *AG* draw *AH* perpendicular to *FG* (*AD*).

Then *AH* is the perpendicular required.

In the plane MN, through H in the first figure and A in the second, draw KL parallel to BC.

Now, since BC is perpendicular to both DA and DE, BC is perpendicular to the plane AG. [XI. 4]

Therefore KL, being parallel to BC, is also perpendicular to the plane AG [XI. 8], and therefore to AH which meets it and is in that plane.

Therefore AH is perpendicular to both FD (AD) and KL at their point of intersection.

Therefore AH is perpendicular to the plane MN.

Thus we have solved the problem in XI. 12 as well as that in XI. 11; and this direct method of drawing a perpendicular to a plane from a point *in* it is obviously preferable to Euclid's method by which the construction of a perpendicular to a plane from a point *without* it is assumed, and a line is merely drawn from a point in the plane parallel to the perpendicular obtained in XI. 11.

PROPOSITION 12.

To set up a straight line at right angles to a given plane from a given point in it.

Let the plane of reference be the given plane, and A the point in it;

thus it is required to set up from the point A a straight line at right angles to the plane of reference.

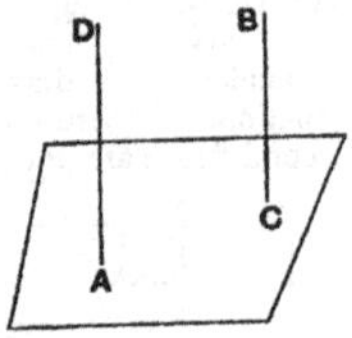

Let any elevated point B be conceived, from B let BC be drawn perpendicular to the plane of reference, [XI. 11]

and through the point A let AD be drawn parallel to BC. [I. 31]

Then, since AD, CB are two parallel straight lines, while one of them, BC, is at right angles to the plane of reference,

therefore the remaining one, AD, is also at right angles to the plane of reference. [XI. 8]

Therefore AD has been set up at right angles to the given plane from the point A in it.

Q. E. F.

PROPOSITION 13.

From the same point two straight lines cannot be set up at right angles to the same plane on the same side.

For, if possible, from the same point A let the two straight lines AB, AC be set up at right angles to the plane of reference and on the same side,

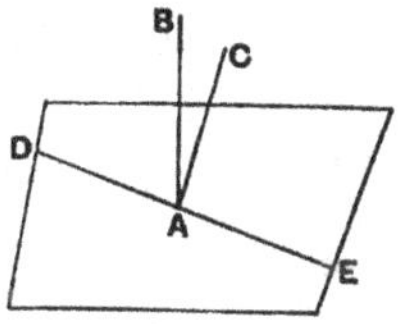

and let a plane be drawn through BA, AC;

it will then make, as section through A in the plane of reference, a straight line. [XI. 3]

Let it make DAE;

therefore the straight lines AB, AC, DAE are in one plane.

And, since CA is at right angles to the plane of reference, it will also make right angles with all the straight lines which meet it and are in the plane of reference. [XI. Def. 3]

But DAE meets it and is in the plane of reference;

therefore the angle CAE is right.

For the same reason

the angle BAE is also right;

therefore the angle CAE is equal to the angle BAE.

And they are in one plane:

which is impossible.

Therefore etc.

Q. E. D.

Simson added words to this as follows:

"Also, from a point above a plane there can be but one perpendicular to that plane; for, if there could be two, they would be parallel to one another [XI. 6], which is absurd."

Euclid does not give this result, but we have already had it in the note above to XI. 4 (*ad fin.*).

PROPOSITION 14.

Planes to which the same straight line is at right angles will be parallel.

For let any straight line AB be at right angles to each of the planes CD, EF;

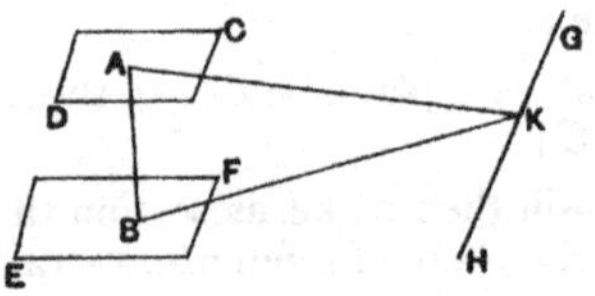

I say that the planes are parallel.

For, if not, they will meet when produced.

Let them meet;

they will then make, as common section, a straight line. [XI. 3]

Let them make GH;

let a point K be taken at random on GH,

and let AK, BK be joined.

Now, since AB is at right angles to the plane EF,

therefore AB is also at right angles to BK which is a straight line in the plane EF produced; [XI. Def. 3]

therefore the angle ABK is right.

For the same reason

the angle BAK is also right.

Thus, in the triangle ABK, the two angles ABK, BAK are equal to two right angles:

which is impossible. [I. 17]

Therefore the planes CD, EF will not meet when produced;

therefore the planes CD, EF are parallel. [XI. Def. 8]

Therefore planes to which the same straight line is at right angles are parallel.

Q. E. D.

PROPOSITION 15.

If two straight lines meeting one another be parallel to two straight lines meeting one another, not being in the same plane, the planes through them are parallel.

For let the two straight lines *AB*, *BC* meeting one another be parallel to the two straight lines *DE*, *EF* meeting one another, not being in the same plane;

I say that the planes produced through *AB*, *BC* and *DE*, *EF* will not meet one another.

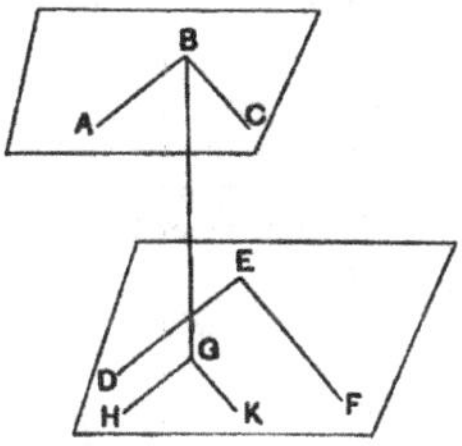

For let *BG* be drawn from the point *B* perpendicular to the plane through *DE*, *EF* [XI. 11], and let it meet the plane at the point *G*;

through *G* let *GH* be drawn parallel to *ED*, and *GK* parallel to *EF*. [I. 31]

Now, since *BG* is at right angles to the plane through *DE*, *EF*,

therefore it will also make right angles with all the straight lines which meet it and are in the plane through *DE*, *EF*. [XI. Def. 3]

But each of the straight lines *GH*, *GK* meets it and is in the plane through *DE*, *EF*;

therefore each of the angles *BGH*, *BGK* is right.

And, since *BA* is parallel to *GH*, [XI. 9]

therefore the angles *GBA*, *BGH* are equal to two right angles. [I. 29]

But the angle *BGH* is right;

therefore the angle *GBA* is also right;

therefore *GB* is at right angles to *BA*.

For the same reason

GB is also at right angles to *BC*.

Since then the straight line *GB* is set up at right angles to the two straight lines *BA*, *BC* which cut one another, therefore *GB* is also at right angles to the plane through *BA*, *BC*. [XI. 4]

But planes to which the same straight line is at right angles are parallel; [XI. 14]
therefore the plane through AB, BC is parallel to the plane through DE, EF.

Therefore, if two straight lines meeting one another be parallel to two straight lines meeting one another, not in the same plane, the planes through them are parallel.

Q. E. D.

This result is arrived at in the American text-books already quoted by starting from the relation between a plane and a straight line parallel to it. The series of propositions is worth giving. A straight line and a plane being parallel if they do not meet however far they may be produced, we have the following propositions.

1. *Any plane containing one, and only one, of two parallel straight lines is parallel to the other.*

For suppose AB, CD to be parallel and CD to lie in the plane MN.

Then AB, CD determine a plane intersecting MN in the straight line CD.

Thus, if AB meets MN, it must meet it at some point in CD.

But this is impossible, since AB is parallel to CD.

Therefore AB will not meet the plane MN, and is therefore parallel to it.

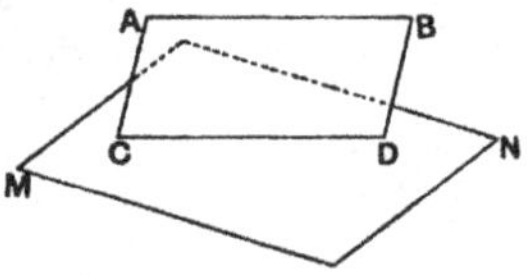

[This proposition and the proof are in Legendre.]

The following theorems follow as corollaries.

2. *Through a given straight line a plane can be drawn parallel to any other given straight line; and, if the lines are not parallel, only one such plane can be drawn.*

We have simply to draw through any point on the first line a straight line parallel to the second line and then pass a plane through these two intersecting lines. This plane is then, by the above proposition, parallel to the second given straight line.

3. *Through a given point a plane can be drawn parallel to any two straight lines in space; and, if the latter are not parallel, only one such plane can be drawn.*

Here we draw through the point straight lines parallel respectively to the given straight lines and then draw a plane through the lines so drawn.

Next we have the partial converse of the first proposition above.

4. *If a straight line is parallel to a plane, it is also parallel to the intersection of any plane through it with the given plane.*

Let AB be parallel to the plane MN, and let any plane through AB intersect MN in CD.

Now AB and CD cannot meet, because, if they did, AB would meet the plane MN.

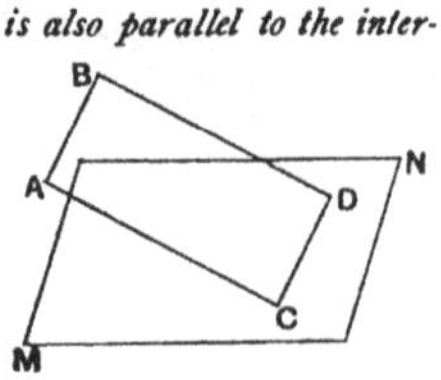

And AB, CD are in one plane.

Therefore AB, CD are parallel.

From this follows as a corollary:

5. *If each of two intersecting straight lines is parallel to a given plane, the plane containing them is parallel to the given plane.*

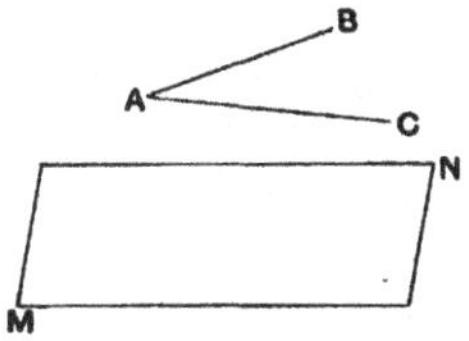

Let AB, AC be parallel to the plane MN.

Then, if the plane ABC were to meet the plane MN, the intersection would be parallel both to AB and to AC: which is impossible.

Lastly, we have Euclid's proposition.

6. *If two straight lines forming an angle are respectively parallel to two other straight lines forming an angle, the plane of the first angle is parallel to the plane of the second.*

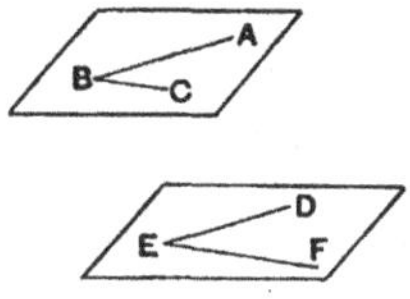

Let ABC, DEF be the angles formed by straight lines parallel to one another respectively.

Then, since AB is parallel to DE,

the plane of DEF is parallel to AB [(1) above].

Similarly the plane of DEF is parallel to BC.

Hence the plane of DEF is parallel to the plane of ABC [(5)].

Legendre arrives at the result by yet another method. He first proves Eucl. XI. 16 to the effect that, *if two parallel planes are cut by a third, the lines of intersection are parallel*, and then deduces from this that, *if two parallel straight lines are terminated by two parallel planes, the straight lines are equal in length.*

(The latter inference is obvious because the plane through the parallels cuts the parallel planes in parallel lines, which therefore, with the given parallel lines, form a parallelogram.)

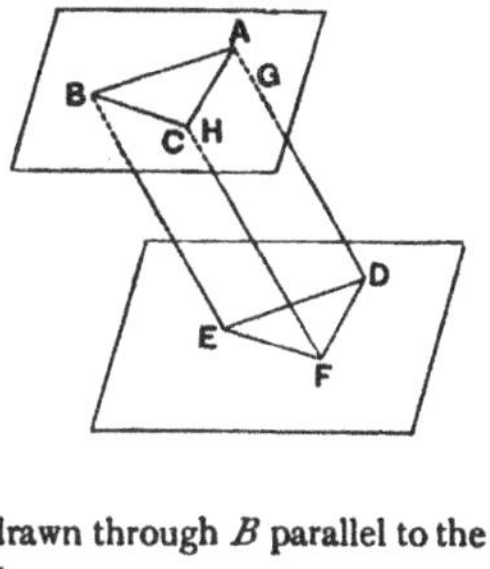

Legendre is now in a position to prove Euclid's proposition XI. 15.

If ABC, DEF be the angles, make AB equal to DE, and BC equal to EF, and join CA, FD, BE, CF, AD.

Then, as in Eucl. XI. 10, the triangles ABC, DEF are equal in all respects;

and AD, BE, CF are all equal.

It is now proved that the planes are parallel by *reductio ad absurdum* from the last preceding result. For, if the plane ABC is not parallel to the plane DEF, let the plane drawn through B parallel to the plane DEF meet CF, AD in H, G respectively.

Then, by the last result BE, HF, GD will all be equal.

But BE, CF, AD are all equal:

which is impossible.

Therefore etc.

PROPOSITION 16.

If two parallel planes be cut by any plane, their common sections are parallel.

For let the two parallel planes *AB*, *CD* be cut by the plane *EFGH*,
and let *EF*, *GH* be their common sections;
I say that *EF* is parallel to *GH*.

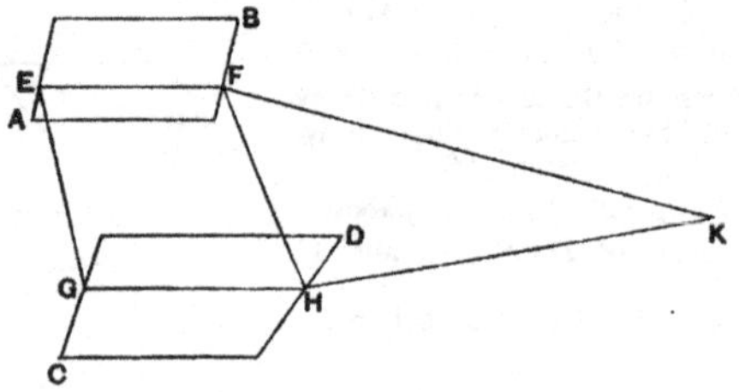

For, if not, *EF*, *GH* will, when produced, meet either in the direction of *F*, *H* or of *E*, *G*.

Let them be produced, as in the direction of *F*, *H*, and let them, first, meet at *K*.

Now, since *EFK* is in the plane *AB*,
therefore all the points on *EFK* are also in the plane *AB*. [XI. 1]

But *K* is one of the points on the straight line *EFK*;
therefore *K* is in the plane *AB*.

For the same reason
K is also in the plane *CD*;
therefore the planes *AB*, *CD* will meet when produced.

But they do not meet, because they are, by hypothesis, parallel;
therefore the straight lines *EF*, *GH* will not meet when produced in the direction of *F*, *H*.

Similarly we can prove that neither will the straight lines *EF*, *GH* meet when produced in the direction of *E*, *G*.

But straight lines which do not meet in either direction are parallel. [I. Def. 23]

Therefore *EF* is parallel to *GH*.

Therefore etc. Q. E. D.

Simson points out that, in here quoting I. Def. 23, Euclid should have said "But straight lines *in one plane* which do not meet in either direction are parallel."

From this proposition is deduced the converse of XI. 14.

If a straight line is perpendicular to one of two parallel planes, it is perpendicular to the other also.

For suppose that *MN*, *PQ* are two parallel planes, and that *AB* is perpendicular to *MN*.

Through *AB* draw any plane, and let it intersect the planes *MN*, *PQ* in *AC*, *BD* respectively.

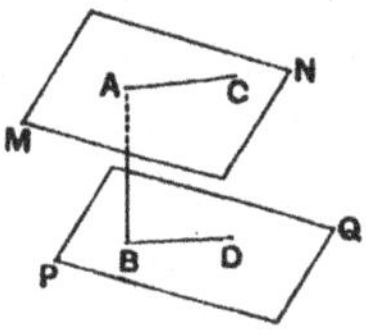

Therefore *AC*, *BD* are parallel. [XI. 16]

But *AC* is perpendicular to *AB*;

therefore *AB* is also perpendicular to *BD*.

That is, *AB* is perpendicular to any line in *PQ* passing through *B*;

therefore *AB* is perpendicular to *PQ*.

It follows as a corollary that

Through a given point one plane, and only one, can be drawn parallel to a given plane.

In the above figure let *A* be the given point and *PQ* the given plane.

Draw *AB* perpendicular to *PQ*.

Through *A* draw a plane *MN* at right angles to *AB* (see note on XI. 5 above).

Then *MN* is parallel to *PQ*. [XI. 14]

If there could pass through *A* a second plane parallel to *PQ*, *AB* would also be perpendicular to it.

That is, *AB* would be perpendicular to two different planes through *A*: which is impossible (see the same note).

Also it is readily proved that,

If two planes are parallel to a third plane, they are parallel to one another.

PROPOSITION 17.

If two straight lines be cut by parallel planes, they will be cut in the same ratios.

For let the two straight lines *AB*, *CD* be cut by the parallel planes *GH*, *KL*, *MN* at the points *A*, *E*, *B* and *C*, *F*, *D*;

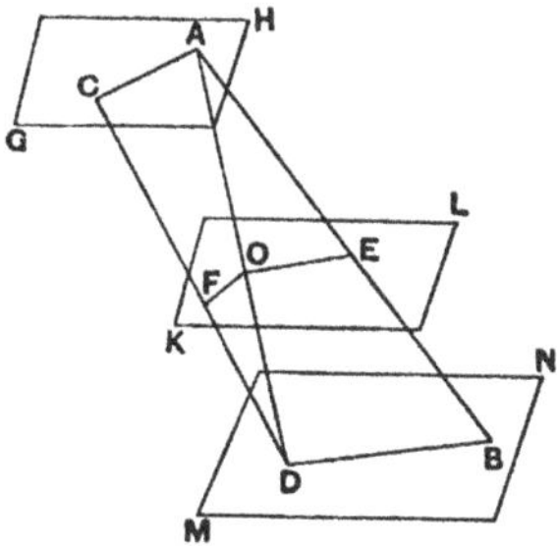

I say that, as the straight line *AE* is to *EB*, so is *CF* to *FD*.

For let *AC*, *BD*, *AD* be joined,

let *AD* meet the plane *KL* at the point *O*,

and let *EO*, *OF* be joined.

Now, since the two parallel planes *KL*, *MN* are cut by the plane *EBDO*,
their common sections *EO*, *BD* are parallel. [XI. 16]

For the same reason, since the two parallel planes *GH*, *KL* are cut by the plane *AOFC*,
their common sections *AC*, *OF* are parallel. [*id.*]

And, since the straight line *EO* has been drawn parallel to *BD*, one of the sides of the triangle *ABD*,
therefore, proportionally, as *AE* is to *EB*, so is *AO* to *OD*. [VI. 2]

Again, since the straight line *OF* has been drawn parallel to *AC*, one of the sides of the triangle *ADC*,
proportionally, as *AO* is to *OD*, so is *CF* to *FD*. [*id.*]

But it was also proved that, as *AO* is to *OD*, so is *AE* to *EB*;
therefore also, as *AE* is to *EB*, so is *CF* to *FD*. [V. 11]

Therefore etc.

Q. E. D.

PROPOSITION 18.

If a straight line be at right angles to any plane, all the planes through it will also be at right angles to the same plane.

For let any straight line *AB* be at right angles to the plane of reference;

I say that all the planes through *AB* are also at right angles to the plane of reference.

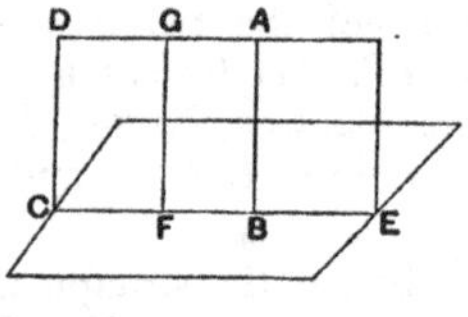

For let the plane *DE* be drawn through *AB*,
let *CE* be the common section of the plane *DE* and the plane of reference,
let a point *F* be taken at random on *CE*,
and from *F* let *FG* be drawn in the plane *DE* at right angles to *CE*. [I. 11]

Now, since *AB* is at right angles to the plane of reference,

AB is also at right angles to all the straight lines which meet it and are in the plane of reference; [XI. Def. 3]
so that it is also at right angles to *CE*;
therefore the angle *ABF* is right.

But the angle *GFB* is also right;
therefore *AB* is parallel to *FG*. [I. 28]

But *AB* is at right angles to the plane of reference;
therefore *FG* is also at right angles to the plane of reference. [XI. 8]

Now a plane is at right angles to a plane, when the straight lines drawn, in one of the planes, at right angles to the common section of the planes are at right angles to the remaining plane. [XI. Def. 4]

And *FG*, drawn in one of the planes *DE* at right angles to *CE*, the common section of the planes, was proved to be at right angles to the plane of reference;
therefore the plane *DE* is at right angles to the plane of reference.

Similarly also it can be proved that all the planes through *AB* are at right angles to the plane of reference.

Therefore etc.

Q. E. D.

Starting as Euclid does from the definition of perpendicular planes as planes such that all straight lines drawn in one of the planes at right angles to the common section are at right angles to the other plane, it is necessary for him to show that, if *F* be *any* point in *CE*, and *FG* be drawn in the plane *DE* at right angles to *CE*, *FG* will be perpendicular to the plane to which *AB* is perpendicular.

It is perhaps more scientific to make the definition, as Legendre makes it, a particular case of the definition of the *inclination of planes*. Perpendicular planes would thus be planes such that the angle which (when it is acute) Euclid calls the inclination of a plane to a plane is a right angle. When to this is added the fact incidentally proved in XI. 10 that the "inclination of a plane to a plane" is the same at whatever point in their common section it is drawn, it is sufficient to prove the perpendicularity of two planes if *one* straight line drawn, in one of them, perpendicular to their common section is perpendicular to the other.

If this point of view is taken, Props. 18, 19 are much simplified (cf. Legendre, H. M. Taylor, Smith and Bryant, Rausenberger, Schultze and Sevenoak, Holgate). The alternative proof is as follows.

Let *AB* be perpendicular to the plane *MN*, and *CE* any plane through *AB*, meeting the plane *MN* in the straight line *CD*.

In the plane *MN* draw *BF* at right angles to *CD*.

Then ABF is the angle which Euclid calls (in the case where it is acute) the "inclination of the plane to the plane."

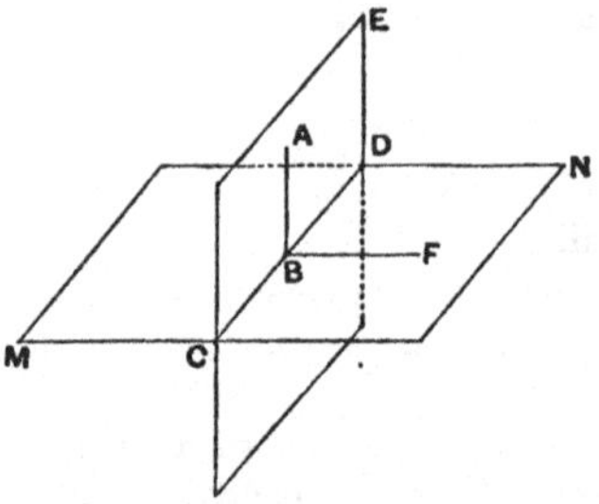

But, since AB is perpendicular to the plane MN, it is perpendicular to BF in it.

Therefore the angle ABF is a right angle;
whence the plane CE is perpendicular to the plane MN.

Proposition 19.

If two planes which cut one another be at right angles to any plane, their common section will also be at right angles to the same plane.

For let the two planes AB, BC be at right angles to the plane of reference,
and let BD be their common section;
I say that BD is at right angles to the plane of reference.

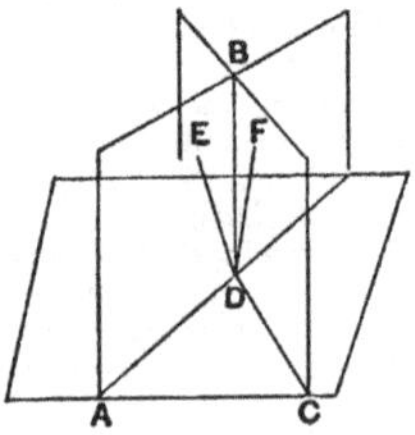

For suppose it is not, and from the point D let DE be drawn in the plane AB at right angles to the straight line AD, and DF in the plane BC at right angles to CD.

Now, since the plane AB is at right angles to the plane of reference,
and DE has been drawn in the plane AB at right angles to AD, their common section,
therefore DE is at right angles to the plane of reference. [XI. Def. 4]

Similarly we can prove that

DF is also at right angles to the plane of reference.

Therefore from the same point *D* two straight lines have been set up at right angles to the plane of reference on the same side:

which is impossible. [XI. 13]

Therefore no straight line except the common section *DB* of the planes *AB*, *BC* can be set up from the point *D* at right angles to the plane of reference.

Therefore etc.

Q. E. D.

Legendre, followed by other writers already quoted, uses a preliminary proposition equivalent to Euclid's definition of planes at right angles to one another.

If two planes are perpendicular to one another, a straight line drawn in one of them perpendicular to their common section will be perpendicular to the other.

Let the perpendicular planes *CE*, *MN* (figure of last note) intersect in *CD*, and let *AB* be drawn in *CE* perpendicular to *CD*.

In the plane *MN* draw *BF* at right angles to *CD*.

Then, since the planes are perpendicular, the angle *ABF* (their *inclination*) is a right angle.

Therefore *AB* is perpendicular to both *CD* and *BF*, and therefore to the plane *MN*.

We are now in a position to prove XI. 19, viz. *If two planes be perpendicular to a third, their intersection is also perpendicular to that third plane.*

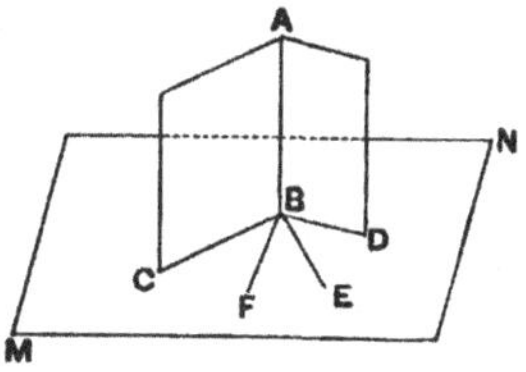

Let each of the two planes *AC*, *AD* intersecting in *AB* be perpendicular to the plane *MN*.

Let *AC*, *AD* intersect *MN* in *BC*, *BD* respectively.

In the plane *MN* draw *BE* at right angles to *BC* and *BF* at right angles to *BD*.

Now, since the planes *AC*, *MN* are at right angles, and *BE* is drawn in the latter perpendicular to *BC*, *BE* is perpendicular to the plane *AC*.

Hence *AB* is perpendicular to *BE*. [XI. 4]

Similarly *AB* is perpendicular to *BF*.

Therefore *AB* is perpendicular to the plane through *BE*, *BF*, i.e. to the plane *MN*.

An useful problem is that of drawing a common perpendicular to two straight lines not in one plane, and in connexion with this the following proposition may be given.

Given a plane and a straight line not perpendicular to it, one plane, and only one, can be drawn through the straight line perpendicular to the plane.

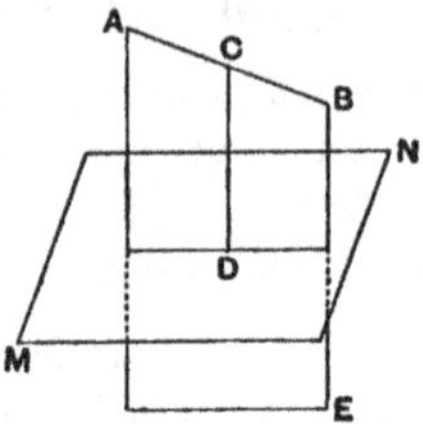

Let AB be the given straight line, MN the given plane.

From any point C in AB draw CD perpendicular to the plane MN.

Through AB and CD draw a plane AE.

Then the plane AE is perpendicular to the plane MN. [XI. 18]

If any other plane could be drawn through AB perpendicular to MN, the intersection AB of the two planes perpendicular to MN would itself be perpendicular to MN: [XI. 19]
which contradicts the hypothesis.

To draw a common perpendicular to two straight lines not in the same plane.

Let AB, CD be the given straight lines.

Through CD draw the plane MN parallel to AB (Prop. 2 in note to XI. 15).

Through AB draw the plane AF perpendicular to the plane MN (see the last preceding proposition).

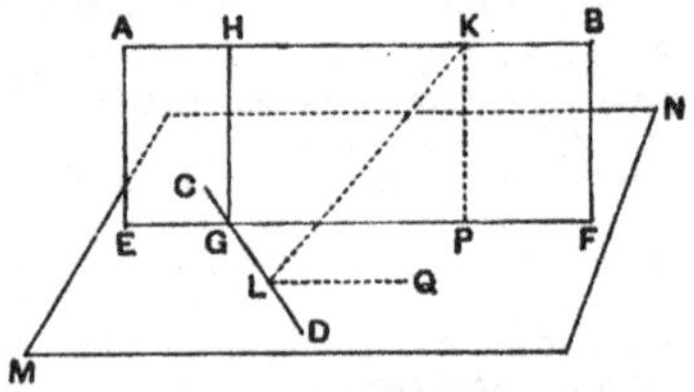

Let the planes AF, MN intersect in EF, and let EF meet CD in G.

From G, in the plane AF, draw GH at right angles to EF, meeting AB in H.

GH is then the required perpendicular.

For AB is parallel to EF (Prop. 4 in note to XI. 15); therefore GH, being perpendicular to EF, is also perpendicular to AB.

But, the plane AF being perpendicular to the plane MN, and GH being perpendicular to EF, their intersection,

GH is perpendicular to the plane MN, and therefore to CD.

Therefore GH is perpendicular to both AB and CD.

Only one common perpendicular can be drawn to two straight lines not in one plane.

For, if possible, let KL also be perpendicular to both AB and CD.

Let the plane through KL, AB meet the plane MN in LQ.

Then AB is parallel to LQ (Prop. 4 in note to XI. 15), so that KL, being perpendicular to AB, is also perpendicular to LQ.

Therefore KL is perpendicular to both CL and LQ, and consequently to the plane MN.

But, if KP be drawn in the plane AF perpendicular to EF, KP is also perpendicular to the plane MN.

Thus there are two perpendiculars from the point *K* to the plane *MN*: which is impossible.

Rausenberger's construction for the same problem is more elegant. Draw, he says, through each straight line a plane parallel to the other. Then draw through each straight line a plane perpendicular to the plane through the other. The two planes last drawn will intersect in a straight line, and this straight line is the common perpendicular required.

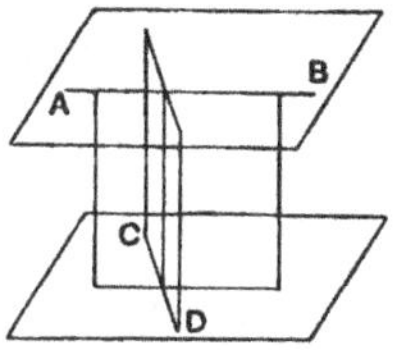

The form of the construction best suited for examination purposes, because the most self-contained, is doubtless that given by Smith and Bryant.

Let *AB*, *CD* be the two given straight lines.

Through any point *E* in *CD* draw *EF* parallel to *AB*.

From any point *G* in *AB* draw *GH* perpendicular to the plane *CDF*, meeting the plane in *H*.

Through *H* in the plane *CDF* draw *HK* parallel to *FE* or *AB*, to cut *CD* in *K*.

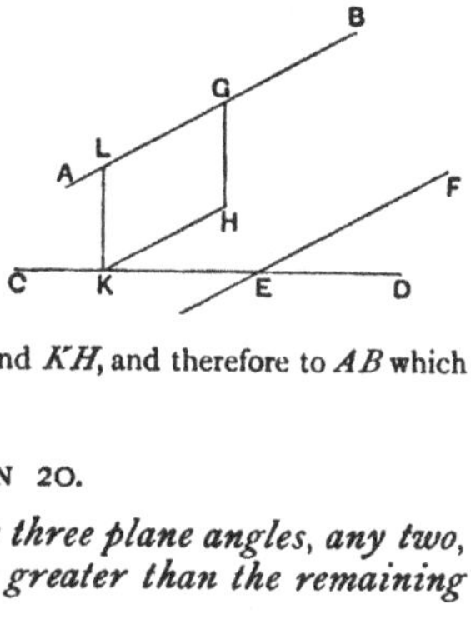

Then, since *AB*, *HK* are parallel, *AGHK* is a plane.

Complete the parallelogram *GHKL*.

Now, since *LK*, *GH* are parallel, and *GH* is perpendicular to the plane *CDF*,

LK is perpendicular to the plane *CDF*.

Therefore *LK* is perpendicular to *CD* and *KH*, and therefore to *AB* which is parallel to *KH*.

PROPOSITION 20.

If a solid angle be contained by three plane angles, any two, taken together in any manner, are greater than the remaining one.

For let the solid angle at *A* be contained by the three plane angles *BAC*, *CAD*, *DAB*;

I say that any two of the angles *BAC*, *CAD*, *DAB*, taken together in any manner, are greater than the remaining one.

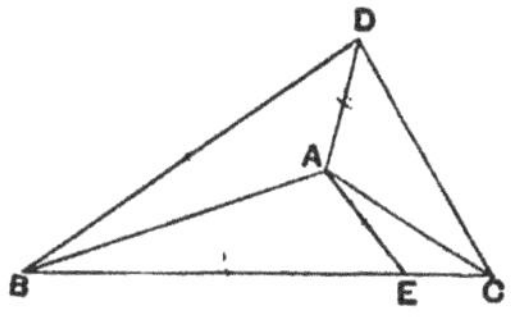

If now the angles *BAC*, *CAD*, *DAB* are equal to one another, it is manifest that any two are greater than the remaining one.

But, if not, let *BAC* be greater,

and on the straight line *AB*, and at the point *A* on it, let the

angle BAE be constructed, in the plane through BA, AC, equal to the angle DAB;
let AE be made equal to AD,
and let BEC, drawn across through the point E, cut the straight lines AB, AC at the points B, C;
let DB, DC be joined.

Now, since DA is equal to AE,
and AB is common,
two sides are equal to two sides;
and the angle DAB is equal to the angle BAE;
therefore the base DB is equal to the base BE. [I. 4]

And, since the two sides BD, DC are greater than BC, [I. 20]
and of these DB was proved equal to BE,
therefore the remainder DC is greater than the remainder EC.

Now, since DA is equal to AE,
and AC is common,
and the base DC is greater than the base EC,
therefore the angle DAC is greater than the angle EAC. [I. 25]

But the angle DAB was also proved equal to the angle BAE;
therefore the angles DAB, DAC are greater than the angle BAC.

Similarly we can prove that the remaining angles also, taken together two and two, are greater than the remaining one.

Therefore etc.

Q. E. D.

After excluding the obvious case in which all three angles are equal, Euclid goes on to say "If not, let the angle BAC be greater," without adding greater than *what*. Heiberg is clearly right in saying that he means greater than BAD, i.e. greater than *one* of the adjacent angles. This is proved by the words at the end "Similarly we can prove," etc. Euclid thus excludes as obvious the case where one of the three angles is not greater than either of the other two, but proves the remaining cases. This is scientific, but he might further have excluded as obvious the case in which one angle is greater than one of the others but equal to or less than the remaining one.

Simson remarks that the angle *BAC* may happen to be ***equal*** to one of the other two and writes accordingly "If they [all three angles] are not [equal], let *BAC* be that angle which is not less than either of the other two, and is greater than one of them *DAB*." He then proves, in the same way as Euclid does, that the angles *DAB*, *DAC* are greater than the angle *BAC*, adding finally: "But *BAC* is not less than either of the angles *DAB*, *DAC*; therefore *BAC*, with either of them, is greater than the other."

It would be better, as indicated by Legendre and Rausenberger, to begin by saying that, "If one of the three angles is either equal to or less than either of the other two, it is evident that the sum of those two is greater than the first. It is therefore only necessary to prove, *for the case in which one angle is greater than each of the others*, that the sum of the two latter is greater than the former.

Accordingly let *BAC* be greater than each of the other angles." We then proceed as in Euclid.

Proposition 21.

Any solid angle is contained by plane angles less than four right angles.

Let the angle at *A* be a solid angle contained by the plane angles *BAC*, *CAD*, *DAB*;

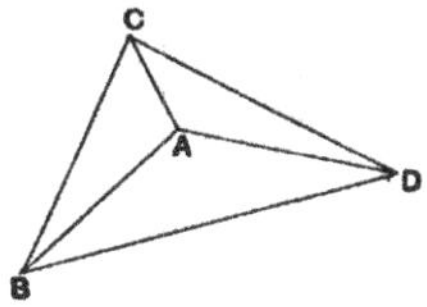

I say that the angles *BAC*, *CAD*, *DAB* are less than four right angles.

For let points *B*, *C*, *D* be taken at random on the straight lines *AB*, *AC*, *AD* respectively,

and let *BC*, *CD*, *DB* be joined.

Now, since the solid angle at *B* is contained by the three plane angles *CBA*, *ABD*, *CBD*,

any two are greater than the remaining one; [XI. 20]

therefore the angles *CBA*, *ABD* are greater than the angle *CBD*.

For the same reason

the angles *BCA*, *ACD* are also greater than the angle *BCD*,

and the angles *CDA*, *ADB* are greater than the angle *CDB*;

therefore the six angles *CBA*, *ABD*, *BCA*, *ACD*, *CDA*, *ADB* are greater than the three angles *CBD*, *BCD*, *CDB*.

But the three angles *CBD*, *BDC*, *BCD* are equal to two right angles; [I. 32]

therefore the six angles *CBA*, *ABD*, *BCA*, *ACD*, *CDA*, *ADB* are greater than two right angles.

And, since the three angles of each of the triangles *ABC*, *ACD*, *ADB* are equal to two right angles,
therefore the nine angles of the three triangles, the angles *CBA*, *ACB*, *BAC*, *ACD*, *CDA*, *CAD*, *ADB*, *DBA*, *BAD* are equal to six right angles;
and of them the six angles *ABC*, *BCA*, *ACD*, *CDA*, *ADB*, *DBA* are greater than two right angles;
therefore the remaining three angles *BAC*, *CAD*, *DAB* containing the solid angle are less than four right angles.

Therefore etc.

Q. E. D.

It will be observed that, although Euclid enunciates this proposition for *any* solid angle, he only proves it for the particular case of a *trihedral* angle. This is in accordance with his manner of proving one case and leaving the others to the reader. The omission of the convex polyhedral angle here corresponds to the omission, after I. 32, of the proposition about the interior angles of a convex polygon given by Proclus and in most books. The proof of the present proposition for any convex polyhedral angle can of course be arranged so as not to assume the proposition that the interior angles of a convex polygon together with four right angles are equal to twice as many right angles as the figure has sides.

Let there be any convex polyhedral angle with *V* as vertex, and let it be cut by any plane meeting its faces in, say, the polygon *ABCDE*.

Take *O* any point within the polygon, and in its plane, and join *OA*, *OB*, *OC*, *OD*, *OE*.

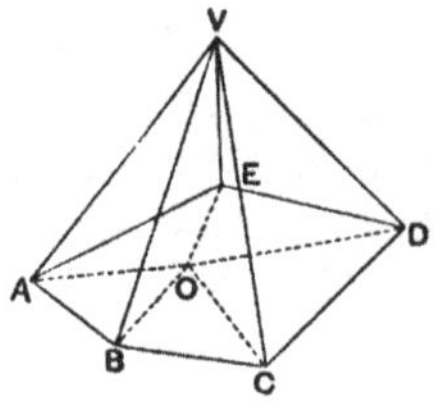

Then all the angles of the triangles with vertex *O* are equal to twice as many right angles as the polygon has sides; [I. 32]
therefore the interior angles of the polygon together with all the angles round *O* are equal to twice as many right angles as the polygon has sides.

Also the sum of the angles of the triangles *VAB*, *VBC*, etc., with vertex *V* are equal to twice as many right angles as the polygon has sides;
and all the said angles are equal to the sum of (1) the plane angles at *V* forming the polyhedral angle and (2) the base angles of the triangles with vertex *V*.

This latter sum is therefore equal to the sum of (3) all the angles round *O* and (4) all the interior angles of the polygon.

Now, by Euclid's proposition, of the three angles forming the solid angle at *A*, the angles *VAE*, *VAB* are together greater than the angle *EAB*.

Similarly, at *B*, the angles *VBA*, *VBC* are together greater than the angle *ABC*.

And so on.

Therefore, by addition, the base angles of the triangles with vertex *V*

[(2) above] are together greater than the sum of the angles of the polygon [(4) above].

Hence, by way of compensation, the sum of the plane angles at V [(1) above] is less than the sum of the angles round O [(3) above].

But the latter sum is equal to four right angles; therefore the plane angles forming the polyhedral angle are together less than four right angles.

The proposition is only true of *convex* polyhedral angles, i.e. those in which the plane of any face cannot, if produced, ever cut the solid angle.

There are certain propositions relating to equal (and symmetrical) trihedral angles which are necessary to the consideration of the polyhedra dealt with by Euclid, all of which (as before remarked) have trihedral angles only.

1. *Two trihedral angles are equal if two face angles and the included dihedral angle of the one are respectively equal to two face angles and the included dihedral angle of the other, the equal parts being arranged in the same order.*

2. *Two trihedral angles are equal if two dihedral angles and the included face angle of the one are respectively equal to two dihedral angles and the included face angle of the other, all equal parts being arranged in the same order.*

These propositions are proved immediately by superposition.

3. *Two trihedral angles are equal if the three face angles of the one are respectively equal to the three face angles of the other, and all are arranged in the same order.*

Let $V—ABC$ and $V'—A'B'C'$ be two trihedral angles such that the angle AVB is equal to the angle $A'V'B'$, the angle BVC to the angle $B'V'C'$, and the angle CVA to the angle $C'V'A'$.

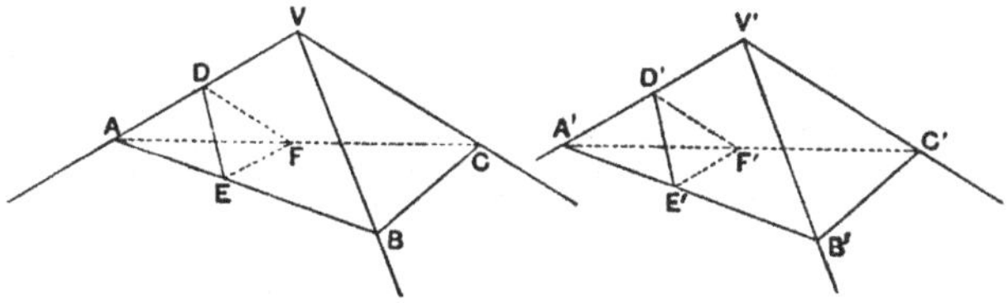

We first prove that *corresponding pairs of face angles include equal dihedral angles.*

E.g., the dihedral angle formed by the plane angles CVA, AVB is equal to that formed by the plane angles $C'V'A'$, $A'V'B'$.

Take points A, B, C on VA, VB, VC and points A', B', C' on $V'A'$, $V'B'$, $V'C'$, such that VA, VB, VC, $V'A'$, $V'B'$, $V'C'$ are all equal.

Join BC, CA, AB, $B'C'$, $C'A'$, $A'B'$.

Take any point D on AV, and measure $A'D'$ along $A'V'$ equal to AD.

From D draw DE in the plane AVB, and DF in the plane CVA, perpendicular to AV. Then DE, DF will meet AB, AC respectively, the angles VAB, VAC, the base angles of two isosceles triangles, being less than right angles.

Join EF.

Draw the triangle $D'E'F'$ in the same way.

Now, by means of the hypothesis and construction, it appears that the triangles VAB, $V'A'B'$ are equal in all respects.

So are the triangles VAC, $V'A'C'$, and the triangles VBC, $V'B'C'$.

Thus BC, CA, AB are respectively equal to $B'C'$, $C'A'$, $A'B'$, and the triangles ABC, $A'B'C'$ are equal in all respects.

Now, in the triangles ADE, $A'D'E'$,
the angles ADE, DAE are equal to the angles $A'D'E'$, $D'A'E'$ respectively,
and AD is equal to $A'D'$.

Therefore the triangles ADE, $A'D'E'$ are equal in all respects.

Similarly the triangles ADF, $A'D'F'$ are equal in all respects.

Thus, in the triangles AEF, $A'E'F'$,
EA, AF are respectively equal to $E'A'$, $A'F'$,
and the angle EAF is equal to the angle $E'A'F'$ (from above);
therefore the triangles AEF, $A'E'F'$ are equal in all respects.

Lastly, in the triangles DEF, $D'E'F'$, the three sides are respectively equal to the three sides;
therefore the triangles are equal in all respects.

Therefore the angles EDF, $E'D'F'$ are equal.

But these angles are the measures of the dihedral angles formed by the planes CVA, AVB and by the planes $C'V'A'$, $A'V'B'$ respectively.

Therefore these dihedral angles are equal.

Similarly for the other two dihedral angles.

Hence the trihedral angles coincide if one is applied to the other;
that is, they are equal.

To understand what is implied by "taken in the same order" we may suppose ourselves to be placed at the vertices, and to take the faces in clockwise direction, or the reverse, for *both* angles.

If the face angles and dihedral angles are *taken in reverse directions*, i.e. in clockwise direction in one and in counterclockwise direction in the other, then, if the other conditions in the above three propositions are fulfilled, the trihedral angles are not equal but *symmetrical.*

If the faces of a trihedral angle be produced beyond the vertex, they form another trihedral angle. It is easily seen that these *vertical trihedral angles* are *symmetrical.*

PROPOSITION 22.

If there be three plane angles of which two, taken together in any manner, are greater than the remaining one, and they are contained by equal straight lines, it is possible to construct a triangle out of the straight lines joining the extremities of the equal straight lines.

Let there be three plane angles ABC, DEF, GHK, of

which two, taken together in any manner, are greater than the remaining one, namely

the angles *ABC*, *DEF* greater than the angle *GHK*,

the angles *DEF*, *GHK* greater than the angle *ABC*,

and, further, the angles *GHK*, *ABC* greater than the angle *DEF*;

let the straight lines *AB*, *BC*, *DE*, *EF*, *GH*, *HK* be equal,

and let *AC*, *DF*, *GK* be joined;

I say that it is possible to construct a triangle out of straight lines equal to *AC*, *DF*, *GK*, that is, that any two of the straight lines *AC*, *DF*, *GK* are greater than the remaining one.

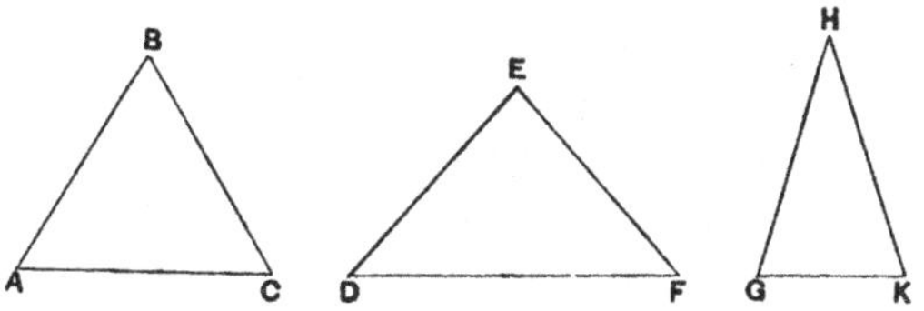

Now, if the angles *ABC*, *DEF*, *GHK* are equal to one another, it is manifest that, *AC*, *DF*, *GK* being equal also, it is possible to construct a triangle out of straight lines equal to *AC*, *DF*, *GK*.

But, if not, let them be unequal,

and on the straight line *HK*, and at the point *H* on it, let the angle *KHL* be constructed equal to the angle *ABC*;

let *HL* be made equal to one of the straight lines *AB*, *BC*, *DE*, *EF*, *GH*, *HK*,

and let *KL*, *GL* be joined.

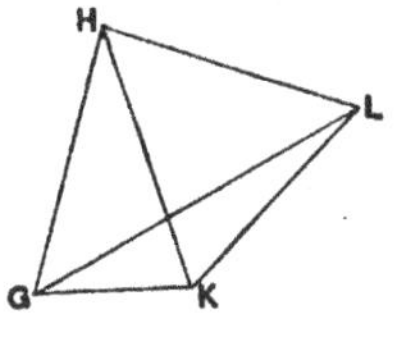

Now, since the two sides *AB*, *BC* are equal to the two sides *KH*, *HL*,

and the angle at *B* is equal to the angle *KHL*,

therefore the base *AC* is equal to the base *KL*. [I. 4]

And, since the angles *ABC*, *GHK* are greater than the angle *DEF*,

while the angle ABC is equal to the angle KHL,
therefore the angle GHL is greater than the angle DEF.

And, since the two sides GH, HL are equal to the two sides DE, EF,
and the angle GHL is greater than the angle DEF,
therefore the base GL is greater than the base DF. [I. 24]

But GK, KL are greater than GL.
Therefore GK, KL are much greater than DF.

But KL is equal to AC;
therefore AC, GK are greater than the remaining straight line DF.

Similarly we can prove that
AC, DF are greater than GK,
and further DF, GK are greater than AC.

Therefore it is possible to construct a triangle out of straight lines equal to AC, DF, GK.

Q. E. D.

The Greek text gives an alternative proof, which is relegated by Heiberg to the Appendix. Simson selected the alternative proof in preference to that given above; he objected however to words near the beginning, "If not, let the angles at the points B, E, H be unequal and that at B greater than either of the angles at E, H," and altered the words so as to take account of the possibility that the angle at B might be equal to one of the other two.

As will be seen, Euclid takes no account of the relative magnitude of the angles except as regards the case when all three are equal. Having proved that *one* base is less than the sum of the two others, he says that "similarly we can prove" the same thing for the other two bases.

If a distinction is to be made according to the relative magnitude of the three angles, we may say, as in the corresponding place in XI. 21, that, if one of the three angles is either equal to or less than *either* of the other two, the bases subtending those two angles must obviously be together greater than the base subtending the first. Thus it is only necessary to prove, for the case in which one angle is *greater than either of the others*, that the sum of the bases subtending those others is greater than that subtending the first. This is practically the course taken in the interpolated alternative proof.

Proposition 23.

To construct a solid angle out of three plane angles two of which, taken together in any manner, are greater than the remaining one: thus the three angles must be less than four right angles.

Let the angles *ABC*, *DEF*, *GHK* be the three given plane angles, and let two of these, taken together in any manner, be greater than the remaining one, while, further, the three are less than four right angles;

thus it is required to construct a solid angle out of angles equal to the angles *ABC*, *DEF*, *GHK*.

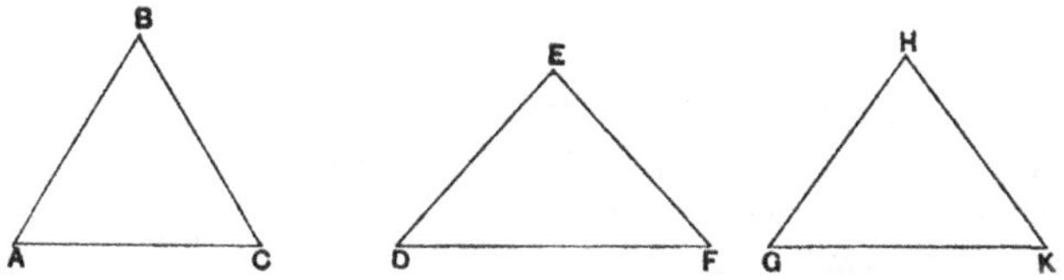

Let *AB*, *BC*, *DE*, *EF*, *GH*, *HK* be cut off equal to one another,

and let *AC*, *DF*, *GK* be joined;

it is therefore possible to construct a triangle out of straight lines equal to *AC*, *DF*, *GK*. [XI. 22]

Let *LMN* be so constructed that *AC* is equal to *LM*, *DF* to *MN*, and further *GK* to *NL*,

let the circle *LMN* be described about the triangle *LMN*,

let its centre be taken, and let it be *O*;

let *LO*, *MO*, *NO* be joined;

I say that *AB* is greater than *LO*.

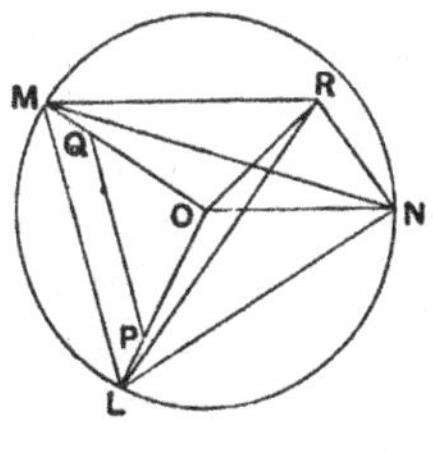

For, if not, *AB* is either equal to *LO*, or less.

First, let it be equal.

Then, since *AB* is equal to *LO*,

while *AB* is equal to *BC*, and *OL* to *OM*,

the two sides *AB*, *BC* are equal to the two sides *LO*, *OM* respectively;

and, by hypothesis, the base *AC* is equal to the base *LM*;

therefore the angle *ABC* is equal to the angle *LOM*. [I. 8]

For the same reason

the angle *DEF* is also equal to the angle *MON*,

and further the angle *GHK* to the angle *NOL*;

therefore the three angles *ABC*, *DEF*, *GHK* are equal to the three angles *LOM*, *MON*, *NOL*.

But the three angles *LOM*, *MON*, *NOL* are equal to four right angles;
therefore the angles *ABC*, *DEF*, *GHK* are equal to four right angles.

But they are also, by hypothesis, less than four right angles: which is absurd.

Therefore *AB* is not equal to *LO*.

I say next that neither is *AB* less than *LO*.

For, if possible, let it be so,
and let *OP* be made equal to *AB*, and *OQ* equal to *BC*,
and let *PQ* be joined.

Then, since *AB* is equal to *BC*,
OP is also equal to *OQ*,
so that the remainder *LP* is equal to *QM*.

Therefore *LM* is parallel to *PQ*, [VI. 2]
and *LMO* is equiangular with *PQO*; [I. 29]
therefore, as *OL* is to *LM*, so is *OP* to *PQ*; [VI. 4]
and alternately, as *LO* is to *OP*, so is *LM* to *PQ*. [V. 16]

But *LO* is greater than *OP*;
therefore *LM* is also greater than *PQ*.

But *LM* was made equal to *AC*;
therefore *AC* is also greater than *PQ*.

Since, then, the two sides *AB*, *BC* are equal to the two sides *PO*, *OQ*,
and the base *AC* is greater than the base *PQ*,
therefore the angle *ABC* is greater than the angle *POQ*. [I. 25]

Similarly we can prove that
the angle *DEF* is also greater than the angle *MON*,
and the angle *GHK* greater than the angle *NOL*.

Therefore the three angles *ABC*, *DEF*, *GHK* are greater than the three angles *LOM*, *MON*, *NOL*.

But, by hypothesis, the angles *ABC*, *DEF*, *GHK* are less than four right angles;
therefore the angles *LOM*, *MON*, *NOL* are much less than four right angles.

But they are also equal to four right angles: which is absurd.

Therefore *AB* is not less than *LO*.

And it was proved that neither is it equal; therefore *AB* is greater than *LO*.

Let then *OR* be set up from the point *O* at right angles to the plane of the circle *LMN*, [XI. 12]

and let the square on *OR* be equal to that area by which the square on *AB* is greater than the square on *LO*; [Lemma]

let *RL*, *RM*, *RN* be joined.

Then, since *RO* is at right angles to the plane of the circle *LMN*,

therefore *RO* is also at right angles to each of the straight lines *LO*, *MO*, *NO*.

And, since *LO* is equal to *OM*,

while *OR* is common and at right angles,

therefore the base *RL* is equal to the base *RM*. [I. 4]

For the same reason

RN is also equal to each of the straight lines *RL*, *RM*;

therefore the three straight lines *RL*, *RM*, *RN* are equal to one another.

Next, since by hypothesis the square on *OR* is equal to that area by which the square on *AB* is greater than the square on *LO*,

therefore the square on *AB* is equal to the squares on *LO*, *OR*.

But the square on *LR* is equal to the squares on *LO*, *OR*,

for the angle *LOR* is right; [I. 47]

therefore the square on *AB* is equal to the square on *RL*;

therefore *AB* is equal to *RL*.

But each of the straight lines *BC*, *DE*, *EF*, *GH*, *HK* is equal to *AB*,

while each of the straight lines *RM*, *RN* is equal to *RL*;

therefore each of the straight lines *AB*, *BC*, *DE*, *EF*, *GH*, *HK* is equal to each of the straight lines *RL*, *RM*, *RN*.

And, since the two sides LR, RM are equal to the two sides AB, BC,
and the base LM is by hypothesis equal to the base AC,
therefore the angle LRM is equal to the angle ABC. [I. 8]

For the same reason
the angle MRN is also equal to the angle DEF,
and the angle LRN to the angle GHK.

Therefore, out of the three plane angles LRM, MRN, LRN, which are equal to the three given angles ABC, DEF, GHK, the solid angle at R has been constructed, which is contained by the angles LRM, MRN, LRN.

Q. E. F.

Lemma.

But how it is possible to take the square on OR equal to that area by which the square on AB is greater than the square on LO, we can show as follows.

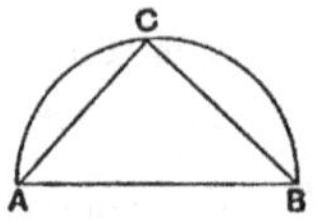

Let the straight lines AB, LO be set out,
and let AB be the greater;
let the semicircle ABC be described on AB,
and into the semicircle ABC let AC be fitted equal to the straight line LO, not being greater than the diameter AB; [IV. 1]
let CB be joined.

Since then the angle ACB is an angle in the semicircle ACB,
therefore the angle ACB is right. [III. 31]

Therefore the square on AB is equal to the squares on AC, CB. [I. 47]

Hence the square on AB is greater than the square on AC by the square on CB.

But AC is equal to LO.

Therefore the square on AB is greater than the square on LO by the square on CB.

If then we cut off OR equal to BC, the square on AB will be greater than the square on LO by the square on OR.

Q. E. F.

The whole difficulty in this proposition is the proof of a fact which makes the construction *possible*, viz. the fact that, if LMN be a triangle with sides respectively equal to the bases of the isosceles triangles which have the given angles as vertical angles and the equal sides all of the same length, then one of these equal sides, as AB, is greater than the radius LO of the circle circumscribing the triangle LMN.

Assuming that AB is greater than LO, we have only to draw from O a perpendicular OR to the plane of the triangle LMN, to make OR of such a length that the sum of the squares on LO, OR is equal to the square on AB, and to join RL, RM, RN. (The manner of finding OR such that the square on it is equal to the difference between the squares on AB and LO is shown in the Lemma at the end of the text of the proposition. We have already had the same construction in the Lemma after X. 13.)

Then clearly RL, RM, RN are equal to AB and to one another [I. 4 and I. 47].

Therefore the triangles LRM, MRN, NRL have their three sides respectively equal to those of the triangles ABC, DEF, GHK respectively.

Hence their vertical angles are equal to the three given angles respectively; and the required solid angle is constructed.

We return now to the proposition to be proved as a preliminary to the construction, viz. that, in the figures, AB is greater than LO.

It will be observed that Euclid, as his manner is, proves it for one case only, that, namely, in which O, the centre of the circle circumscribing the triangle LMN, falls *within* the triangle, leaving the other cases for the reader to prove. As usual, however, the two other cases are found in the Greek text, after the formal conclusion of the proposition, as above, ending with the words ὅπερ ἔδει ποιῆσαι. This position for the proofs itself suggests that they are not Euclid's but are interpolated; and this is rendered certain by the fact that words distinguishing three cases at the point where the centre O of the circumscribing circle is found, "It [the centre] will then be either within the triangle LMN or on one of its sides or without. First let it be within," are found in the MSS. B and V only and are manifestly interpolated. Nevertheless the additional two cases must have been inserted very early, as they are found in all the best MSS.

In order to give a clear view of the proof of all three cases as given in the text, we will reproduce all three (Euclid's as well as the others) with abbreviations to make them catch the eye better.

In all three cases the proof is by *reductio ad absurdum*, and it is proved first that AB cannot be *equal* to LO, and secondly that AB cannot be *less* than LO.

Case I.

(1) Suppose, if possible, that $AB = LO$.

Then AB, BC are respectively equal to LO, OM;

and $AC = LM$ (by construction).

Therefore $\angle ABC = \angle LOM$.

Similarly $\angle DEF = \angle MON$,

$\angle GHK = \angle NOL$.

Adding, we have

$$\angle ABC + \angle DEF + \angle GHK = \angle LOM + \angle MON + \angle NOL$$

$$= \text{four right angles}:$$

which contradicts the hypothesis.

Therefore $AB \neq LO$.

(2) Suppose that $AB < LO$.

Make OP, OQ (measured along OL, OM) each equal to AB.

Thus, OL, OM being equal also, it follows that

$$PQ \text{ is } \parallel \text{ to } LM.$$

Hence $$LM : PQ = LO : OP;$$

and, since $LO > OP$,

LM, i.e. AC, $> PQ$.

Thus, in $\triangle$s POQ, ABC, two sides are equal to two sides, and base $AC >$ base PQ;

therefore $$\angle ABC > \angle POQ, \quad \text{i.e. } \angle LOM.$$

Similarly $$\angle DEF > \angle MON,$$
$$\angle GHK > \angle NOL,$$

and it follows by addition that

$$\angle ABC + \angle DEF + \angle GHK > \text{(four right angles)}:$$

which again contradicts the hypothesis.

Case II.

(1) Suppose, if possible, that $AB = LO$.

Then $(AB + BC)$, or $(DE + EF) = MO + OL$
$$= MN$$
$$= DF:$$

which contradicts the hypothesis.

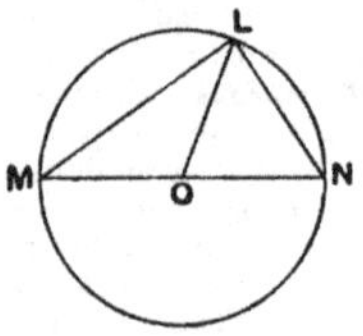

(2) The supposition that $AB < LO$ is even more impossible; for in this case it would result that

$$DE + EF < DF.$$

Case III.

(1) Suppose, if possible, that $AB = LO$.

Then, in the triangles ABC, LOM, two sides AB, BC are respectively equal to two sides LO, OM, and the bases AC, LM are equal;

therefore $\angle ABC = \angle LOM$.

Similarly $\angle GHK = \angle NOL$.

Therefore, by addition,

$$\angle MON = \angle ABC + \angle GHK$$
$$> \angle DEF \text{ (by hypothesis)}.$$

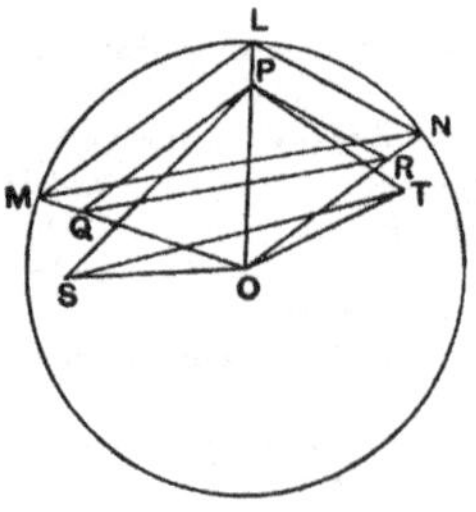

But, in the triangles DEF, MON, which are equal in all respects,

$$\angle MON = \angle DEF.$$

But it was proved that $\angle MON > \angle DEF$: which is impossible.

(2) Suppose, if possible, that $AB < LO$.

Along OL, OM measure OP, OQ each equal to AB.

Then LM, PQ are parallel, and

$$LM : PQ = LO : OP,$$

whence, since $LO > OP$,

$$LM, \text{ or } AC, > PQ.$$

Thus, in the triangles ABC, POQ,

$$\angle ABC > \angle POQ, \text{ i.e. } \angle LOM.$$

Similarly, by taking OR along ON equal to AB, we prove that

$$\angle GHK > \angle LON.$$

Now, at O, make $\angle POS$ equal to $\angle ABC$, and $\angle POT$ equal to $\angle GHK$.

Make OS, OT each equal to OP, and join ST, SP, TP.

Then, in the equal triangles ABC, POS,

$$AC = PS,$$

so that $$LM = PS.$$

Similarly $$LN = PT.$$

Therefore in the triangles MLN, SPT, since $\angle MLN > \angle SPT$ [this is assumed, but should have been explained],

$$MN > ST,$$

or $$DF > ST.$$

Lastly, in Δs DEF, SOT, which have two sides equal to two sides, since $DF > ST$,

$$\angle DEF > \angle SOT$$
$$> \angle ABC + \angle GHK \text{ (by construction)}:$$

which contradicts the hypothesis.

Simson gives rather different proofs for all three cases; but the essence of them can be put, I think, a little more shortly than in his text, as well as more clearly.

Case I. (O within ΔLMN.)

(1) Let AB be, if possible, equal to LO.

Then the Δs ABC, DEF, GHK must be identically equal to the Δs LOM, MON, NOL respectively.

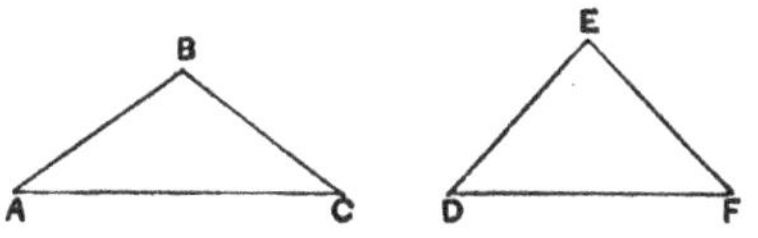

Therefore the vertical angles at O in the latter triangles are equal respectively to the angles at B, E, H.

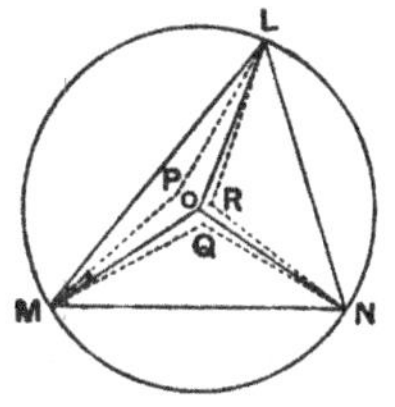

The latter are therefore together equal to four right angles:

which is impossible.

(2) If AB be less than LO, construct on the bases LM, MN, NL triangles with vertices P, Q, R and identically equal to the Δs ABC, DEF, GHK respectively.

Then P, Q, R will fall within the respective angles at O, since $PL = PM$ and $< LO$, and similarly in the other cases.

Thus [I. 21] the angles at P, Q, R are respectively greater than the angles at O in which they lie.

Therefore the sum of the angles at P, Q, R, i.e. the sum of the angles at B, E, H, is greater than four right angles:

which again contradicts the hypothesis.

Case II. (O lying on MN.)

In this case, whether (1) $AB = LO$, or (2) $AB < LO$, a triangle cannot be formed with MN as base and each of the other sides equal to AB. In other words, the triangle DEF either reduces to a straight line or is impossible.

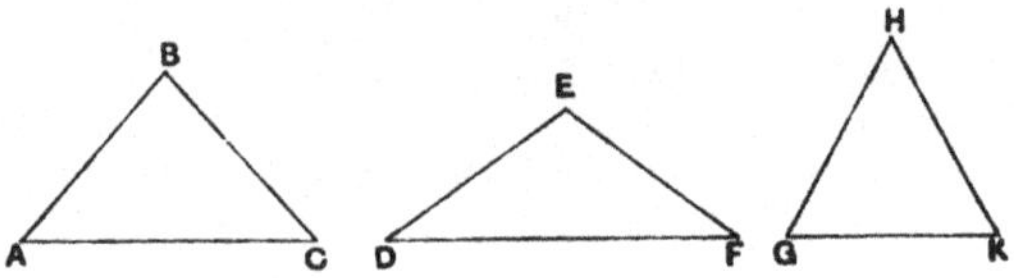

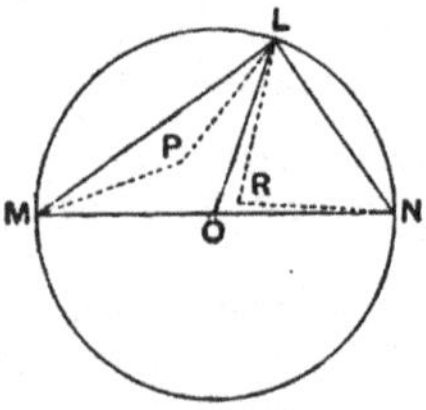

Case III. (O lying outside the $\triangle LMN$.)

(1) Suppose, if possible, that $AB = LO$.

Then the triangles LOM, MON, NOL are identically equal to the triangles ABC, DEF, GHK.

Since $\angle LOM + \angle LON = \angle MON$,

$\angle ABC + \angle GHK = \angle DEF$:

which contradicts the hypothesis.

(2) Suppose that $AB < OL$.

Draw, as before, on LM, MN, NL as bases triangles with vertices P, Q, R and identically equal to the $\triangle$s ABC, DEF, GHK.

Next, at N on the straight line NR, make $\angle RNS$ equal to the angle PLM, cut off NS equal to LM and join RS, LS.

Then $\triangle NRS$ is identically equal to $\triangle LPM$ or $\triangle ABC$.

Now $(\angle LNR + \angle RNS) < (\angle NLO + \angle OLM)$,

that is, $\angle LNS < \angle NLM$.

Thus, in $\triangle$s LNS, NLM, two sides are equal to two sides, and the included angle in the former is less than the included angle in the other.

Therefore $LS < MN$.

Hence, in the triangles MQN, LRS, two sides are equal to two sides, and $MN > LS$.

Therefore $\angle MQN > \angle LRS$

$> (\angle LRN + \angle SRN)$

$> (\angle LRN + \angle LPM)$.

That is, $\angle DEF > (\angle GHK + \angle ABC)$:

which is impossible.

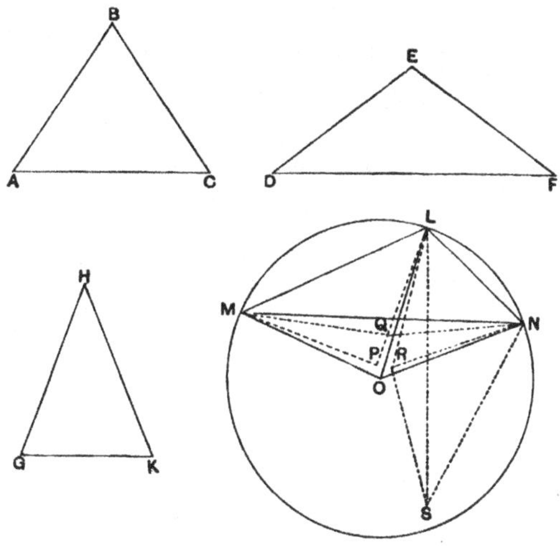

PROPOSITION 24.

If a solid be contained by parallel planes, the opposite planes in it are equal and parallelogrammic.

For let the solid $CDHG$ be contained by the parallel planes AC, GF, AH, DF, BF, AE;

I say that the opposite planes in it are equal and parallelogrammic.

For, since the two parallel planes BG, CE are cut by the plane AC,

their common sections are parallel. [XI. 16]

Therefore *AB* is parallel to *DC*.

Again, since the two parallel planes *BF*, *AE* are cut by the plane *AC*,
their common sections are parallel. [XI. 16]

Therefore *BC* is parallel to *AD*.

But *AB* was also proved parallel to *DC*;
therefore *AC* is a parallelogram.

Similarly we can prove that each of the planes *DF*, *FG*, *GB*, *BF*, *AE* is a parallelogram.

Let *AH*, *DF* be joined.

Then, since *AB* is parallel to *DC*, and *BH* to *CF*,
the two straight lines *AB*, *BH* which meet one another are parallel to the two straight lines *DC*, *CF* which meet one another, not in the same plane;
therefore they will contain equal angles; [XI. 10]
therefore the angle *ABH* is equal to the angle *DCF*.

And, since the two sides *AB*, *BH* are equal to the two sides *DC*, *CF*, [I. 34]
and the angle *ABH* is equal to the angle *DCF*,
therefore the base *AH* is equal to the base *DF*,
and the triangle *ABH* is equal to the triangle *DCF*. [I. 4]

And the parallelogram *BG* is double of the triangle *ABH*,
and the parallelogram *CE* double of the triangle *DCF*; [I. 34]
therefore the parallelogram *BG* is equal to the parallelogram *CE*.

Similarly we can prove that
AC is also equal to *GF*,
and *AE* to *BF*.

Therefore etc.

Q. E. D.

As Heiberg says, this proposition is carelessly enunciated. Euclid means a solid contained by *six* planes and not more, the planes are parallel *two and two*, and the opposite faces are equal in the sense of *identically* equal, or, as Simson puts it, equal *and similar*. The *similarity* is necessary in order to enable the equality of the parallelepipeds in the next proposition to be inferred from the 10th definition of Book XI. Hence a better enunciation would be:

If a solid be contained by six planes parallel two and two, the opposite faces respectively are equal and similar parallelograms.

The proof is simple and requires no elucidation.

PROPOSITION 25.

If a parallelepipedal solid be cut by a plane which is parallel to the opposite planes, then, as the base is to the base, so will the solid be to the solid.

For let the parallelepipedal solid *ABCD* be cut by the plane *FG* which is parallel to the opposite planes *RA*, *DH*;
I say that, as the base *AEFV* is to the base *EHCF*, so is the solid *ABFU* to the solid *EGCD*.

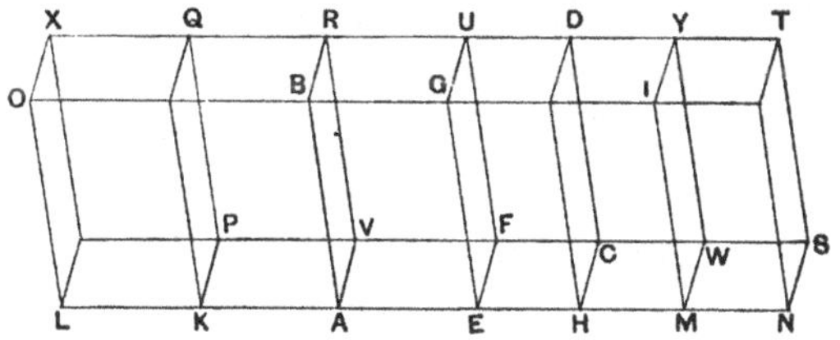

For let *AH* be produced in each direction,
let any number of straight lines whatever, *AK*, *KL*, be made equal to *AE*,
and any number whatever, *HM*, *MN*, equal to *EH*;
and let the parallelograms *LP*, *KV*, *HW*, *MS* and the solids *LQ*, *KR*, *DM*, *MT* be completed.

Then, since the straight lines *LK*, *KA*, *AE* are equal to one another,
the parallelograms *LP*, *KV*, *AF* are also equal to one another,
KO, *KB*, *AG* are equal to one another,
and further *LX*, *KQ*, *AR* are equal to one another, for they are opposite. [XI. 24]

For the same reason
the parallelograms *EC*, *HW*, *MS* are also equal to one another,
HG, *HI*, *IN* are equal to one another,
and further *DH*, *MY*, *NT* are equal to one another.

Therefore in the solids *LQ*, *KR*, *AU* three planes are equal to three planes.

But the three planes are equal to the three opposite; therefore the three solids *LQ*, *KR*, *AU* are equal to one another.

For the same reason
the three solids *ED*, *DM*, *MT* are also equal to one another.

Therefore, whatever multiple the base *LF* is of the base *AF*, the same multiple also is the solid *LU* of the solid *AU*.

For the same reason,
whatever multiple the base *NF* is of the base *FH*, the same multiple also is the solid *NU* of the solid *HU*.

And, if the base *LF* is equal to the base *NF*, the solid *LU* is also equal to the solid *NU*;
if the base *LF* exceeds the base *NF*, the solid *LU* also exceeds the solid *NU*;
and, if one falls short, the other falls short.

Therefore, there being four magnitudes, the two bases *AF*, *FH*, and the two solids *AU*, *UH*,
equimultiples have been taken of the base *AF* and the solid *AU*, namely the base *LF* and the solid *LU*,
and equimultiples of the base *HF* and the solid *HU*, namely the base *NF* and the solid *NU*,
and it has been proved that, if the base *LF* exceeds the base *FN*, the solid *LU* also exceeds the solid *NU*,
if the bases are equal, the solids are equal,
and if the base falls short, the solid falls short.

Therefore, as the base *AF* is to the base *FH*, so is the solid *AU* to the solid *UH*. [v. Def. 5]

Q. E. D.

It is to be observed that, as the word *parallelogrammic* was used in Book I. without any definition of its meaning, so παραλληλεπίπεδος, *parallelepipedal*, is here used without explanation. While it means simply "with parallel planes," i.e. "faces," the term is appropriated to the particular solid which has *six* plane faces parallel two and two. The proper translation of στερεὸν παραλληλεπίπεδον is *parallelepipedal solid*, not *solid parallelepiped*, as it is usually translated. Still less is the solid a parallel*o*piped, as the word is not uncommonly written.

The opposite faces in each set of parallelepipedal solids in this proposition are not only equal but equal *and similar*. Euclid infers that the solids in each set are equal from Def. 10; but, as we have seen in the note on Deff. 9, 10, though it is true, where no solid angle in the figures is contained by more than three plane angles, that two solid figures are equal and similar which are contained by the same number of equal and similar faces, similarly arranged, the fact should have been proved. To do this, we have only to prove the proposition, given above in the note on XI. 21, that *two trihedral angles are equal if the three face angles of the one are respectively equal to the three face angles in the other, and all are arranged in the same order*, and then to prove equality by applying one figure to the other as is done by Simson in his proposition C.

Application will also, of course, establish what is assumed by Euclid of the solids formed by the multiples of the original solids, namely that, if $LF \gtreqless NF$, the solid $LU \gtreqless$ the solid NU.

PROPOSITION 26.

On a given straight line, and at a given point on it, to construct a solid angle equal to a given solid angle.

Let AB be the given straight line, A the given point on it, and the angle at D, contained by the angles EDC, EDF, FDC, the given solid angle;

thus it is required to construct on the straight line AB, and at the point A on it, a solid angle equal to the solid angle at D.

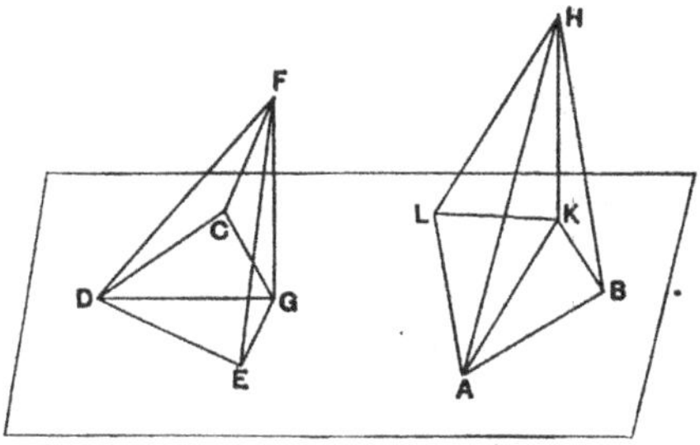

For let a point F be taken at random on DF,

let FG be drawn from F perpendicular to the plane through ED, DC, and let it meet the plane at G, [XI. 11]

let DG be joined,

let there be constructed on the straight line AB and at the point A on it the angle BAL equal to the angle EDC, and the angle BAK equal to the angle EDG, [I. 23]

let AK be made equal to DG,

let KH be set up from the point K at right angles to the plane through BA, AL, [XI. 12]

let KH be made equal to GF,

and let HA be joined;

I say that the solid angle at A, contained by the angles BAL, BAH, HAL is equal to the solid angle at D contained by the angles EDC, EDF, FDC.

For let AB, DE be cut off equal to one another,

and let HB, KB, FE, GE be joined.

Then, since FG is at right angles to the plane of reference, it will also make right angles with all the straight lines which meet it and are in the plane of reference; [XI. Def. 3]

therefore each of the angles FGD, FGE is right.

For the same reason

each of the angles HKA, HKB is also right.

And, since the two sides KA, AB are equal to the two sides GD, DE respectively,

and they contain equal angles,

therefore the base KB is equal to the base GE. [I. 4]

But KH is also equal to GF,

and they contain right angles;

therefore HB is also equal to FE. [I. 4]

Again, since the two sides AK, KH are equal to the two sides DG, GF,

and they contain right angles,

therefore the base AH is equal to the base FD. [I. 4]

But AB is also equal to DE;

therefore the two sides HA, AB are equal to the two sides DF, DE.

And the base HB is equal to the base FE;

therefore the angle BAH is equal to the angle EDF. [I. 8]

For the same reason

the angle HAL is also equal to the angle FDC.

And the angle BAL is also equal to the angle EDC.

Therefore on the straight line AB, and at the point A on it, a solid angle has been constructed equal to the given solid angle at D.

Q. E. F.

This proposition again assumes the equality of two trihedral angles which have the three plane angles of the one respectively equal to the three plane angles of the other taken in the same order.

Proposition 27.

On a given straight line to describe a parallelepipedal solid similar and similarly situated to a given parallelepipedal solid.

Let AB be the given straight line and CD the given parallelepipedal solid;

thus it is required to describe on the given straight line AB a parallelepipedal solid similar and similarly situated to the given parallelepipedal solid CD.

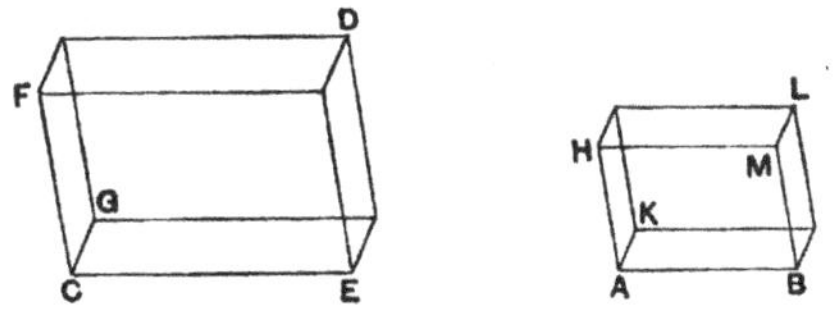

For on the straight line AB and at the point A on it let the solid angle, contained by the angles BAH, HAK, KAB, be constructed equal to the solid angle at C, so that the angle BAH is equal to the angle ECF, the angle BAK equal to the angle ECG, and the angle KAH to the angle GCF;

and let it be contrived that,

as EC is to CG, so is BA to AK,

and, as GC is to CF, so is KA to AH. [VI. 12]

Therefore also, *ex aequali*,

as EC is to CF, so is BA to AH. [V. 22]

Let the parallelogram HB and the solid AL be completed.

Now since, as EC is to CG, so is BA to AK,

and the sides about the equal angles ECG, BAK are thus proportional, therefore the parallelogram *GE* is similar to the parallelogram *KB*.

For the same reason
the parallelogram *KH* is also similar to the parallelogram *GF*, and further *FE* to *HB*;
therefore three parallelograms of the solid *CD* are similar to three parallelograms of the solid *AL*.

But the former three are both equal and similar to the three opposite parallelograms,
and the latter three are both equal and similar to the three opposite parallelograms;
therefore the whole solid *CD* is similar to the whole solid *AL*. [XI. Def. 9]

Therefore on the given straight line *AB* there has been described *AL* similar and similarly situated to the given parallelepipedal solid *CD*.

Q. E. F.

PROPOSITION 28.

If a parallelepipedal solid be cut by a plane through the diagonals of the opposite planes, the solid will be bisected by the plane.

For let the parallelepipedal solid *AB* be cut by the plane *CDEF* through the diagonals *CF*, *DE* of opposite planes;
I say that the solid *AB* will be bisected by the plane *CDEF*.

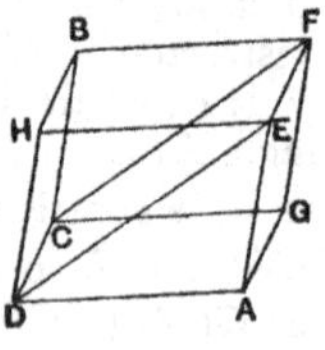

For, since the triangle *CGF* is equal to the triangle *CFB*, [I. 34]
and *ADE* to *DEH*,
while the parallelogram *CA* is also equal to the parallelogram *EB*, for they are opposite,
and *GE* to *CH*,
therefore the prism contained by the two triangles *CGF*, *ADE* and the three parallelograms *GE*, *AC*, *CE* is also equal to the prism contained by the two triangles *CFB*, *DEH* and the three parallelograms *CH*, *BE*, *CE*;

for they are contained by planes equal both in multitude and in magnitude. [XI. Def. 10]

Hence the whole solid *AB* is bisected by the plane *CDEF*.

Q. E. D.

Simson properly observes that it ought to be proved that the diagonals of two opposite faces *are* in one plane, before we speak of drawing a plane through them. Clavius supplied the proof, which is of course simple enough.

Since *EF*, *CD* are both parallel to *AG* or *BH*, they are parallel to one another.

Consequently a plane can be drawn through *CD*, *EF* and the diagonals *DE*, *CF* are in that plane [XI. 7]. Moreover *CD*, *EF* are equal as well as parallel; so that *CF*, *DE* are also equal and parallel.

Simson does not, however, seem to have noticed a more serious difficulty. The two prisms are shown by Euclid to be contained by equal faces—the faces are in fact equal and similar—and Euclid then infers at once that the prisms are *equal*. But they are not equal in the only sense in which we have, at present, a right to speak of solids being equal, namely in the sense that they can be *applied*, the one to the other. They cannot be so applied because the faces, though equal respectively, are not *similarly arranged*; consequently the prisms are *symmetrical*, and it ought to be proved that they are, though not equal *and similar*, equal in content, or *equivalent*, as Legendre has it.

Legendre addressed himself to proving that the two prisms are equivalent, and his method has been adopted, though his name is not mentioned, by Schultze and Sevenoak and by Holgate. Certain preliminary propositions are necessary.

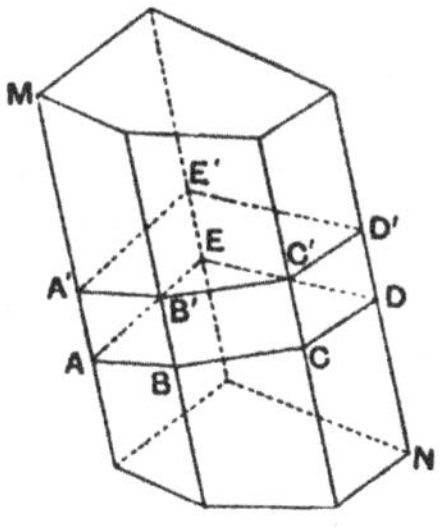

1. *The sections of a prism made by parallel planes cutting all the lateral edges are equal polygons.*

Suppose a prism *MN* cut by parallel planes which make sections *ABCDE*, *A'B'C'D'E'*.

Now *AB*, *BC*, *CD*, ... are respectively parallel to *A'B'*, *B'C'*, *C'D'*, ... [XI. 16]

Therefore the angles *ABC*, *BCD*, ... are equal to the angles *A'B'C'*, *B'C'D'*, ... respectively. [XI. 10]

Also *AB*, *BC*, *CD*, ... are respectively equal to *A'B'*, *B'C'*, *C'D'*, ... [I. 34]

Thus the polygons *ABCDE*, *A'B'C'D'E'* are equilateral and equiangular to one another.

2. *Two prisms are equal when they have a solid angle in each contained by three faces equal each to each and similarly arranged.*

Let the faces *ABCDE*, *AG*, *AL* be equal and similarly placed to the faces *A'B'C'D'E'*, *A'G'*, *A'L'*.

Since the three plane angles at *A*, *A'* are equal respectively and are similarly placed, the trihedral angle at *A* is equal to the trihedral angle at *A'*. [(3) in note to XI. 21]

Place the trihedral angle at A on that at A'.

Then the face $ABCDE$ coincides with the face $A'B'C'D'E'$, the face AG with the face $A'G'$, and the face AL with the face $A'L'$.

The point C falls on C' and D on D'.

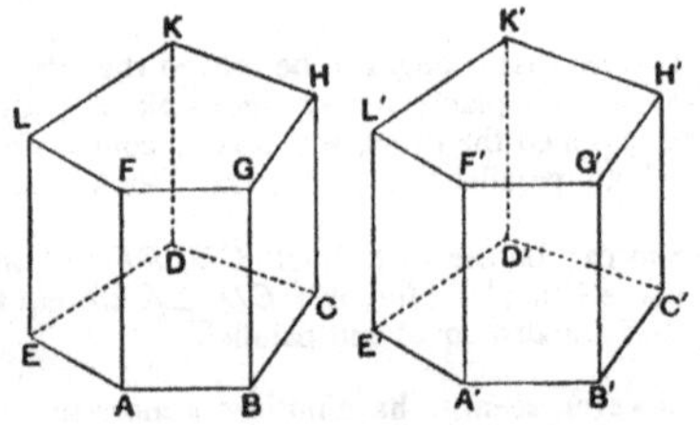

Since the lateral edges of a prism are parallel, CH will fall an $C'H'$, and DK on $D'K'$.

And the points F, G, L coincide respectively with F', G', L', so that

the planes GK, $G'K'$ coincide.

Hence H, K coincide with H', K' respectively.

Thus the prisms coincide throughout and are equal.

In the same way we can prove that two *truncated* prisms with three faces forming a solid angle related to one another as in the above proposition are identically equal.

In particular,

Cor. *Two right prisms having equal bases and equal heights are equal.*

3. *An oblique prism is equivalent to a right prism whose base is a right section of the oblique prism and whose height is equal to a lateral edge of the oblique prism.*

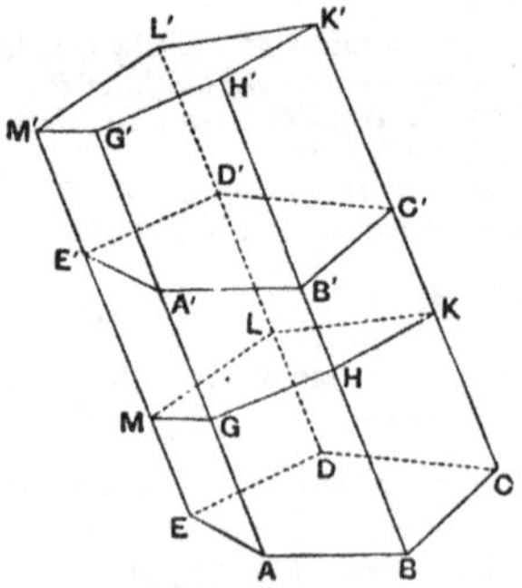

Suppose GL to be a right section of the oblique prism AD', and let GL' be a right prism on GL as base and with height equal to a lateral edge of AD'.

Now the lateral edges of GL' are equal to the lateral edges of AD'.

Therefore $AG = A'G'$, $BH = B'H'$, $CK = C'K'$, etc.

Thus the faces AH, BK, CL are equal respectively to the faces $A'H'$, $B'K'$, $C'L'$.

Therefore [by the proposition above]

(truncated prism AL) = (truncated prism $A'L'$).

Subtracting each from the whole solid AL', we see that

the prisms AD', GL' are equivalent.

Now suppose the parallelepiped of Euclid's proposition to be cut by the plane through *AG, DF*.

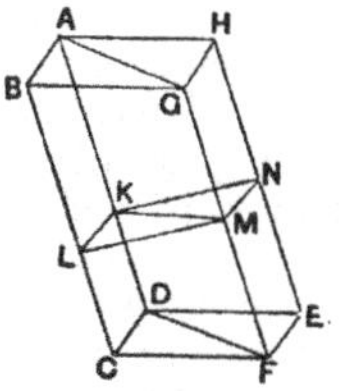

Let *KLMN* be a right section of the parallelepiped cutting the edges *AD, BC, GF, HE*.

Then *KLMN* is a parallelogram; and, if the diagonal *KM* be drawn,

$$\triangle KLM = \triangle MNK.$$

Now the prism of which the $\triangle$s *ABG, DCF* are the bases is equal to the right prism on $\triangle KLM$ as base and of height *AD*.

Similarly the prism of which the $\triangle$s *AGH, DFE* are the bases is equal to the right prism on $\triangle MNK$ as base and with height *AD*. [(3) above]

And the right prisms on $\triangle$s *KLM, MNK* as bases and of equal height *AD* are equal. [(2), Cor. above]

Consequently the two prisms into which the parallelepiped is divided are *equivalent*.

PROPOSITION 29.

Parallelepipedal solids which are on the same base and of the same height, and in which the extremities of the sides which stand up are on the same straight lines, are equal to one another.

Let *CM, CN* be parallelepipedal solids on the same base *AB* and of the same height,

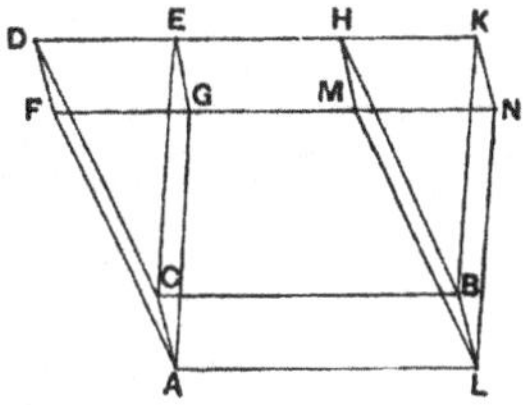

and let the extremities of their sides which stand up, namely *AG, AF, LM, LN, CD, CE, BH, BK*, be on the same straight lines *FN, DK*;

I say that the solid *CM* is equal to the solid *CN*.

For, since each of the figures *CH, CK* is a parallelogram, *CB* is equal to each of the straight lines *DH, EK*; [I. 34]

hence *DH* is also equal to *EK*.

Let *EH* be subtracted from each;

therefore the remainder *DE* is equal to the remainder *HK*.

Hence the triangle *DCE* is also equal to the triangle *HBK*, [I. 8, 4]

and the parallelogram *DG* to the parallelogram *HN*. [I. 36]

For the same reason
the triangle AFG is also equal to the triangle MLN.

But the parallelogram CF is equal to the parallelogram BM, and CG to BN, for they are opposite;
therefore the prism contained by the two triangles AFG, DCE and the three parallelograms AD, DG, CG is equal to the prism contained by the two triangles MLN, HBK and the three parallelograms BM, HN, BN.

Let there be added to each the solid of which the parallelogram AB is the base and $GEHM$ its opposite;
therefore the whole parallelepipedal solid CM is equal to the whole parallelepipedal solid CN.

Therefore etc.

Q. E. D.

As usual, Euclid takes one case only and leaves the reader to prove for himself the two other possible cases shown in the subjoined figures. Euclid's proof holds with a very slight change in each case. With the first figure, the

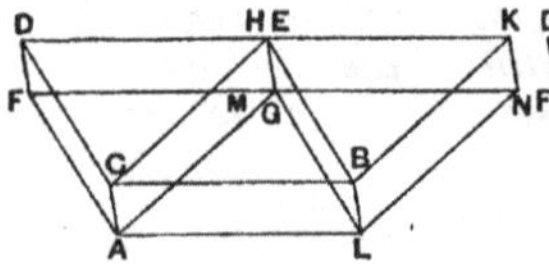

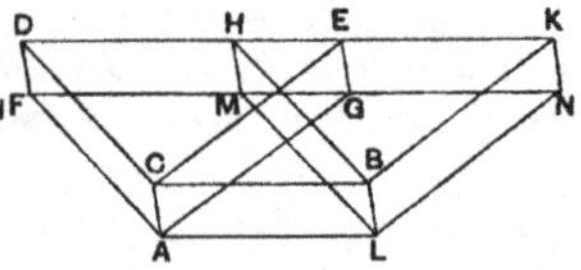

only difference is that the prism of which the $\triangle$s GAL, ECB are the bases takes the place of "the solid of which the parallelogram AB is the base and $GEHM$ its opposite"; while with the second figure we have to subtract the prisms which are proved equal successively from the solid of which the parallelogram AB is the base and $FDKN$ its opposite.

Simson, as usual, suspects mutilation by "some unskilful editor," but gives a curious reason why the case in which the two parallelograms opposite to AB have a side common ought not to have been omitted, namely that this case "is immediately deduced from the preceding 28th Prop. which seems for this purpose to have been premised to the 29th." But, apart from the fact that Euclid's Prop. 28 does *not* prove the theorem which it enunciates (as we have seen), that theorem is not in the least necessary for the proof of this case of Prop. 29, as Euclid's proof applies to it perfectly well.

Proposition 30.

Parallelepipedal solids which are on the same base and of the same height, and in which the extremities of the sides which stand up are not on the same straight lines, are equal to one another.

Let *CM*, *CN* be parallelepipedal solids on the same base *AB* and of the same height,

and let the extremities of their sides which stand up, namely *AF*, *AG*, *LM*, *LN*, *CD*, *CE*, *BH*, *BK*, not be on the same straight lines;

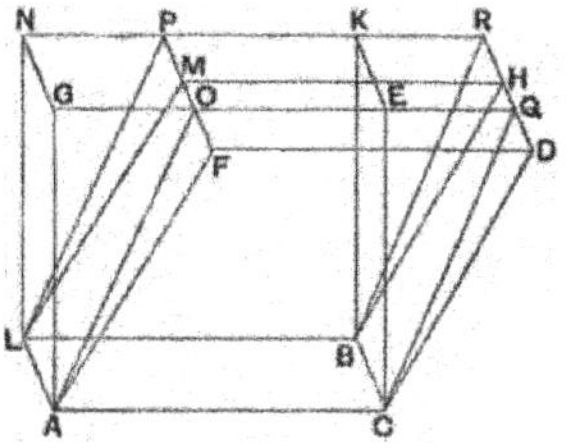

I say that the solid *CM* is equal to the solid *CN*.

For let *NK*, *DH* be produced and meet one another at *R*,

and further let *FM*, *GE* be produced to *P*, *Q*;

let *AO*, *LP*, *CQ*, *BR* be joined.

Then the solid *CM*, of which the parallelogram *ACBL* is the base, and *FDHM* its opposite, is equal to the solid *CP*, of which the parallelogram *ACBL* is the base, and *OQRP* its opposite;

for they are on the same base *ACBL* and of the same height, and the extremities of their sides which stand up, namely *AF*, *AO*, *LM*, *LP*, *CD*, *CQ*, *BH*, *BR*, are on the same straight lines *FP*, *DR*. [XI. 29]

But the solid *CP*, of which the parallelogram *ACBL* is the base, and *OQRP* its opposite, is equal to the solid *CN*, of which the parallelogram *ACBL* is the base and *GEKN* its opposite;

for they are again on the same base *ACBL* and of the same height, and the extremities of their sides which stand up, namely *AG*, *AO*, *CE*, *CQ*, *LN*, *LP*, *BK*, *BR*, are on the same straight lines *GQ*, *NR*.

Hence the solid *CM* is also equal to the solid *CN*.

Therefore etc.

Q. E. D.

This proposition completes the proof of the theorem that

Two parallelepipeds on the same base and of the same height are equivalent.

Legendre deduced the useful theorem that

Every parallelepiped can be changed into an equivalent rectangular parallelepiped having the same height and an equivalent base.

For suppose we have a parallelepiped on the base *ABCD* with *EFGH* for the opposite face.

Draw *AI*, *BK*, *CL*, *DM* perpendicular to the plane through *EFGH* and all equal to the height of the parallelepiped *AG*. Then, on joining *IK*, *KL*, *LM*, *MI*, we have a parallelepiped equivalent to the original one and having its lateral faces *AK*, *BL*, *CM*, *DI* rectangles.

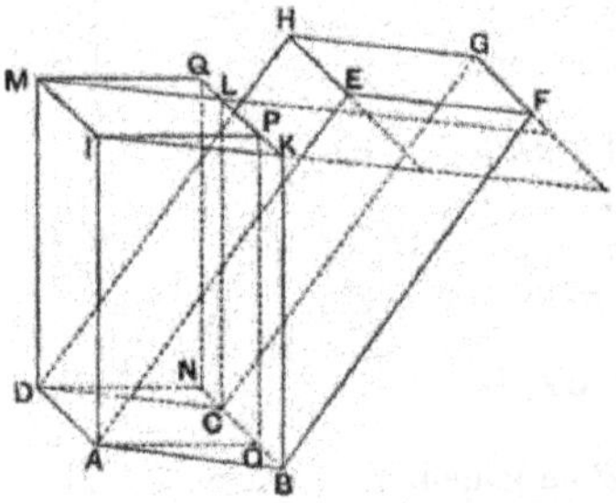

If *ABCD* is not a rectangle, draw *AO*, *DN* in the plane *AC* perpendicular to *BC*, and *IP*, *MQ* in the plane *IL* perpendicular to *KL*.

Joining *OP*, *NQ*, we have a *rectangular* parallelepiped on *AOND* as base which is equivalent to the parallelepiped with *ABCD* as base and *IKLM* as opposite face, since we may regard these parallelepipeds as being on the same base *ADMI* and of the same height (*AO*).

That is, a rectangular parallelepiped has been constructed which is equivalent to the given parallelepiped and has (1) the same height, (2) an equivalent base.

The American text-books which I have quoted adopt a somewhat different construction shown in the subjoined figure.

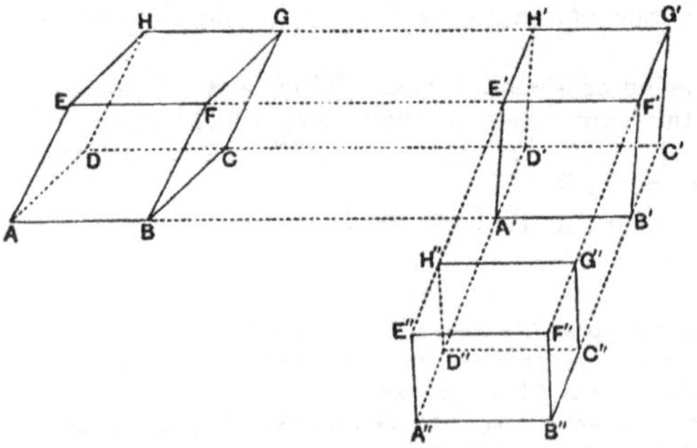

The edges *AB*, *DC*, *EF*, *HG* of the original parallelepiped are produced and cut at right angles by two parallel planes at a distance apart *A'B'* equal to *AB*.

Thus a parallelepiped is formed in which all the faces are rectangles except *A'H'*, *B'G'*.

Next produce $D'A'$, $C'B'$, $G'F'$, $H'E'$ and cut them perpendicularly by two parallel planes at a distance apart $B''C''$ equal to $B'C'$.

The points of section determine a *rectangular* parallelepiped.

The equivalence of the three parallelepipeds is proved, not by Eucl. XI. 29, 30, but by the proposition about a right section of a prism given above in the note to XI. 28 (3 in that note).

PROPOSITION 31.

Parallelepipedal solids which are on equal bases and of the same height are equal to one another.

Let the parallelepipedal solids AE, CF, of the same height, be on equal bases AB, CD.

I say that the solid AE is equal to the solid CF.

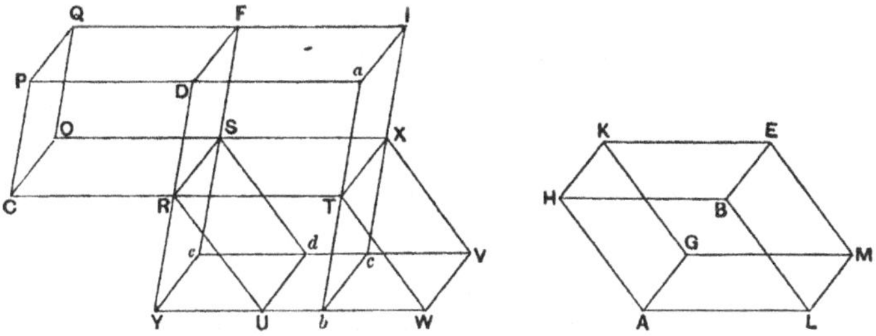

First, let the sides which stand up, HK, BE, AG, LM, PQ, DF, CO, RS, be at right angles to the bases AB, CD;
let the straight line RT be produced in a straight line with CR;
on the straight line RT, and at the point R on it, let the angle TRU be constructed equal to the angle ALB, [I. 23]
let RT be made equal to AL, and RU equal to LB,
and let the base RW and the solid XU be completed.

Now, since the two sides TR, RU are equal to the two sides AL, LB,
and they contain equal angles,
therefore the parallelogram RW is equal and similar to the parallelogram HL.

Since again AL is equal to RT, and LM to RS,
and they contain right angles,

therefore the parallelogram *RX* is equal and similar to the parallelogram *AM*.

For the same reason

LE is also equal and similar to *SU*;

therefore three parallelograms of the solid *AE* are equal and similar to three parallelograms of the solid *XU*.

But the former three are equal and similar to the three opposite, and the latter three to the three opposite; [XI. 24]

therefore the whole parallelepipedal solid *AE* is equal to the whole parallelepipedal solid *XU*. [XI. Def. 10]

Let *DR*, *WU* be drawn through and meet one another at *Y*,

let *aTb* be drawn through *T* parallel to *DY*,

let *PD* be produced to *a*,

and let the solids *YX*, *RI* be completed.

Then the solid *XY*, of which the parallelogram *RX* is the base and *Yc* its opposite, is equal to the solid *XU* of which the parallelogram *RX* is the base and *UV* its opposite,

for they are on the same base *RX* and of the same height, and the extremities of their sides which stand up, namely *RY*, *RU*, *Tb*, *TW*, *Se*, *Sd*, *Xc*, *XV*, are on the same straight lines *YW*, *eV*. [XI. 29]

But the solid *XU* is equal to *AE*;

therefore the solid *XY* is also equal to the solid *AE*.

And, since the parallelogram *RUWT* is equal to the parallelogram *YT*,

for they are on the same base *RT* and in the same parallels *RT*, *YW*, [I. 35]

while *RUWT* is equal to *CD*, since it is also equal to *AB*,

therefore the parallelogram *YT* is also equal to *CD*.

But *DT* is another parallelogram;

therefore, as the base *CD* is to *DT*, so is *YT* to *DT*. [V. 7]

And, since the parallelepipedal solid *CI* has been cut by the plane *RF* which is parallel to opposite planes,

as the base *CD* is to the base *DT*, so is the solid *CF* to the solid *RI*. [XI. 25]

For the same reason,

since the parallelepipedal solid YI has been cut by the plane RX which is parallel to opposite planes,

as the base YT is to the base TD, so is the solid YX to the solid RI. [XI. 25]

But, as the base CD is to DT, so is YT to DT;

therefore also, as the solid CF is to the solid RI, so is the solid YX to RI. [V. 11]

Therefore each of the solids CF, YX has to RI the same ratio;

therefore the solid CF is equal to the solid YX. [V. 9]

But YX was proved equal to AE;

therefore AE is also equal to CF.

Next, let the sides standing up, AG, HK, BE, LM, CN, PQ, DF, RS, not be at right angles to the bases AB, CD;

I say again that the solid AE is equal to the solid CF.

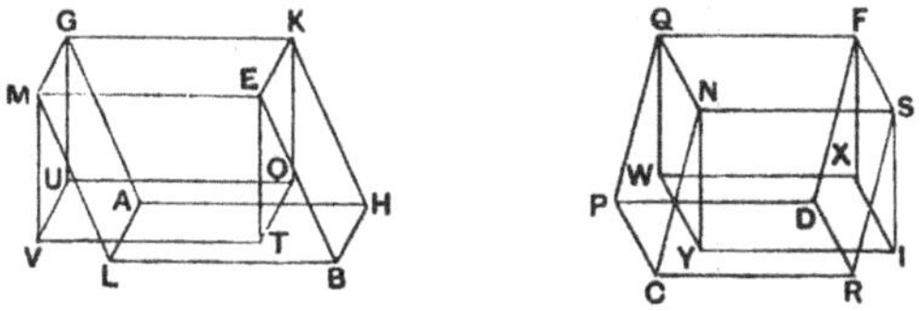

For from the points K, E, G, M, Q, F, N, S let KO, ET, GU, MV, QW, FX, NY, SI be drawn perpendicular to the plane of reference, and let them meet the plane at the points O, T, U, V, W, X, Y, I,

and let OT, OU, UV, TV, WX, WY, YI, IX be joined.

Then the solid KV is equal to the solid QI,

for they are on the equal bases KM, QS and of the same height, and their sides which stand up are at right angles to their bases. [First part of this Prop.]

But the solid KV is equal to the solid AE,

and QI to CF;

for they are on the same base and of the same height, while the extremities of their sides which stand up are not on the same straight lines. [XI. 30]

Therefore the solid AE is also equal to the solid CF.

Therefore etc.

Q. E. D.

It is interesting to observe that, in the figure of this proposition, the bases are represented as lying "in the plane of the paper," as it were, and the third dimension as "standing up" from that plane. The figure is that of the manuscript P slightly corrected as regards the solid AE.

Nothing could well be more ingenious than the proof of this proposition, which recalls the brilliant proposition I. 44 and the proofs of VI. 14 and 23.

As the proof occupies considerable space in the text, it will no doubt be well to give a summary.

I. First, suppose that the edges terminating at the angular points of the bases are ***perpendicular*** to the bases.

AB, CD being the bases, Euclid constructs a solid identically equal to AE (he might simply have ***moved*** AE itself), placing it so that RS is the edge corresponding to HK ($RS = HK$ because the heights are equal), and the face RX corresponding to HE is in the plane of CS.

The faces CD, RW are in one plane because both are perpendicular to RS. Thus DR, WU meet, if produced, in Y say.

Complete the parallelograms YT, DT and the solids YX, FT.

Then (solid YX) = (solid UX),

because they are on the same base ST and of the same height. [XI. 29]

Also, CI, YI being parallelepipeds cut by planes RF, RX parallel to pairs of opposite faces respectively,

(solid CF) : (solid RI) = $\square$ CD : $\square$ DT, [XI. 25]

and (solid YX) : (solid RI) = $\square$ YT : $\square$ DT.

But [I. 35] $\square$ YT = $\square$ UT

= $\square$ AB

= $\square$ CD, by hypothesis.

Therefore (solid CF) = (solid YX)

= (solid UX)

= (solid AE).

II. If the edges terminating at the base are ***not*** perpendicular to it, turn each solid into an equivalent one on the same base with edges perpendicular to it (by drawing four perpendiculars from the angular points of the base to the plane of the opposite face). (XI. 29, 30 prove the equivalence.)

Then the equivalent solids are equal, by Part I.; so that the original solids are also equal.

Simson observes that Euclid has made no mention of the case in which the bases of the two solids are ***equiangular***, and he prefixes this case to Part I. in the text. This is surely unnecessary, as Part I. covers it well enough: the only difference in the figure is that UW would coincide with Yb and dV with ec.

Simson further remarks that in the demonstration of Part II. it is not proved that the new solids constructed in the manner described ***are*** parallelepipeds. The proof is, however, so simple that it scarcely needed insertion into the text. He is correct in his remark that the words "while the extremities of their sides which stand up are not on the same straight lines" just before the end of the proposition would be better absent, since they *may* be "on the same straight lines."

Proposition 32.

Parallelepipedal solids which are of the same height are to one another as their bases.

Let *AB*, *CD* be parallelepipedal solids of the same height; I say that the parallelepipedal solids *AB*, *CD* are to one another as their bases, that is, that, as the base *AE* is to the base *CF*, so is the solid *AB* to the solid *CD*.

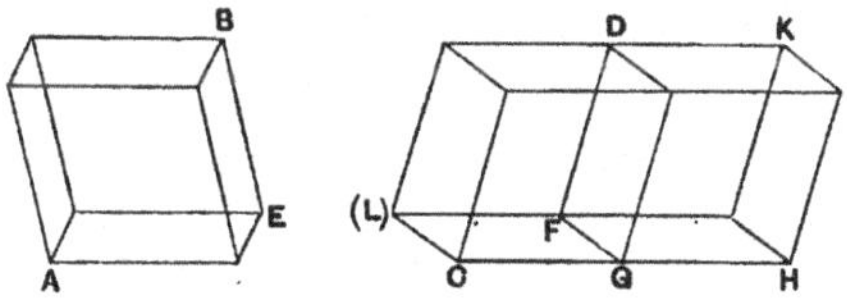

For let *FH* equal to *AE* be applied to *FG*, [I. 45]
and on *FH* as base, and with the same height as that of *CD*, let the parallelepipedal solid *GK* be completed.

Then the solid *AB* is equal to the solid *GK*;
for they are on equal bases *AE*, *FH* and of the same height. [XI. 31]

And, since the parallelepipedal solid *CK* is cut by the plane *DG* which is parallel to opposite planes,
therefore, as the base *CF* is to the base *FH*, so is the solid *CD* to the solid *DH*. [XI. 25]

But the base *FH* is equal to the base *AE*,
and the solid *GK* to the solid *AB*;
therefore also, as the base *AE* is to the base *CF*, so is the solid *AB* to the solid *CD*.

Therefore etc.

Q. E. D.

As Clavius observed, Euclid should have said, in applying the parallelogram *FH* to *FG*, that it should be applied "*in the angle* FGH *equal to the angle* LCG." Simson is however, I think, hypercritical when he states as regards the completion of the solid *GK* that it ought to be said, "complete the solid of which the base is *FH, and one of its insisting straight lines is FD.*" Surely, when we have two faces *DG, FH* meeting in an edge, to say "*complete* the solid" is quite sufficient, though the words "on *FH* as base" might perhaps as well be left out. The same "completion" of a parallelepipedal solid occurs in XI. 31 and 33.

Proposition 33.

Similar parallelepipedal solids are to one another in the triplicate ratio of their corresponding sides.

Let *AB, CD* be similar parallelepipedal solids, and let *AE* be the side corresponding to *CF*;
I say that the solid *AB* has to the solid *CD* the ratio triplicate of that which *AE* has to *CF*.

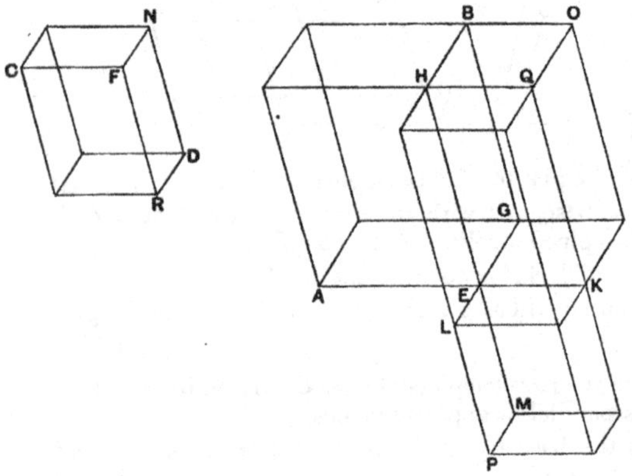

For let *EK, EL, EM* be produced in a straight line with *AE, GE, HE*,
let *EK* be made equal to *CF*, *EL* equal to *FN*, and further *EM* equal to *FR*,
and let the parallelogram *KL* and the solid *KP* be completed.

Now, since the two sides *KE, EL* are equal to the two sides *CF, FN*,
while the angle *KEL* is also equal to the angle *CFN*, inasmuch as the angle *AEG* is also equal to the angle *CFN* because of the similarity of the solids *AB, CD*,

therefore the parallelogram *KL* is equal < and similar > to the parallelogram *CN*.

For the same reason
the parallelogram *KM* is also equal and similar to *CR*,
and further *EP* to *DF*;
therefore three parallelograms of the solid *KP* are equal and similar to three parallelograms of the solid *CD*.

But the former three parallelograms are equal and similar to their opposites, and the latter three to their opposites; [XI. 24]
therefore the whole solid *KP* is equal and similar to the whole solid *CD*. [XI. Def. 10]

Let the parallelogram *GK* be completed,
and on the parallelograms *GK*, *KL* as bases, and with the same height as that of *AB*, let the solids *EO*, *LQ* be completed.

Then since, owing to the similarity of the solids *AB*, *CD*,
as *AE* is to *CF*, so is *EG* to *FN*, and *EH* to *FR*,
while *CF* is equal to *EK*, *FN* to *EL*, and *FR* to *EM*,
therefore, as *AE* is to *EK*, so is *GE* to *EL*, and *HE* to *EM*.

But, as *AE* is to *EK*, so is *AG* to the parallelogram *GK*,
as *GE* is to *EL*, so is *GK* to *KL*,
and, as *HE* is to *EM*, so is *QE* to *KM*; [VI. 1]
therefore also, as the parallelogram *AG* is to *GK*, so is *GK* to *KL*, and *QE* to *KM*.

But, as *AG* is to *GK*, so is the solid *AB* to the solid *EO*,
as *GK* is to *KL*, so is the solid *OE* to the solid *QL*,
and, as *QE* is to *KM*, so is the solid *QL* to the solid *KP*; [XI. 32]
therefore also, as the solid *AB* is to *EO*, so is *EO* to *QL*, and *QL* to *KP*.

But, if four magnitudes be continuously proportional, the first has to the fourth the ratio triplicate of that which it has to the second; [V. Def. 10]
therefore the solid *AB* has to *KP* the ratio triplicate of that which *AB* has to *EO*.

But, as *AB* is to *EO*, so is the parallelogram *AG* to *GK*, and the straight line *AE* to *EK* [VI. 1];

hence the solid AB has also to KP the ratio triplicate of that which AE has to EK.

But the solid KP is equal to the solid CD,
and the straight line EK to CF;
therefore the solid AB has also to the solid CD the ratio triplicate of that which the corresponding side of it, AE, has to the corresponding side CF.

Therefore etc.

Q. E. D.

PORISM. From this it is manifest that, if four straight lines be <continuously> proportional, as the first is to the fourth, so will a parallelepipedal solid on the first be to the similar and similarly described parallelepipedal solid on the second, inasmuch as the first has to the fourth the ratio triplicate of that which it has to the second.

The proof may be summarised as follows.

The three edges AE, GE, HE of the parallelepiped AB which meet at E, the vertex corresponding to R in the other parallelepiped, are produced, and lengths EK, EL, EM are marked off equal respectively to the edges CF, FN, FR of CD.

The parallelograms and solids are then completed as shown in the figure.

Euclid first shows that the solid CD and the new solid PK are equal and similar according to the criterion in XI. Def. 10, viz. that they are contained by the same number of equal and similar planes. (They are arranged in the same order, and it would be easy to prove equality by proving the equality of a pair of solid angles and then applying one solid to the other.)

We have now, by hypothesis,

$$AE : CF = EG : FN = EH : FR:$$

that is,

$$AE : EK = EG : EL = EH : EM.$$

But

$$AE : EK = \square AG : \square GK, \qquad \text{[VI. 1]}$$

$$EG : EL = \square GK : \square KL,$$

$$EH : EM = \square HK : \square KM.$$

Again, by XI. 25 or 32,

$$\square AG : \square GK = (\text{solid } AB) : (\text{solid } EO),$$

$$\square GK : \square KL = (\text{solid } EO) : (\text{solid } QL).$$

$$\square HK : \square KM = (\text{solid } QL) : (\text{solid } KP).$$

Therefore

$$(\text{solid } AB) : (\text{solid } EO) = (\text{solid } EO) : (\text{solid } QL) = (\text{solid } QL) : (\text{solid } KP),$$

or the solid AB is to the solid KP (that is, CD) in the ratio triplicate of that which the solid AB has to the solid EO, i.e. the ratio triplicate of that which AE has to EK (or CF).

Heiberg doubts whether the Porism appended to this proposition is genuine.

Simson adds, as Prop. D, a useful theorem which we should have expected to find here, on the analogy of VI. 23 following VI. 19, 20, viz. that *Solid parallelepipeds contained by parallelograms equiangular to one another, each to each, that is, of which the solid angles are equal, each to each, have to one another the ratio compounded of the ratios of their sides.*

The proof follows the method of the proposition XI. 33, and we can use the same figure. In order to obtain one ratio between lines to represent the ratio compounded of the ratios of the sides, after the manner of VI. 23, we take any straight line a, and then determine three other straight lines b, c, d, such that

$$AE : CF = a : b,$$
$$EG : FN = b : c,$$
$$EH : FR = c : d,$$

whence $a : d$ represents the ratio compounded of the ratios of the sides.

We obtain, in the same manner as above,

$$(\text{solid } AB) : (\text{solid } EO) = \square\, AG : \square\, GK = AE : EK = AE : CF$$
$$= a : b,$$
$$(\text{solid } EO) : (\text{solid } QL) = \square\, GK : \square\, KL = GE : EL = GE : FN$$
$$= b : c,$$
$$(\text{solid } QL) : (\text{solid } KP) = \square\, HK : \square\, KM = EH : EM = EH : FR$$
$$= c : d,$$

whence, by composition [V. 22],

$$(\text{solid } AB) : (\text{solid } KP) = a : d,$$

or

$$(\text{solid } AB) : (\text{solid } CD) = a : d.$$

Proposition 34.

In equal parallelepipedal solids the bases are reciprocally proportional to the heights; and those parallelepipedal solids in which the bases are reciprocally proportional to the heights are equal.

Let *AB*, *CD* be equal parallelepipedal solids;

I say that in the parallelepipedal solids *AB*, *CD* the bases are reciprocally proportional to the heights,

that is, as the base *EH* is to the base *NQ*, so is the height of the solid *CD* to the height of the solid *AB*.

First, let the sides which stand up, namely *AG*, *EF*, *LB*, *HK*, *CM*, *NO*, *PD*, *QR*, be at right angles to their bases;

I say that, as the base *EH* is to the base *NQ*, so is *CM* to *AG*.

If now the base *EH* is equal to the base *NQ*,

while the solid *AB* is also equal to the solid *CD*,

CM will also be equal to *AG*.

For parallelepipedal solids of the same height are to one another as the bases; [XI. 32]
and, as the base *EH* is to *NQ*, so will *CM* be to *AG*,
and it is manifest that in the parallelepipedal solids *AB*, *CD* the bases are reciprocally proportional to the heights.

Next, let the base *EH* not be equal to the base *NQ*, but let *EH* be greater.

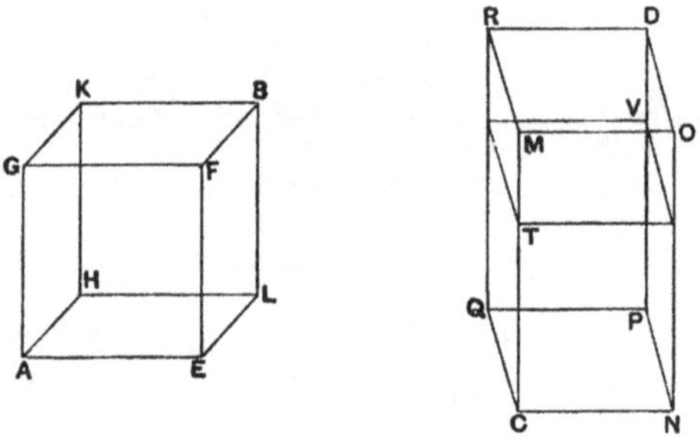

Now the solid *AB* is equal to the solid *CD*;
therefore *CM* is also greater than *AG*.

Let then *CT* be made equal to *AG*,
and let the parallelepipedal solid *VC* be completed on *NQ* as base and with *CT* as height.

Now, since the solid *AB* is equal to the solid *CD*,
and *CV* is outside them,
while equals have to the same the same ratio, [V. 7]
therefore, as the solid *AB* is to the solid *CV*, so is the solid *CD* to the solid *CV*.

But, as the solid *AB* is to the solid *CV*, so is the base *EH* to the base *NQ*,
for the solids *AB*, *CV* are of equal height; [XI. 32]
and, as the solid *CD* is to the solid *CV*, so is the base *MQ* to the base *TQ* [XI. 25] and *CM* to *CT* [VI. 1];
therefore also, as the base *EH* is to the base *NQ*, so is *MC* to *CT*.

But *CT* is equal to *AG*;
therefore also, as the base *EH* is to the base *NQ*, so is *MC* to *AG*.

Therefore in the parallelepipedal solids AB, CD the bases are reciprocally proportional to the heights.

Again, in the parallelepipedal solids AB, CD let the bases be reciprocally proportional to the heights, that is, as the base EH is to the base NQ, so let the height of the solid CD be to the height of the solid AB;

I say that the solid AB is equal to the solid CD.

Let the sides which stand up be again at right angles to the bases.

Now, if the base EH is equal to the base NQ,

and, as the base EH is to the base NQ, so is the height of the solid CD to the height of the solid AB,

therefore the height of the solid CD is also equal to the height of the solid AB.

But parallelepipedal solids on equal bases and of the same height are equal to one another; [XI. 31]

therefore the solid AB is equal to the solid CD.

Next, let the base EH not be equal to the base NQ,

but let EH be greater;

therefore the height of the solid CD is also greater than the height of the solid AB,

that is, CM is greater than AG.

Let CT be again made equal to AG,

and let the solid CV be similarly completed.

Since, as the base EH is to the base NQ, so is MC to AG,

while AG is equal to CT,

therefore, as the base EH is to the base NQ, so is CM to CT.

But, as the base EH is to the base NQ, so is the solid AB to the solid CV,

for the solids AB, CV are of equal height; [XI. 32]

and, as CM is to CT, so is the base MQ to the base QT [VI. 1]

and the solid CD to the solid CV. [XI. 25]

Therefore also, as the solid AB is to the solid CV, so is the solid CD to the solid CV;

therefore each of the solids AB, CD has to CV the same ratio.

Therefore the solid *AB* is equal to the solid *CD*. [V. 9]

Now let the sides which stand up, *FE*, *BL*, *GA*, *HK*, *ON*, *DP*, *MC*, *RQ*, not be at right angles to their bases;
let perpendiculars be drawn from the points *F*, *G*, *B*, *K*, *O*, *M*, *D*, *R* to the planes through *EH*, *NQ*, and let them meet the planes at *S*, *T*, *U*, *V*, *W*, *X*, *Y*, *a*,
and let the solids *FV*, *Oa* be completed;
I say that, in this case too, if the solids *AB*, *CD* are equal, the bases are reciprocally proportional to the heights, that is, as the base *EH* is to the base *NQ*, so is the height of the solid *CD* to the height of the solid *AB*.

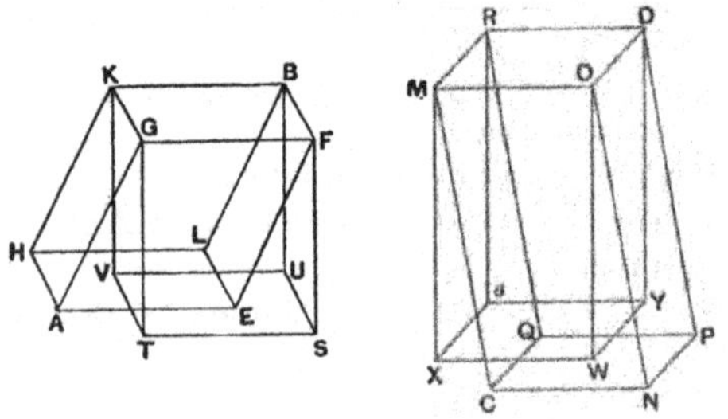

Since the solid *AB* is equal to the solid *CD*,
while *AB* is equal to *BT*,
for they are on the same base *FK* and of the same height; [XI. 29, 30]
and the solid *CD* is equal to *DX*,
for they are again on the same base *RO* and of the same height; [*id.*]
therefore the solid *BT* is also equal to the solid *DX*.

Therefore, as the base *FK* is to the base *OR*, so is the height of the solid *DX* to the height of the solid *BT*. [Part I.]

But the base *FK* is equal to the base *EH*,
and the base *OR* to the base *NQ*;
therefore, as the base *EH* is to the base *NQ*, so is the height of the solid *DX* to the height of the solid *BT*.

But the solids *DX*, *BT* and the solids *DC*, *BA* have the same heights respectively;
therefore, as the base *EH* is to the base *NQ*, so is the height of the solid *DC* to the height of the solid *AB*.

Therefore in the parallelepipedal solids *AB*, *CD* the bases are reciprocally proportional to the heights.

Again, in the parallelepipedal solids *AB*, *CD* let the bases be reciprocally proportional to the heights,
that is, as the base *EH* is to the base *NQ*, so let the height of the solid *CD* be to the height of the solid *AB*;
I say that the solid *AB* is equal to the solid *CD*.

For, with the same construction,
since, as the base *EH* is to the base *NQ*, so is the height of the solid *CD* to the height of the solid *AB*,
while the base *EH* is equal to the base *FK*,
and *NQ* to *OR*,
therefore, as the base *FK* is to the base *OR*, so is the height of the solid *CD* to the height of the solid *AB*.

But the solids *AB*, *CD* and *BT*, *DX* have the same heights respectively;
therefore, as the base *FK* is to the base *OR*, so is the height of the solid *DX* to the height of the solid *BT*.

Therefore in the parallelepipedal solids *BT*, *DX* the bases are reciprocally proportional to the heights;
therefore the solid *BT* is equal to the solid *DX*. [Part I.]

But *BT* is equal to *BA*,
for they are on the same base *FK* and of the same height; [XI. 29, 30]
and the solid *DX* is equal to the solid *DC*. [*id.*]

Therefore the solid *AB* is also equal to the solid *CD*.

Q. E. D.

In this proposition Euclid makes two assumptions which require notice, (1) that, if two parallelepipeds are equal, and have equal bases, their heights are equal, and (2) that, if the bases of two equal parallelepipeds are unequal, that which has the lesser base has the greater height. In justification of the former statement Euclid says, according to what Heiberg holds to be the genuine reading, "for parallelepipedal solids of the same height are to one another as their bases" [XI. 32]. This apparently struck some very early editor as not being sufficient, and he added the explanation appearing in Simson's text, "For if, the bases *EH*, *NQ* being equal, the heights *AG*, *CM* were not equal, neither would the solid AB be equal to CD. But it is by hypothesis equal. Therefore the height CM is not unequal to the height AG; therefore it is equal." Then, it being perceived that there ought not to be two explanations, the genuine one was erased from the inferior MSS. While the interpolated explanation does not take us very far, the truth of the statement may be deduced with perhaps greater ease from XI. 31 than from XI. 32 quoted by Euclid. For, assuming one height greater than the other, while the bases are equal, we have only to cut from the higher solid so much as will make its height equal to that of the other. Then this *part* of the higher solid is equal to the whole of the other solid which is by hypothesis equal to the higher solid itself. That is, the whole is equal to its part: which is impossible.

The genuine text contains no explanation of the second assumption that, if the base EH be greater than the base NQ, while the solids are equal, the height CM is greater than the height AG; for the added words "for, if not, neither again will the solids AB, CD be equal; but they are equal by hypothesis" are no doubt interpolated. In this case the truth of the assumption is easily deduced from XI. 32 by *reductio ad absurdum*. If the height CM were *equal* to the height AG, the solid AB would be to the solid CD as the base EH is to the base NQ, i.e. as a greater to a less, so that the solids would not be equal, as they are by hypothesis. Again, if the height CM were *less* than the height AG, we could increase the height of CD till it was equal to that of AB, and it would then appear that AB is greater than the heightened solid and *a fortiori* greater than CD: which contradicts the hypothesis.

Clavius rather ingeniously puts the first assumption the other way, saying that, if the heights are equal in the equal parallelepipeds, the bases must be equal. This follows *directly* from XI. 32, which proves that the parallelepipeds are to one another as their bases; though Clavius deduces it indirectly from XI. 31. The advantage of Clavius' alternative is that it makes the second assumption unnecessary. He merely says, if the *heights* be not equal, let CM be the greater, and then proceeds with Euclid's construction.

It is also to be observed that, when Euclid comes to the corresponding proposition for cones and cylinders [XII. 15], he begins by supposing the *heights* equal, inferring by XII. 11 (corresponding to XI. 32) that, the solids being equal, the bases are also equal, and then proceeds to the case where the heights are unequal without making any preliminary inference about the bases. The analogy then of XII. 15, and the fact that he quotes XI. 32 here (which directly proves that, if the solids are equal, and also their heights, their bases are also equal), make Clavius' form the more convenient to adopt.

The two assumptions being proved as above, the proposition can be put shortly as follows.

I. Suppose the edges terminating at the corners of the base to be *perpendicular* to it.

Then (*a*), if the base EH be equal to the base NQ, the parallelepipeds being also equal, the heights must be equal (converse of XI. 31), so that the bases are reciprocally proportional to the heights, the ratio of the bases and the ratio of the heights being both ratios of equality.

(*b*) If the base EH be greater than the base NQ, and consequently (by deduction from XI. 32) the height CM greater than the height AG, cut off CT from CM equal to AG, and draw the plane TV through T parallel to the base NQ, making the parallelepiped CV, with CT (= AG) for its height.

Then, since the solids AB, CD are equal,

$$(\text{solid } AB) : (\text{solid } CV) = (\text{solid } CD) : (\text{solid } CV). \qquad [\text{v. } 7]$$

But (solid AB) : (solid CV) = $\square HE : \square NQ$, [XI. 32]
and (solid CD) : (solid CV) = $\square MQ : \square TQ$ [XI. 25]
$= CM : CT.$ [VI. 1]

Therefore $\square HE : \square NQ = CM : CT$
$= CM : AG.$

Conversely (*a*), if the bases EH, NQ be equal and reciprocally proportional to the heights, the heights must be equal.

Consequently (solid AB) = (solid CD). [XI. 31]

(*b*) If the bases EH, NQ be unequal, if, e.g. $\square EH > \square NQ$,
then, since $\square EH : \square NQ = CM : AG$,
$CM > AG.$

Make the same construction as before.

Then $\square EH : \square NQ$ = (solid AB) : (solid CV), [XI. 32]
and $CM : AG = CM : CT$
$= \square MQ : \square TQ$ [VI. 1]
= (solid CD) : (solid CV). [XI. 25]

Therefore

(solid AB) : (solid CV) = (solid CD) : (solid CV),

whence (solid AB) = solid CD. [V. 9]

II. Suppose that the edges terminating at the corners of the bases are *not* perpendicular to it.

Drop perpendiculars on the bases from the corners of the faces opposite to the bases.

We thus have two parallelepipeds equal to AB, CD respectively, since they are on the same bases FK, RO and of the same height respectively. [XI. 29, 30]

If then (1) the solid AB is equal to the solid CD,

(solid BT) = (solid DX),

and, by the first part of this proposition,

$\square KF : \square OR = MX : GT,$
or $\square HE : \square NQ = MX : GT.$

(2) If $\square HE : \square NQ = MX : GT,$
then $\square KF : \square OR = MX : GT,$

so that, by the first half of the proposition, the solids BT, DX are equal, and consequently

(solid AB) = (solid CD).

The text of the second part of the proposition four times contains, after the words "of the same height," the words "in which the sides which stand up are not on the same straight lines." As Simson observed, they are inept, as the extremities of the edges may or may not be "on the same straight lines"; cf. the similar words incorrectly inserted at the end of XI. 31.

Words purporting to quote the result of the first part of the proposition are also twice inserted; but they are rejected as unnecessary and as containing an absurd expression—"(solids) in which *the heights* are at right angles to their bases," as if the *heights* could be otherwise than perpendicular to the bases.

PROPOSITION 35.

If there be two equal plane angles, and on their vertices there be set up elevated straight lines containing equal angles with the original straight lines respectively, if on the elevated straight lines points be taken at random and perpendiculars be drawn from them to the planes in which the original angles are, and if from the points so arising in the planes straight lines be joined to the vertices of the original angles, they will contain, with the elevated straight lines, equal angles.

Let the angles *BAC*, *EDF* be two equal rectilineal angles, and from the points *A*, *D* let the elevated straight lines *AG*, *DM* be set up containing, with the original straight lines, equal angles respectively, namely, the angle *MDE* to the angle *GAB* and the angle *MDF* to the angle *GAC*,

let points *G*, *M* be taken at random on *AG*, *DM*,

let *GL*, *MN* be drawn from the points *G*, *M* perpendicular to the planes through *BA*, *AC* and *ED*, *DF*, and let them meet the planes at *L*, *N*,

and let *LA*, *ND* be joined;

I say that the angle *GAL* is equal to the angle *MDN*.

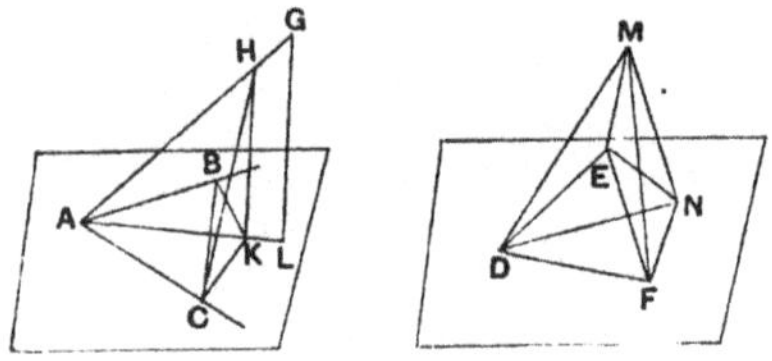

Let *AH* be made equal to *DM*,

and let *HK* be drawn through the point *H* parallel to *GL*.

But *GL* is perpendicular to the plane through *BA*, *AC*; therefore *HK* is also perpendicular to the plane through *BA*, *AC*. [XI. 8]

From the points *K*, *N* let *KC*, *NF*, *KB*, *NE* be drawn perpendicular to the straight lines *AC*, *DF*, *AB*, *DE*,

and let *HC*, *CB*, *MF*, *FE* be joined.

Since the square on HA is equal to the squares on HK, KA,
and the squares on KC, CA are equal to the square on KA, [I. 47]
therefore the square on HA is also equal to the squares on HK, KC, CA.

But the square on HC is equal to the squares on HK, KC; [I. 47]
therefore the square on HA is equal to the squares on HC, CA.

Therefore the angle HCA is right. [I. 48]

For the same reason
the angle DFM is also right.

Therefore the angle ACH is equal to the angle DFM.

But the angle HAC is also equal to the angle MDF.

Therefore MDF, HAC are two triangles which have two angles equal to two angles respectively, and one side equal to one side, namely, that subtending one of the equal angles, that is, HA equal to MD;
therefore they will also have the remaining sides equal to the remaining sides respectively. [I. 26]

Therefore AC is equal to DF.

Similarly we can prove that AB is also equal to DE.

Since then AC is equal to DF, and AB to DE,
the two sides CA, AB are equal to the two sides FD, DE.

But the angle CAB is also equal to the angle FDE;
therefore the base BC is equal to the base EF, the triangle to the triangle, and the remaining angles to the remaining angles; [I. 4]
therefore the angle ACB is equal to the angle DFE.

But the right angle ACK is also equal to the right angle DFN;
therefore the remaining angle BCK is also equal to the remaining angle EFN.

For the same reason
the angle CBK is also equal to the angle FEN.

Therefore *BCK*, *EFN* are two triangles which have two angles equal to two angles respectively, and one side equal to one side, namely, that adjacent to the equal angles, that is, *BC* equal to *EF*;
therefore they will also have the remaining sides equal to the remaining sides. [I. 26]

Therefore *CK* is equal to *FN*.

But *AC* is also equal to *DF*;
therefore the two sides *AC*, *CK* are equal to the two sides *DF*, *FN*;
and they contain right angles.

Therefore the base *AK* is equal to the base *DN*. [I. 4]

And, since *AH* is equal to *DM*,
the square on *AH* is also equal to the square on *DM*.

But the squares on *AK*, *KH* are equal to the square on *AH*,
for the angle *AKH* is right; [I. 47]
and the squares on *DN*, *NM* are equal to the square on *DM*,
for the angle *DNM* is right; [I. 47]
therefore the squares on *AK*, *KH* are equal to the squares on *DN*, *NM*;
and of these the square on *AK* is equal to the square on *DN*;
therefore the remaining square on *KH* is equal to the square on *NM*;
therefore *HK* is equal to *MN*.

And, since the two sides *HA*, *AK* are equal to the two sides *MD*, *DN* respectively,
and the base *HK* was proved equal to the base *MN*,
therefore the angle *HAK* is equal to the angle *MDN*. [I. 8]

Therefore etc.

PORISM. From this it is manifest that, if there be two equal plane angles, and if there be set up on them elevated straight lines which are equal and contain equal angles with the original straight lines respectively, the perpendiculars drawn from their extremities to the planes in which are the original angles are equal to one another.

Q. E. D.

This proposition is required for the next, where it is necessary to know that, if in two equiangular parallelepipeds equal angles, one in each, be contained by three plane angles respectively, one of which is an angle of the parallelogram forming the *base* in one parallelepiped, while its equal is likewise in the *base* of the other, and the edges in which the two remaining angles forming the solid angles meet are *equal*, the parallelepipeds *are of the same height*.

Bearing in mind the definition of *the inclination of a straight line to a plane*, we might enunciate the proposition more shortly thus.

If there be two trihedral angles identically equal to one another, corresponding edges in each are equally inclined to the planes through the other two edges respectively.

The proof, which is necessarily somewhat long, may be summarised thus.

It is required to prove that the angles GAL, MDN in the figure are equal, G, M being any points on AG, DM, and GL, MN perpendicular to the planes BAC, EDF respectively.

If AH is made equal to DM, and HK is drawn in the plane GAL parallel to GL,

HK is also perpendicular to the plane BAC. [XI. 8]

Draw KB, KC perpendicular to AB, AC respectively and NE, NF perpendicular to DE, DF respectively, and complete the figures.

Now (1)
$$\left.\begin{aligned} HA^2 &= HK^2 + KA^2 \\ &= HK^2 + KC^2 + CA^2 \\ &= HC^2 + CA^2 \end{aligned}\right\}. \qquad [\text{I. } 47]$$

Therefore $\angle HCA =$ a right angle.

Similarly $\angle MFD =$ a right angle.

(2) $\triangle$s HAC, MDF have therefore two angles equal and one side.

Therefore $\triangle HAC \equiv \triangle MDF$, and $AC = DF$. [I. 26]

(3) Similarly $\triangle HAB \equiv \triangle MDE$, and $AB = DE$.

(4) Hence $\triangle$s ABC, DEF are equal in all respects, so that $BC = EF$,

and
$$\angle ABC = \angle DEF,$$
$$\angle ACB = \angle DFE.$$

(5) Therefore the complements of these angles are equal,

i.e. $\angle KBC = \angle NEF$,

and $\angle KCB = \angle NFE$.

(6) The $\triangle$s KBC, NEF have two angles equal and one side, and are therefore equal in all respects, so that

$$KB = NE,$$
$$KC = NF.$$

(7) The right-angled triangles KAC, NDF are equal in all respects, since $AC = DF$ [(2) above], $KC = NF$.

Consequently $AK = DN$.

(8) In $\triangle$s HAK, MDN,

$$\begin{aligned} HK^2 + KA^2 &= HA^2 \\ &= MD^2, \text{ by hypothesis,} \\ &= MN^2 + ND^2. \end{aligned}$$

Subtracting the equals KA^2, ND^2,

we have $$HK^2 = MN^2,$$

or $$HK = MN.$$

(9) $\triangle$s HAK, MDN are now equal in all respects, by I. 8 and I. 4, and therefore

$$\angle HAK = \angle MDN.$$

The Porism is merely a statement of the result arrived at in (8).

Legendre uses, practically, the construction and argument of this proposition to prove the theorem given under (3) of the note on XI. 21 above that *In two equal trihedral angles, corresponding pairs of face angles include equal dihedral angles.* This fact is readily deduced from the above proposition.

Since [(1)] HC, KC are both perpendicular to AC, and MF, NF both perpendicular to DF, the angles HCK, MFN are the measures of the *dihedral* angles between the planes HAC, BAC, and MDF, EDF respectively. [XI. Def. 6]

By (6), $$KC = NF,$$

and, by (8), $$HK = MN,$$

while the angles HKC, MNF, both being right, are equal.

Consequently the $\triangle$s HCK, MFN are equal in all respects, [I. 4]

so that $$\angle HCK = \angle MFN.$$

Simson substituted a different proof of (1) in the above summary, as follows.

Since HK is perpendicular to the plane BAC, the plane HBK, passing through HK, is also perpendicular to the plane BAC. [XI. 18]

And AB, being drawn in the plane BAC perpendicular to BK, the common section of the planes HBK, BAC, is perpendicular to the plane HBK [XI. Def. 4], and is therefore perpendicular to every straight line meeting it in that plane [XI. Def. 3].

Hence the angle ABH is a right angle.

I think Euclid's proof much preferable to this with its references to definitions which are more of the nature of theorems.

PROPOSITION 36.

If three straight lines be proportional, the parallelepipedal solid formed out of the three is equal to the parallelepipedal solid on the mean which is equilateral, but equiangular with the aforesaid solid.

Let A, B, C be three straight lines in proportion, so that, as A is to B, so is B to C;

I say that the solid formed out of A, B, C is equal to the solid on B which is equilateral, but equiangular with the aforesaid solid.

Let there be set out the solid angle at E contained by the angles DEG, GEF, FED, let each of the straight lines *DE*, *GE*, *EF* be made equal to *B*, and let the parallelepipedal solid *EK* be completed,

let *LM* be made equal to *A*,

and on the straight line *LM*, and at the point *L* on it, let there be constructed a solid angle equal to the solid angle at *E*, namely that contained by *NLO*, *OLM*, *MLN*;

let *LO* be made equal to *B*, and *LN* equal to *C*.

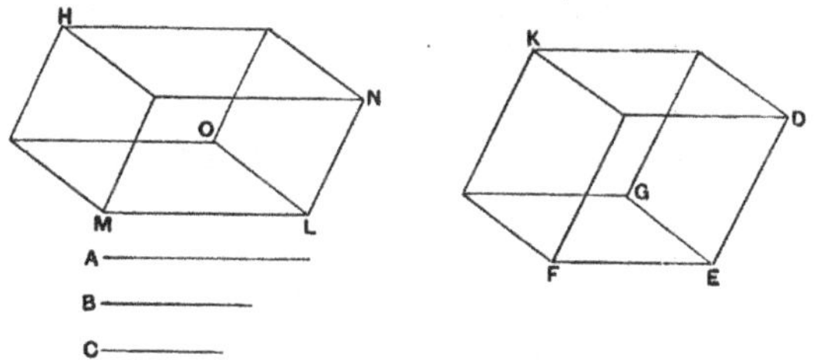

Now, since, as *A* is to *B*, so is *B* to *C*,

while *A* is equal to *LM*, *B* to each of the straight lines *LO*, *ED*, and *C* to *LN*,

therefore, as *LM* is to *EF*, so is *DE* to *LN*.

Thus the sides about the equal angles *NLM*, *DEF* are reciprocally proportional;

therefore the parallelogram *MN* is equal to the parallelogram *DF*. [VI. 14]

And, since the angles *DEF*, *NLM* are two plane rectilineal angles, and on them the elevated straight lines *LO*, *EG* are set up which are equal to one another and contain equal angles with the original straight lines respectively,

therefore the perpendiculars drawn from the points *G*, *O* to the planes through *NL*, *LM* and *DE*, *EF* are equal to one another; [XI. 35, Por.]

hence the solids *LH*, *EK* are of the same height.

But parallelepipedal solids on equal bases and of the same height are equal to one another; [XI. 31]

therefore the solid *HL* is equal to the solid *EK*.

And *LH* is the solid formed out of *A*, *B*, *C*, and *EK* the solid on *B*;

therefore the parallelepipedal solid formed out of A, B, C is equal to the solid on B which is equilateral, but equiangular with the aforesaid solid.

Q. E. D.

The edges of the parallelepiped HL being respectively equal to A, B, C, and those of the equiangular parallelepiped KE being all equal to B, we regard MN (*not* containing the edge OL equal to B) as the base of the first parallelepiped, and consequently FD, equiangular to MN, as the base of KE.

Then the solids have the same height. [XI. 35, Por.]

Hence (solid HL) : (solid KE) = $\square MN : \square FD$. [XI. 32]

But, since A, B, C are in continued proportion,

$$A : B = B : C,$$

or

$$LM : EF = DE : LN.$$

Thus the sides of the equiangular $\square$s MN, FD are reciprocally proportional, whence

$$\square MN = \square FD, \quad \text{[VI. 14]}$$

and therefore

$$(\text{solid } HL) = (\text{solid } KE).$$

PROPOSITION 37.

If four straight lines be proportional, the parallelepipedal solids on them which are similar and similarly described will also be proportional; and, if the parallelepipedal solids on them which are similar and similarly described be proportional, the straight lines will themselves also be proportional.

Let AB, CD, EF, GH be four straight lines in proportion, so that, as AB is to CD, so is EF to GH;

and let there be described on AB, CD, EF, GH the similar and similarly situated parallelepipedal solids KA, LC, ME, NG;

I say that, as KA is to LC, so is ME to NG.

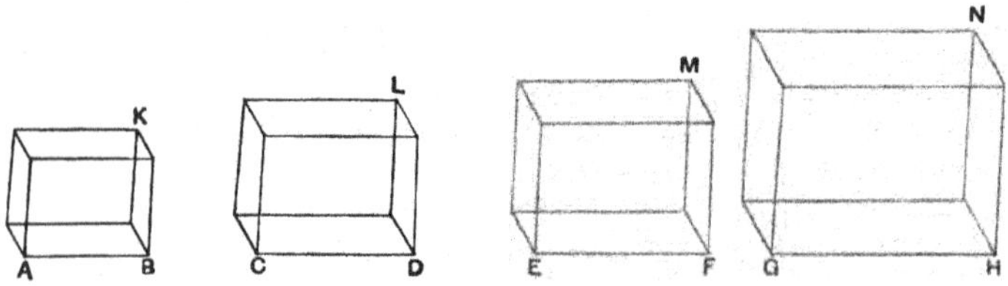

For, since the parallelepipedal solid KA is similar to LC, therefore KA has to LC the ratio triplicate of that which AB has to CD. [XI. 33]

For the same reason

ME also has to *NG* the ratio triplicate of that which *EF* has to *GH*. [*id.*]

And, as *AB* is to *CD*, so is *EF* to *GH*.

Therefore also, as *AK* is to *LC*, so is *ME* to *NG*.

Next, as the solid *AK* is to the solid *LC*, so let the solid *ME* be to the solid *NG*;

I say that, as the straight line *AB* is to *CD*, so is *EF* to *GH*.

For since, again, *KA* has to *LC* the ratio triplicate of that which *AB* has to *CD*, [XI. 33]

and *ME* also has to *NG* the ratio triplicate of that which *EF* has to *GH*, [*id.*]

and, as *KA* is to *LC*, so is *ME* to *NG*,

therefore also, as *AB* is to *CD*, so is *EF* to *GH*.

Therefore etc.

Q. E. D.

In this proposition it is assumed that, if two ratios be equal, the ratio triplicate of one is equal to the ratio triplicate of the other and, conversely, that, if ratios which are the triplicate of two other ratios are equal, those other ratios are themselves equal.

To avoid the necessity for these assumptions Simson adopts the alternative proof found in the MS. which Heiberg calls b, and also adopted by Clavius, who, however, gives Euclid's proof as well, attributing it to Theon. The alternative proof proceeds after the manner of VI. 22, thus.

Make *AB*, *CD*, *O*, *P* continuous proportionals, and also *EF*, *GH*, *Q*, *R*.

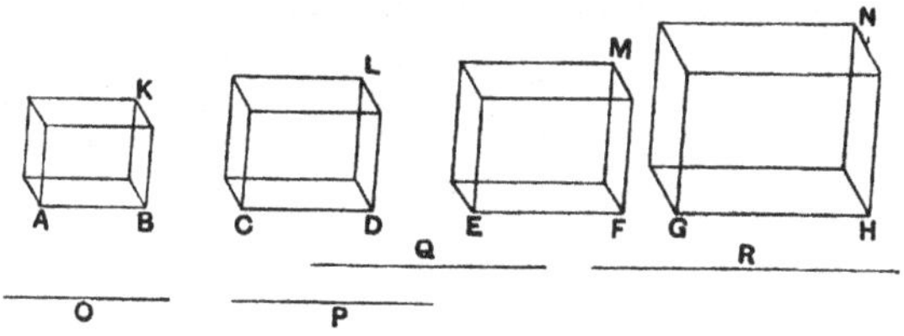

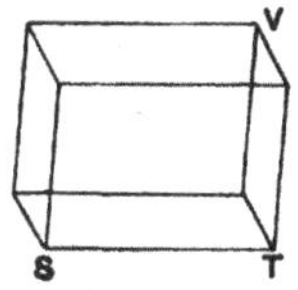

I. Then, since

$$AB : CD = EF : GH,$$

we have, *ex aequali*,

$$AB : P = EF : R. \qquad [\text{V. } 22]$$

But (solid *AK*) : (solid *CL*) = *AB* : *P*, [XI. 33 and Por.]

and (solid *EM*) : (solid *GN*) = *EF* : *R*.

Therefore

(solid *AK*) : (solid *CL*) = (solid *EM*) : (solid *GN*).

II. If the *solids* are proportional, take ST such that

$$AB : CD = EF : ST,$$

and on ST describe the parallelepiped SV similar and similarly situated to either of the parallelepipeds EM, GN.

Then, by the first part,

$$(\text{solid } AK) : (\text{solid } CL) = (\text{solid } EM) : (\text{solid } SV),$$

whence it follows that

$$(\text{solid } GN) = (\text{solid } SV).$$

But these solids are similar and similarly situated;

therefore their faces are similar and equal; [XI. Def. 10]

therefore the corresponding sides GH, ST are equal.

[For this inference cf. note on VI. 22. The equality of GH, ST may readily be proved by application of the two parallelepipeds to one another, since, being similar, they are equiangular.]

Hence $AB : CD = EF : GH$.

The text of the MSS. has here a proposition which is as badly placed as it is unnecessary. *If a plane be at right angles to a plane, and from any one of the points in one of the planes a perpendicular be drawn to the other plane, the perpendicular so drawn will fall on the common section of the planes.* It is of the nature of a lemma to XII. 17, where alone the fact is made use of. Heiberg observes that it is omitted in b and that the copyist of P knew other texts which did not contain it. From these facts it is fairly concluded that the proposition was interpolated. The truth of it is of course immediately obvious by *reductio ad absurdum*. Let the plane CAD be perpendicular to the plane AB, and let a perpendicular be drawn to the latter from any point E in the former.

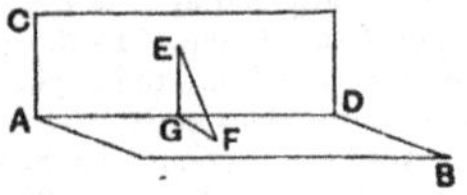

If it does not fall on AD, the common section, let it meet the plane AB in F.

Draw FG in AB perpendicular to AD, and join EG.

Then FG is perpendicular to the plane CAD [XI. Def. 4], and therefore to GE [XI. Def. 3]. Therefore $\angle EGF$ is right.

Also, since EF is perpendicular to AB,

the angle EFG is right.

That is, the triangle EGF has two right angles:

which is impossible.

PROPOSITION 38.

If the sides of the opposite planes of a cube be bisected, and planes be carried through the points of section, the common section of the planes and the diameter of the cube bisect one another.

For let the sides of the opposite planes CF, AH of the cube AF be bisected at the points K, L, M, N, O, Q, P, R, and through the points of section let the planes *KN*, *OR* be carried;

let *US* be the common section of the planes, and *DG* the diameter of the cube *AF*.

I say that *UT* is equal to *TS*, and *DT* to *TG*.

For let *DU*, *UE*, *BS*, *SG* be joined.

Then, since *DO* is parallel to *PE*,

the alternate angles *DOU*, *UPE* are equal to one another. [I. 29]

And, since *DO* is equal to *PE*, and *OU* to *UP*,

and they contain equal angles,

therefore the base *DU* is equal to the base *UE*,

the triangle *DOU* is equal to the triangle *PUE*,

and the remaining angles are equal to the remaining angles; [I. 4]

therefore the angle *OUD* is equal to the angle *PUE*.

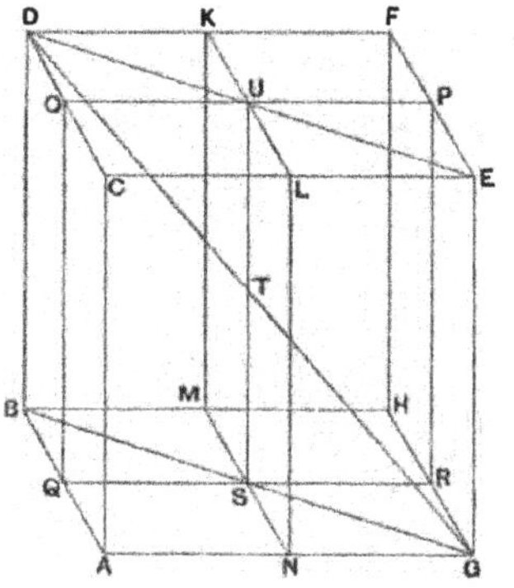

For this reason *DUE* is a straight line. [I. 14]

For the same reason, *BSG* is also a straight line,

and *BS* is equal to *SG*.

Now, since *CA* is equal and parallel to *DB*,

while *CA* is also equal and parallel to *EG*,

therefore *DB* is also equal and parallel to *EG*. [XI. 9]

And the straight lines DE, BG join their extremities; therefore DE is parallel to BG. [I. 33]

Therefore the angle EDT is equal to the angle BGT, for they are alternate; [I. 29]
and the angle DTU is equal to the angle GTS. [I. 15]

Therefore DTU, GTS are two triangles which have two angles equal to two angles, and one side equal to one side, namely that subtending one of the equal angles, that is, DU equal to GS,
for they are the halves of DE, BG;
therefore they will also have the remaining sides equal to the remaining sides. [I. 26]

Therefore DT is equal to TG, and UT to TS.

Therefore etc.

Q. E. D.

Euclid enunciates this proposition of a *cube* only, though it is true of any parallelepiped, no doubt because its truth for a cube is all that was wanted for the only proposition where it is needed, viz. XIII. 17.

Simson remarks that it should be proved that the straight lines bisecting the corresponding opposite sides of opposite planes *are* in one plane. This is, however, clear because e.g. since DK, CL are equal and parallel, KL is equal and parallel to CD. And, since KL, AB are both parallel to DC, KL is parallel to AB. And lastly, since KL, MN are both parallel to AB, KL is parallel to MN and therefore in one plane with it.

The essential thing to be proved is that the plane passing through the opposite edges DB, EG passes through the straight line US, since, only if this be the case, can US, DG intersect one another.

To prove this we have only to prove that, if DU, UE and BS, SG be joined, DUE and BSG are both straight lines.

Now, since DO is parallel to PE,

$$\angle DOU = \angle EPU.$$

Thus, in the $\triangle$s DUO, EUP, two sides DO, OU are equal to two sides EP, PU, and the included angles are equal.

Therefore $$\triangle DUO \equiv \triangle EUP,$$
$$DU = UE,$$
and $$\angle DUO = \angle EUP,$$
so that DUE is a straight line, bisected at U. Similarly BSG is a straight line, bisected at S.

Thus the plane through DB, EG (DB, EG being equal and parallel) contains the straight lines DUE, BSG (which are therefore equal and parallel also) and also [XI. 7] the straight lines US, DG (which accordingly intersect).

In $\triangle$s DTU, GTS, the angles UDT, SGT are equal (being alternate), and the angles UTD, STG are also equal (being vertically opposite), while DU (half of DE) is equal to GS (half of BG).

Therefore [I. 26] the triangles DTU, GTS are equal in all respects, so that

$$DT = TG,$$
$$UT = TS.$$

Proposition 39.

If there be two prisms of equal height, and one have a parallelogram as base and the other a triangle, and if the parallelogram be double of the triangle, the prisms will be equal.

Let $ABCDEF$, $GHKLMN$ be two prisms of equal height,

let one have the parallelogram AF as base, and the other the triangle GHK,

and let the parallelogram AF be double of the triangle GHK;

I say that the prism $ABCDEF$ is equal to the prism $GHKLMN$.

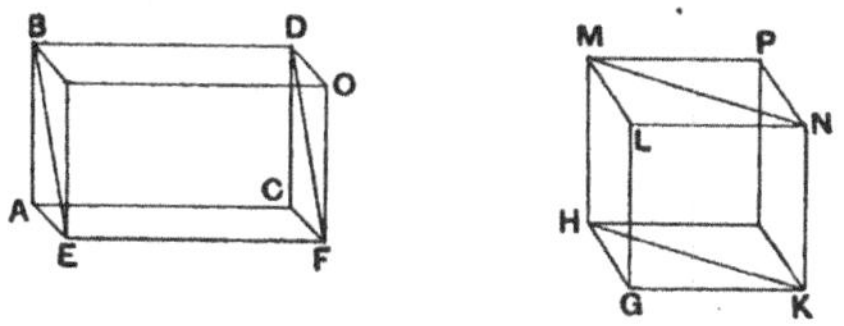

For let the solids AO, GP be completed.

Since the parallelogram AF is double of the triangle GHK, while the parallelogram HK is also double of the triangle GHK, [I. 34]

therefore the parallelogram AF is equal to the parallelogram HK.

But parallelepipedal solids which are on equal bases and of the same height are equal to one another; [XI. 31]

therefore the solid AO is equal to the solid GP.

And the prism $ABCDEF$ is half of the solid AO,

and the prism $GHKLMN$ is half of the solid GP; [XI. 28]

therefore the prism $ABCDEF$ is equal to the prism $GHKLMN$.

Therefore etc.

Q. E. D.

This proposition is made use of in XII. 3, 4. The phraseology is interesting because we find one of the *parallelogrammic* faces of one of the triangular prisms called its *base*, and the perpendicular on this plane from that vertex of either *triangular* face which is not in this plane the *height*.

The proof is simple because we have only to complete parallelepipeds which are double the prisms respectively and then use XI. 31. It has to be borne in mind, however, that, if the parallelepipeds are not rectangular, the proof in XI. 28 is not sufficient to establish the fact that the parallelepipeds are double of the prisms, but has to be supplemented as shown in the note on that proposition. XII. 4 does, however, require the theorem in its general form.

BOOK XII.

HISTORICAL NOTE.

The predominant feature of Book XII. is the use of the *method of exhaustion*, which is applied in Propositions 2, 3—5, 10, 11, 12, and (in a slightly different form) in Propositions 16—18. We conclude therefore that for the content of this Book Euclid was greatly indebted to Eudoxus, to whom the discovery of the method of exhaustion is attributed. The evidence for this attribution comes mainly from Archimedes. (1) In the preface to *On the Sphere and Cylinder* I., after stating the main results obtained by himself regarding the surface of a sphere or a segment thereof, and the volume and surface of a right cylinder with height equal to its diameter as compared with those of a sphere with the same diameter, Archimedes adds: "Having now discovered that the properties mentioned are true of these figures, I cannot feel any hesitation in setting them side by side both with my former investigations and *with those of the theorems of Eudoxus on solids* which are held to be most irrefragably established, namely that *any pyramid is one third part of the prism which has the same base with the pyramid and equal height* [i.e. Eucl. XII. 7], and that *any cone is one third part of the cylinder which has the same base with the cone and equal height* [i.e. Eucl. XII. 10]. For, though these properties also were naturally inherent in the figures all along, yet they were in fact unknown to all the many able geometers who lived before Eudoxus and had not been observed by any one." (2) In the preface to the treatise known as the *Quadrature of the Parabola* Archimedes states the "lemma" assumed by him and known as the "Axiom of Archimedes" (see note on X. 1 above) and proceeds: "Earlier geometers (οἱ πρότερον γεωμέτραι) have also used this lemma; for it is by the use of this same lemma that they have shown that *circles are to one another in the duplicate ratio of their diameters* [Eucl. XII. 2], and that *spheres are to one another in the triplicate ratio of their diameters* [Eucl. XII. 18], and further that *every pyramid is one third part of the prism which has the same base with the pyramid and equal height* [Eucl. XII. 7]; also, that *every cone is one third part of the cylinder which has the same base with the cone and equal height* [Eucl. XII. 10] they proved by assuming *a certain lemma similar to that aforesaid.*" Thus in the first passage two theorems of Eucl. XII. are definitely attributed to Eudoxus; and, when Archimedes says, in the second passage, that "earlier geometers" proved these two theorems by means of the lemma known as the "Axiom of Archimedes" and of a lemma similar to it respectively, we can hardly suppose him to be alluding to any other proof than that given by Eudoxus. As a matter of fact, the lemma used by Euclid to prove both propositions (XII. 3—5 and 7, and XII. 10) is the theorem of Eucl. X. 1. As regards the connexion between the two "lemmas" see note on X. 1.

We are not, however, to suppose that none of the *results* obtained by the method of exhaustion had been discovered before the time of Eudoxus (fl. about 368—5 B.C.). Two at least are of earlier date, those of Eucl. XII. 2 and XII. 7.

(*a*) Simplicius (*Comment. in Aristot. Phys.* p. 61, ed. Diels) quotes Eudemus as saying, in his *History of Geometry*, that Hippocrates of Chios (fl. say 430 B.C.) first laid it down (ἔθετο) that similar segments of circles are in the ratio of the squares on their bases and that he proved this (ἐδείκνυεν) by proving (ἐκ τοῦ δεῖξαι) that the squares on the diameters have the same ratio as the (whole) circles. We know nothing of the method by which Hippocrates proved this proposition; but, having regard to the evidence from Archimedes quoted above, it is not permissible to suppose that the method was the fully developed *method of exhaustion* as we know it.

(*b*) As regards the two theorems about the volume of a pyramid and of a cone respectively, which Eudoxus was the first to prove, a new piece of evidence is now forthcoming in the fragment of Archimedes recently brought to light at Constantinople and published by Heiberg (for the Greek text see *Hermes* XLII., 1907, pp. 235—303; for Heiberg's translation and Zeuthen's notes see *Bibliotheca Mathematica* VII$_3$, 1907, pp. 321—363). This is nothing less than a considerable portion of a work under the title Ἀρχιμήδους περὶ τῶν μηχανικῶν θεωρημάτων πρὸς Ἐρατοσθένην ἔφοδος, which "Method," addressed to Eratosthenes, is the ἐφόδιον on which, according to Suidas, Theodosius wrote a commentary, and which is several times cited by Heron in his *Metrica*; and it adds a new and important chapter to the history of the integral calculus. In the preface to this work (*Hermes l.c.* p. 245, *Bibliotheca Mathematica l.c.* p. 323) Archimedes alludes to the theorems which he first discovered by means of mechanical considerations, but proved afterwards by geometry because the investigation by means of mechanics did not constitute a rigid proof; he observes, however, that the mechanical method is of great use for the *discovery* of theorems, and it is much easier to provide the rigid proof when the fact to be proved has once been discovered than it would be if nothing were known to begin with. He goes on: "Hence too, in the case of those theorems the proof of which was first discovered by Eudoxus, namely those relating to the cone and the pyramid, that the cone is one third part of the cylinder, and the pyramid one third part of the prism, having the same base and equal height, no small part of the credit will naturally be assigned to Democritus, who was the first to make the statement (of the fact) regarding the said figure [i.e. property], though without proving it." Hence the *discovery* of the two theorems must now be attributed to Democritus (fl. towards the end of 5th cent. B.C.). The words "without proving it" (χωρὶς ἀποδείξεως) do not mean that Democritus gave no sort of proof, but only that he did not give a proof on the rigorous lines required later; for the same words are used by Archimedes of his own investigations by means of mechanics, which, however, do constitute a reasoned argument. The character of Archimedes' mechanical arguments combined with a passage of Plutarch about a particular question in infinitesimals said to have been raised by Democritus may perhaps give a clue to the line of Democritus' argument as regards the prism. The essential feature of Archimedes' mechanical arguments in this tract is that he regards an area as the sum of an infinite number of *straight lines* parallel to one another and terminated by the boundary or boundaries of the closed figure the area of which is to be found, and a volume as the sum of an infinite number of *plane sections* parallel to one another: which is of course the same thing as taking (as we do in the integral calculus) the sum of an infinite number of strips of breadth dx (say), when dx becomes indefinitely small, or the sum of an infinite number of parallel laminae of depth dz (say), when dz becomes indefinitely small. To give only one instance, we may take the case of the area of a segment of a parabola cut off by a chord.

Let CBA be the parabolic segment, CE the tangent at C meeting the

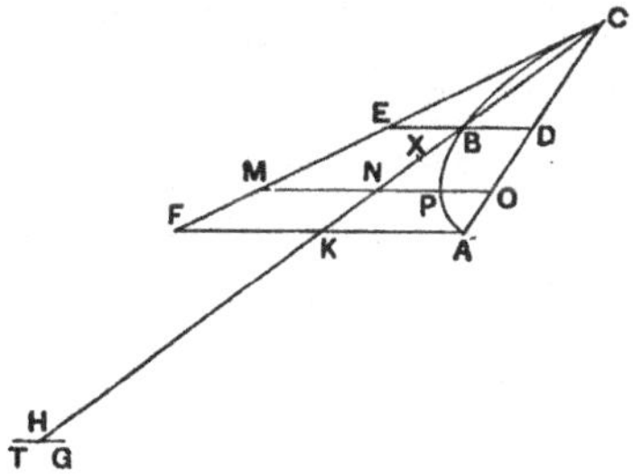

diameter EBD through the middle point of the chord CA in E, so that

$$EB = BD.$$

Draw AF parallel to ED meeting CE produced in F. Produce CB to H so that $CK = KH$, where K is the point in which CH meets AF; and suppose CH to be a lever.

Let any diameter $MNPO$ be drawn meeting the curve in P and CF, CK, CA in M, N, O respectively.

Archimedes then observes that

$$CA : AO = MO : OP$$

("for this is proved in a lemma"),

whence $$HK : KN = MO : OP,$$

so that, if a straight line TG equal to PO be placed with its middle point at H, the straight line MO with centre of gravity at N, and the straight line TG with centre of gravity at H, will balance about K.

Taking all other parts of diameters like PO intercepted between the curve and CA, and placing equal straight lines with their centres of gravity at H, these straight lines collected at H will balance (about K) all the lines like MO parallel to FA intercepted within the triangle CFA in the positions in which they severally lie in the figure.

Hence Archimedes infers that an area equal to that of the parabolic segment hung at H will balance (about K) the triangle CFA hung at its centre of gravity, the point X (a point on CK such that $CK = 3XK$), and therefore that

$$\begin{aligned}(\text{area of triangle } CFA) : (\text{area of segment}) &= HK : KX \\ &= 3 : 1,\end{aligned}$$

from which it follows that

$$\text{area of parabolic segment} = \tfrac{4}{3} \triangle ABC.$$

The same sort of argument is used for solids, ***plane sections*** taking the place of ***straight lines.***

Archimedes is careful to state once more that this method of argument does not constitute a ***proof.*** Thus, at the end of the above proposition about the parabolic segment, he adds: "This property is of course not proved by what has just been said; but it has furnished a sort of ***indication*** (ἔμφασίν τινα) that the conclusion is true."

Let us now turn to the passage of Plutarch (*De Comm. Not. adv. Stoicos* XXXIX. 3) about Democritus above referred to. Plutarch speaks of Democritus as having raised the question in natural philosophy (φυσικῶς): "if a cone were cut by a plane parallel to the base [by which is clearly meant a plane indefinitely near to the base], what must we think of the surfaces of the sections, that they are equal or unequal? For, if they are unequal, they will make the cone irregular, as having many indentations, like steps, and unevennesses; but, if they are equal, the sections will be equal, and the cone will appear to have the property of the cylinder and to be made up of equal, not unequal circles, which is very absurd." The phrase "***made up*** of equal...circles" (ἐξ ἴσων συγκείμενος...κύκλων) shows that Democritus already had the idea of a solid being the sum of an infinite number of parallel planes, or indefinitely thin laminae, indefinitely near together: a most important anticipation of the same thought which led to such fruitful results in Archimedes. If then one may hazard a conjecture as to Democritus' argument with regard to a pyramid, it seems probable that he would notice that, if two pyramids of the same height and equal triangular bases are respectively cut by planes parallel to the base and dividing the heights in the same ratio, the corresponding sections of the two pyramids are equal, whence he would infer that the pyramids are equal as being the sum of the same infinite number of equal plane sections or indefinitely thin laminae. (This would be a particular anticipation of Cavalieri's proposition that the areal or solid content of two figures are equal if two sections of them taken at the same height, whatever the height may be, always give equal straight lines or equal surfaces respectively.) And Democritus would of course see that the three pyramids into which a prism on the same base and of equal height with the original pyramid is divided (as in Eucl. XII. 7) satisfy this test of equality, so that the pyramid would be one third part of the prism. The extension to a pyramid with a polygonal base would be easy. And Democritus may have stated the proposition for the cone (of course without an absolute proof) as a natural inference from the result of increasing indefinitely the number of sides in a regular polygon forming the base of a pyramid.

PROPOSITION 1.

Similar polygons inscribed in circles are to one another as the squares on the diameters.

Let ABC, FGH be circles,
let $ABCDE$, $FGHKL$ be similar polygons inscribed in them, and let BM, GN be diameters of the circles;
I say that, as the square on BM is to the square on GN, so is the polygon $ABCDE$ to the polygon $FGHKL$.

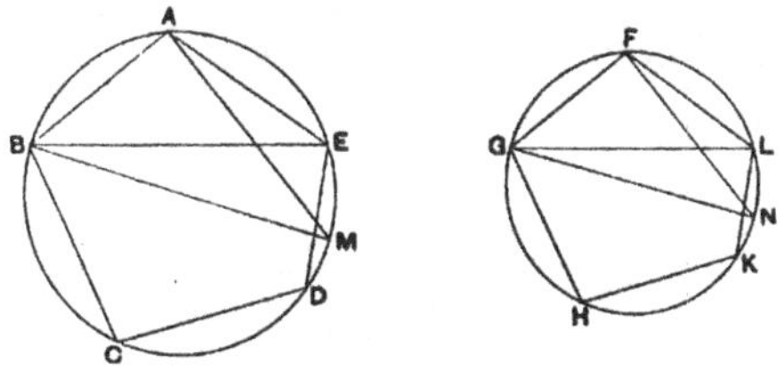

For let BE, AM, GL, FN be joined.

Now, since the polygon $ABCDE$ is similar to the polygon $FGHKL$,
the angle BAE is equal to the angle GFL,
and, as BA is to AE, so is GF to FL. [VI. Def. 1]

Thus BAE, GFL are two triangles which have one angle equal to one angle, namely the angle BAE to the angle GFL, and the sides about the equal angles proportional;
therefore the triangle ABE is equiangular with the triangle FGL. [VI. 6]

Therefore the angle AEB is equal to the angle FLG.

But the angle AEB is equal to the angle AMB, for they stand on the same circumference; [III. 27]
and the angle FLG to the angle FNG;
therefore the angle AMB is also equal to the angle FNG.

But the right angle BAM is also equal to the right angle GFN; [III. 31]
therefore the remaining angle is equal to the remaining angle. [I. 32]

Therefore the triangle ABM is equiangular with the triangle FGN.

Therefore, proportionally, as BM is to GN, so is BA to GF. [VI. 4]

But the ratio of the square on BM to the square on GN is duplicate of the ratio of BM to GN,
and the ratio of the polygon $ABCDE$ to the polygon $FGHKL$ is duplicate of the ratio of BA to GF; [VI. 20]
therefore also, as the square on BM is to the square on GN, so is the polygon $ABCDE$ to the polygon $FGHKL$.

Therefore etc.

Q. E. D.

As, from this point onward, the text of each proposition usually occupies considerable space, I shall generally give in the notes a summary of the argument, to enable it to be followed more easily.

Here we have to prove that a pair of corresponding sides are in the ratio of the corresponding diameters.

Since ∠s BAE, GFL are equal, and the sides about those angles proportional,

$$\triangle\text{s } ABE, FGL \text{ are equiangular,}$$

so that

$$\angle AEB = \angle FLG.$$

Hence their equals in the same segments, ∠s AMB, FNG, are equal.

And the right angles BAM, GFN are equal.

Therefore △s ABM, FGN are equiangular, so that

$$BM : GN = BA : GF.$$

The duplicates of these ratios are therefore equal,

whence

$$\begin{aligned}(\text{polygon } ABCDE) : (\text{polygon } FGHKL) &= \text{duplicate ratio of } BA \text{ to } GF \\ &= \text{duplicate ratio of } BM \text{ to } GN \\ &= BM^2 : GN^2.\end{aligned}$$

Proposition 2.

Circles are to one another as the squares on the diameters.

Let *ABCD*, *EFGH* be circles, and *BD*, *FH* their diameters;

I say that, as the circle *ABCD* is to the circle *EFGH*, so is the square on *BD* to the square on *FH*.

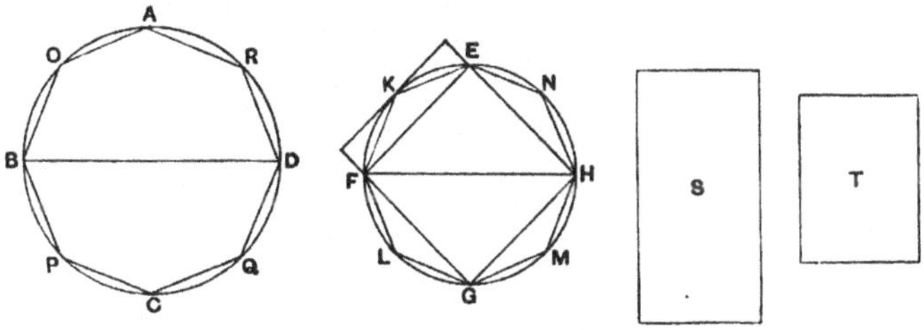

For, if the square on *BD* is not to the square on *FH* as the circle *ABCD* is to the circle *EFGH*,

then, as the square on *BD* is to the square on *FH*, so will the circle *ABCD* be either to some less area than the circle *EFGH*, or to a greater.

First, let it be in that ratio to a less area *S*.

Let the square *EFGH* be inscribed in the circle *EFGH*; then the inscribed square is greater than the half of the circle *EFGH*, inasmuch as, if through the points *E*, *F*, *G*, *H* we draw tangents to the circle, the square *EFGH* is half the square circumscribed about the circle, and the circle is less than the circumscribed square;

hence the inscribed square *EFGH* is greater than the half of the circle *EFGH*.

Let the circumferences *EF*, *FG*, *GH*, *HE* be bisected at the points *K*, *L*, *M*, *N*,

and let *EK*, *KF*, *FL*, *LG*, *GM*, *MH*, *HN*, *NE* be joined;

therefore each of the triangles *EKF*, *FLG*, *GMH*, *HNE* is also greater than the half of the segment of the circle about it, inasmuch as, if through the points *K*, *L*, *M*, *N* we draw tangents to the circle and complete the parallelograms on the straight lines *EF*, *FG*, *GH*, *HE*, each of the triangles *EKF*,

FLG, GMH, HNE will be half of the parallelogram about it,
while the segment about it is less than the parallelogram;
hence each of the triangles *EKF, FLG, GMH, HNE* is greater than the half of the segment of the circle about it.

Thus, by bisecting the remaining circumferences and joining straight lines, and by doing this continually, we shall leave some segments of the circle which will be less than the excess by which the circle *EFGH* exceeds the area *S*.

For it was proved in the first theorem of the tenth book that, if two unequal magnitudes be set out, and if from the greater there be subtracted a magnitude greater than the half, and from that which is left a greater than the half, and if this be done continually, there will be left some magnitude which will be less than the lesser magnitude set out.

Let segments be left such as described, and let the segments of the circle *EFGH* on *EK, KF, FL, LG, GM, MH, HN, NE* be less than the excess by which the circle *EFGH* exceeds the area *S*.

Therefore the remainder, the polygon *EKFLGMHN*, is greater than the area *S*.

Let there be inscribed, also, in the circle *ABCD* the polygon *AOBPCQDR* similar to the polygon *EKFLGMHN*;
therefore, as the square on *BD* is to the square on *FH*, so is the polygon *AOBPCQDR* to the polygon *EKFLGMHN*. [XII. 1]

But, as the square on *BD* is to the square on *FH*, so also is the circle *ABCD* to the area *S*;
therefore also, as the circle *ABCD* is to the area *S*, so is the polygon *AOBPCQDR* to the polygon *EKFLGMHN*; [V. 11]
therefore, alternately, as the circle *ABCD* is to the polygon inscribed in it, so is the area *S* to the polygon *EKFLGMHN*. [V. 16]

But the circle *ABCD* is greater than the polygon inscribed in it;
therefore the area *S* is also greater than the polygon *EKFLGMHN*.

But it is also less:
which is impossible.

Therefore, as the square on *BD* is to the square on *FH*, so is not the circle *ABCD* to any area less than the circle *EFGH*.

Similarly we can prove that neither is the circle *EFGH* to any area less than the circle *ABCD* as the square on *FH* is to the square on *BD*.

I say next that neither is the circle *ABCD* to any area greater than the circle *EFGH* as the square on *BD* is to the square on *FH*.

For, if possible, let it be in that ratio to a greater area *S*.

Therefore, inversely, as the square on *FH* is to the square on *DB*, so is the area *S* to the circle *ABCD*.

But, as the area *S* is to the circle *ABCD*, so is the circle *EFGH* to some area less than the circle *ABCD*;
therefore also, as the square on *FH* is to the square on *BD*, so is the circle *EFGH* to some area less than the circle *ABCD*: [V. 11]
which was proved impossible.

Therefore, as the square on *BD* is to the square on *FH*, so is not the circle *ABCD* to any area greater than the circle *EFGH*.

And it was proved that neither is it in that ratio to any area less than the circle *EFGH*;
therefore, as the square on *BD* is to the square on *FH*, so is the circle *ABCD* to the circle *EFGH*.

Therefore etc.

Q. E. D.

Lemma.

I say that, the area *S* being greater than the circle *EFGH*, as the area *S* is to the circle *ABCD*, so is the circle *EFGH* to some area less than the circle *ABCD*.

For let it be contrived that, as the area *S* is to the circle *ABCD*, so is the circle *EFGH* to the area *T*.

I say that the area *T* is less than the circle *ABCD*.

For since, as the area *S* is to the circle *ABCD*, so is the circle *EFGH* to the area *T*, therefore, alternately, as the area S is to the circle $EFGH$, so is the circle $ABCD$ to the area T. [V. 16]

But the area S is greater than the circle $EFGH$;
therefore the circle $ABCD$ is also greater than the area T.

Hence, as the area S is to the circle $ABCD$, so is the circle $EFGH$ to some area less than the circle $ABCD$.

Q. E. D.

Though this theorem is said to have been proved by Hippocrates, we may with tolerable certainty attribute the proof of it given by Euclid to Eudoxus, to whom XII. 7 Por. and XII. 10 (which Euclid proves in exactly the same manner) are specifically attributed by Archimedes. As regards the lemma used herein (Eucl. X. 1) and the somewhat different lemma by means of which Archimedes says that the theorems of XII. 2, XII. 7 Por. and XII. 18 were proved, see my note on X. 1 above.

The first essential in this proposition is to prove that we can ***exhaust*** a circle, in the sense of X. 1, by successively inscribing in it regular polygons, each of which has twice as many sides as the preceding one. We take first an inscribed square, then bisect the arcs subtended by the sides and so form an equilateral polygon of eight sides, then do the same with the latter, forming a polygon of 16 sides, and so on. And we have to prove that what is left over when any one of these polygons is taken away from the circle is ***more than half exhausted*** when the next polygon is made and subtracted from the circle.

Euclid proves that the inscribed square is greater than half the circle and that the regular octagon when subtracted takes away more than half of what was left by the square. He then infers that the same thing will happen whenever the number of sides is doubled.

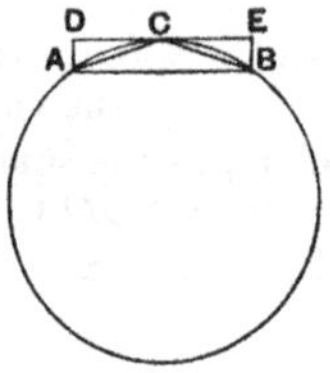

This can be seen generally by taking *any* arc of a circle cut off by a chord AB. Bisect the arc in C. Draw a tangent to the circle at C, and let AD, BE be drawn perpendicular to the tangent. Join AC, CB.

Then DE is parallel to AB, since

$\angle ECB = \angle CAB$, in alternate segment, [III. 32]

$= \angle CBA$. [III. 29, I. 5]

Thus $ABED$ is a ▱;
and it is greater than the segment ACB.

Therefore its half, the $\triangle ACB$, is greater than half the segment.

Thus, by X. 1, Euclid's construction of successive regular polygons in a circle, if continued far enough, will at length leave segments which are together less than any given area.

Now let X, X' be the areas of the circles, d, d' their diameters, respectively.

Then, if $$X : X' \neq d^2 : d'^2,$$
$$d^2 : d'^2 = X : S,$$
where S is some area either greater or less than X'.

I. Suppose $S < X'$.

Continue the construction of polygons in X' until we arrive at one which leaves over segments together less than the excess of X' over S, i.e. a polygon such that

$$X' > (\text{polygon in } X') > S.$$

Inscribe in the circle X a polygon similar to that in X'.

Then $(\text{polygon in } X) : (\text{polygon in } X') = d^2 : d'^2$ [XII. 1]

$= X : S$, by hypothesis;

and, alternately,

$$(\text{polygon in } X) : X = (\text{polygon in } X') : S.$$

But $(\text{polygon in } X) < X$;

therefore $(\text{polygon in } X') < S$.

But, by construction, $(\text{polygon in } X') > S$:

which is impossible.

Hence S cannot be *less* than X' as supposed.

II. Suppose $S > X'$.

Since $d^2 : d'^2 = X : S$,

we have, inversely, $d'^2 : d^2 = S : X$.

Suppose that $S : X = X' : T$,

whence, since $S > X'$, $X > T$. [V. 14]

Consequently $d'^2 : d^2 = X' : T$,

where $T < X$.

This can be proved impossible in exactly the same way as shown in Part I.

Hence S cannot be *greater* than X' as supposed.

Since then S is neither greater nor less than X',

$$S = X',$$

and therefore $d^2 : d'^2 = X : X'$.

With reference to the assumption that there *is* some space S such that

$$d^2 : d'^2 = X : S,$$

i.e. that there is a fourth proportional to the areas d^2, d'^2, X, Simson observes that it is sufficient, in this and the like cases, that a thing made use of in the reasoning can possibly *exist*, though it cannot be exhibited by a geometrical construction. As regards the assumption see note on V. 18 above.

There is grave reason for suspecting the genuineness of the Lemma at the end of the proposition; though, if it be rejected, it will be necessary to delete the words "as was before proved" in corresponding places in XII. 5, 18.

It will be observed that Euclid proves the impossibility in the second case by reducing it to the first. If it is desired to prove the second case independently, we must *circumscribe* successive polygons to the circles instead of inscribing them, in the way shown by Archimedes in his first proposition on the *Measurement of a circle*. Of course we require, as a preliminary, the proposition corresponding to XII. 1, that *Similar polygons circumscribed about circles are to one another as the squares on the diameters.*

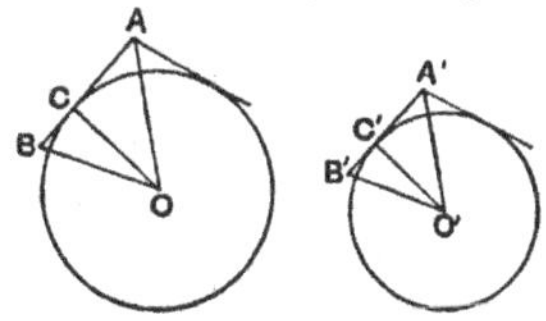

Let AB, $A'B'$ be corresponding sides of the two similar polygons. Then $\angle$s OAB, $O'A'B'$ are equal, since AO, $A'O'$ bisect equal angles.

Similarly $\angle ABO = \angle A'B'O'$.

Therefore $\triangle$s AOB, $A'O'B'$ are similar, so that their areas are in the duplicate ratio of AB to $A'B'$.

The radii OC, $O'C'$ drawn to the points of contact are perpendicular to AB, $A'B'$, and it follows that

$$AB : A'B' = CO : C'O'.$$

Thus the polygons are to one another in the duplicate ratio of the radii, and therefore of the diameters.

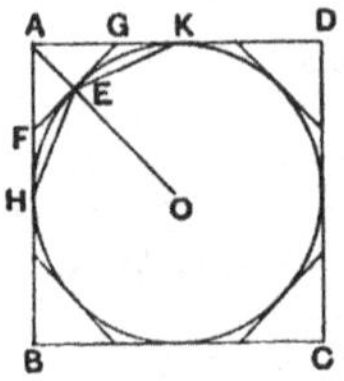

Now suppose a square $ABCD$ described about a circle.

Make an octagon described about the circle by drawing tangents at the points E etc., where OA etc. meet the circle.

Then shall the tangent at E cut off more than half of the area between AK, AH and the arc HEK.

For the angle AEG is right, and is therefore $> \angle EAG$.

Therefore $$AG > EG$$
$$> GK.$$

Therefore $$\triangle AGE > \triangle EGK.$$

Similarly $$\triangle AFE > \triangle EFH.$$

Hence $\triangle AFG > \frac{1}{2}$ (re-entrant quadrilateral $AHEK$),

and *a fortiori*, $\triangle AFG > \frac{1}{2}$ (area between AH, AK and the arc).

Thus the octagon takes from the square more than half the space between the square and the circle.

Similarly, if a figure of 16 equal sides be circumscribed by cutting off symmetrically the corners of the octagon, it will take away more than half of the space between the octagon and circle.

Suppose now, with the original notation, that

$$d^2 : d'^2 = X : S,$$

where S is greater than X'.

Continue the construction of circumscribed polygons about X' until the total area between the polygon and the circle is less than the difference between S and X', i.e. till

$$S > (\text{polygon about } X') > X'.$$

Circumscribe a similar polygon about X.

Then $(\text{polygon about } X) : (\text{polygon about } X') = d^2 : d'^2$
$= X : S$, by hypothesis,

and, alternately,

$$(\text{polygon about } X) : X = (\text{polygon about } X') : S.$$

But $(\text{polygon about } X) > X.$

Therefore $(\text{polygon about } X') > S.$

But $S > (\text{polygon about } X')$: [above]

which is impossible.

Hence S cannot be *greater* than X'.

Legendre proves this proposition by a method equally rigorous but not, I think, possessing any advantages over Euclid's. It depends on a lemma corresponding to Eucl. XII. 16, but with another part added to it.

Two concentric circles being given, we can always inscribe in the greater a regular polygon such that its sides do not meet the circumference of the lesser, and we can also circumscribe about the lesser a regular polygon such that its sides do not meet the circumference of the greater.

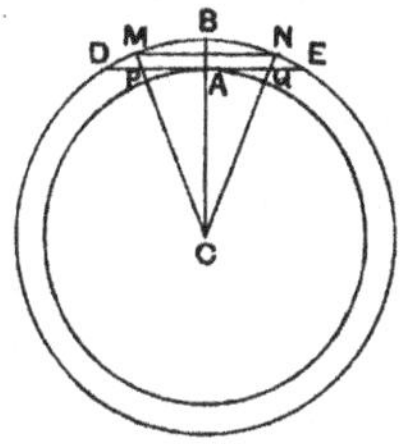

Let CA, CB be the radii of the circles.

I. At A on the inner circle draw the tangent DE meeting the outer circle in D, E.

Inscribe in the outer circle any of the regular polygons which we can inscribe, e.g. a square.

Bisect the arc subtended by a side, bisect the half, bisect that again, and so on, until we arrive at an arc less than the arc DBE.

Let this arc be MN, and suppose it so placed that B is its middle point.

Then the chord MN is clearly more distant from the centre C than DE is; and the regular polygon, of which MN is a side, does not anywhere meet the circumference of the inner circle.

II. Join CM, CN, meeting DE in P, Q.

Then PQ will be the side of a polygon circumscribed about the inner circle and similar to the polygon inscribed in the outer;

and the circumscribed polygon of which PQ is a side will not anywhere meet the outer circle.

Legendre now proves XII. 2 after the following manner.

For brevity, let us denote the area of the circle with radius CA by (circ. CA).

Then it is required to prove that, if OB be the radius of a second circle,

$$(\text{circ. } CA) : (\text{circ. } OB) = CA^2 : OB^2.$$

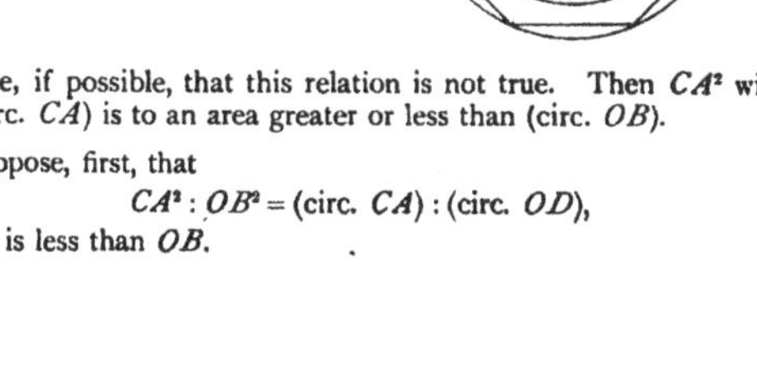

Suppose, if possible, that this relation is not true. Then CA^2 will be to OB^2 as (circ. CA) is to an area greater or less than (circ. OB).

I. Suppose, first, that

$$CA^2 : OB^2 = (\text{circ. } CA) : (\text{circ. } OD),$$

where OD is less than OB.

Inscribe in the circle with radius OB a regular polygon such that its sides do not anywhere meet the circumference of the circle with centre OD; [Lemma]

and inscribe a similar polygon in the other circle.

The areas of the polygons will then be in the duplicate ratio of CA to OB, or [XII. 1]

$$\text{(polygon in circ. } CA) : \text{(polygon in circ. } OB)$$
$$= CA^2 : OB^2$$
$$= \text{(circ. } CA) : \text{(circ. } OD\text{), by hypothesis.}$$

But this is impossible, because the polygon in (circ. CA) is *less* than (circ. CA), but the polygon in (circ. OB) is *greater* than (circ. OD).

Therefore CA^2 cannot be to OB^2 as (circ. CA) is to a *less* circle than (circ. OB).

II. Suppose, if possible, that

$$CA^2 : OB^2 = \text{(circ. } CA) : \text{(some circle} > \text{circ. } OB).$$

Then inversely

$$OB^2 : CA^2 = \text{(circ. } OB) : \text{(some circle} < \text{circ. } CA),$$

and this is proved impossible exactly as in Part I.

Therefore $$CA^2 : OB^2 = \text{(circ. } CA) : \text{(circ. } OB).$$

Proposition 3.

Any pyramid which has a triangular base is divided into two pyramids equal and similar to one another, similar to the whole and having triangular bases, and into two equal prisms; and the two prisms are greater than the half of the whole pyramid.

Let there be a pyramid of which the triangle ABC is the base and the point D the vertex;

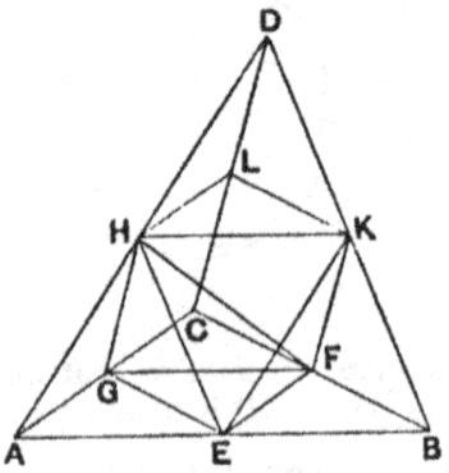

I say that the pyramid $ABCD$ is divided into two pyramids equal to one another, having triangular bases and similar to the whole pyramid, and into two equal prisms; and the two prisms are greater than the half of the whole pyramid.

For let AB, BC, CA, AD, DB, DC be bisected at the points E, F, G, H, K, L, and let HE, EG, GH, HK, KL, LH, KF, FG be joined.

Since AE is equal to EB, and AH to DH, therefore EH is parallel to DB. [VI. 2]

For the same reason
HK is also parallel to *AB*.

Therefore *HEBK* is a parallelogram;
therefore *HK* is equal to *EB*. [I. 34]

But *EB* is equal to *EA*;
therefore *AE* is also equal to *HK*.

But *AH* is also equal to *HD*;
therefore the two sides *EA*, *AH* are equal to the two sides *KH*, *HD* respectively;
and the angle *EAH* is equal to the angle *KHD*;
therefore the base *EH* is equal to the base *KD*. [I. 4]

Therefore the triangle *AEH* is equal and similar to the triangle *HKD*.

For the same reason
the triangle *AHG* is also equal and similar to the triangle *HLD*.

Now, since two straight lines *EH*, *HG* meeting one another are parallel to two straight lines *KD*, *DL* meeting one another, and are not in the same plane, they will contain equal angles. [XI. 10]

Therefore the angle *EHG* is equal to the angle *KDL*.

And, since the two straight lines *EH*, *HG* are equal to the two *KD*, *DL* respectively,
and the angle *EHG* is equal to the angle *KDL*,
therefore the base *EG* is equal to the base *KL*; [I. 4]
therefore the triangle *EHG* is equal and similar to the triangle *KDL*.

For the same reason
the triangle *AEG* is also equal and similar to the triangle *HKL*.

Therefore the pyramid of which the triangle *AEG* is the base and the point *H* the vertex is equal and similar to the pyramid of which the triangle *HKL* is the base and the point *D* the vertex. [XI. Def. 10]

And, since *HK* has been drawn parallel to *AB*, one of the sides of the triangle *ADB*,

the triangle *ADB* is equiangular to the triangle *DHK*, [I. 29]
and they have their sides proportional;
therefore the triangle *ADB* is similar to the triangle *DHK*. [VI. Def. 1]

For the same reason
the triangle *DBC* is also similar to the triangle *DKL*, and the triangle *ADC* to the triangle *DLH*.

Now, since the two straight lines *BA*, *AC* meeting one another are parallel to the two straight lines *KH*, *HL* meeting one another, not in the same plane, they will contain equal angles. [XI. 10]

Therefore the angle *BAC* is equal to the angle *KHL*.

And, as *BA* is to *AC*, so is *KH* to *HL*;
therefore the triangle *ABC* is similar to the triangle *HKL*.

Therefore also the pyramid of which the triangle *ABC* is the base and the point *D* the vertex is similar to the pyramid of which the triangle *HKL* is the base and the point *D* the vertex.

But the pyramid of which the triangle *HKL* is the base and the point *D* the vertex was proved similar to the pyramid of which the triangle *AEG* is the base and the point *H* the vertex.

Therefore each of the pyramids *AEGH*, *HKLD* is similar to the whole pyramid *ABCD*.

Next, since *BF* is equal to *FC*,
the parallelogram *EBFG* is double of the triangle *GFC*.

And since, if there be two prisms of equal height, and one have a parallelogram as base, and the other a triangle, and if the parallelogram be double of the triangle, the prisms are equal, [XI. 39]
therefore the prism contained by the two triangles *BKF*, *EHG*, and the three parallelograms *EBFG*, *EBKH*, *HKFG* is equal to the prism contained by the two triangles *GFC*, *HKL* and the three parallelograms *KFCL*, *LCGH*, *HKFG*.

And it is manifest that each of the prisms, namely that in which the parallelogram *EBFG* is the base and the straight line *HK* is its opposite, and that in which the triangle *GFC* is the base and the triangle *HKL* its opposite, is greater than each of the pyramids of which the triangles *AEG*, *HKL* are the bases and the points *H*, *D* the vertices, inasmuch as, if we join the straight lines *EF*, *EK*, the prism in which the parallelogram *EBFG* is the base and the straight line *HK* its opposite is greater than the pyramid of which the triangle *EBF* is the base and the point *K* the vertex.

But the pyramid of which the triangle *EBF* is the base and the point *K* the vertex is equal to the pyramid of which the triangle *AEG* is the base and the point *H* the vertex;

for they are contained by equal and similar planes.

Hence also the prism in which the parallelogram *EBFG* is the base and the straight line *HK* its opposite is greater than the pyramid of which the triangle *AEG* is the base and the point *H* the vertex.

But the prism in which the parallelogram *EBFG* is the base and the straight line *HK* its opposite is equal to the prism in which the triangle *GFC* is the base and the triangle *HKL* its opposite,

and the pyramid of which the triangle *AEG* is the base and the point *H* the vertex is equal to the pyramid of which the triangle *HKL* is the base and the point *D* the vertex.

Therefore the said two prisms are greater than the said two pyramids of which the triangles *AEG*, *HKL* are the bases and the points *H*, *D* the vertices.

Therefore the whole pyramid, of which the triangle *ABC* is the base and the point *D* the vertex, has been divided into two pyramids equal to one another and into two equal prisms, and the two prisms are greater than the half of the whole pyramid.

Q. E. D.

We will denote a pyramid with vertex *D* and base *ABC* by *D* (*ABC*) or *D-ABC* and the triangular prism with triangles *GCF*, *HLK* for bases by (*GCF*, *HLK*).

The following are the steps of the proof.

I. To prove pyramid $H(AEG)$ equal and similar to pyramid $D(HKL)$.

Since sides of $\triangle DAB$ are bisected at *H*, *E*, *K*,

$$HE \parallel DB, \text{ and } HK \parallel AB.$$

Hence
$$HK = EB = EA,$$
$$HE = KB = DK.$$

Therefore (1) $\triangle$s *HAE*, *DHK* are equal and similar.

Similarly (2) $\triangle$s *HAG*, *DHL* are equal and similar.

Again, *LH*, *HK* are respectively $\parallel$ to *GA*, *AE* in a different plane;

therefore
$$\angle GAE = \angle LHK.$$

And LH, HK are respectively equal to GA, AE.

Therefore (3) △s GAE, LHK are equal and similar.

Similarly (4) △s HGE, DLK are equal and similar.

Therefore [XI. Def. 10] the pyramids $H(AEG)$ and $D(HKL)$ are equal and similar.

II. To prove the pyramid $D(HKL)$ similar to the pyramid $D(ABC)$.

(1) The △s DHK, DAB are equiangular and therefore similar.

Similarly (2) △s DLH, DCA are similar, as also (3) the △s DLK, DCB.

Again, BA, AC are respectively parallel to KH, HL in a different plane; therefore

$$\angle BAC = \angle KHL.$$

And

$$BA : AC = KH : HL.$$

Therefore (4) △s BAC, KHL are similar.

Consequently the pyramid $D(ABC)$ is similar to the pyramid $D(HKL)$, and therefore also to the pyramid $H(AEG)$.

III. To prove prism (GCF, HLK) equal to prism (HGE, KFB).

The prisms may be regarded as having the same *height* (the distance between the planes HKL, ABC) and having for bases (1) the △ CGF and (2) the ▱ $EBFG$, which is the double of the △ CGF.

Therefore, by XI. 39, the prisms are equal.

IV. To prove the prisms greater than the small pyramids.

Prism (HGE, KFB) is clearly greater than pyramid $K(EFB)$ and therefore greater than pyramid $H(AEG)$.

Therefore each of the prisms is greater than each of the small pyramids; and the sum of the two prisms is greater than the sum of the two small pyramids, which, with the two prisms, make up the whole pyramid.

Proposition 4.

If there be two pyramids of the same height which have triangular bases, and each of them be divided into two pyramids equal to one another and similar to the whole, and into two equal prisms, then, as the base of the one pyramid is to the base of the other pyramid, so will all the prisms in the one pyramid be to all the prisms, being equal in multitude, in the other pyramid.

Let there be two pyramids of the same height which have the triangular bases ABC, DEF, and vertices the points G, H,

and let each of them be divided into two pyramids equal to one another and similar to the whole and into two equal prisms; [XII. 3]

I say that, as the base ABC is to the base DEF, so are all the prisms in the pyramid $ABCG$ to all the prisms, being equal in multitude, in the pyramid $DEFH$,

For, since BO is equal to OC, and AL to LC,
therefore LO is parallel to AB,
and the triangle ABC is similar to the triangle LOC.

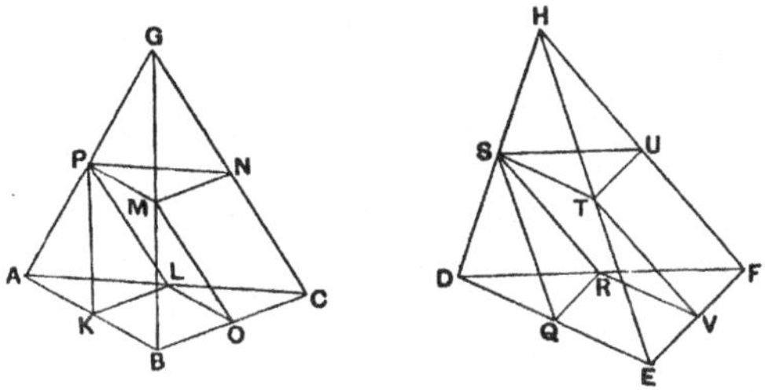

For the same reason
the triangle DEF is also similar to the triangle RVF.

And, since BC is double of CO, and EF of FV,
therefore, as BC is to CO, so is EF to FV.

And on BC, CO are described the similar and similarly situated rectilineal figures ABC, LOC,
and on EF, FV the similar and similarly situated figures DEF, RVF;
therefore, as the triangle ABC is to the triangle LOC, so is the triangle DEF to the triangle RVF; [VI. 22]
therefore, alternately, as the triangle ABC is to the triangle DEF, so is the triangle LOC to the triangle RVF. [V. 16]

But, as the triangle LOC is to the triangle RVF, so is the prism in which the triangle LOC is the base and PMN its opposite to the prism in which the triangle RVF is the base and STU its opposite; [Lemma following]
therefore also, as the triangle ABC is to the triangle DEF, so is the prism in which the triangle LOC is the base and PMN its opposite to the prism in which the triangle RVF is the base and STU its opposite.

But, as the said prisms are to one another, so is the prism in which the parallelogram $KBOL$ is the base and the straight line PM its opposite to the prism in which the parallelogram $QEVR$ is the base and the straight line ST its opposite. [XI. 39; cf. XII. 3]

Therefore also the two prisms, that in which the parallelogram *KBOL* is the base and *PM* its opposite, and that in which the triangle *LOC* is the base and *PMN* its opposite, are to the prisms in which *QEVR* is the base and the straight line *ST* its opposite and in which the triangle *RVF* is the base and *STU* its opposite in the same ratio. [V. 12]

Therefore also, as the base *ABC* is to the base *DEF*, so are the said two prisms to the said two prisms.

And similarly, if the pyramids *PMNG*, *STUH* be divided into two prisms and two pyramids,
as the base *PMN* is to the base *STU*, so will the two prisms in the pyramid *PMNG* be to the two prisms in the pyramid *STUH*.

But, as the base *PMN* is to the base *STU*, so is the base *ABC* to the base *DEF*;
for the triangles *PMN*, *STU* are equal to the triangles *LOC*, *RVF* respectively.

Therefore also, as the base *ABC* is to the base *DEF*, so are the four prisms to the four prisms.

And similarly also, if we divide the remaining pyramids into two pyramids and into two prisms, then, as the base *ABC* is to the base *DEF*, so will all the prisms in the pyramid *ABCG* be to all the prisms, being equal in multitude, in the pyramid *DEFH*.

Q. E. D.

Lemma.

But that, as the triangle *LOC* is to the triangle *RVF*, so is the prism in which the triangle *LOC* is the base and *PMN* its opposite to the prism in which the triangle *RVF* is the base and *STU* its opposite, we must prove as follows.

For in the same figure let perpendiculars be conceived drawn from *G*, *H* to the planes *ABC*, *DEF*; these are of course equal because, by hypothesis, the pyramids are of equal height.

Now, since the two straight lines *GC* and the perpendicular from *G* are cut by the parallel planes *ABC*, *PMN*,
they will be cut in the same ratios. [XI. 17]

And GC is bisected by the plane PMN at N;
therefore the perpendicular from G to the plane ABC will also be bisected by the plane PMN.

For the same reason
the perpendicular from H to the plane DEF will also be bisected by the plane STU.

And the perpendiculars from G, H to the planes ABC, DEF are equal;
therefore the perpendiculars from the triangles PMN, STU to the planes ABC, DEF are also equal.

Therefore the prisms in which the triangles LOC, RVF are bases, and PMN, STU their opposites, are of equal height.

Hence also the parallelepipedal solids described from the said prisms are of equal height and are to one another as their bases; [XI. 32]
therefore their halves, namely the said prisms, are to one another as the base LOC is to the base RVF.

Q. E. D.

We can incorporate the lemma at the end of the proposition and summarise the proof thus.

Since LO is parallel to AB,

$\triangle$s ABC, LOC are similar.

In like manner $\triangle$s DEF, RVF are similar.

And, since $BC : CO = EF : FV$,

$$\triangle ABC : \triangle LOC = \triangle DEF : \triangle RVF, \qquad \text{[VI. 22]}$$

and, alternately,

$$\triangle ABC : \triangle DEF = \triangle LOC : \triangle RVF.$$

Now the prisms (LOC, PMN) and (RVF, STU) are equal in height:
for the perpendiculars from G, H on the bases ABC, DEF are divided by the planes PMN, STU (parallel to the bases) in the same proportion as GC, HF are divided by those planes [XI. 17], i.e. they are bisected;
hence the heights of the prisms, being half the equal heights of the pyramids, are equal.

And the prisms are the halves respectively of parallelepipeds of the same height on parallelogrammic bases double of the $\triangle$s LOC, RVF respectively; [XI. 28 and note]
hence they are in the same ratio as those parallelepipeds, and therefore as their bases [XI. 32].

Therefore

$$(\text{prism } LOC, PMN) : (\text{prism } RVF, STU) = \triangle LOC : \triangle RVF$$
$$= \triangle ABC : \triangle DEF.$$

And since the other prisms in the pyramids are equal to these prisms respectively,

(sum of prisms in $GABC$) : (sum of prisms in $HDEF$) $= \triangle ABC : \triangle DEF$.

Similarly, if the pyramids $GPMN$, $HSTU$ be divided in like manner, and also the pyramids $PAKL$, $SDQR$, we shall have e.g.

(sum of prisms in $GPMN$) : (sum of prisms in $HSTU$) $= \triangle PMN : \triangle STU$
$= \triangle ABC : \triangle DEF$,

and similarly for the second pair of pyramids.

The process may be continued indefinitely, and we shall always have

(sum of prisms in $GABC$) : (sum of prisms in $HDEF$) $= \triangle ABC : \triangle DEF$.

Proposition 5.

Pyramids which are of the same height and have triangular bases are to one another as the bases.

Let there be pyramids of the same height, of which the triangles ABC, DEF are the bases and the points G, H the vertices;

I say that, as the base ABC is to the base DEF, so is the pyramid $ABCG$ to the pyramid $DEFH$.

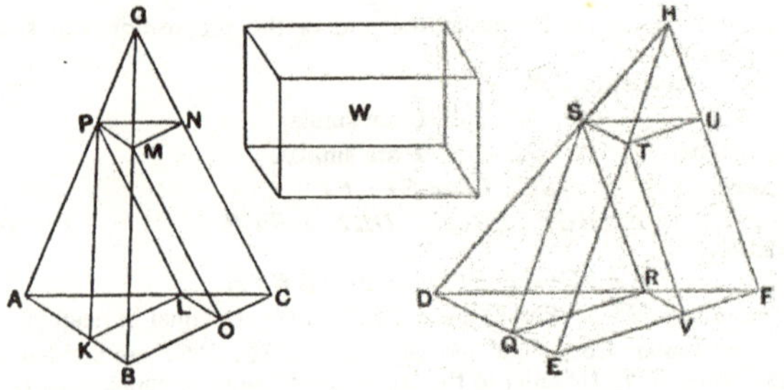

For, if the pyramid $ABCG$ is not to the pyramid $DEFH$ as the base ABC is to the base DEF,

then, as the base ABC is to the base DEF, so will the pyramid $ABCG$ be either to some solid less than the pyramid $DEFH$ or to a greater.

Let it, first, be in that ratio to a less solid W, and let the pyramid $DEFH$ be divided into two pyramids equal to one another and similar to the whole and into two equal prisms;

then the two prisms are greater than the half of the whole pyramid. [XII. 3]

Again, let the pyramids arising from the division be similarly divided,
and let this be done continually until there are left over from the pyramid *DEFH* some pyramids which are less than the excess by which the pyramid *DEFH* exceeds the solid *W*. [X. 1]

Let such be left, and let them be, for the sake of argument, *DQRS*, *STUH*;
therefore the remainders, the prisms in the pyramid *DEFH*, are greater than the solid *W*.

Let the pyramid *ABCG* also be divided similarly, and a similar number of times, with the pyramid *DEFH*;
therefore, as the base *ABC* is to the base *DEF*, so are the prisms in the pyramid *ABCG* to the prisms in the pyramid *DEFH*. [XII. 4]

But, as the base *ABC* is to the base *DEF*, so also is the pyramid *ABCG* to the solid *W*;
therefore also, as the pyramid *ABCG* is to the solid *W*, so are the prisms in the pyramid *ABCG* to the prisms in the pyramid *DEFH*; [V. 11]
therefore, alternately, as the pyramid *ABCG* is to the prisms in it, so is the solid *W* to the prisms in the pyramid *DEFH*. [V. 16]

But the pyramid *ABCG* is greater than the prisms in it;
therefore the solid *W* is also greater than the prisms in the pyramid *DEFH*.

But it is also less:
which is impossible.

Therefore the prism *ABCG* is not to any solid less than the pyramid *DEFH* as the base *ABC* is to the base *DEF*.

Similarly it can be proved that neither is the pyramid *DEFH* to any solid less than the pyramid *ABCG* as the base *DEF* is to the base *ABC*.

I say next that neither is the pyramid *ABCG* to any solid greater than the pyramid *DEFH* as the base *ABC* is to the base *DEF*.

For, if possible, let it be in that ratio to a greater solid *W*;
therefore, inversely, as the base *DEF* is to the base *ABC*, so is the solid *W* to the pyramid *ABCG*.

But, as the solid W is to the solid $ABCG$, so is the pyramid $DEFH$ to some solid less than the pyramid $ABCG$, as was before proved; [XII. 2, Lemma]

therefore also, as the base DEF is to the base ABC, so is the pyramid $DEFH$ to some solid less than the pyramid $ABCG$: [V. 11]

which was proved absurd.

Therefore the pyramid $ABCG$ is not to any solid greater than the pyramid $DEFH$ as the base ABC is to the base DEF.

But it was proved that neither is it in that ratio to a less solid.

Therefore, as the base ABC is to the base DEF, so is the pyramid $ABCG$ to the pyramid $DEFH$.

Q. E. D.

In the two preceding propositions it has been shown how we can divide a pyramid with a triangular base into (1) two equal prisms which are together greater than half the pyramid and (2) two equal pyramids similar to the original one, and that, if this process be continued with the two pyramids, then with the four resulting pyramids, and so on, and if, further, another pyramid of the same height as the original one be similarly divided, the subdivision being made the same number of times, the sum of all the prisms in one pyramid is to the sum of all the prisms in the other as the base of the first is to the base of the second.

We can now prove in the manner of XII. 2 that the volumes of the pyramids themselves are as the bases.

Let us call the pyramids P, P' and their respective bases B, B'.

If $$P : P' \neq B : B',$$

suppose that $$B : B' = P : W.$$

I. Let W be $< P'$.

Divide P' into two prisms and two pyramids, subdivide the latter similarly, and so on, until the sum of the *pyramids* remaining is less than the difference between P' and W [X. 1], so that

$$P' > (\text{prisms in } P') > W.$$

Then divide P similarly, the same number of times.

Now $$(\text{prisms in } P) : (\text{prisms in } P') = B : B' \quad \text{[XII. 4]}$$
$$= P : W, \text{ by hypothesis,}$$

and, alternately,

$$(\text{prisms in } P) : P = (\text{prisms in } P') : W.$$

But $$(\text{prisms in } P) < P;$$

therefore $$(\text{prisms in } P') < W.$$

But, by construction, $$(\text{prisms in } P') > W.$$

Hence W cannot be less than P'.

II. Suppose, if possible, that $W > P'$.

Then, inversely, $B' : B = W : P$.

$= P' : V$,

where V is *some solid less than P*. [Cf. XII. 2, Lemma, and note.]

But this can be proved impossible exactly as in Part I.

Therefore W is neither less nor greater than P',

so that $B : B' = P : P'$.

Legendre, followed by the American editors already mentioned, and by others, approaches the subject by a different route, proving the following propositions.

1. *If a pyramid be cut by a plane parallel to the base, (a) the lateral edges and the height will be cut in the same proportion, (b) the section by the plane will be a polygon similar to the base.*

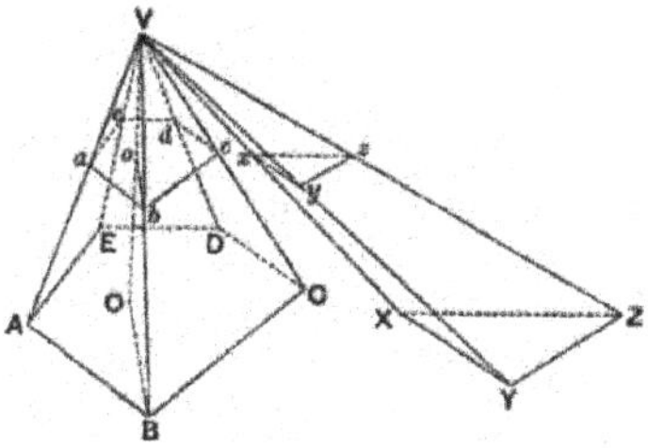

(*a*) Since a lateral face VAB of the pyramid $V(ABCDE)$ is cut by two parallel planes in AB, ab,

$$AB \parallel ab;$$

Similarly $BC \parallel bc$, and so on.

Therefore $VA : Va = VB : Vb = VC : Vc = \ldots$.

And, if VO the height be cut in O, o,

$BO \parallel bo$; and each of the above ratios is equal to $VO : Vo$.

(*b*) Since $BA \parallel ba$, and $BC \parallel bc$,

$\angle ABC = \angle abc$. [XI. 10]

Similarly for all the other angles of the polygons, which are therefore equiangular.

Also, by similar triangles,

$$VA : Va = AB : ab,$$

and so on.

Therefore, by the ratios above,

$$AB : ab = BC : bc = \ldots.$$

Therefore the polygons are similar.

2. *If two pyramids of the same height be cut by planes which are at the same perpendicular distance from the vertices, the sections are as the respective bases.*

For, if we place the pyramids so that the vertices coincide and the bases are in one plane, the planes of the sections will coincide.

If, e.g., the base of the second pyramid be XYZ and the section xyz, we shall have, by the argument of the last proposition,

$$VX : Vx = VY : Vy = VZ : Vz = VO : Vo = VA : Va = \ldots,$$

and XYZ, xyz will be similar.

Now (polygon $ABCDE$) : (polygon $abcde$) $= AB^2 : ab^2$

$$= VA^2 : Va^2,$$

and

$$\triangle XYZ : \triangle xyz = XY^2 : xy^2$$
$$= VX^2 : Vx^2$$
$$= VA^2 : Va^2.$$

Therefore

$$(\text{polygon } ABCDE) : (\text{polygon } abcde) = \triangle XYZ : \triangle xyz.$$

As a particular case, *if the bases of the two pyramids are equivalent, the sections are also equivalent.*

3. *Two triangular pyramids which have equivalent bases and equal heights are equivalent.*

Let $VABC$, $vabc$ be pyramids with equivalent bases ABC, abc, which for convenience we will suppose placed in one plane, and let TA be the common height.

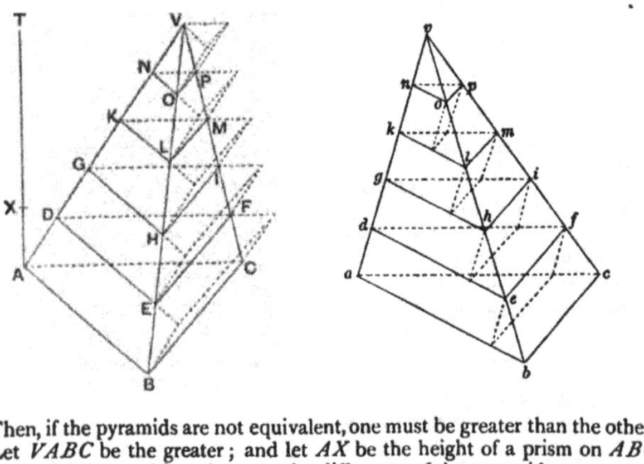

Then, if the pyramids are not equivalent, one must be greater than the other.

Let $VABC$ be the greater; and let AX be the height of a prism on ABC as base which is equal in volume to the difference of the pyramids.

Divide the height AT into equal parts such that each is less than AX, and let each part be equal to z.

Through the points of division draw planes parallel to the bases cutting both pyramids in the sections DEF, GHI,... and def, ghi, ...

The sections DEF, def will then be equivalent; so will the sections GHI, ghi, and so on. [(2) above]

On the triangles ABC, DEF, GHI, ... as bases draw *exterior* prisms having for edges the parts AD, DG, GK, ... of the edge AV;

and on the triangles *def*, *ghi*, ... as bases draw *interior* prisms having for edges the parts *ad*, *dg*, ... of *av*.

All the partial prisms will then have the same height *z*.

Now the sum of the exterior prisms of the pyramid *VABC* is *greater* than that pyramid;

and the sum of the interior prisms in the pyramid *vabc* is *less* than that pyramid.

Consequently the difference between the sum of the first set of prisms and the sum of the second set of prisms is greater than the difference between the two pyramids.

Again, if we start from the bases *ABC*, *abc*, the *second* exterior prism *DEFG* is equivalent to the *first* interior prism *defa*, since their bases are equivalent and they have the same height *z*. [XI. 28 and note; XI. 32]

Similarly the third exterior prism is equivalent to the second interior prism, and so on, until we arrive at the last of each.

Therefore the prism *ABCD*, the *first* exterior prism, is the difference between the sums of the exterior and interior prisms respectively.

Therefore the difference between the two pyramids is *less* than the prism *ABCD*, which should therefore be greater than the prism with base *ABC* and height *AX*.

But the prism *ABCD* is, by hypothesis, less than the latter prism:

which is impossible.

Consequently the pyramid *VABC* cannot be greater than the pyramid *vabc*.

Similarly it may be proved that *vabc* cannot be greater than *VABC*.

Therefore the pyramids are *equivalent*.

Legendre next establishes a proposition corresponding to Eucl. XII. 7, viz.

4. *Any triangular pyramid is one third of the triangular prism on the same base and of the same height*,

and from this he deduces that

COR. *The volume of a triangular pyramid is equal to a third of the product of its base by its height.*

He has previously proved that the volume of a triangular prism is equal to the product of its base and height, since (1) the prism is half of a parallelepiped of the same height and with a parallelogram for base which is double of the base of the prism, and (2) this parallelepiped can be transformed into an equivalent *rectangular* parallelepiped with the same height and an equivalent base.

The theorem (4) is then extended to *any* pyramid in the proposition

5. *Any pyramid has for its measure the third part of the product of its base and its height*, from which follow

COR. I. *Any pyramid is the third part of the prism on the same base and of the same height.*

COR. II. *Two pyramids of the same height are to one another as their bases, and two pyramids on the same base are to one another as their heights.*

The first part of the second corollary corresponds to the present proposition as extended by the next, XII. 6.

PROPOSITION 6.

Pyramids which are of the same height and have polygonal bases are to one another as the bases.

Let there be pyramids of the same height of which the polygons *ABCDE*, *FGHKL* are the bases and the points *M*, *N* the vertices;

I say that, as the base *ABCDE* is to the base *FGHKL*, so is the pyramid *ABCDEM* to the pyramid *FGHKLN*.

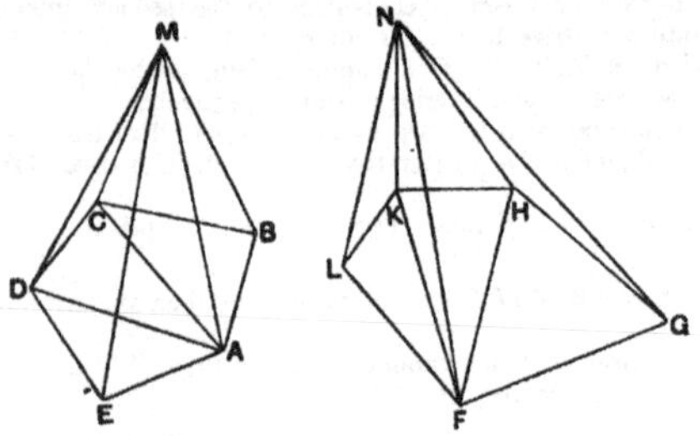

For let *AC*, *AD*, *FH*, *FK* be joined.

Since then *ABCM*, *ACDM* are two pyramids which have triangular bases and equal height,

they are to one another as the bases; [XII. 5]

therefore, as the base *ABC* is to the base *ACD*, so is the pyramid *ABCM* to the pyramid *ACDM*.

And, *componendo*, as the base *ABCD* is to the base *ACD*, so is the pyramid *ABCDM* to the pyramid *ACDM*. [V. 18]

But also, as the base *ACD* is to the base *ADE*, so is the pyramid *ACDM* to the pyramid *ADEM*. [XII. 5]

Therefore, *ex aequali*, as the base *ABCD* is to the base *ADE*, so is the pyramid *ABCDM* to the pyramid *ADEM*. [V. 22]

And, again *componendo*, as the base *ABCDE* is to the base *ADE*, so is the pyramid *ABCDEM* to the pyramid *ADEM*. [V. 18]

Similarly also it can be proved that, as the base *FGHKL* is to the base *FGH*, so is the pyramid *FGHKLN* to the pyramid *FGHN*.

And, since $ADEM$, $FGHN$ are two pyramids which have triangular bases and equal height,

therefore, as the base ADE is to the base FGH, so is the pyramid $ADEM$ to the pyramid $FGHN$. [XII. 5]

But, as the base ADE is to the base $ABCDE$, so was the pyramid $ADEM$ to the pyramid $ABCDEM$.

Therefore also, *ex aequali*, as the base $ABCDE$ is to the base FGH, so is the pyramid $ABCDEM$ to the pyramid $FGHN$. [V. 22]

But further, as the base FGH is to the base $FGHKL$, so also was the pyramid $FGHN$ to the pyramid $FGHKLN$.

Therefore also, *ex aequali*, as the base $ABCDE$ is to the base $FGHKL$, so is the pyramid $ABCDEM$ to the pyramid $FGHKLN$. [V. 22]

Q. E. D.

It will be seen that, in order to obtain the proportion

$$(\text{base } ABCDE) : \triangle ADE = (\text{pyramid } MABCDE) : (\text{pyramid } MADE),$$

Euclid employs V. 18 (*componendo*) twice over, with an *ex aequali* step [V. 22] intervening.

We might arrive at it more concisely by using V. 24 extended to any number of antecedents.

Thus

$$\triangle ABC : \triangle ADE = (\text{pyramid } MABC) : (\text{pyramid } MADE),$$

$$\triangle ACD : \triangle ADE = (\text{pyramid } MACD) : (\text{pyramid } MADE),$$

and lastly

$$\triangle ADE : \triangle ADE = (\text{pyramid } MADE) : (\text{pyramid } MADE).$$

Therefore, adding the antecedents [V. 24], we have

$$(\text{polygon } ABCDE) : \triangle ADE = (\text{pyramid } MABCDE) : (\text{pyramid } MADE).$$

Again, since the pyramids $MADE$, $NFGH$ are of the same height,

$$\triangle ADE : \triangle FGH = (\text{pyramid } MADE) : (\text{pyramid } NFGH).$$

Lastly, using the same argument for the pyramid $NFGHKL$ as for $MABCDE$, and inverting, we have

$$\triangle FGH : (\text{polygon } FGHKL) = (\text{pyramid } NFGH) : (\text{pyramid } NFGHKL).$$

Thus from the three proportions, *ex aequali*,

$$(\text{polygon } ABCDE) : (\text{polygon } FGHKL)$$
$$= (\text{pyramid } MABCDE) : (\text{pyramid } NFGHKL).$$

PROPOSITION 7.

Any prism which has a triangular base is divided into three pyramids equal to one another which have triangular bases.

Let there be a prism in which the triangle ABC is the base and DEF its opposite;

I say that the prism $ABCDEF$ is divided into three pyramids equal to one another, which have triangular bases.

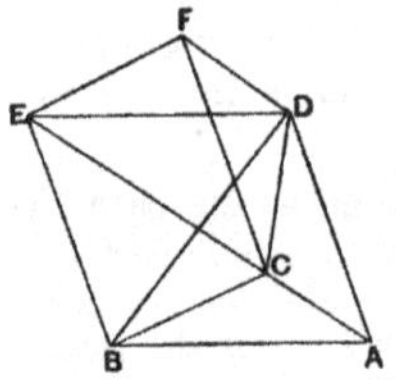

For let BD, EC, CD be joined.

Since $ABED$ is a parallelogram, and BD is its diameter, therefore the triangle ABD is equal to the triangle EBD; [I. 34]

therefore also the pyramid of which the triangle ABD is the base and the point C the vertex is equal to the pyramid of which the triangle DEB is the base and the point C the vertex. [XII. 5]

But the pyramid of which the triangle DEB is the base and the point C the vertex is the same with the pyramid of which the triangle EBC is the base and the point D the vertex;

for they are contained by the same planes.

Therefore the pyramid of which the triangle ABD is the base and the point C the vertex is also equal to the pyramid of which the triangle EBC is the base and the point D the vertex.

Again, since $FCBE$ is a parallelogram, and CE is its diameter, the triangle CEF is equal to the triangle CBE. [I. 34]

Therefore also the pyramid of which the triangle BCE is the base and the point D the vertex is equal to the pyramid of which the triangle ECF is the base and the point D the vertex. [XII. 5]

But the pyramid of which the triangle BCE is the base and the point D the vertex was proved equal to the pyramid of which the triangle ABD is the base and the point C the vertex;

therefore also the pyramid of which the triangle *CEF* is the base and the point *D* the vertex is equal to the pyramid of which the triangle *ABD* is the base and the point *C* the vertex;

therefore the prism *ABCDEF* has been divided into three pyramids equal to one another which have triangular bases.

And, since the pyramid of which the triangle *ABD* is the base and the point *C* the vertex is the same with the pyramid of which the triangle *CAB* is the base and the point *D* the vertex,

for they are contained by the same planes,

while the pyramid of which the triangle *ABD* is the base and the point *C* the vertex was proved to be a third of the prism in which the triangle *ABC* is the base and *DEF* its opposite,

therefore also the pyramid of which the triangle *ABC* is the base and the point *D* the vertex is a third of the prism which has the same base, the triangle *ABC*, and *DEF* as its opposite.

PORISM. From this it is manifest that any pyramid is a third part of the prism which has the same base with it and equal height.

Q. E. D.

If we denote by *C-ABD* a pyramid with vertex *C* and base *ABD*, Euclid's argument is easily followed thus.

The ▱ *ABED* being bisected by *BD*,

(pyramid *C-ABD*) = (pyramid *C-DEB*) [XII. 5]

≡ (pyramid *D-EBC*).

And, the ▱ *EBCF* being bisected by *EC*,

(pyramid *D-EBC*) = (pyramid *D-ECF*).

Thus (pyramid *C-ABD*) = (pyramid *D-EBC*) = (pyramid *D-ECF*), and these three pyramids make up the whole prism, so that each is one-third of the prism.

And, since (pyramid *C-ABD*) ≡ (pyramid *D-ABC*),

(pyramid *D-ABC*) = $\frac{1}{3}$ (prism *ABC*, *DEF*).

PROPOSITION 8.

Similar pyramids which have triangular bases are in the triplicate ratio of their corresponding sides.

Let there be similar and similarly situated pyramids of

which the triangles *ABC, DEF* are the bases and the points *G, H* the vertices;

I say that the pyramid *ABCG* has to the pyramid *DEFH* the ratio triplicate of that which *BC* has to *EF*.

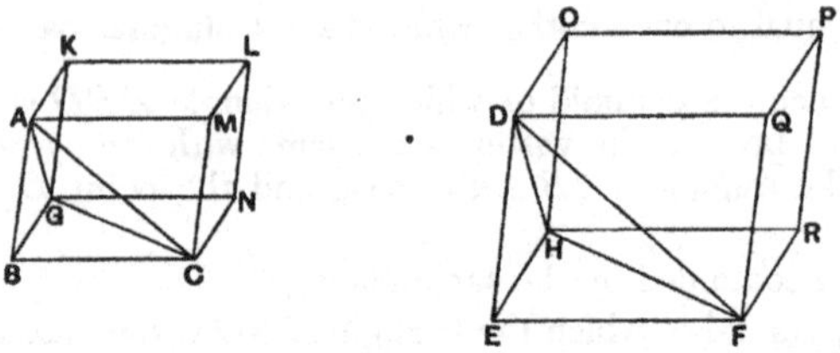

For let the parallelepipedal solids *BGML, EHQP* be completed.

Now, since the pyramid *ABCG* is similar to the pyramid *DEFH*,

therefore the angle *ABC* is equal to the angle *DEF*,

the angle *GBC* to the angle *HEF*,

and the angle *ABG* to the angle *DEH*;

and, as *AB* is to *DE*, so is *BC* to *EF*, and *BG* to *EH*.

And since, as *AB* is to *DE*, so is *BC* to *EF*,

and the sides are proportional about equal angles,

therefore the parallelogram *BM* is similar to the parallelogram *EQ*.

For the same reason

BN is also similar to *ER*, and *BK* to *EO*;

therefore the three parallelograms *MB, BK, BN* are similar to the three *EQ, EO, ER*.

But the three parallelograms *MB, BK, BN* are equal and similar to their three opposites,

and the three *EQ, EO, ER* are equal and similar to their three opposites. [XI. 24]

Therefore the solids *BGML, EHQP* are contained by similar planes equal in multitude.

Therefore the solid *BGML* is similar to the solid *EHQP*.

But similar parallelepipedal solids are in the triplicate ratio of their corresponding sides. [XI. 33]

Therefore the solid *BGML* has to the solid *EHQP* the ratio triplicate of that which the corresponding side *BC* has to the corresponding side *EF*.

But, as the solid *BGML* is to the solid *EHQP*, so is the pyramid *ABCG* to the pyramid *DEFH*,
inasmuch as the pyramid is a sixth part of the solid, because the prism which is half of the parallelepipedal solid [XI. 28] is also triple of the pyramid. [XII. 7].

Therefore the pyramid *ABCG* also has to the pyramid *DEFH* the ratio triplicate of that which *BC* has to *EF*.

Q. E. D.

PORISM. From this it is manifest that similar pyramids which have polygonal bases are also to one another in the triplicate ratio of their corresponding sides.

For, if they are divided into the pyramids contained in them which have triangular bases, by virtue of the fact that the similar polygons forming their bases are also divided into similar triangles equal in multitude and corresponding to the wholes [VI. 20],
then, as the one pyramid which has a triangular base in the one complete pyramid is to the one pyramid which has a triangular base in the other complete pyramid, so also will all the pyramids which have triangular bases contained in the one pyramid be to all the pyramids which have triangular bases contained in the other pyramid [V. 12], that is, the pyramid itself which has a polygonal base to the pyramid which has a polygonal base.

But the pyramid which has a triangular base is to the pyramid which has a triangular base in the triplicate ratio of the corresponding sides;
therefore also the pyramid which has a polygonal base has to the pyramid which has a similar base the ratio triplicate of that which the side has to the side.

It is at once proved that, the pyramids being similar, the parallelepipeds constructed as shown in the figure are also similar.

Consequently, as these latter are in the triplicate ratio of their corresponding sides [XI. 33], so are the pyramids which are their sixth parts respectively (being one third of the respective *prisms* on the same bases, i.e. of the halves of the respective parallelepipeds, XI. 28).

As the Porism is not used where Euclid might have been expected to use it (see note on XII. 12, p. 416), there is some reason to doubt its genuineness. P only has it in the margin, though in the first hand.

Proposition 9.

In equal pyramids which have triangular bases the bases are reciprocally proportional to the heights; and those pyramids in which the bases are reciprocally proportional to the heights are equal.

For let there be equal pyramids which have the triangular bases *ABC*, *DEF* and vertices the points *G*, *H*;
I say that in the pyramids *ABCG*, *DEFH* the bases are reciprocally proportional to the heights, that is, as the base *ABC* is to the base *DEF*, so is the height of the pyramid *DEFH* to the height of the pyramid *ABCG*.

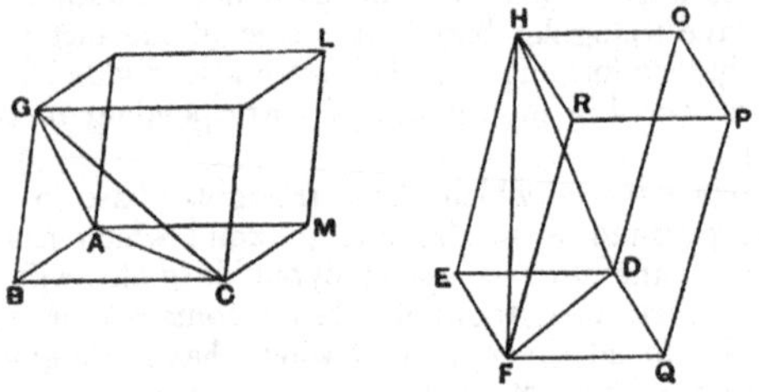

For let the parallelepipedal solids *BGML*, *EHQP* be completed.

Now, since the pyramid *ABCG* is equal to the pyramid *DEFH*,
and the solid *BGML* is six times the pyramid *ABCG*,
and the solid *EHQP* six times the pyramid *DEFH*,
therefore the solid *BGML* is equal to the solid *EHQP*.

But in equal parallelepipedal solids the bases are reciprocally proportional to the heights; [XI. 34]
therefore, as the base *BM* is to the base *EQ*, so is the height of the solid *EHQP* to the height of the solid *BGML*.

But, as the base *BM* is to *EQ*, so is the triangle *ABC* to the triangle *DEF*. [I. 34]

Therefore also, as the triangle *ABC* is to the triangle *DEF*, so is the height of the solid *EHQP* to the height of the solid *BGML*. [V. 11]

But the height of the solid *EHQP* is the same with the height of the pyramid *DEFH*,

and the height of the solid *BGML* is the same with the height of the pyramid *ABCG*,

therefore, as the base *ABC* is to the base *DEF*, so is the height of the pyramid *DEFH* to the height of the pyramid *ABCG*.

Therefore in the pyramids *ABCG*, *DEFH* the bases are reciprocally proportional to the heights.

Next, in the pyramids *ABCG*, *DEFH* let the bases be reciprocally proportional to the heights;

that is, as the base *ABC* is to the base *DEF*, so let the height of the pyramid *DEFH* be to the height of the pyramid *ABCG*;

I say that the pyramid *ABCG* is equal to the pyramid *DEFH*.

For, with the same construction,

since, as the base *ABC* is to the base *DEF*, so is the height of the pyramid *DEFH* to the height of the pyramid *ABCG*,

while, as the base *ABC* is to the base *DEF*, so is the parallelogram *BM* to the parallelogram *EQ*,

therefore also, as the parallelogram *BM* is to the parallelogram *EQ*, so is the height of the pyramid *DEFH* to the height of the pyramid *ABCG*. [v. 11]

But the height of the pyramid *DEFH* is the same with the height of the parallelepiped *EHQP*,

and the height of the pyramid *ABCG* is the same with the height of the parallelepiped *BGML*;

therefore, as the base *BM* is to the base *EQ*, so is the height of the parallelepiped *EHQP* to the height of the parallelepiped *BGML*.

But those parallelepipedal solids in which the bases are reciprocally proportional to the heights are equal; [XI. 34]

therefore the parallelepipedal solid *BGML* is equal to the parallelepipedal solid *EHQP*.

And the pyramid *ABCG* is a sixth part of *BGML*, and the pyramid *DEFH* a sixth part of the parallelepiped *EHQP*;

therefore the pyramid *ABCG* is equal to the pyramid *DEFH*.

Therefore etc.

Q. E. D.

The volumes of the pyramids are respectively one sixth part of the volumes of the parallelepipeds described, as in the figure, on double the bases and with the same heights as the pyramids.

I. Thus the parallelepipeds are equal if the pyramids are equal.

And, the parallelepipeds being equal, their bases are reciprocally proportional to their heights; [XI. 34]

hence the bases of the equal pyramids (which are the halves of the bases of the parallelepipeds) are proportional to their heights.

II. If the bases of the pyramids are reciprocally proportional to their heights, so are the bases of the parallelepipeds to their heights (since the bases of the parallelepipeds are double of the bases of the pyramids respectively).

Consequently the parallelepipeds are equal. [XI. 34]

Therefore their sixth parts, the pyramids, are also equal.

Proposition 10.

Any cone is a third part of the cylinder which has the same base with it and equal height.

For let a cone have the same base, namely the circle *ABCD*, with a cylinder and equal height;

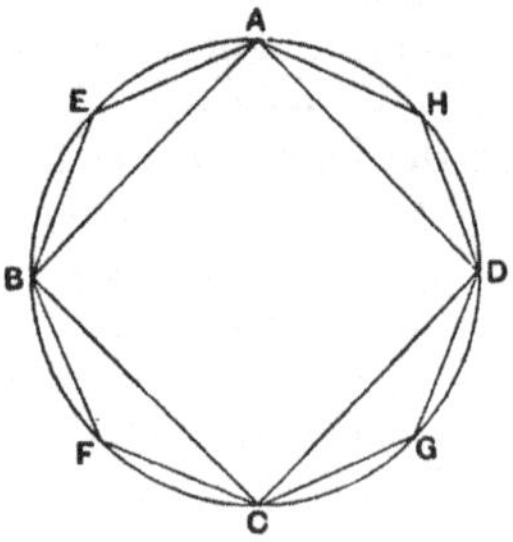

I say that the cone is a third part of the cylinder, that is, that the cylinder is triple of the cone.

For if the cylinder is not triple of the cone, the cylinder will be either greater than triple or less than triple of the cone.

First let it be greater than triple,

and let the square *ABCD* be inscribed in the circle *ABCD*; [IV. 6]

then the square *ABCD* is greater than the half of the circle *ABCD*.

From the square *ABCD* let there be set up a prism of equal height with the cylinder.

Then the prism so set up is greater than the half of the cylinder,

inasmuch as, if we also circumscribe a square about the circle *ABCD* [IV. 7], the square inscribed in the circle *ABCD* is half of that circumscribed about it,

and the solids set up from them are parallelepipedal prisms of equal height,

while parallelepipedal solids which are of the same height are to one another as their bases; [XI. 32]

therefore also the prism set up on the square *ABCD* is half of the prism set up from the square circumscribed about the circle *ABCD*; [cf. XI. 28, or XII. 6 and 7, Por.]

and the cylinder is less than the prism set up from the square circumscribed about the circle *ABCD*;

therefore the prism set up from the square *ABCD* and of equal height with the cylinder is greater than the half of the cylinder.

Let the circumferences *AB*, *BC*, *CD*, *DA* be bisected at the points *E*, *F*, *G*, *H*,

and let *AE*, *EB*, *BF*, *FC*, *CG*, *GD*, *DH*, *HA* be joined;

then each of the triangles *AEB*, *BFC*, *CGD*, *DHA* is greater than the half of that segment of the circle *ABCD* which is about it, as we proved before. [XII. 2]

On each of the triangles *AEB*, *BFC*, *CGD*, *DHA* let prisms be set up of equal height with the cylinder;

then each of the prisms so set up is greater than the half part of that segment of the cylinder which is about it,

inasmuch as, if we draw through the points *E*, *F*, *G*, *H* parallels to *AB*, *BC*, *CD*, *DA*, complete the parallelograms on *AB*, *BC*, *CD*, *DA*, and set up from them parallelepipedal solids of equal height with the cylinder, the prisms on the triangles *AEB*, *BFC*, *CGD*, *DHA* are halves of the several solids set up;

and the segments of the cylinder are less than the parallelepipedal solids set up;

hence also the prisms on the triangles *AEB*, *BFC*, *CGD*, *DHA* are greater than the half of the segments of the cylinder about them.

Thus, bisecting the circumferences that are left, joining

straight lines, setting up on each of the triangles prisms of equal height with the cylinder,

and doing this continually,

we shall leave some segments of the cylinder which will be less than the excess by which the cylinder exceeds the triple of the cone. [X. 1]

Let such segments be left, and let them be *AE*, *EB*, *BF*, *FC*, *CG*, *GD*, *DH*, *HA*;

therefore the remainder, the prism of which the polygon *AEBFCGDH* is the base and the height is the same as that of the cylinder, is greater than triple of the cone.

But the prism of which the polygon *AEBFCGDH* is the base and the height the same as that of the cylinder is triple of the pyramid of which the polygon *AEBFCGDH* is the base and the vertex is the same as that of the cone; [XII. 7, Por.]

therefore also the pyramid of which the polygon *AEBFCGDH* is the base and the vertex is the same as that of the cone is greater than the cone which has the circle *ABCD* as base.

But it is also less, for it is enclosed by it:

which is impossible.

Therefore the cylinder is not greater than triple of the cone.

I say next that neither is the cylinder less than triple of the cone,

For, if possible, let the cylinder be less than triple of the cone;

therefore, inversely, the cone is greater than a third part of the cylinder.

Let the square *ABCD* be inscribed in the circle *ABCD*;

therefore the square *ABCD* is greater than the half of the circle *ABCD*.

Now let there be set up from the square *ABCD* a pyramid having the same vertex with the cone;

therefore the pyramid so set up is greater than the half part of the cone,

seeing that, as we proved before, if we circumscribe a square about the circle, the square *ABCD* will be half of the square circumscribed about the circle,

and if we set up from the squares parallelepipedal solids of equal height with the cone, which are also called prisms, the solid set up from the square *ABCD* will be half of that set up from the square circumscribed about the circle;

for they are to one another as their bases. [XI. 32]

Hence also the thirds of them are in that ratio;

therefore also the pyramid of which the square *ABCD* is the base is half of the pyramid set up from the square circumscribed about the circle.

And the pyramid set up from the square about the circle is greater than the cone,

for it encloses it.

Therefore the pyramid of which the square *ABCD* is the base and the vertex is the same with that of the cone is greater than the half of the cone.

Let the circumferences *AB*, *BC*, *CD*, *DA* be bisected at the points *E*, *F*, *G*, *H*,

and let *AE*, *EB*, *BF*, *FC*, *CG*, *GD*, *DH*, *HA* be joined;

therefore also each of the triangles *AEB*, *BFC*, *CGD*, *DHA* is greater than the half part of that segment of the circle *ABCD* which is about it.

Now, on each of the triangles *AEB*, *BFC*, *CGD*, *DHA* let pyramids be set up which have the same vertex as the cone;

therefore also each of the pyramids so set up is, in the same manner, greater than the half part of that segment of the cone which is about it.

Thus, by bisecting the circumferences that are left, joining straight lines, setting up on each of the triangles a pyramid which has the same vertex as the cone,

and doing this continually,

we shall leave some segments of the cone which will be less than the excess by which the cone exceeds the third part of the cylinder. [X. 1]

Let such be left, and let them be the segments on *AE*, *EB*, *BF*, *FC*, *CG*, *GD*, *DH*, *HA*;

therefore the remainder, the pyramid of which the polygon *AEBFCGDH* is the base and the vertex the same with that of the cone, is greater than a third part of the cylinder.

But the pyramid of which the polygon *AEBFCGDH* is the base and the vertex the same with that of the cone is a third part of the prism of which the polygon *AEBFCGDH* is the base and the height is the same with that of the cylinder;
therefore the prism of which the polygon *AEBFCGDH* is the base and the height is the same with that of the cylinder is greater than the cylinder of which the circle *ABCD* is the base.

But it is also less, for it is enclosed by it:
which is impossible.

Therefore the cylinder is not less than triple of the cone.

But it was proved that neither is it greater than triple;
therefore the cylinder is triple of the cone;
hence the cone is a third part of the cylinder.

Therefore etc.

Q. E. D.

We observe the use in this proposition of the term "parallelepipedal prism," which recalls Heron's "parallelogrammic" or "parallel-sided prism."

The course of the proof is exactly the same as in XII. 2, except that an arithmetical fraction takes the place of a ratio which, being incommensurable, could only be expressed as a ratio. Consequently we do not need *proportions* in this proposition, as we did in XII. 2, and shall again in XII. 11, etc.

Euclid *exhausts* the cylinder and cone respectively by setting up prisms and pyramids of the same height on the successive regular polygons inscribed in the circle which is the common base, viz. the square, the regular polygon of 8 sides, that of 16 sides, etc.

If *AB* be the side of one polygon, we obtain two sides of the next by bisecting the arc *ACB* and joining *AC*, *CB*. Draw the tangent *DE* at *C* and complete the parallelogram *ABED*.

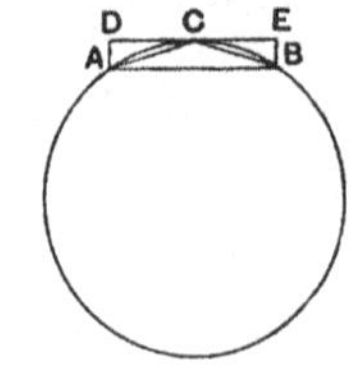

Now suppose a prism erected on the polygon of which *AB* is a side, and of the same height as that of the cylinder.

To obtain the prism of the same height on the next polygon we add all the triangular prisms of the same height on the bases *ACB* and the rest.

Now the prism on *ACB* is half the prism of the same height on the ▱ *ABED* as base.

[cf. XI. 28]

And the prism on □ $ABED$ includes, and is greater than, the portion of the cylinder standing on the segment ACB of the circle.

The same thing is true in regard to the other sides of the polygon of which AB is one side.

Thus the process begins with a prism on the square inscribed in the circle, which is more than half the cylinder, the next prism (with eight lateral faces) takes away more than half the remainder, and so on;

hence [x. 1], if we proceed far enough, we shall ultimately arrive at a prism leaving over portions of the cylinder together less than any assigned volume.

The construction of pyramids on the successive polygons exhausts the cone in exactly the same way.

Now, if the cone is not equal to one-third of the cylinder, it must be either greater or less.

I. Suppose, if possible, that, V, O being their volumes respectively,

$$O > 3V.$$

Construct successive inscribed polygons in the bases and prisms on them until we arrive at a prism P leaving over portions of the cylinder together less than $(O - 3V)$, i.e. such that

$$O > P > 3V.$$

But P is triple of the pyramid on the same base and of the same height; and this pyramid is included by, and is therefore less than, V;

therefore $P < 3V$.

But, by construction, $P > 3V$:

which is impossible.

Therefore $O \not> 3V$.

II. Suppose, if possible, that $O < 3V$.

Therefore $V > \frac{1}{3}O$.

Construct successive pyramids in the cone in the manner described until we arrive at a pyramid Π leaving over portions of the cone together less than $(V - \frac{1}{3}O)$, i.e. such that

$$V > \Pi > \tfrac{1}{3}O.$$

Now Π is one-third of the prism on the same base and of the same height; and this prism is included by, and is therefore less than, the cylinder;

therefore $\Pi < \frac{1}{3}O$.

But, by construction, $\Pi > \frac{1}{3}O$:

which is impossible.

Therefore O is neither greater nor less than $3V$, so that

$$O = 3V.$$

It will be observed that here, as in XII. 2, Euclid always *exhausts* the solid by (as it were) building up to it from inside. Hence the solid to be exhausted must, with him, be supposed *greater* than the solid to which it is to be proved equal; and this is the reason why, in the second part, the initial supposition is turned round.

In this case too Euclid might have approximated to the cone and cylinder by *circumscribing* successive pyramids and prisms in the way shown, after Archimedes, in the note on XII. 2.

PROPOSITION 11.

Cones and cylinders which are of the same height are to one another as their bases.

Let there be cones and cylinders of the same height, let the circles *ABCD*, *EFGH* be their bases, *KL*, *MN* their axes and *AC*, *EG* the diameters of their bases;

I say that, as the circle *ABCD* is to the circle *EFGH*, so is the cone *AL* to the cone *EN*.

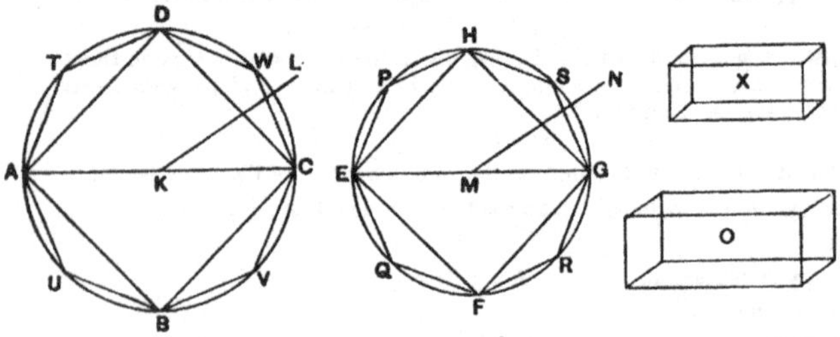

For, if not, then, as the circle *ABCD* is to the circle *EFGH*, so will the cone *AL* be either to some solid less than the cone *EN* or to a greater.

First, let it be in that ratio to a less solid *O*, and let the solid *X* be equal to that by which the solid *O* is less than the cone *EN*;

therefore the cone *EN* is equal to the solids *O*, *X*.

Let the square *EFGH* be inscribed in the circle *EFGH*; therefore the square is greater than the half of the circle.

Let there be set up from the square *EFGH* a pyramid of equal height with the cone;

therefore the pyramid so set up is greater than the half of the cone,

inasmuch as, if we circumscribe a square about the circle, and set up from it a pyramid of equal height with the cone, the inscribed pyramid is half of the circumscribed pyramid,

for they are to one another as their bases, [XII. 6]

while the cone is less than the circumscribed pyramid.

Let the circumferences *EF, FG, GH, HE* be bisected at the points *P, Q, R, S,*
and let *HP, PE, EQ, QF, FR, RG, GS, SH* be joined.

Therefore each of the triangles *HPE, EQF, FRG, GSH* is greater than the half of that segment of the circle which is about it.

On each of the triangles *HPE, EQF, FRG, GSH* let there be set up a pyramid of equal height with the cone;
therefore, also, each of the pyramids so set up is greater than the half of that segment of the cone which is about it.

Thus, bisecting the circumferences which are left, joining straight lines, setting up on each of the triangles pyramids of equal height with the cone,
and doing this continually,
we shall leave some segments of the cone which will be less than the solid *X*. [X. 1]

Let such be left, and let them be the segments on *HPE, EQF, FRG, GSH*;
therefore the remainder, the pyramid of which the polygon *HPEQFRGS* is the base and the height the same with that of the cone, is greater than the solid *O*.

Let there also be inscribed in the circle *ABCD* the polygon *DTAUBVCW* similar and similarly situated to the polygon *HPEQFRGS*,
and on it let a pyramid be set up of equal height with the cone *AL*.

Since then, as the square on *AC* is to the square on *EG*, so is the polygon *DTAUBVCW* to the polygon *HPEQFRGS*, [XII. 1]
while, as the square on *AC* is to the square on *EG*, so is the circle *ABCD* to the circle *EFGH*, [XII. 2]
therefore also, as the circle *ABCD* is to the circle *EFGH*, so is the polygon *DTAUBVCW* to the polygon *HPEQFRGS*.

But, as the circle *ABCD* is to the circle *EFGH*, so is the cone *AL* to the solid *O*,
and, as the polygon *DTAUBVCW* is to the polygon *HPEQFRGS*, so is the pyramid of which the polygon *DTAUBVCW* is the base and the point *L* the vertex to the pyramid of which the polygon *HPEQFRGS* is the base and the point *N* the vertex. [XII. 6]

Therefore also, as the cone *AL* is to the solid *O*, so is the pyramid of which the polygon *DTAUBVCW* is the base and the point *L* the vertex to the pyramid of which the polygon *HPEQFRGS* is the base and the point *N* the vertex; [v. 11]
therefore, alternately, as the cone *AL* is to the pyramid in it, so is the solid *O* to the pyramid in the cone *EN*. [v. 16]

But the cone *AL* is greater than the pyramid in it;
therefore the solid *O* is also greater than the pyramid in the cone *EN*.

But it is also less:
which is absurd.

Therefore the cone *AL* is not to any solid less than the cone *EN* as the circle *ABCD* is to the circle *EFGH*.

Similarly we can prove that neither is the cone *EN* to any solid less than the cone *AL* as the circle *EFGH* is to the circle *ABCD*.

I say next that neither is the cone *AL* to any solid greater than the cone *EN* as the circle *ABCD* is to the circle *EFGH*.

For, if possible, let it be in that ratio to a greater solid *O*;
therefore, inversely, as the circle *EFGH* is to the circle *ABCD*, so is the solid *O* to the cone *AL*.

But, as the solid *O* is to the cone *AL*, so is the cone *EN* to some solid less than the cone *AL*;
therefore also, as the circle *EFGH* is to the circle *ABCD*, so is the cone *EN* to some solid less than the cone *AL*:
which was proved impossible.

Therefore the cone *AL* is not to any solid greater than the cone *EN* as the circle *ABCD* is to the circle *EFGH*.

But it was proved that neither is it in this ratio to a less solid;
therefore, as the circle *ABCD* is to the circle *EFGH*, so is the cone *AL* to the cone *EN*.

But, as the cone is to the cone, so is the cylinder to the cylinder,
for each is triple of each; [XII. 10]

Therefore also, as the circle $ABCD$ is to the circle $EFGH$, so are the cylinders on them which are of equal height.

Therefore etc.

Q. E. D.

We need not again repeat the preliminary construction of successive pyramids and prisms exhausting the cones and cylinders.

Let Z, Z' be the volumes of the two cones, β, β' their respective bases.

If $$\beta : \beta' \neq Z : Z',$$
then must $$\beta : \beta' = Z : O,$$
where O is either less or greater than Z'.

I. Suppose, if possible, that O is *less* than Z'.

Inscribe in Z' a pyramid (Π') leaving over portions of it together less than $(Z' - O)$, i.e. such that
$$Z' > \Pi' > O.$$

Inscribe in Z a pyramid Π on a polygon inscribed in the circular base of Z similar to the polygon which is the base of Π'.

Now, if d, d' be the diameters of the bases,
$$\beta : \beta' = d^2 : d'^2 \quad \text{[XII. 2]}$$
$$= (\text{polygon in } \beta) : (\text{polygon in } \beta') \quad \text{[XII. 1]}$$
$$= \Pi : \Pi'. \quad \text{[XII. 6]}$$

Therefore $$Z : O = \Pi : \Pi',$$
and, alternately, $$Z : \Pi = O : \Pi'.$$

But $Z > \Pi$, since it includes it;

therefore $$O > \Pi'.$$

But, by construction, $$O < \Pi':$$
which is impossible.

Therefore $$O \nless Z.$$

II. Suppose, if possible, that
$$\beta : \beta' = Z : O,$$
where O is *greater* than Z'.

Therefore $$\beta : \beta' = O' : Z',$$
where O' is some solid less than Z.

That is, $$\beta' : \beta = Z' : O',$$
where $O' < Z$.

This is proved impossible exactly in the same way as the assumption in Part I. was proved impossible.

Therefore Z has not either to a less solid than Z' or to a greater solid than Z' the ratio of β to β';

therefore $$\beta : \beta' = Z : Z'.$$

The same is true of the cylinders which are equal to $3Z$, $3Z'$ respectively.

PROPOSITION 12.

Similar cones and cylinders are to one another in the triplicate ratio of the diameters in their bases.

Let there be similar cones and cylinders,

let the circles *ABCD*, *EFGH* be their bases, *BD*, *FH* the diameters of the bases, and *KL*, *MN* the axes of the cones and cylinders;

I say that the cone of which the circle *ABCD* is the base and the point *L* the vertex has to the cone of which the circle *EFGH* is the base and the point *N* the vertex the ratio triplicate of that which *BD* has to *FH*.

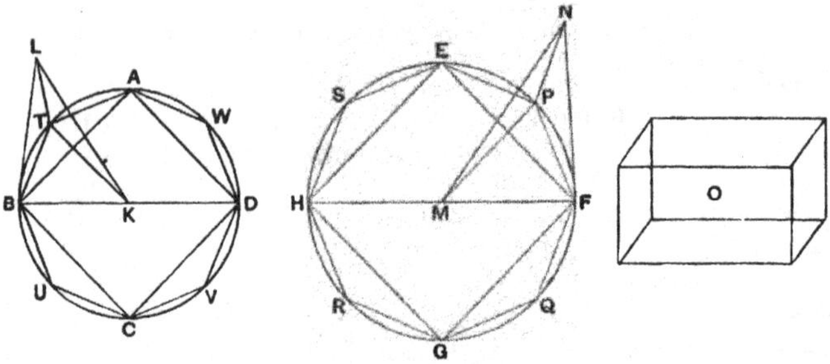

For, if the cone *ABCDL* has not to the cone *EFGHN* the ratio triplicate of that which *BD* has to *FH*,

the cone *ABCDL* will have that triplicate ratio either to some solid less than the cone *EFGHN* or to a greater.

First, let it have that triplicate ratio to a less solid *O*.

Let the square *EFGH* be inscribed in the circle *EFGH*; [IV. 6]

therefore the square *EFGH* is greater than the half of the circle *EFGH*.

Now let there be set up on the square *EFGH* a pyramid having the same vertex with the cone;

therefore the pyramid so set up is greater than the half part of the cone.

Let the circumferences *EF, FG, GH, HE* be bisected at the points *P, Q, R, S,*

and let *EP, PF, FQ, QG, GR, RH, HS, SE* be joined.

Therefore each of the triangles *EPF, FQG, GRH, HSE* is also greater than the half part of that segment of the circle *EFGH* which is about it.

Now on each of the triangles *EPF, FQG, GRH, HSE* let a pyramid be set up having the same vertex with the cone;

therefore each of the pyramids so set up is also greater than the half part of that segment of the cone which is about it.

Thus, bisecting the circumferences so left, joining straight lines, setting up on each of the triangles pyramids having the same vertex with the cone,

and doing this continually,

we shall leave some segments of the cone which will be less than the excess by which the cone *EFGHN* exceeds the solid *O*. [x. 1]

Let such be left, and let them be the segments on *EP, PF, FQ, QG, GR, RH, HS, SE*;

therefore the remainder, the pyramid of which the polygon *EPFQGRHS* is the base and the point *N* the vertex, is greater than the solid *O*.

Let there be also inscribed in the circle *ABCD* the polygon *ATBUCVDW* similar and similarly situated to the polygon *EPFQGRHS*,

and let there be set up on the polygon *ATBUCVDW* a pyramid having the same vertex with the cone;

of the triangles containing the pyramid of which the polygon *ATBUCVDW* is the base and the point *L* the vertex let *LBT* be one,

and of the triangles containing the pyramid of which the polygon *EPFQGRHS* is the base and the point *N* the vertex let *NFP* be one;

and let *KT, MP* be joined.

Now, since the cone *ABCDL* is similar to the cone *EFGHN*,

therefore, as *BD* is to *FH*, so is the axis *KL* to the axis *MN*. [xi. Def. 24]

But, as BD is to FH, so is BK to FM;
therefore also, as BK is to FM, so is KL to MN.

And, alternately, as BK is to KL, so is FM to MN. [V. 16]

And the sides are proportional about equal angles, namely the angles BKL, FMN;
therefore the triangle BKL is similar to the triangle FMN. [VI. 6]

Again, since, as BK is to KT, so is FM to MP,
and they are about equal angles, namely the angles BKT, FMP,
inasmuch as, whatever part the angle BKT is of the four right angles at the centre K, the same part also is the angle FMP of the four right angles at the centre M;
since then the sides are proportional about equal angles,
therefore the triangle BKT is similar to the triangle FMP. [VI. 6]

Again, since it was proved that, as BK is to KL, so is FM to MN,
while BK is equal to KT, and FM to PM,
therefore, as TK is to KL, so is PM to MN;
and the sides are proportional about equal angles, namely the angles TKL, PMN, for they are right;
therefore the triangle LKT is similar to the triangle NMP. [VI. 6]

And since, owing to the similarity of the triangles LKB, NMF,
as LB is to BK, so is NF to FM,
and, owing to the similarity of the triangles BKT, FMP,
as KB is to BT, so is MF to FP,
therefore, *ex aequali*, as LB is to BT, so is NF to FP. [V. 22]

Again since, owing to the similarity of the triangles LTK, NPM,
as LT is to TK, so is NP to PM,
and, owing to the similarity of the triangles TKB, PMF,
as KT is to TB, so is MP to PF;
therefore, *ex aequali*, as LT is to TB, so is NP to PF. [V. 22]

But it was also proved that, as TB is to BL, so is PF to FN.

Therefore, *ex aequali*, as TL is to LB, so is PN to NF. [V. 22]

Therefore in the triangles LTB, NPF the sides are proportional;

therefore the triangles LTB, NPF are equiangular; [VI. 5]

hence they are also similar. [VI. Def. 1]

Therefore the pyramid of which the triangle BKT is the base and the point L the vertex is also similar to the pyramid of which the triangle FMP is the base and the point N the vertex,

for they are contained by similar planes equal in multitude. [XI. Def. 9]

But similar pyramids which have triangular bases are to one another in the triplicate ratio of their corresponding sides. [XII. 8]

Therefore the pyramid $BKTL$ has to the pyramid $FMPN$ the ratio triplicate of that which BK has to FM.

Similarly, by joining straight lines from A, W, D, V, C, U to K, and from E, S, H, R, G, Q to M, and setting up on each of the triangles pyramids which have the same vertex with the cones,

we can prove that each of the similarly arranged pyramids will also have to each similarly arranged pyramid the ratio triplicate of that which the corresponding side BK has to the corresponding side FM, that is, which BD has to FH.

And, as one of the antecedents is to one of the consequents, so are all the antecedents to all the consequents; [V. 12]

therefore also, as the pyramid $BKTL$ is to the pyramid $FMPN$, so is the whole pyramid of which the polygon $ATBUCVDW$ is the base and the point L the vertex to the whole pyramid of which the polygon $EPFQGRHS$ is the base and the point N the vertex;

hence also the pyramid of which $ATBUCVDW$ is the base and the point L the vertex has to the pyramid of which the polygon $EPFQGRHS$ is the base and the point N the vertex the ratio triplicate of that which BD has to FH.

But, by hypothesis, the cone of which the circle $ABCD$

is the base and the point *L* the vertex has also to the solid *O* the ratio triplicate of that which *BD* has to *FH*;

therefore, as the cone of which the circle *ABCD* is the base and the point *L* the vertex is to the solid *O*, so is the pyramid of which the polygon *ATBUCVDW* is the base and *L* the vertex to the pyramid of which the polygon *EPFQGRHS* is the base and the point *N* the vertex;

therefore, alternately, as the cone of which the circle *ABCD* is the base and *L* the vertex is to the pyramid contained in it of which the polygon *ATBUCVDW* is the base and *L* the vertex, so is the solid *O* to the pyramid of which the polygon *EPFQGRHS* is the base and *N* the vertex. [v. 16]

But the said cone is greater than the pyramid in it;

for it encloses it.

Therefore the solid *O* is also greater than the pyramid of which the polygon *EPFQGRHS* is the base and *N* the vertex.

But it is also less:

which is impossible.

Therefore the cone of which the circle *ABCD* is the base and *L* the vertex has not to any solid less than the cone of which the circle *EFGH* is the base and the point *N* the vertex the ratio triplicate of that which *BD* has to *FH*.

Similarly we can prove that neither has the cone *EFGHN* to any solid less than the cone *ABCDL* the ratio triplicate of that which *FH* has to *BD*.

I say next that neither has the cone *ABCDL* to any solid greater than the cone *EFGHN* the ratio triplicate of that which *BD* has to *FH*.

For, if possible, let it have that ratio to a greater solid *O*.

Therefore, inversely, the solid *O* has to the cone *ABCDL* the ratio triplicate of that which *FH* has to *BD*.

But, as the solid *O* is to the cone *ABCDL*, so is the cone *EFGHN* to some solid less than the cone *ABCDL*.

Therefore the cone *EFGHN* also has to some solid less than the cone *ABCDL* the ratio triplicate of that which *FH* has to *BD*:

which was proved impossible.

Therefore the cone *ABCDL* has not to any solid greater than the cone *EFGHN* the ratio triplicate of that which *BD* has to *FH*.

But it was proved that neither has it this ratio to a less solid than the cone *EFGHN*.

Therefore the cone *ABCDL* has to the cone *EFGHN* the ratio triplicate of that which *BD* has to *FH*.

But, as the cone is to the cone, so is the cylinder to the cylinder,

for the cylinder which is on the same base as the cone and of equal height with it is triple of the cone; [XII. 10]

therefore the cylinder also has to the cylinder the ratio triplicate of that which *BD* has to *FH*.

Therefore etc.

Q. E. D.

The method of proof is precisely that of the previous proposition. The only addition is caused by the necessity of proving that, if similar equilateral polygons be inscribed in the bases of two similar cones, and pyramids be erected on them with the same vertices as those of the cones, the pyramids (are similar and) are to one another in the triplicate ratio of corresponding edges.

Let *KL*, *MN* be the axes of the cones, *L*, *N* the vertices, and let *BT*, *FP* be sides of similar polygons inscribed in the bases. Join *BK*, *TK*, *BL*, *TL*, *PM*, *FM*, *PN*, *FN*.

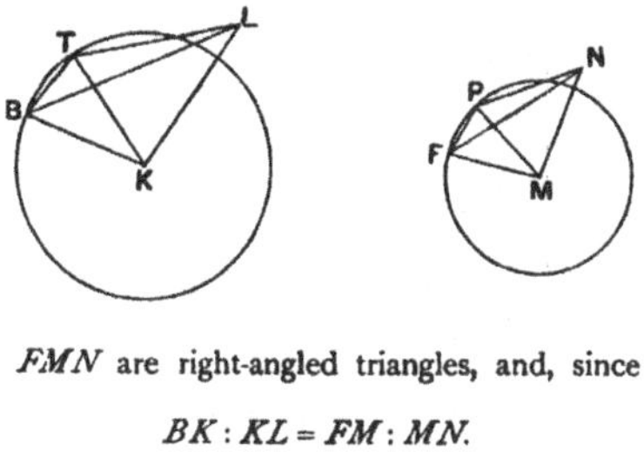

Now *BKL*, *FMN* are right-angled triangles, and, since the cones are similar,

$$BK : KL = FM : MN. \qquad \text{[XI. Def. 24]}$$

Therefore (1) △s *BKL*, *FMN* are similar. [VI. 6]

Similarly (2) △s *TKL*, *PMN* are similar.

Next, in △s *BKT*, *FMP*, the angles *BKT*, *FMP* are equal, since each is the same fraction of four right angles; and the sides about the equal angles are proportional;

therefore (3) △s *BKT*, *FMP* are similar.

Again, since from the similar △s BKL, FMN, and the similar △s BKT, FMP respectively,

$$LB : BK = NF : FM,$$
$$BK : BT = MF : FP,$$

ex aequali, $LB : BT = NF : FP.$

Similarly $LT : TB = NP : PF.$

Inverting the latter ratio and compounding it with the preceding one, we have, *ex aequali*,

$$LB : LT = NF : NP.$$

Thus in △s LTB, NPF the sides are proportional in pairs;

therefore (4) △s LTB, NPF are similar.

Thus the partial pyramids $L\text{-}BKT$, $N\text{-}FMP$ are similar.

In exactly the same way it is proved that all the other partial pyramids are similar.

Now

(pyramid $L\text{-}BKT$) : (pyramid $N\text{-}FMP$) = ratio triplicate of ($BK : FM$).

The other partial pyramids are to one another in the same triplicate ratio.

The sum of the antecedents is therefore to the sum of the consequents in the same triplicate ratio,

i.e. (pyramid $L\text{-}ATBU\ldots$) : (pyramid $N\text{-}EPFQ\ldots$)
= ratio triplicate of ratio ($BK : FM$)
= ratio triplicate of ratio ($BD : FH$).

[The fact that Euclid makes this transition from the partial pyramids to the whole pyramids in the body of this proposition seems to me to suggest grave doubts as to the genuineness of the Porism to XII. 8, which contains a similar but rather more general extension from the case of triangular pyramids to pyramids with polygonal bases. Were that Porism genuine, Euclid would have been more likely to refer to it than to repeat here the same arguments which it contains.]

Now we are in a position to apply the method of exhaustion.

If X, X' be the volumes of the cones, d, d' the diameters of their bases, and if

$$(\text{ratio triplicate of } d : d') \neq X : X',$$

then must (ratio triplicate of $d : d'$) = $X : O$,

where O is either less or greater than X'.

I. Suppose that O is *less* than X'.

Construct in the way described a pyramid (Π') in X' leaving over portions of X' together less than $(X' - O)$, so that $X' > \Pi' > O$,

and construct in X a pyramid (Π), with the same vertex as X has, on a polygon inscribed in its base similar to the base of Π'.

Then, by what has just been proved,

$$\Pi : \Pi' = (\text{ratio triplicate of } d : d')$$
$$= X : O, \text{ by hypothesis,}$$

and, alternately, $\Pi : X = \Pi' : O.$

But X includes, and is therefore greater than, Π;

therefore $O > \Pi'$.

But, by construction, $O < \Pi'$:

which is impossible.

Therefore O cannot be less than X'.

II. Suppose, if possible, that

(ratio triplicate of $d : d'$) $= X : O$,

where O is *greater* than X';

then (ratio triplicate of $d : d'$) $= Z : X'$,

or, inversely, (ratio triplicate of $d' : d$) $= X' : Z$,

where Z is *some solid less than* X.

This is proved impossible by the exact method of Part I.

Hence O cannot be either greater or less than X',

and $X : X' =$ (ratio triplicate of ratio $d : d'$).

Proposition 13.

If a cylinder be cut by a plane which is parallel to its opposite planes, then, as the cylinder is to the cylinder, so will the axis be to the axis.

For let the cylinder AD be cut by the plane GH which is parallel to the opposite planes AB, CD,

and let the plane GH meet the axis at the point K;

I say that, as the cylinder BG is to the cylinder GD, so is the axis EK to the axis KF.

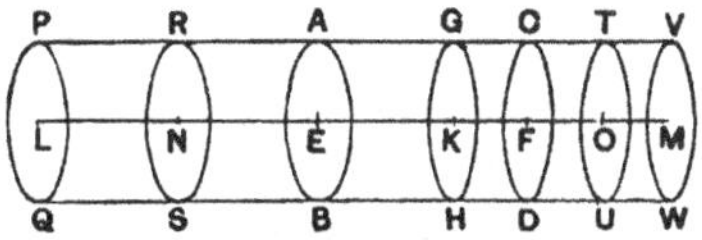

For let the axis EF be produced in both directions to the points L, M,

and let there be set out any number whatever of axes EN, NL equal to the axis EK,

and any number whatever FO, OM equal to FK;

and let the cylinder PW on the axis LM be conceived of which the circles PQ, VW are the bases.

Let planes be carried through the points N, O parallel to AB, CD and to the bases of the cylinder PW,

and let them produce the circles RS, TU about the centres N, O.

Then, since the axes LN, NE, EK are equal to one another, therefore the cylinders *QR*, *RB*, *BG* are to one another as their bases. [XII. 11]

But the bases are equal;

therefore the cylinders *QR*, *RB*, *BG* are also equal to one another.

Since then the axes *LN*, *NE*, *EK* are equal to one another,

and the cylinders *QR*, *RB*, *BG* are also equal to one another,

and the multitude of the former is equal to the multitude of the latter,

therefore, whatever multiple the axis *KL* is of the axis *EK*, the same multiple also will the cylinder *QG* be of the cylinder *GB*.

For the same reason, whatever multiple the axis *MK* is of the axis *KF*, the same multiple also is the cylinder *WG* of the cylinder *GD*.

And, if the axis *KL* is equal to the axis *KM*, the cylinder *QG* will also be equal to the cylinder *GW*,

if the axis is greater than the axis, the cylinder will also be greater than the cylinder,

and if less, less.

Thus, there being four magnitudes, the axes *EK*, *KF* and the cylinders *BG*, *GD*,

there have been taken equimultiples of the axis *EK* and of the cylinder *BG*, namely the axis *LK* and the cylinder *QG*,

and equimultiples of the axis *KF* and of the cylinder *GD*, namely the axis *KM* and the cylinder *GW*;

and it has been proved that,

if the axis *KL* is in excess of the axis *KM*, the cylinder *QG* is also in excess of the cylinder *GW*,

if equal, equal,

and if less, less.

Therefore, as the axis *EK* is to the axis *KF*, so is the cylinder *BG* to the cylinder *GD*. [V. Def. 5]

Q. E. D.

It is not necessary to reproduce the proof, as it follows exactly the method of VI. 1 and XI. 25.

The fact that cylinders described about axes of equal length and having equal bases are equal is inferred from XII. 11 to the effect that cylinders of equal height are to one another as their bases.

That, of two cylinders with unequal axes but equal bases, the greater is that which has the longer axis is of course obvious either by application or by cutting off from the cylinder with the longer axis a cylinder with an axis of the same length as that of the other given cylinder.

PROPOSITION 14.

Cones and cylinders which are on equal bases are to one another as their heights.

For let *EB*, *FD* be cylinders on equal bases, the circles *AB*, *CD*;

I say that, as the cylinder *EB* is to the cylinder *FD*, so is the axis *GH* to the axis *KL*.

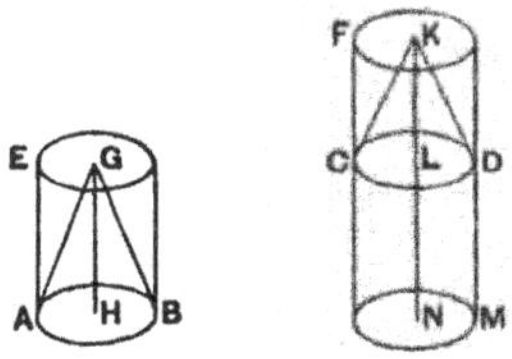

For let the axis *KL* be produced to the point *N*,

let *LN* be made equal to the axis *GH*,

and let the cylinder *CM* be conceived about *LN* as axis.

Since then the cylinders *EB*, *CM* are of the same height, they are to one another as their bases. [XII. 11]

But the bases are equal to one another;

therefore the cylinders *EB*, *CM* are also equal.

And, since the cylinder *FM* has been cut by the plane *CD* which is parallel to its opposite planes,

therefore, as the cylinder *CM* is to the cylinder *FD*, so is the axis *LN* to the axis *KL*. [XII. 13]

But the cylinder *CM* is equal to the cylinder *EB*,

and the axis *LN* to the axis *GH*;

therefore, as the cylinder *EB* is to the cylinder *FD*, so is the axis *GH* to the axis *KL*.

But, as the cylinder *EB* is to the cylinder *FD*, so is the cone *ABG* to the cone *CDK*. [XII. 10]

Therefore also, as the axis *GH* is to the axis *KL*, so is the cone *ABG* to the cone *CDK* and the cylinder *EB* to the cylinder *FD*. Q. E. D.

No separate proposition corresponding to this is necessary in the case of parallelepipeds, for XI. 25 really contains the property corresponding to that in this proposition as well as the property corresponding to that in XII. 13.

Proposition 15.

In equal cones and cylinders the bases are reciprocally proportional to the heights; and those cones and cylinders in which the bases are reciprocally proportional to the heights are equal.

Let there be equal cones and cylinders of which the circles *ABCD*, *EFGH* are the bases;
let *AC*, *EG* be the diameters of the bases,
and *KL*, *MN* the axes, which are also the heights of the cones or cylinders;
let the cylinders *AO*, *EP* be completed.

I say that in the cylinders *AO*, *EP* the bases are reciprocally proportional to the heights,
that is, as the base *ABCD* is to the base *EFGH*, so is the height *MN* to the height *KL*.

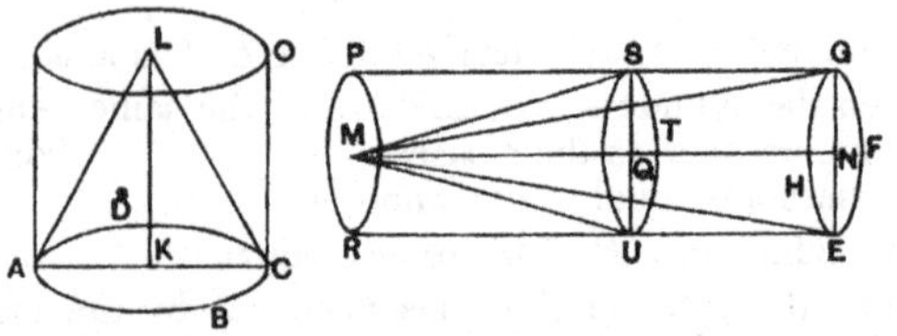

For the height *LK* is either equal to the height *MN* or not equal.

First, let it be equal.

Now the cylinder *AO* is also equal to the cylinder *EP*.

But cones and cylinders which are of the same height are to one another as their bases; [XII. 11]
therefore the base *ABCD* is also equal to the base *EFGH*.

Hence also, reciprocally, as the base *ABCD* is to the base *EFGH*, so is the height *MN* to the height *KL*.

Next, let the height *LK* not be equal to *MN*,
but let *MN* be greater;
from the height *MN* let *QN* be cut off equal to *KL*,
through the point *Q* let the cylinder *EP* be cut by the plane *TUS* parallel to the planes of the circles *EFGH*, *RP*,

and let the cylinder ES be conceived erected from the circle $EFGH$ as base and with height NQ.

Now, since the cylinder AO is equal to the cylinder EP, therefore, as the cylinder AO is to the cylinder ES, so is the cylinder EP to the cylinder ES. [V. 7]

But, as the cylinder AO is to the cylinder ES, so is the base $ABCD$ to the base $EFGH$,

for the cylinders AO, ES are of the same height; [XII. 11]

and, as the cylinder EP is to the cylinder ES, so is the height MN to the height QN,

for the cylinder EP has been cut by a plane which is parallel to its opposite planes. [XII. 13]

Therefore also, as the base $ABCD$ is to the base $EFGH$, so is the height MN to the height QN. [V. 11]

But the height QN is equal to the height KL;

therefore, as the base $ABCD$ is to the base $EFGH$, so is the height MN to the height KL.

Therefore in the cylinders AO, EP the bases are reciprocally proportional to the heights.

Next, in the cylinders AO, EP let the bases be reciprocally proportional to the heights,

that is, as the base $ABCD$ is to the base $EFGH$, so let the height MN be to the height KL;

I say that the cylinder AO is equal to the cylinder EP.

For, with the same construction,

since, as the base $ABCD$ is to the base $EFGH$, so is the height MN to the height KL,

while the height KL is equal to the height QN,

therefore, as the base $ABCD$ is to the base $EFGH$, so is the height MN to the height QN.

But, as the base $ABCD$ is to the base $EFGH$, so is the cylinder AO to the cylinder ES,

for they are of the same height; [XII. 11]

and, as the height MN is to QN, so is the cylinder EP to the cylinder ES; [XII. 13]

therefore, as the cylinder AO is to the cylinder ES, so is the cylinder EP to the cylinder ES. [V. 11]

Therefore the cylinder AO is equal to the cylinder EP. [V. 9]

And the same is true for the cones also.

Q. E. D.

I. If the heights of the two cylinders are equal, and their volumes are equal, the bases are equal, since the latter are proportional to the volumes. [XII. 11]

If the heights are *not* equal, cut off from the higher cylinder a cylinder of the same height as the lower.

Then, if LK, QN be the equal heights,

we have, by XII. 11,

$$(\text{base } ABCD) : (\text{base } EFGH) = (\text{cylinder } AO) : (\text{cylinder } ES)$$
$$= (\text{cylinder } EP) : (\text{cylinder } ES),$$

by hypothesis,

$$= MN : QN \quad \text{[XII. 13]}$$
$$= MN : KL.$$

II. In the converse part of the proposition, Euclid omits the case where the cylinders have equal heights. In this case of course the reciprocal ratios are both ratios of equality; the bases are therefore equal, and consequently the cylinders.

If the heights are *not* equal, we have, with the same construction as before,

$$(\text{base } ABCD) : (\text{base } EFGH) = MN : KL.$$

But [XII. 11]

$$(\text{base } ABCD) : (\text{base } EFGH) = (\text{cylinder } AO) : (\text{cylinder } ES),$$

and $MN : KL = MN : QN$

$$= (\text{cylinder } EP) : (\text{cylinder } ES). \quad \text{[XII. 13]}$$

Therefore

$$(\text{cylinder } AO) : (\text{cylinder } ES) = (\text{cylinder } EP) : (\text{cylinder } ES),$$

and consequently $(\text{cylinder } AO) = (\text{cylinder } EP)$.

Similarly for the cones, which are equal to one-third of the cylinders respectively.

Legendre deduces these propositions about cones and cylinders from two others which he establishes by a method similar to that adopted by him for the theorem of XII. 2 (see note on that proposition).

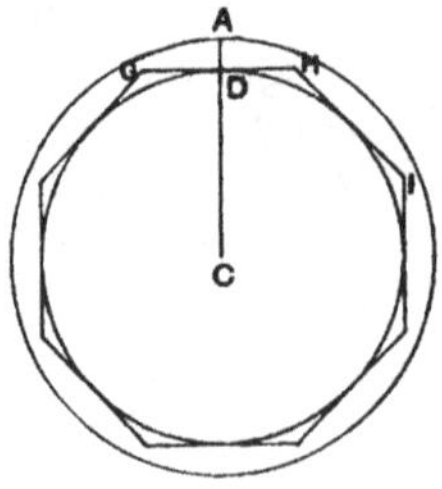

The first (for the cylinder) is as follows.

The volume of a cylinder is equal to the product of its base by its height.

Suppose CA to be the radius of the base of the given cylinder, h its height.

For brevity let us denote by (surf. CA) the area of the circle of which CA is the radius.

If (surf. CA) $\times h$ is not the measure of the given cylinder, it will be the measure of a cylinder greater or less than it.

I. First let it be the measure of a less cylinder, that, for example, of which the circle with radius CD is the base, and h is the height.

Circumscribe about the circle with radius CD a regular polygon $GHI\ldots$ such that its sides do not anywhere meet the circle with radius CA. [See note on XII. 2, p. 393 above, for Legendre's lemma relating to this construction.]

Imagine a prism erected on the polygon as base and with height h.

Then (volume of prism) = (polygon $GHI\ldots$) $\times h$.

[Legendre has previously proved this proposition, first for a parallelepiped (by transforming it into a rectangular one), then for a triangular prism (half of a parallelepiped of the same height), and lastly for a prism with a polygonal base.]

But (polygon $GHI\ldots$) < (surf. CA).

Therefore (volume of prism) < (surf. CA) $\times h$

< (cylinder on circle of rad. CD),

by hypothesis.

But the prism is *greater* than the latter cylinder, since it includes it: which is impossible.

II. In order not to multiply figures let us, in this second case, suppose that CD is the radius of the base of the given cylinder, and that (surf. CD) $\times h$ is the measure of a cylinder greater than it, e.g. a cylinder on the circle with radius CA as base and of height h.

Then, with the same construction,

(volume of prism) = (polygon $GHI\ldots$) $\times h$.

And (polygon $GHI\ldots$) > (surf. CD).

Therefore (volume of prism) > (surf. CD) $\times h$

> (cylinder on surf. CA), by hypothesis.

But the volume of the prism is also *less* than that cylinder, being included by it:

which is impossible.

Therefore (volume of cylinder) = (its base) $\times$ (its height).

It follows as a corollary that

Cylinders of the same height are to one another as their bases [XII. 13], *and cylinders on the same base are to one another as their heights* [XII. 14].

Also

Similar cylinders are as the cubes of their heights, or as the cubes of the diameters of their bases [Eucl. XII. 12].

For the bases are as the squares on their diameters; and, since the cylinders are similar, the diameters of the bases are as their heights.

Therefore the bases are as the squares on the heights, and the bases multiplied by the heights, or the cylinders themselves, are as the cubes of the heights.

I need not reproduce Legendre's proofs of the corresponding propositions for the cone.

PROPOSITION 16.

Given two circles about the same centre, to inscribe in the greater circle an equilateral polygon with an even number of sides which does not touch the lesser circle.

Let $ABCD$, $EFGH$ be the two given circles about the same centre K;

thus it is required to inscribe in the greater circle $ABCD$ an equilateral polygon with an even number of sides which does not touch the circle $EFGH$.

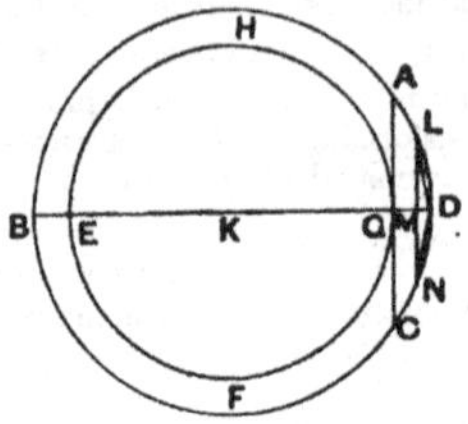

For let the straight line BKD be drawn through the centre K,

and from the point G let GA be drawn at right angles to the straight line BD and carried through to C;

therefore AC touches the circle $EFGH$. [III. 16, Por.]

Then, bisecting the circumference BAD, bisecting the half of it, and doing this continually, we shall leave a circumference less than AD. [x. 1]

Let such be left, and let it be LD;

from L let LM be drawn perpendicular to BD and carried through to N,

and let LD, DN be joined;

therefore LD is equal to DN. [III. 3, I. 4]

Now, since LN is parallel to AC,

and AC touches the circle $EFGH$,

therefore LN does not touch the circle $EFGH$;

therefore LD, DN are far from touching the circle $EFGH$.

If then we fit into the circle $ABCD$ straight lines equal to the straight line LD and placed continuously, there will be inscribed in the circle $ABCD$ an equilateral polygon with an even number of sides which does not touch the lesser circle $EFGH$. Q. E. F.

It must be carefully observed that the polygon inscribed in the outer circle in this proposition is such that not only do its own sides not touch the inner circle, but also *the chords*, as *LN*, *joining angular points next but one to each other do not touch the inner circle either.* In other words, the polygon is the second in order, not the first, which satisfies the condition of the enunciation. This is important, because such a polygon is wanted in the next proposition; hence in that proposition the *exact* construction here given must be followed.

PROPOSITION 17.

Given two spheres about the same centre, to inscribe in the greater sphere a polyhedral solid which does not touch the lesser sphere at its surface.

Let two spheres be conceived about the same centre *A*; thus it is required to inscribe in the greater sphere a polyhedral solid which does not touch the lesser sphere at its surface.

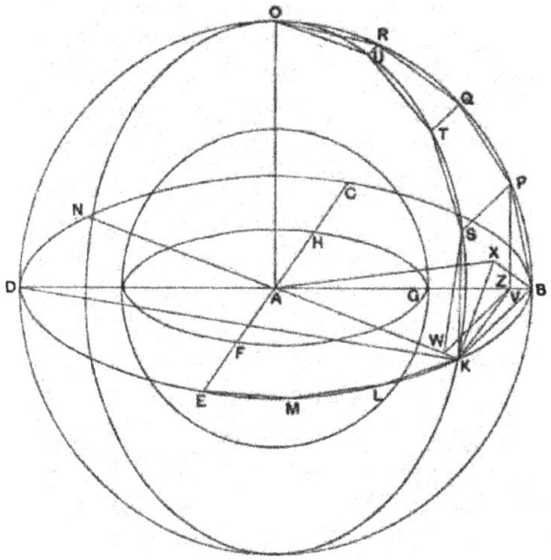

Let the spheres be cut by any plane through the centre; then the sections will be circles,

inasmuch as the sphere was produced by the diameter remaining fixed and the semicircle being carried round it;

[XI. Def. 14]

hence, in whatever position we conceive the semicircle to be, the plane carried through it will produce a circle on the circumference of the sphere.

And it is manifest that this circle is the greatest possible, inasmuch as the diameter of the sphere, which is of course the diameter both of the semicircle and of the circle, is greater than all the straight lines drawn across in the circle or the sphere.

Let then *BCDE* be the circle in the greater sphere,
and *FGH* the circle in the lesser sphere;
let two diameters in them, *BD*, *CE*, be drawn at right angles to one another;
then, given the two circles *BCDE*, *FGH* about the same centre, let there be inscribed in the greater circle *BCDE* an equilateral polygon with an even number of sides which does not touch the lesser circle *FGH*,
let *BK*, *KL*, *LM*, *ME* be its sides in the quadrant *BE*,
let *KA* be joined and carried through to *N*,
let *AO* be set up from the point *A* at right angles to the plane of the circle *BCDE*, and let it meet the surface of the sphere at *O*,
and through *AO* and each of the straight lines *BD*, *KN* let planes be carried;
they will then make greatest circles on the surface of the sphere, for the reason stated.

Let them make such,
and in them let *BOD*, *KON* be the semicircles on *BD*, *KN*.

Now, since *OA* is at right angles to the plane of the circle *BCDE*,
therefore all the planes through *OA* are also at right angles to the plane of the circle *BCDE*; [XI. 18]
hence the semicircles *BOD*, *KON* are also at right angles to the plane of the circle *BCDE*.

And, since the semicircles *BED*, *BOD*, *KON* are equal,
for they are on the equal diameters *BD*, *KN*,
therefore the quadrants *BE*, *BO*, *KO* are also equal to one another.

Therefore there are as many straight lines in the quadrants *BO*, *KO* equal to the straight lines *BK*, *KL*, *LM*, *ME* as there are sides of the polygon in the quadrant *BE*.

Let them be inscribed, and let them be *BP*, *PQ*, *QR*, *RO* and *KS*, *ST*, *TU*, *UO*,
let *SP*, *TQ*, *UR* be joined,

and from P, S let perpendiculars be drawn to the plane of the circle $BCDE$; [XI. 11]

these will fall on BD, KN, the common sections of the planes, inasmuch as the planes of BOD, KON are also at right angles to the plane of the circle $BCDE$. [cf. XI. Def. 4]

Let them so fall, and let them be PV, SW,

and let WV be joined.

Now since, in the equal semicircles BOD, KON, equal straight lines BP, KS have been cut off,

and the perpendiculars PV, SW have been drawn,

therefore PV is equal to SW, and BV to KW. [III. 27, I. 26]

But the whole BA is also equal to the whole KA;

therefore the remainder VA is also equal to the remainder WA;

therefore, as BV is to VA, so is KW to WA;

therefore WV is parallel to KB. [VI. 2]

And, since each of the straight lines PV, SW is at right angles to the plane of the circle $BCDE$,

therefore PV is parallel to SW. [XI. 6]

But it was also proved equal to it;

therefore WV, SP are also equal and parallel. [I. 33]

And, since WV is parallel to SP,

while WV is parallel to KB,

therefore SP is also parallel to KB. [XI. 9]

And BP, KS join their extremities;

therefore the quadrilateral $KBPS$ is in one plane,

inasmuch as, if two straight lines be parallel, and points be taken at random on each of them, the straight line joining the points is in the same plane with the parallels. [XI. 7]

For the same reason

each of the quadrilaterals $SPQT$, $TQRU$ is also in one plane.

But the triangle URO is also in one plane. [XI. 2]

If then we conceive straight lines joined from the points P, S, Q, T, R, U to A, there will be constructed a certain polyhedral solid figure between the circumferences BO, KO, consisting of pyramids of which the quadrilaterals $KBPS$, $SPQT$, $TQRU$ and the triangle URO are the bases and the point A the vertex.

And, if we make the same construction in the case of each of the sides *KL*, *LM*, *ME* as in the case of *BK*, and further in the case of the remaining three quadrants,

there will be constructed a certain polyhedral figure inscribed in the sphere and contained by pyramids, of which the said quadrilaterals and the triangle *URO*, and the others corresponding to them, are the bases and the point *A* the vertex.

I say that the said polyhedron will not touch the lesser sphere at the surface on which the circle *FGH* is.

Let *AX* be drawn from the point *A* perpendicular to the plane of the quadrilateral *KBPS*, and let it meet the plane at the point *X*; [XI. 11]

let *XB*, *XK* be joined.

Then, since *AX* is at right angles to the plane of the quadrilateral *KBPS*,

therefore it is also at right angles to all the straight lines which meet it and are in the plane of the quadrilateral. [XI. Def. 3]

Therefore *AX* is at right angles to each of the straight lines *BX*, *XK*.

And, since *AB* is equal to *AK*,

the square on *AB* is also equal to the square on *AK*.

And the squares on *AX*, *XB* are equal to the square on *AB*,

for the angle at *X* is right; [I. 47]

and the squares on *AX*, *XK* are equal to the square on *AK*. [*id.*]

Therefore the squares on *AX*, *XB* are equal to the squares on *AX*, *XK*.

Let the square on *AX* be subtracted from each;

therefore the remainder, the square on *BX*, is equal to the remainder, the square on *XK*;

therefore *BX* is equal to *XK*.

Similarly we can prove that the straight lines joined from *X* to *P*, *S* are equal to each of the straight lines *BX*, *XK*.

Therefore the circle described with centre X and distance one of the straight lines XB, XK will pass through P, S also, and $KBPS$ will be a quadrilateral in a circle.

Now, since KB is greater than WV,
while WV is equal to SP,
therefore KB is greater than SP.

But KB is equal to each of the straight lines KS, BP;
therefore each of the straight lines KS, BP is greater than SP.

And, since $KBPS$ is a quadrilateral in a circle,
and KB, BP, KS are equal, and PS less,
and BX is the radius of the circle,
therefore the square on KB is greater than double of the square on BX.

Let KZ be drawn from K perpendicular to BV.

Then, since BD is less than double of DZ,
and, as BD is to DZ, so is the rectangle DB, BZ to the rectangle DZ, ZB,
if a square be described upon BZ and the parallelogram on ZD be completed,
then the rectangle DB, BZ is also less than double of the rectangle DZ, ZB.

And, if KD be joined,
the rectangle DB, BZ is equal to the square on BK,
and the rectangle DZ, ZB equal to the square on KZ;
[III. 31, VI. 8 and Por.]
therefore the square on KB is less than double of the square on KZ.

But the square on KB is greater than double of the square on BX;
therefore the square on KZ is greater than the square on BX.

And, since BA is equal to KA,
the square on BA is equal to the square on AK.

And the squares on BX, XA are equal to the square on BA,
and the squares on KZ, ZA equal to the square on KA;
[I. 47]
therefore the squares on BX, XA are equal to the squares on KZ, ZA,

and of these the square on KZ is greater than the square on BX;

therefore the remainder, the square on ZA, is less than the square on XA.

Therefore AX is greater than AZ;

therefore AX is much greater than AG.

And AX is the perpendicular on one base of the polyhedron,

and AG on the surface of the lesser sphere;

hence the polyhedron will not touch the lesser sphere on its surface.

Therefore, given two spheres about the same centre, a polyhedral solid has been inscribed in the greater sphere which does not touch the lesser sphere at its surface.

Q. E. F.

PORISM. But if in another sphere also a polyhedral solid be inscribed similar to the solid in the sphere $BCDE$,

the polyhedral solid in the sphere $BCDE$ has to the polyhedral solid in the other sphere the ratio triplicate of that which the diameter of the sphere $BCDE$ has to the diameter of the other sphere.

For, the solids being divided into their pyramids similar in multitude and arrangement, the pyramids will be similar.

But similar pyramids are to one another in the triplicate ratio of their corresponding sides; [XII. 8, Por.]

therefore the pyramid of which the quadrilateral $KBPS$ is the base, and the point A the vertex, has to the similarly arranged pyramid in the other sphere the ratio triplicate of that which the corresponding side has to the corresponding side, that is, of that which the radius AB of the sphere about A as centre has to the radius of the other sphere.

Similarly also each pyramid of those in the sphere about A as centre has to each similarly arranged pyramid of those in the other sphere the ratio triplicate of that which AB has to the radius of the other sphere.

And, as one of the antecedents is to one of the consequents, so are all the antecedents to all the consequents;

[V. 12]

hence the whole polyhedral solid in the sphere about A as centre has to the whole polyhedral solid in the other sphere the ratio triplicate of that which AB has to the radius of the other sphere, that is, of that which the diameter BD has to the diameter of the other sphere.

Q. E. D.

This proposition is of great length and therefore requires summarising in order to make it easier to grasp. Moreover there are some assumptions in it which require to be proved, and some omissions to be supplied. The figure also is one of some complexity, and, in addition, the text and the figure treat two points Z and V, which are really one and the same, as different.

The first thing needed is to know that all sections of a sphere by planes through the centre are circles and equal to one another (great circles or "greatest circles" as Euclid calls them, more appropriately). Euclid uses his definition of a sphere as the figure described by a semicircle revolving about its diameter. This of course establishes that all planes through the particular diameter make equal circular sections; but it is also assumed that the same sphere is generated by *any other* semicircle of the same size and with its centre at the same point.

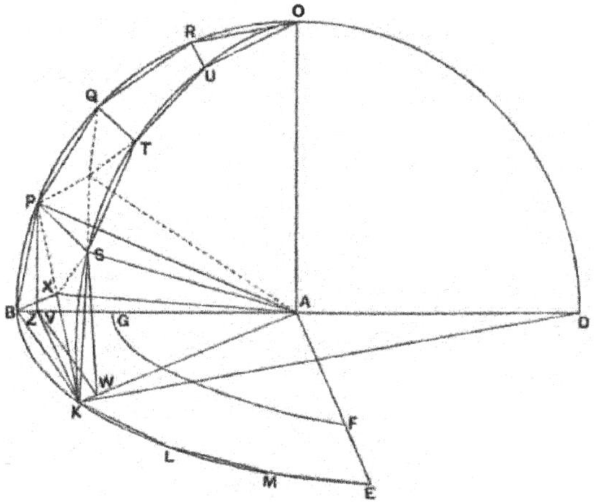

The construction and argument of the proposition may be shortly given as follows.

A plane through the centre of two concentric spheres cuts them in great circles of which BE, GF are quadrants.

A regular polygon with an even number of sides is inscribed (*exactly* as in Prop. 16) to the outer circle such that its sides do not touch the inner circle. BK, KL, LM, ME are the sides in the quadrant BE.

AO is drawn at right angles to the plane ABE, and through AO are drawn planes passing through B, K, L, M, E, etc., cutting the sphere in great circles.

OB, OK are quadrants of two of these great circles.

As these quadrants are equal to the quadrant BE, they will be divisible into arcs equal in number and magnitude to the arcs BK, KL, LM, ME.

Dividing the other quadrants of these circles, and also all the quadrants of the other circles through OA, in this way we shall have in all the circles a polygon equal to that in the circle of which BE is a quadrant.

BP, PQ, QR, RO and KS, ST, TU, UO are the sides of these polygons in the quadrants BO, KO.

Joining PS, QT, RU, and making the same construction all round the circles through AO, we have a certain polyhedron inscribed in the outer sphere.

Draw PV perpendicular to AB and therefore (since the planes OAB, BAE are at right angles) perpendicular to the plane BAE; [XI. Def. 4]
draw SW perpendicular to AK and therefore (for a like reason) perpendicular to the plane BAE.

Draw KZ perpendicular to BA. (Since $BK = BP$, and $DB \cdot BV = BP^2$, $DB \cdot BZ = BK^2$, it follows that $BV = BZ$, and Z, V coincide.)

Now, since $\angle$s PAV, SAW, being angles subtended at the centre by equal arcs of equal circles, are equal,

and since $\angle$s PVA, SWA are right,

while $AS = AP$,

$\triangle$s PAV, SAW are equal in all respects, [I. 26]

and $AV = AW$.

Consequently $AB : AV = AK : AW$;

and VW, BK are parallel.

But PV, SW are parallel (being both perpendicular to one plane) and equal (by the equal $\triangle$s PAV, SAW),

therefore VW, PS are equal and parallel.

Therefore BK (being parallel to VW) is parallel to PS.

Consequently (1) $BPSK$ is a quadrilateral in *one plane.*

Similarly the other quadrilaterals $PQTS$, $QRUT$ are in one plane; and the triangle ORU is in one plane.

In order now to prove that the plane BPSK *does not anywhere touch the inner sphere we have to prove that the shortest distance from* A *to the plane is greater than* AZ, *which by the construction in* XII. 16 *is greater than* AG.

Draw AX perpendicular to the plane $BPSK$.

Then $AX^2 + XB^2 = AX^2 + XK^2 = AX^2 + XS^2 = AX^2 + XP^2 = AB^2$,

whence $XB = XK = XS = XP$,

or (2) the quadrilateral $BPSK$ is inscribable in a circle with X as centre and radius XB.

Now $BK > VW$
$> PS$;

therefore in the quadrilateral $BPSK$ three sides BK, BP, KS are equal, but PS is less.

Consequently the angles about X are three equal angles and one smaller angle;

therefore any one of the equal angles is greater than a right angle, i.e. $\angle BXK$ is obtuse.

Therefore (3) $BK^2 > 2BX^2$. [II. 12]

Next, consider the semicircle BKD with KZ drawn perpendicular to BD.

We have $BD < 2DZ$,

so that $DB \cdot BZ < 2DZ \cdot ZB$,

or $BK^2 < 2KZ^2$;

therefore, *a fortiori*, [by (3) above]

(4) $BX^2 < KZ^2$.

Now $AK^2 = AB^2$;

therefore $AZ^2 + ZK^2 = AX^2 + XB^2$.

And $BX^2 < KZ^2$;

therefore $AX^2 > AZ^2$,

or (5) $AX > AZ$.

But, by the construction in XII. 16, $AZ > AG$; therefore, *a fortiori*, $AX > AG$.

And, since the perpendicular AX is the shortest distance from A to the plane $BPSK$,

(6) the plane $BPSK$ does not anywhere meet the inner sphere.

Euclid omits to prove that, *a fortiori*, the other quadrilaterals $PQTS$, $QRUT$, and the triangle ROU, do not anywhere meet the inner sphere.

For this purpose it is only necessary to show that the radii of the circles circumscribing $BPSK$, $PQTS$, $QRUT$ and ROU are in descending order of magnitude.

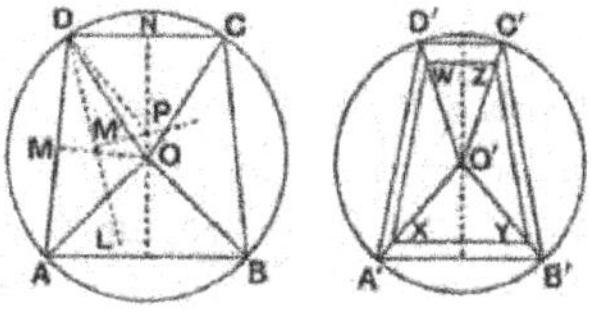

We have therefore to prove that, if $ABCD$, $A'B'C'D'$ are two quadrilaterals inscribable in circles, and

$$AD = BC = A'D' = B'C',$$

while AB is not greater than AD, $A'B' = CD$, and $AB > CD > C'D'$,

then the radius OA of the circle circumscribing the first quadrilateral is greater than the radius $O'A'$ of the circle circumscribing the second.

Clavius, and Simson after him, prove this by *reductio ad absurdum*.

(1) If $OA = O'A'$,

it follows that $\angle$s AOD, BOC, $A'O'D'$, $B'O'C'$ are all equal.

Also $\angle AOB > \angle A'O'B'$,

$\angle COD > \angle C'O'D'$,

whence the four angles about O are together greater than the four angles about O', i.e. greater than four right angles;

which is impossible.

(2) If $O'A' > OA$,

cut off from $O'A'$, $O'B'$, $O'C'$, $O'D'$ lengths equal to OA, and draw the inner quadrilateral as shown in the figure ($XYZW$).

Then
$$AB > A'B' > XY,$$
$$CD > C'D' > ZW,$$
$$AD = A'D' > WX,$$
$$BC = B'C' > YZ.$$

Consequently the same absurdity as in (1) follows *a fortiori*.

Therefore, since OA is neither equal to nor less than $O'A'$,
$$OA > O'A'.$$

The fact is also sufficiently clear if we draw MO, NO bisecting DA, DC perpendicularly and therefore meeting in O, the centre of the circumscribed circle, and then suppose the side DA with the perpendicular MO to turn inwards about D as centre. Then the intersection of MO and NO, as P, will gradually move towards N.

Simson gives his proof as "Lemma II." immediately before XII. 17. He adds to the Porism some words explaining how we may construct a similar polyhedron in another sphere and how we may prove that the polyhedra are similar.

The Porism is of course of the essence of the matter because it is the porism which as much as the construction is wanted in the next proposition. It would therefore not have been amiss to include the Porism in the enunciation of XII. 17 so as to call attention to it.

PROPOSITION 18.

Spheres are to one another in the triplicate ratio of their respective diameters.

Let the spheres ABC, DEF be conceived,

and let BC, EF be their diameters;

I say that the sphere ABC has to the sphere DEF the ratio triplicate of that which BC has to EF.

For, if the sphere ABC has not to the sphere DEF the ratio triplicate of that which BC has to EF,

then the sphere ABC will have either to some less sphere than the sphere DEF, or to a greater, the ratio triplicate of that which BC has to EF.

First, let it have that ratio to a less sphere GHK,

let DEF be conceived about the same centre with GHK,

let there be inscribed in the greater sphere DEF a polyhedral solid which does not touch the lesser sphere GHK at its surface, [XII. 17] and let there also be inscribed in the sphere *ABC* a polyhedral solid similar to the polyhedral solid in the sphere *DEF*;

therefore the polyhedral solid in *ABC* has to the polyhedral solid in *DEF* the ratio triplicate of that which *BC* has to *EF*. [XII. 17, Por.]

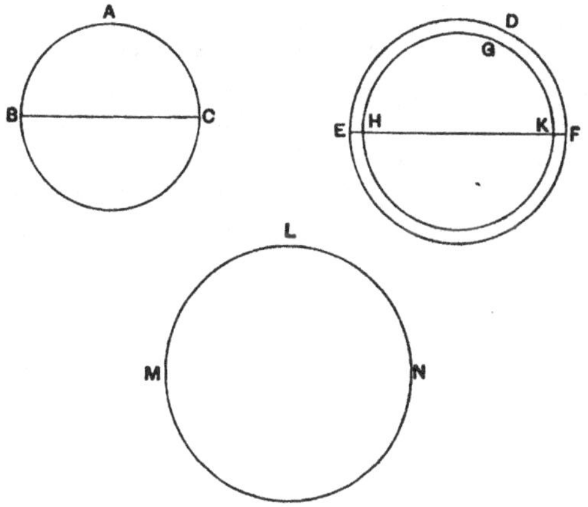

But the sphere *ABC* also has to the sphere *GHK* the ratio triplicate of that which *BC* has to *EF*;

therefore, as the sphere *ABC* is to the sphere *GHK*, so is the polyhedral solid in the sphere *ABC* to the polyhedral solid in the sphere *DEF*;

and, alternately, as the sphere *ABC* is to the polyhedron in it, so is the sphere *GHK* to the polyhedral solid in the sphere *DEF*. [v. 16]

But the sphere *ABC* is greater than the polyhedron in it;

therefore the sphere *GHK* is also greater than the polyhedron in the sphere *DEF*.

But it is also less,

for it is enclosed by it.

Therefore the sphere *ABC* has not to a less sphere than the sphere *DEF* the ratio triplicate of that which the diameter *BC* has to *EF*.

Similarly we can prove that neither has the sphere *DEF* to a less sphere than the sphere *ABC* the ratio triplicate of that which *EF* has to *BC*.

I say next that neither has the sphere *ABC* to any greater sphere than the sphere *DEF* the ratio triplicate of that which *BC* has to *EF*.

For, if possible, let it have that ratio to a greater, *LMN*;

therefore, inversely, the sphere *LMN* has to the sphere *ABC* the ratio triplicate of that which the diameter *EF* has to the diameter *BC*.

But, inasmuch as *LMN* is greater than *DEF*,

therefore, as the sphere *LMN* is to the sphere *ABC*, so is the sphere *DEF* to some less sphere than the sphere *ABC*, as was before proved. [XII. 2, Lemma]

Therefore the sphere *DEF* also has to some less sphere than the sphere *ABC* the ratio triplicate of that which *EF* has to *BC*:

which was proved impossible.

Therefore the sphere *ABC* has not to any sphere greater than the sphere *DEF* the ratio triplicate of that which *BC* has to *EF*.

But it was proved that neither has it that ratio to a less sphere.

Therefore the sphere *ABC* has to the sphere *DEF* the ratio triplicate of that which *BC* has to *EF*.

Q. E. D.

It is the method of this proposition which Legendre adopted for his proof of XII. 2 (see note on that proposition).

The argument can be put very shortly. We will suppose S, S' to be the volumes of the spheres, and d, d' to be their diameters; and we will for brevity express the triplicate ratio of d to d' by $d^3 : d'^3$.

If $$d^3 : d'^3 \neq S : S',$$

then $$d^3 : d'^3 = S : T,$$

where T is the volume of some sphere either greater or less than S'.

I. Suppose, if possible, that $T < S'$.

Let T be supposed concentric with S'.

As in XII. 17, inscribe a polyhedron in S' such that its faces do not anywhere touch T;

and inscribe in S a polyhedron similar to that in S'.

Then $S : T = d^3 : d'^3$

$= (\text{polyhedron in } S) : (\text{polyhedron in } S')$;

or, alternately,

$$S : (\text{polyhedron in } S) = T : (\text{polyhedron in } S').$$

And $S > (\text{polyhedron in } S)$;

therefore $T > (\text{polyhedron in } S')$.

But, by construction, $T < (\text{polyhedron in } S')$:

which is impossible.

Therefore $T \nless S'$.

II. Suppose, if possible, that $T > S'$.

Now $d^3 : d'^3 = S : T$

$= X : S'$,

where X is the volume of some sphere less than S, [XII. 2, Lemma]

or, inversely, $d'^3 : d^3 = S' : X$,

where $X < S$.

This is proved impossible exactly as in Part I.

Therefore $T \ngtr S'$.

Hence T, not being greater or less than S', is equal to it, and

$$d^3 : d'^3 = S : S'.$$

BOOK XIII.

HISTORICAL NOTE.

I have already given, in the note to IV. 10, the evidence upon which the construction of the five regular solids is attributed to the Pythagoreans. Some of them, the cube, the tetrahedron (which is nothing but a pyramid), and the octahedron (which is only a double pyramid with a square base), cannot but have been known to the Egyptians. And it appears that dodecahedra have been found, of bronze or other material, which may belong to periods earlier than Pythagoras' time by some centuries (for references see Cantor's *Geschichte der Mathematik* I_3, pp. 175—6).

It is true that the author of the scholium No. 1 to Eucl. XIII. says that the Book is about "the five so-called Platonic figures, which however do not belong to Plato, three of the aforesaid five figures being due to the Pythagoreans, namely the cube, the pyramid and the dodecahedron, while the octahedron and the icosahedron are due to Theaetetus." This statement (taken probably from Geminus) may perhaps rest on the fact that Theaetetus was the first to write at any length about the two last-mentioned solids. We are told indeed by Suidas (s. v. Θεαίτητος) that Theaetetus "first wrote on the 'five solids' as they are called." This no doubt means that Theaetetus was the first to write a complete and systematic treatise on all the regular solids; it does not exclude the possibility that Hippasus or others had already written on the dodecahedron. The fact that Theaetetus wrote upon the regular solids agrees very well with the evidence which we possess of his contributions to the theory of irrationals, the connexion between which and the investigation of the regular solids is seen in Euclid's Book XIII.

Theaetetus flourished about 380 B.C., and his work on the regular solids was soon followed by another, that of Aristaeus, an elder contemporary of Euclid, who also wrote an important book on *Solid Loci*, i.e. on conics treated as loci. This Aristaeus (known as "the elder") wrote in the period about 320 B.C. We hear of his *Comparison of the five regular solids* from Hypsicles (2nd cent. B.C.), the writer of the short book commonly included in the editions of the *Elements* as Book XIV. Hypsicles gives in this Book some six propositions supplementing Eucl. XIII.; and he introduces the second of the propositions (Heiberg's Euclid, Vol. V. p. 6) as follows:

"*The same circle circumscribes both the pentagon of the dodecahedron and the triangle of the icosahedron when both are inscribed in the same sphere.* This is proved by Aristaeus in the book entitled *Comparison of the five figures*."

Hypsicles proceeds (pp. 7 sqq.) to give a proof of this theorem. Allman pointed out (*Greek Geometry from Thales to Euclid*, 1889, pp. 201—2) that this proof depends on eight theorems, six of which appear in Euclid's Book XIII. (in Propositions 8, 10, 12, 15, 16 with Por., 17); two other propositions not mentioned by Allman are also used, namely XIII. 4 and 9. This seems, as Allman says, to confirm the inference of Bretschneider (p. 171) that, as Aristaeus' work was the newest and latest in which, before Euclid's time, this subject was treated, we have in Eucl. XIII. at least a partial recapitulation of the contents of the treatise of Aristaeus.

After Euclid, Apollonius wrote on the comparison of the dodecahedron and the icosahedron inscribed in one and the same sphere. This we also learn from Hypsicles, who says in the next words following those about Aristaeus above quoted: "But it is proved by Apollonius in the second edition of his *Comparison of the dodecahedron with the icosahedron* that, as the surface of the dodecahedron is to the surface of the icosahedron [inscribed in the same sphere], so is the dodecahedron itself [i.e. its volume] to the icosahedron, because the perpendicular is the same from the centre of the sphere to the pentagon of the dodecahedron and to the triangle of the icosahedron."

Proposition 1.

If a straight line be cut in extreme and mean ratio, the square on the greater segment added to the half of the whole is five times the square on the half.

For let the straight line AB be cut in extreme and mean ratio at the point C,

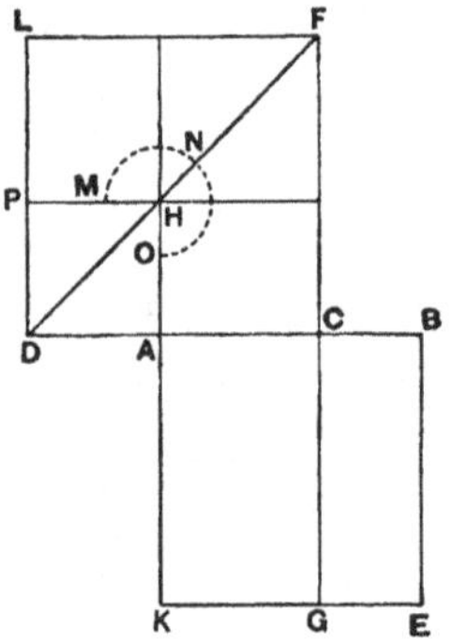

and let AC be the greater segment; let the straight line AD be produced in a straight line with CA, and let AD be made half of AB; I say that the square on CD is five times the square on AD.

For let the squares AE, DF be described on AB, DC, and let the figure in DF be drawn; let FC be carried through to G.

Now, since AB has been cut in extreme and mean ratio at C, therefore the rectangle AB, BC is equal to the square on AC. [VI. Def. 3, VI. 17]

And CE is the rectangle AB, BC, and FH the square on AC;

therefore CE is equal to FH.

And, since BA is double of AD,

while BA is equal to KA, and AD to AH,

therefore KA is also double of AH.

But, as KA is to AH, so is CK to CH; [VI. 1]

therefore CK is double of CH.

But LH, HC are also double of CH.

Therefore KC is equal to LH, HC.

But CE was also proved equal to HF;
therefore the whole square AE is equal to the gnomon MNO.

And, since BA is double of AD,
the square on BA is quadruple of the square on AD,
that is, AE is quadruple of DH.

But AE is equal to the gnomon MNO;
therefore the gnomon MNO is also quadruple of AP;
therefore the whole DF is five times AP.

And DF is the square on DC, and AP the square on DA;
therefore the square on CD is five times the square on DA.

Therefore etc.

Q. E. D.

The first five propositions are in the nature of lemmas, which are required for later propositions but are not in themselves of much importance.

It will be observed that, while the method of the propositions is that of Book II., being strictly geometrical and not algebraical, none of the results of that Book are made use of (except indeed in the Lemma to XIII. 2, which is probably not genuine). It would therefore appear as though these propositions were taken from an earlier treatise without being revised or rewritten in the light of Book II. It will be remembered that, according to Proclus (p. 67, 6), Eudoxus "greatly added to the number of the theorems which originated with Plato regarding *the section*" (i.e. presumably the "*golden* section"); and it is therefore probable that the five theorems are due to Eudoxus.

That, if AB is divided at C in extreme and mean ratio, the rectangle AB, BC is equal to the square on AC is inferred from VI. 17.

AD is made equal to half AB, and we have to prove that

$$(\text{sq. on } CD) = 5\,(\text{sq. on } AD).$$

The figure shows at once that

$$\square\, CH = \square\, HL,$$

so that $\square\, CH + \square\, HL = 2\,(\square\, CH)$
$= \square\, AG.$

Also sq. HF = (sq. on AC)
= rect. AB, BC
= CE.

By addition,

(gnomon MNO) = sq. on AB
= 4 (sq. on AD);

whence, adding the sq. on AD to each, we have

$$(\text{sq. on } CD) = 5\,(\text{sq. on } AD).$$

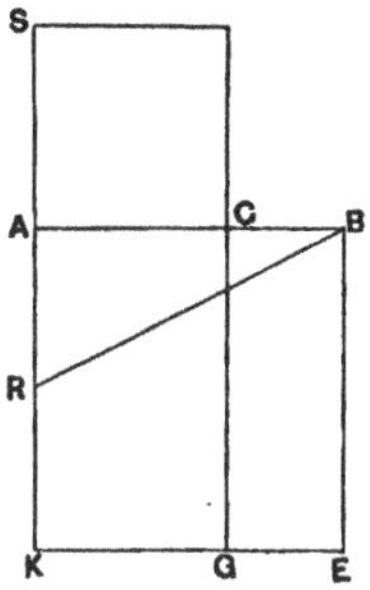

The result here, and in the next propositions, is really seen more readily by means of the figure of II. 11.

In this figure $SR = AC + \frac{1}{2}AB$, by construction; and we have therefore to prove that

$$(\text{sq. on } SR) = 5\,(\text{sq. on } AR).$$

This is obvious, for

$$(\text{sq. on } SR) = (\text{sq. on } RB)$$
$$= \text{sum of sqs. on } AB, AR$$
$$= 5\,(\text{sq. on } AR).$$

The MSS. contain a curious addition to XIII. 1—5 in the shape of analyses and syntheses for each proposition prefaced by the heading:

"What is analysis and what is synthesis.

"Analysis is the assumption of that which is sought as if it were admitted <and the arrival> by means of its consequences at something admitted to be true.

"Synthesis is an assumption of that which is admitted <and the arrival> by means of its consequences at something admitted to be true."

There must apparently be some corruption in the text; it does not, in the case of synthesis, give what is wanted. B and V have, instead of "something admitted to be true," the words "the end or attainment of what is sought."

The whole of this addition is evidently interpolated. To begin with, the analyses and syntheses of the five propositions are placed all together in four MSS.; in P, q they come after an alternative proof of XIII. 5 (which alternative proof P gives after XIII. 6, while q gives it instead of XIII. 6), in B (which has not the alternative proof of XIII. 5) after XIII. 6, and in b (in which XIII. 6 is wanting, and the alternative proof of XIII. 5 is in the margin, in the first hand) after XIII. 5, while V has the analyses of 1—3 in the text after XIII. 6 and those of 4—5 in the same place in the margin, by the second hand. Further, the addition is altogether alien from the plan and manner of the *Elements*. The interpolation took place before Theon's time, and the probability is that it was originally in the margin, whence it crept into the text of P after XIII. 5. Heiberg (after Bretschneider) suggested in his edition (Vol. V. p. lxxxiv.) that it might be a relic of analytical investigations by Theaetetus or Eudoxus, and he cited the remark of Pappus (V. p. 410) at the beginning of his "comparisons of the five [regular solid] figures which have an equal surface," to the effect that he will not use "the so-called analytical investigation by means of which some of the ancients effected their demonstrations." More recently (*Paralipomena zu Euklid* in *Hermes* XXXVIII., 1903) Heiberg conjectures that the author is Heron, on the ground that the sort of analysis and synthesis recalls Heron's remarks on analysis and synthesis in his commentary on the beginning of Book II. (quoted by an-Nairīzī, ed. Curtze, p. 89) and his quasi-algebraical alternative proofs of propositions in that Book.

To show the character of the interpolated matter I need only give the analysis and synthesis of one proposition. In the case of XIII. 1 it is in substance as follows. The figure is a mere straight line.

Let AB be divided in extreme and mean ratio at C, AC being the greater segment;

and let $$AD = \tfrac{1}{2}AB.$$

I say that $$(\text{sq. on } CD) = 5\,(\text{sq. on } AD).$$

(Analysis.)

"For, since $(\text{sq. on } CD) = 5\,(\text{sq. on } AD)$,"

and $(\text{sq. on } CD) = (\text{sq. on } CA) + (\text{sq. on } AD) + 2\,(\text{rect. } CA, AD)$,

therefore $(\text{sq. on } CA) + 2\,(\text{rect. } CA, AD) = 4\,(\text{sq. on } AD)$.

But $$\text{rect. } BA \,.\, AC = 2\,(\text{rect. } CA \,.\, AD),$$

and $$(\text{sq. on } CA) = (\text{rect. } AB, BC).$$

Therefore

$$(\text{rect. } BA, AC) + (\text{rect. } AB, BC) = 4\,(\text{sq. on } AD),$$

or

$$(\text{sq. on } AB) = 4\,(\text{sq. on } AD):$$

and this is true, since $AD = \frac{1}{2}AB$.

(Synthesis.)

Since $(\text{sq. on } AB) = 4\,(\text{sq. on } AD)$,

and $(\text{sq. on } AB) = (\text{rect. } BA, AC) + (\text{rect. } AB, BC)$,

therefore $4\,(\text{sq. on } AD) = 2\,(\text{rect. } DA, AC) + \text{sq. on } AC$.

Adding to each the square on AD, we have

$$(\text{sq. on } CD) = 5\,(\text{sq. on } AD).$$

Proposition 2.

If the square on a straight line be five times the square on a segment of it, then, when the double of the said segment is cut in extreme and mean ratio, the greater segment is the remaining part of the original straight line.

For let the square on the straight line AB be five times the square on the segment AC of it,

and let CD be double of AC;

I say that, when CD is cut in extreme and mean ratio, the greater segment is CB.

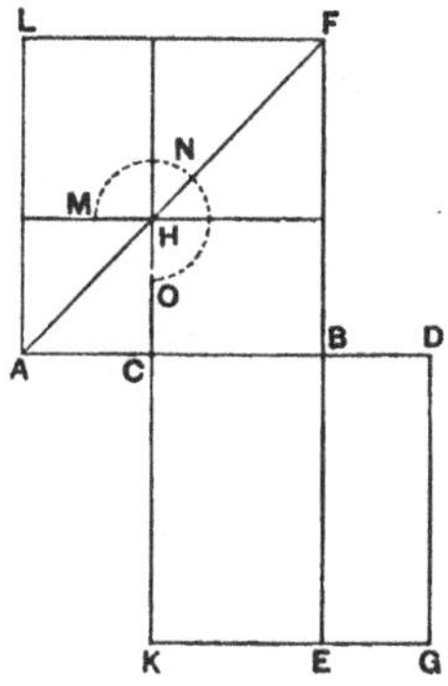

Let the squares AF, CG be described on AB, CD respectively, let the figure in AF be drawn, and let BE be drawn through.

Now, since the square on BA is five times the square on AC, AF is five times AH.

Therefore the gnomon MNO is quadruple of AH.

And, since DC is double of CA, therefore the square on DC is quadruple of the square on CA, that is, CG is quadruple of AH.

But the gnomon MNO was also proved quadruple of AH; therefore the gnomon MNO is equal to CG.

And, since DC is double of CA, while DC is equal to CK, and AC to CH, therefore KB is also double of BH. [VI. 1]

But LH, HB are also double of HB;
therefore KB is equal to LH, HB.

But the whole gnomon MNO was also proved equal to the whole CG;
therefore the remainder HF is equal to BG.

And BG is the rectangle CD, DB,
for CD is equal to DG;
and HF is the square on CB;
therefore the rectangle CD, DB is equal to the square on CB.

Therefore, as DC is to CB, so is CB to BD.

But DC is greater than CB;
therefore CB is also greater than BD.

Therefore, when the straight line CD is cut in extreme and mean ratio, CB is the greater segment.

Therefore etc.

Q. E. D.

LEMMA.

That the double of AC is greater than BC is to be proved thus.

If not, let BC be, if possible, double of CA.

Therefore the square on BC is quadruple of the square on CA;
therefore the squares on BC, CA are five times the square on CA.

But, by hypothesis, the square on BA is also five times the square on CA;
therefore the square on BA is equal to the squares on BC, CA:
which is impossible. [II. 4]

Therefore CB is not double of AC.

Similarly we can prove that neither is a straight line less than CB double of CA;
for the absurdity is much greater.

Therefore the double of AC is greater than CB.

Q. E. D.

This proposition is the converse of Prop. 1. We have to prove that, if AB be so divided at C that

$$(\text{sq. on } AB) = 5\,(\text{sq. on } AC),$$

and if $CD = 2AC$,

then $$(\text{rect. } CD, DB) = (\text{sq. on } CB).$$

Subtract from each side the sq. on AC;

then (gnomon MNO) = 4 (sq. on AC)

= (sq. on CD).

Now, as in the last proposition,

$$\square CE = 2(\square BH)$$
$$= \square BH + \square HL.$$

Subtracting these equals from the equals, the square on CD and the gnomon MNO respectively, we have

$$\square BG = (\text{square } HF),$$

i.e. (rect. CD, DB) = (sq. on CB).

Here again the proposition can readily be proved by means of a figure similar to that of II. 11.

Draw CA through C *at right angles* to CB and of length equal to CA in the original figure; make CD double of CA; produce AC to R so that $CR = CB$.

Complete the squares on CB and CD, and join AD.

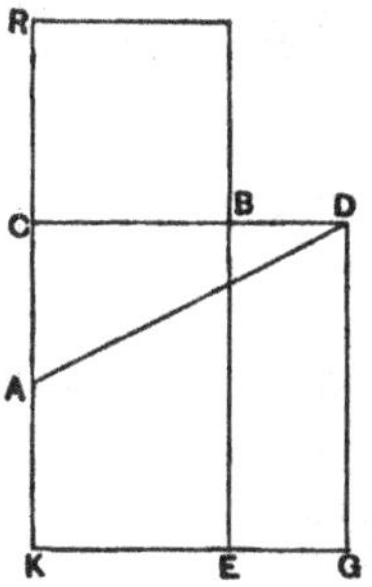

Now we are given the fact that

(sq. on AR) = 5 (sq. on CA).

But

5 (sq. on AC) = (sq. on AC) + (sq. on CD)

= (sq. on AD).

Therefore

(sq. on AR) = (sq. on AD),

or $AR = AD$.

Now

(rect. KR, RC) + (sq. on AC) = (sq. on AR)

= (sq. on AD)

= (sq. on AC) + (sq. on CD).

Therefore (rect. $KR . RC$) = (sq. on CD).

That is, (rectangle RE) = (square CG).

Subtract the common part CE,

and (rect. BG) = (sq. RB),

or rect. CD, DB = (sq. on CB).

Heiberg, with reason, doubts the genuineness of the Lemma following this proposition.

PROPOSITION 3.

If a straight line be cut in extreme and mean ratio, the square on the lesser segment added to the half of the greater segment is five times the square on the half of the greater segment.

For let any straight line *AB* be cut in extreme and mean ratio at the point *C*,

let *AC* be the greater segment, and let *AC* be bisected at *D*;

I say that the square on *BD* is five times the square on *DC*.

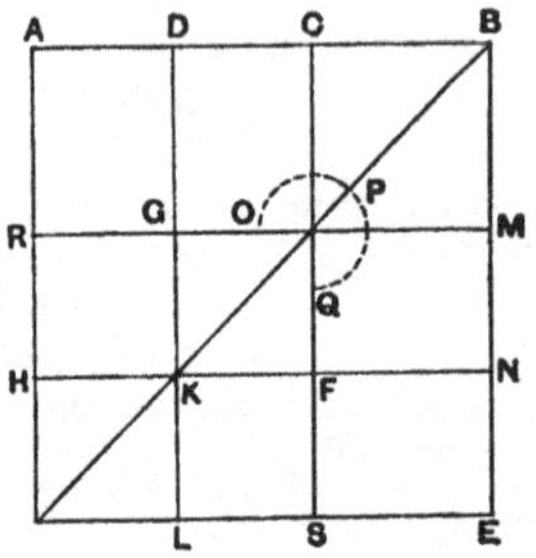

For let the square *AE* be described on *AB*,

and let the figure be drawn double.

Since *AC* is double of *DC*,

therefore the square on *AC* is quadruple of the square on *DC*,

that is, *RS* is quadruple of *FG*.

And, since the rectangle *AB*, *BC* is equal to the square on *AC*,

and *CE* is the rectangle *AB*, *BC*,

therefore *CE* is equal to *RS*.

But *RS* is quadruple of *FG*;

therefore *CE* is also quadruple of *FG*.

Again, since *AD* is equal to *DC*,

HK is also equal to *KF*.

Hence the square *GF* is also equal to the square *HL*.

Therefore *GK* is equal to *KL*, that is, *MN* to *NE*;

hence *MF* is also equal to *FE*.

But *MF* is equal to *CG*;

therefore *CG* is also equal to *FE*.

Let *CN* be added to each;

therefore the gnomon *OPQ* is equal to *CE*.

But *CE* was proved quadruple of *GF*;

therefore the gnomon *OPQ* is also quadruple of the square *FG*.

Therefore the gnomon *OPQ* and the square *FG* are five times *FG*.

But the gnomon *OPQ* and the square *FG* are the square *DN*.

And *DN* is the square on *DB*, and *GF* the square on *DC*.

Therefore the square on *DB* is five times the square on *DC*.

Q. E. D.

In this case we have

(sq. on BD) = (sq. FG) + (rect. CG) + (rect. CN)
= (sq. FG) + (rect. FE) + (rect. CN)
= (sq. FG) + (rect. CE)
= (sq. FG) + (rect. AB, BC)
= (sq. FG) + (sq. on AC), by hypothesis,
= 5 (sq. on DC).

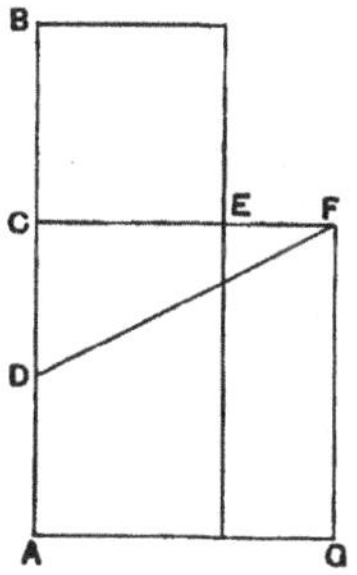

The theorem is still more obvious if the figure of II. 11 be used. Let CF be divided in extreme and mean ratio at E, by the method of II. 11.

Then, since

(rect. AB, BC) + (sq. on CD)
= sq. on BD
= sqs. on CD, CF,
(rect. AB, BC) = (sq. on CF)
= (sq. on CA),

and AB is divided at C in extreme and mean ratio.

And (sq. on BD) = (sq. on DF)
= 5 (sq. on CD).

Proposition 4.

If a straight line be cut in extreme and mean ratio, the square on the whole and the square on the lesser segment together are triple of the square on the greater segment.

Let AB be a straight line,
let it be cut in extreme and mean ratio at C,
and let AC be the greater segment;
I say that the squares on AB, BC are triple of the square on CA.

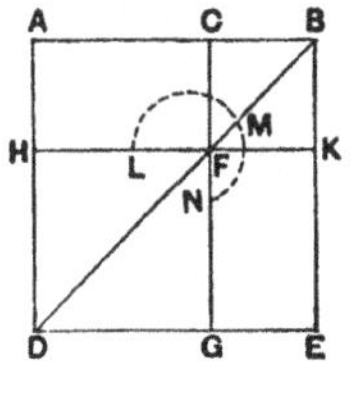

For let the square $ADEB$ be described on AB,
and let the figure be drawn.

Since then AB has been cut in extreme and mean ratio at C,
and AC is the greater segment,
therefore the rectangle AB, BC is equal to the square on AC. [VI. Def. 3, VI. 17]

And AK is the rectangle AB, BC, and HG the square on AC;
therefore AK is equal to HG.

And, since AF is equal to FE,
let CK be added to each;
therefore the whole AK is equal to the whole CE;
therefore AK, CE are double of AK.

But AK, CE are the gnomon LMN and the square CK;
therefore the gnomon LMN and the square CK are double of AK.

But, further, AK was also proved equal to HG;
therefore the gnomon LMN and the squares CK, HG are triple of the square HG.

And the gnomon LMN and the squares CK, HG are the whole square AE and CK, which are the squares on AB, BC,
while HG is the square on AC.

Therefore the squares on AB, BC are triple of the square on AC.

Q. E. D.

Here, as in the preceding propositions, the results are proved *de novo* by the method of Book II., without reference to that Book. Otherwise the proof might have been shorter.

For, by II. 7,

$$(\text{sq. on } AB) + (\text{sq. on } BC) = 2\,(\text{rect. } AB, BC) + (\text{sq. on } AC)$$
$$= 3\,(\text{sq. on } AC).$$

PROPOSITION 5.

If a straight line be cut in extreme and mean ratio, and there be added to it a straight line equal to the greater segment, the whole straight line has been cut in extreme and mean ratio, and the original straight line is the greater segment.

For let the straight line AB be cut in extreme and mean ratio at the point C,
let AC be the greater segment,
and let AD be equal to AC.

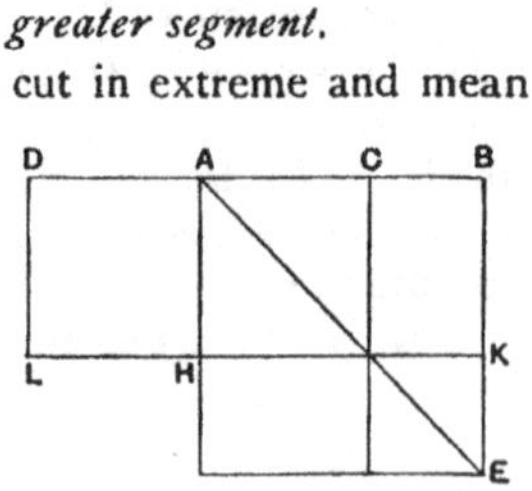

I say that the straight line DB has been cut in extreme and mean ratio at A, and the original straight line AB is the greater segment.

For let the square AE be described on AB, and let the figure be drawn.

Since AB has been cut in extreme and mean ratio at C, therefore the rectangle AB, BC is equal to the square on AC. [VI. Def. 3, VI. 17]

And CE is the rectangle AB, BC, and CH the square on AC;

therefore CE is equal to HC.

But HE is equal to CE,

and DH is equal to HC;

therefore DH is also equal to HE.

Therefore the whole DK is equal to the whole AE.

And DK is the rectangle BD, DA,

for AD is equal to DL;

and AE is the square on AB;

therefore the rectangle BD, DA is equal to the square on AB.

Therefore, as DB is to BA, so is BA to AD. [VI. 17]

And DB is greater than BA;

therefore BA is also greater than AD. [V. 14]

Therefore DB has been cut in extreme and mean ratio at A, and AB is the greater segment.

Q. E. D.

We have (sq. DH) = (sq. HC)
= (rect. CE), by hypothesis,
= (rect. HE).

Add to each side the rectangle AK, and
(rect. DK) = (sq. AE),

or (rect. BD, DA) = (sq. on AB).

The result is of course obvious from II. 11.

There is an alternative proof given in P after XIII. 6, which depends on Book V.

By hypothesis, $BA : AC = AC : CB$,

or, inversely, $AC : AB = CB : AC$.

Componendo, $(AB + AC) : AB = AB : AC$,

or $DB : BA = BA : AD$.

Proposition 6.

If a rational straight line be cut in extreme and mean ratio, each of the segments is the irrational straight line called apotome.

Let AB be a rational straight line, let it be cut in extreme and mean ratio at C,

and let AC be the greater segment;

I say that each of the straight lines AC, CB is the irrational straight line called apotome.

For let BA be produced, and let AD be made half of BA.

Since then the straight line AB has been cut in extreme and mean ratio,

and to the greater segment AC is added AD which is half of AB,

therefore the square on CD is five times the square on DA. [XIII. 1]

Therefore the square on CD has to the square on DA the ratio which a number has to a number;

therefore the square on CD is commensurable with the square on DA. [X. 6]

But the square on DA is rational,

for DA is rational, being half of AB which is rational;

therefore the square on CD is also rational; [X. Def. 4]

therefore CD is also rational.

And, since the square on CD has not to the square on DA the ratio which a square number has to a square number,

therefore CD is incommensurable in length with DA; [X. 9]

therefore CD, DA are rational straight lines commensurable in square only;

therefore AC is an apotome. [X. 73]

Again, since AB has been cut in extreme and mean ratio, and AC is the greater segment,

therefore the rectangle AB, BC is equal to the square on AC. [VI. Def. 3, VI. 17]

Therefore the square on the apotome AC, if applied to the rational straight line AB, produces BC as breadth.

But the square on an apotome, if applied to a rational straight line, produces as breadth a first apotome; [X. 97]

therefore CB is a first apotome.

And *CA* was also proved to be an apotome.
Therefore etc.

Q. E. D.

It seems certain that this proposition is an interpolation. P has it, but the copyist (or rather the copyist of its archetype) says that "this theorem is not found in most copies of the new recension, but is found in those of the old." In the first place, there is a scholium to XIII. 17 in P itself which proves the same thing as XIII. 6, and which would therefore have been useless if XIII. 6 had preceded. Hence, when the scholium was written, this proposition had not yet been interpolated. Secondly, P has it before the alternative proof of XIII. 5; this proof is considered, on general grounds, to be interpolated, and it would appear that it must have been a *later* interpolation (XIII. 6) which divorced it from the proposition to which it belonged. Thirdly, there is cause for suspicion in the proposition itself, for, while the enunciation states that each segment of the straight line is an *apotome*, the proposition adds that the lesser segment is a *first* apotome. The scholium in P referred to has not this blot. What is actually wanted in XIII. 17 is the fact that the *greater* segment is an apotome. It is probable that Euclid *assumed* this fact as evident enough from XIII. 1 without further proof, and that he neither wrote XIII. 6 nor the quotation of its enunciation in XIII. 17.

PROPOSITION 7.

If three angles of an equilateral pentagon, taken either in order or not in order, be equal, the pentagon will be equiangular.

For in the equilateral pentagon *ABCDE* let, first, three angles taken in order, those at *A*, *B*, *C*, be equal to one another;

I say that the pentagon *ABCDE* is equiangular.

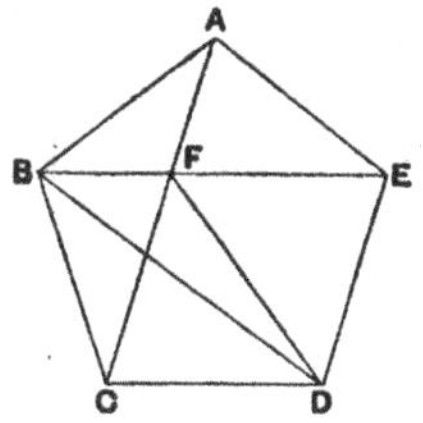

For let *AC*, *BE*, *FD* be joined.

Now, since the two sides *CB*, *BA* are equal to the two sides *BA*, *AE* respectively,

and the angle *CBA* is equal to the angle *BAE*,

therefore the base *AC* is equal to the base *BE*,

the triangle *ABC* is equal to the triangle *ABE*,

and the remaining angles will be equal to the remaining angles, namely those which the equal sides subtend, [I. 4]

that is, the angle *BCA* to the angle *BEA*, and the angle *ABE* to the angle *CAB*;

hence the side *AF* is also equal to the side *BF*. [I. 6]

But the whole *AC* was also proved equal to the whole *BE*;
therefore the remainder *FC* is also equal to the remainder *FE*.

But *CD* is also equal to *DE*.

Therefore the two sides *FC*, *CD* are equal to the two sides *FE*, *ED*;
and the base *FD* is common to them;
therefore the angle *FCD* is equal to the angle *FED*. [I. 8]

But the angle *BCA* was also proved equal to the angle *AEB*;
therefore the whole angle *BCD* is also equal to the whole angle *AED*.

But, by hypothesis, the angle *BCD* is equal to the angles at *A*, *B*;
therefore the angle *AED* is also equal to the angles at *A*, *B*.

Similarly we can prove that the angle *CDE* is also equal to the angles at *A*, *B*, *C*;
therefore the pentagon *ABCDE* is equiangular.

Next, let the given equal angles not be angles taken in order, but let the angles at the points *A*, *C*, *D* be equal;
I say that in this case too the pentagon *ABCDE* is equiangular.

For let *BD* be joined.

Then, since the two sides *BA*, *AE* are equal to the two sides *BC*, *CD*,
and they contain equal angles,
therefore the base *BE* is equal to the base *BD*,
the triangle *ABE* is equal to the triangle *BCD*,
and the remaining angles will be equal to the remaining angles, namely those which the equal sides subtend; [I. 4]
therefore the angle *AEB* is equal to the angle *CDB*.

But the angle *BED* is also equal to the angle *BDE*,
since the side *BE* is also equal to the side *BD*. [I. 5]

Therefore the whole angle *AED* is equal to the whole angle *CDE*.

But the angle *CDE* is, by hypothesis, equal to the angles at *A*, *C*;
therefore the angle *AED* is also equal to the angles at *A*, *C*.

For the same reason

the angle ABC is also equal to the angles at A, C, D.

Therefore the pentagon $ABCDE$ is equiangular.

Q. E. D.

This proposition is required in XIII. 17.

The steps of the proof may be shown thus.

I. Suppose that the angles at A, B, C are all equal.

Then the isosceles triangles BAE, ABC are equal in all respects;

thus $BE = AC$, $\angle BCA = \angle BEA$, $\angle CAB = \angle EBA$.

By the last equality, $FA = FB$,

so that, since $BE = AC$, $FC = FE$.

The $\triangle$s FED, FCD are now equal in all respects, [I. 8, 4]

and $\angle FCD = \angle FED$.

But $\angle ACB = \angle AEB$, from above,

whence, by addition, $\angle BCD = \angle AED$.

Similarly it may be proved that $\angle CDE$ is also equal to any one of the angles at A, B, C.

II. Suppose the angles at A, C, D to be equal.

Then the isosceles triangles ABE, CBD are equal in all respects, and hence $BE = BD$ (so that $\angle BDE = \angle BED$),

and $\angle CDB = \angle AEB$.

By addition of the equal angles,

$\angle CDE = \angle DEA$.

Similarly it may be proved that $\angle ABC$ is also equal to each of the angles at A, C, D.

PROPOSITION 8.

If in an equilateral and equiangular pentagon straight lines subtend two angles taken in order, they cut one another in extreme and mean ratio, and their greater segments are equal to the side of the pentagon.

For in the equilateral and equiangular pentagon $ABCDE$ let the straight lines AC, BE, cutting one another at the point H, subtend two angles taken in order, the angles at A, B;

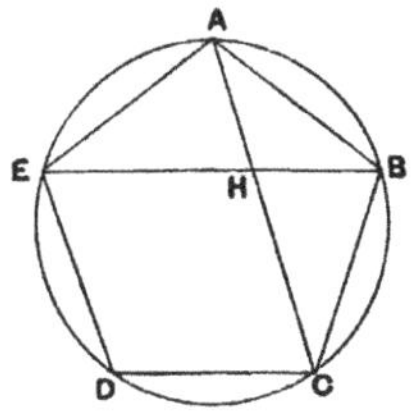

I say that each of them has been cut in extreme and mean ratio at the point H, and their greater segments are equal to the side of the pentagon.

For let the circle $ABCDE$ be circumscribed about the pentagon $ABCDE$. [IV. 14]

Then, since the two straight lines *EA*, *AB* are equal to the two *AB*, *BC*,
and they contain equal angles,
therefore the base *BE* is equal to the base *AC*,
the triangle *ABE* is equal to the triangle *ABC*,
and the remaining angles will be equal to the remaining angles respectively, namely those which the equal sides subtend. [I. 4]

Therefore the angle *BAC* is equal to the angle *ABE*;
therefore the angle *AHE* is double of the angle *BAH*. [I. 32]

But the angle *EAC* is also double of the angle *BAC*,
inasmuch as the circumference *EDC* is also double of the circumference *CB*; [III. 28, VI. 33]
therefore the angle *HAE* is equal to the angle *AHE*;
hence the straight line *HE* is also equal to *EA*, that is, to *AB*. [I. 6]

And, since the straight line *BA* is equal to *AE*,
the angle *ABE* is also equal to the angle *AEB*. [I. 5]

But the angle *ABE* was proved equal to the angle *BAH*;
therefore the angle *BEA* is also equal to the angle *BAH*.

And the angle *ABE* is common to the two triangles *ABE* and *ABH*;
therefore the remaining angle *BAE* is equal to the remaining angle *AHB*; [I. 32]
therefore the triangle *ABE* is equiangular with the triangle *ABH*;
therefore, proportionally, as *EB* is to *BA*, so is *AB* to *BH*. [VI. 4]

But *BA* is equal to *EH*;
therefore, as *BE* is to *EH*, so is *EH* to *HB*.

And *BE* is greater than *EH*;
therefore *EH* is also greater than *HB*. [V. 14]

Therefore *BE* has been cut in extreme and mean ratio at *H*, and the greater segment *HE* is equal to the side of the pentagon.

Similarly we can prove that *AC* has also been cut in extreme and mean ratio at *H*, and its greater segment *CH* is equal to the side of the pentagon.

Q. E. D.

In order to prove this theorem we have to show (1) that the △s AEB, HAB are similar, and (2) that $EH = EA$ (= AB).

To prove (2) we have

△s AEB, BAC equal in all respects,

whence $EB = AC$,

and ∠ BAC = ∠ ABE.

Therefore ∠ AHE = 2 ∠ BAC

= ∠ EAC,

so that $EH = EA$

$= AB$.

To prove (1) we have, in the △s AEB, HAB,

∠ BAH = ∠ EBA

= ∠ AEB,

and ∠ ABE is common;

therefore the third ∠s AHB, EAB are equal,

and △s AEB, HAB are similar.

Now, since these triangles are similar,

$EB : BA = BA : BH$,

or (rect. EB, BH) = (sq. on BA)

= (sq. on EH),

so that EB is divided in extreme and mean ratio at H.

Similarly its equal, CA, is divided in extreme and mean ratio at H.

Proposition 9.

If the side of the hexagon and that of the decagon inscribed in the same circle be added together, the whole straight line has been cut in extreme and mean ratio, and its greater segment is the side of the hexagon.

Let ABC be a circle;

of the figures inscribed in the circle ABC let BC be the side of a decagon, CD that of a hexagon,

and let them be in a straight line;

I say that the whole straight line BD has been cut in extreme and mean ratio, and CD is its greater segment.

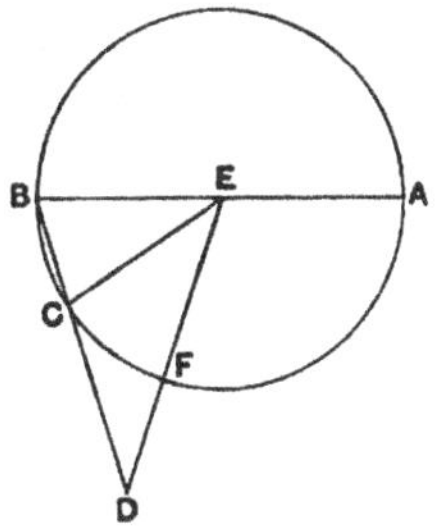

For let the centre of the circle, the point E, be taken,

let EB, EC, ED be joined,

and let BE be carried through to A.

Since BC is the side of an equilateral decagon,

therefore the circumference *ACB* is five times the circumference *BC*;
therefore the circumference *AC* is quadruple of *CB*.

But, as the circumference *AC* is to *CB*, so is the angle *AEC* to the angle *CEB*; [VI. 33]
therefore the angle *AEC* is quadruple of the angle *CEB*.

And, since the angle *EBC* is equal to the angle *ECB*, [I. 5]
therefore the angle *AEC* is double of the angle *ECB*. [I. 32]

And, since the straight line *EC* is equal to *CD*,
for each of them is equal to the side of the hexagon inscribed in the circle *ABC*, [IV. 15, Por.]
the angle *CED* is also equal to the angle *CDE*; [I. 5]
therefore the angle *ECB* is double of the angle *EDC*. [I. 32]

But the angle *AEC* was proved double of the angle *ECB*;
therefore the angle *AEC* is quadruple of the angle *EDC*.

But the angle *AEC* was also proved quadruple of the angle *BEC*;
therefore the angle *EDC* is equal to the angle *BEC*.

But the angle *EBD* is common to the two triangles *BEC* and *BED*;
therefore the remaining angle *BED* is also equal to the remaining angle *ECB*; [I. 32]
therefore the triangle *EBD* is equiangular with the triangle *EBC*.

Therefore, proportionally, as *DB* is to *BE*, so is *EB* to *BC*. [VI. 4]

But *EB* is equal to *CD*.

Therefore, as *BD* is to *DC*, so is *DC* to *CB*.

And *BD* is greater than *DC*;
therefore *DC* is also greater than *CB*.

Therefore the straight line *BD* has been cut in extreme and mean ratio, and *DC* is its greater segment.

Q. E. D.

BC is the side of a regular decagon inscribed in the circle; *CD* is the side of the inscribed regular hexagon, and is therefore equal to the radius *BE* or *EC*.

Therefore, in order to prove our theorem, we have only to show that ***the triangles* EBC, DBE *are similar.***

Since BC is the side of a regular decagon,

$$(\text{arc } BCA) = 5\,(\text{arc } BC),$$

so that $$(\text{arc } CFA) = 4\,(\text{arc } BC),$$

whence $$\angle CEA = 4 \angle BEC.$$

But $$\angle CEA = 2 \angle ECB.$$

Therefore $$\angle ECB = 2 \angle BEC \quad \ldots\ldots\ldots\ldots\ldots\ldots\ldots\ldots(1).$$

But, since $CD = CE$,

$$\angle CDE = \angle CED,$$

so that $$\angle ECB = 2 \angle CDE.$$

It follows from (1) that $\angle BEC = \angle CDE$.

Now, in the $\triangle$s EBC, DBE,

$$\angle BEC = \angle BDE,$$

and $\angle EBC$ is common,

so that $$\angle ECB = \angle DEB,$$

and $\triangle$s EBC, DBE are similar.

Hence $$DB : BE = EB : BC,$$

or $$(\text{rect. } DB, BC) = (\text{sq. on } EB)$$
$$= (\text{sq. on } CD),$$

and DB is divided at C in extreme and mean ratio.

To find the side of the decagon algebraically in terms of the radius we have, if x be the side required,

$$(r + x)x = r^2,$$

whence $$x = \frac{r}{2}(\sqrt{5} - 1).$$

Proposition 10.

If an equilateral pentagon be inscribed in a circle, the square on the side of the pentagon is equal to the squares on the side of the hexagon and on that of the decagon inscribed in the same circle.

Let $ABCDE$ be a circle,

and let the equilateral pentagon $ABCDE$ be inscribed in the circle $ABCDE$.

I say that the square on the side of the pentagon $ABCDE$ is equal to the squares on the side of the hexagon and on that of the decagon inscribed in the circle $ABCDE$.

For let the centre of the circle, the point F, be taken,

let AF be joined and carried through to the point G,

let FB be joined,

let FH be drawn from F perpendicular to AB and be carried through to K, let *AK*, *KB* be joined,

let *FL* be again drawn from *F* perpendicular to *AK*, and be carried through to *M*,

and let *KN* be joined.

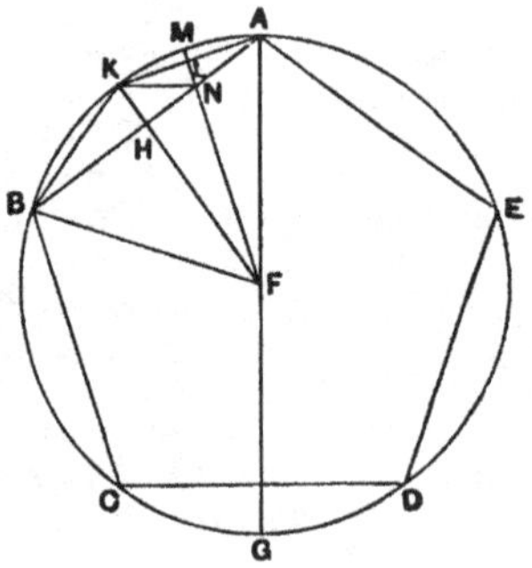

Since the circumference *ABCG* is equal to the circumference *AEDG*,

and in them *ABC* is equal to *AED*,

therefore the remainder, the circumference *CG*, is equal to the remainder *GD*.

But *CD* belongs to a pentagon;

therefore *CG* belongs to a decagon.

And, since *FA* is equal to *FB*,

and *FH* is perpendicular,

therefore the angle *AFK* is also equal to the angle *KFB*. [I. 5, I. 26]

Hence the circumference *AK* is also equal to *KB*; [III. 26]

therefore the circumference *AB* is double of the circumference *BK*;

therefore the straight line *AK* is a side of a decagon.

For the same reason

AK is also double of *KM*.

Now, since the circumference *AB* is double of the circumference *BK*,

while the circumference *CD* is equal to the circumference *AB*,

therefore the circumference *CD* is also double of the circumference *BK*.

But the circumference *CD* is also double of *CG*;

therefore the circumference *CG* is equal to the circumference *BK*.

But *BK* is double of *KM*, since *KA* is so also;

therefore *CG* is also double of *KM*.

But, further, the circumference CB is also double of the circumference BK,

for the circumference CB is equal to BA.

Therefore the whole circumference GB is also double of BM;

hence the angle GFB is also double of the angle BFM. [VI. 33]

But the angle GFB is also double of the angle FAB,

for the angle FAB is equal to the angle ABF.

Therefore the angle BFN is also equal to the angle FAB.

But the angle ABF is common to the two triangles ABF and BFN;

therefore the remaining angle AFB is equal to the remaining angle BNF; [I. 32]

therefore the triangle ABF is equiangular with the triangle BFN.

Therefore, proportionally, as the straight line AB is to BF, so is FB to BN; [VI. 4]

therefore the rectangle AB, BN is equal to the square on BF. [VI. 17]

Again, since AL is equal to LK,

while LN is common and at right angles,

therefore the base KN is equal to the base AN; [I. 4]

therefore the angle LKN is also equal to the angle LAN.

But the angle LAN is equal to the angle KBN;

therefore the angle LKN is also equal to the angle KBN.

And the angle at A is common to the two triangles AKB and AKN.

Therefore the remaining angle AKB is equal to the remaining angle KNA; [I. 32]

therefore the triangle KBA is equiangular with the triangle KNA.

Therefore, proportionally, as the straight line BA is to AK, so is KA to AN; [VI. 4]

therefore the rectangle BA, AN is equal to the square on AK. [VI. 17]

But the rectangle AB, BN was also proved equal to the square on BF;

therefore the rectangle AB, BN together with the rectangle BA, AN, that is, the square on BA [II. 2], is equal to the square on BF together with the square on AK.

And BA is a side of the pentagon, BF of the hexagon [IV. 15, Por.], and AK of the decagon.

Therefore etc.

Q. E. D.

$ABCDE$ being a regular pentagon inscribed in a circle, and AG the diameter through A, it follows that

$$(\text{arc } CG) = (\text{arc } GD),$$

and CG, GD are sides of an inscribed regular decagon.

FHK being drawn perpendicular to AB, it follows, by I. 26, that $\angle$s AFK, BFK are equal, and BK, KA are sides of the regular decagon.

Similarly it may be proved that, FLM being perpendicular to AK,

$$AL = LK,$$

and $$(\text{arc } AM) = (\text{arc } MK).$$

The main facts to prove are that

(1) the triangles ABF, FBN are similar, and (2) the triangles ABK, AKN are similar.

(1) $$2\,(\text{arc } CG) = (\text{arc } CD)$$
$$= (\text{arc } AB)$$
$$= 2\,(\text{arc } BK),$$

or $$(\text{arc } CG) = (\text{arc } BK) = (\text{arc } AK)$$
$$= 2\,(\text{arc } KM).$$

And $$(\text{arc } CB) = 2\,(\text{arc } BK).$$

Therefore, by addition,

$$(\text{arc } BCG) = 2\,(\text{arc } BKM).$$

Therefore $$\angle BFG = 2\angle BFN.$$

But $$\angle BFG = 2\angle FAB,$$

so that $$\angle FAB = \angle BFN.$$

Hence, in the $\triangle$s ABF, FBN,

$$\angle FAB = \angle BFN,$$

and $\angle ABF$ is common;

therefore $$\angle AFB = \angle BNF,$$

and $\triangle$s ABF, FBN are similar.

(2) Since $AL = LK$, and the angles at L are right,

$$AN = NK,$$

and $$\angle NKA = \angle NAK$$
$$= \angle KBA.$$

Hence, in the $\triangle$s ABK, AKN,

$$\angle ABK = \angle AKN,$$

and $\angle KAN$ is common,

whence the third angles are equal;

therefore the triangles ABK, AKN are similar.

Now from the similarity of △s ABF, FBN it follows that

$$AB : BF = BF : BN,$$

or $$(\text{rect. } AB, BN) = (\text{sq. on } BF).$$

And, from the similarity of ABK, AKN,

$$BA : AK = AK : AN,$$

or $$(\text{rect. } BA, AN) = (\text{sq. on } AK).$$

Therefore, by addition,

$$(\text{rect. } AB, BN) + (\text{rect. } BA, AN) = (\text{sq. on } BF) + (\text{sq. on } AK),$$

that is, $$(\text{sq. on } AB) = (\text{sq. on } BF) + (\text{sq. on } AK).$$

If r be the radius of the circle, we have seen (XIII. 9, note) that

$$AK = \frac{r}{2}(\sqrt{5} - 1).$$

Therefore $$(\text{side of pentagon})^2 = r^2 + \frac{r^2}{4}(6 - 2\sqrt{5})$$

$$= \frac{r^2}{4}(10 - 2\sqrt{5}),$$

so that $$(\text{side of pentagon}) = \frac{r}{2}\sqrt{10 - 2\sqrt{5}}.$$

PROPOSITION 11.

If in a circle which has its diameter rational an equilateral pentagon be inscribed, the side of the pentagon is the irrational straight line called minor.

For in the circle $ABCDE$ which has its diameter rational let the equilateral pentagon $ABCDE$ be inscribed;

I say that the side of the pentagon is the irrational straight line called minor.

For let the centre of the circle, the point F, be taken,

let AF, FB be joined and carried through to the points, G, H,

let AC be joined,

and let FK be made a fourth part of AF.

Now AF is rational;

therefore FK is also rational.

But BF is also rational;

therefore the whole BK is rational.

And, since the circumference ACG is equal to the circumference ADG,

and in them ABC is equal to AED,

therefore the remainder CG is equal to the remainder GD.

And, if we join AD, we conclude that the angles at L are right,
and CD is double of CL.

For the same reason
the angles at M are also right,
and AC is double of CM.

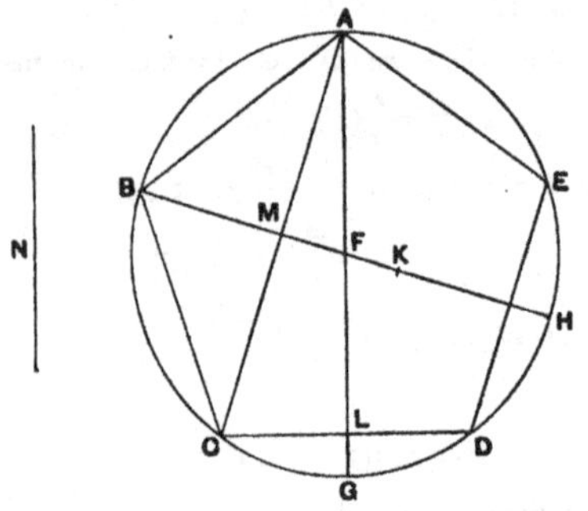

Since then the angle ALC is equal to the angle AMF,
and the angle LAC is common to the two triangles ACL and AMF,
therefore the remaining angle ACL is equal to the remaining angle MFA; [I. 32]
therefore the triangle ACL is equiangular with the triangle AMF;
therefore, proportionally, as LC is to CA, so is MF to FA.

And the doubles of the antecedents may be taken;
therefore, as the double of LC is to CA, so is the double of MF to FA.

But, as the double of MF is to FA, so is MF to the half of FA;
therefore also, as the double of LC is to CA, so is MF to the half of FA.

And the halves of the consequents may be taken;
therefore, as the double of LC is to the half of CA, so is MF to the fourth of FA.

And DC is double of LC, CM is half of CA, and FK a fourth part of FA;

therefore, as DC is to CM, so is MF to FK.

Componendo also, as the sum of DC, CM is to CM, so is MK to KF; [v. 18]

therefore also, as the square on the sum of DC, CM is to the square on CM, so is the square on MK to the square on KF.

And since, when the straight line subtending two sides of the pentagon, as AC, is cut in extreme and mean ratio, the greater segment is equal to the side of the pentagon, that is, to DC, [XIII. 8]

while the square on the greater segment added to the half of the whole is five times the square on the half of the whole, [XIII. 1]

and CM is half of the whole AC,

therefore the square on DC, CM taken as one straight line is five times the square on CM.

But it was proved that, as the square on DC, CM taken as one straight line is to the square on CM, so is the square on MK to the square on KF;

therefore the square on MK is five times the square on KF.

But the square on KF is rational,

for the diameter is rational;

therefore the square on MK is also rational;

therefore MK is rational.

And, since BF is quadruple of FK,

therefore BK is five times KF;

therefore the square on BK is twenty-five times the square on KF.

But the square on MK is five times the square on KF;

therefore the square on BK is five times the square on KM;

therefore the square on BK has not to the square on KM the ratio which a square number has to a square number;

therefore BK is incommensurable in length with KM. [x. 9]

And each of them is rational.

Therefore BK, KM are rational straight lines commensurable in square only.

But, if from a rational straight line there be subtracted a rational straight line which is commensurable with the whole in square only, the remainder is irrational, namely an apotome; therefore MB is an apotome and MK the annex to it. [X. 73]

I say next that MB is also a fourth apotome.

Let the square on N be equal to that by which the square on BK is greater than the square on KM; therefore the square on BK is greater than the square on KM by the square on N.

And, since KF is commensurable with FB, *componendo* also, KB is commensurable with FB. [X. 15]

But BF is commensurable with BH; therefore BK is also commensurable with BH. [X. 12]

And, since the square on BK is five times the square on KM, therefore the square on BK has to the square on KM the ratio which 5 has to 1.

Therefore, *convertendo*, the square on BK has to the square on N the ratio which 5 has to 4 [V. 19, Por.], and this is not the ratio which a square number has to a square number; therefore BK is incommensurable with N; [X. 9] therefore the square on BK is greater than the square on KM by the square on a straight line incommensurable with BK.

Since then the square on the whole BK is greater than the square on the annex KM by the square on a straight line incommensurable with BK, and the whole BK is commensurable with the rational straight line, BH, set out, therefore MB is a fourth apotome. [X. Deff. III. 4]

But the rectangle contained by a rational straight line and a fourth apotome is irrational, and its square root is irrational, and is called minor. [X. 94]

But the square on AB is equal to the rectangle HB, BM, because, when AH is joined, the triangle ABH is equiangular with the triangle ABM, and, as HB is to BA, so is AB to BM.

Therefore the side AB of the pentagon is the irrational straight line called minor.

Q. E. D.

Here we require certain definitions and propositions of Book X.

First we require the definition of an *apotome* [see X. 73], which is a straight line of the form $(\rho \sim \sqrt{k} \cdot \rho)$, where ρ is a "rational" straight line and k is any integer or numerical fraction, the square root of which is not integral or expressible in integers. The lesser of the straight lines ρ, $\sqrt{k} \cdot \rho$ is the *annex*.

Next we require the definition of the *fourth apotome* [X. Deff. III. (after X. 84)], which is a straight line of the form $(x - y)$, where x, y (being both rational and commensurable in square only) are also such that $\sqrt{x^2 - y^2}$ is incommensurable with x, while x is commensurable with a given rational straight line ρ. As shown on X. 88 (note), the *fourth apotome* is of the form

$$\left(k\rho - \frac{k\rho}{\sqrt{1 + \lambda}}\right).$$

Lastly the *minor* (straight line) is the irrational straight line defined in X. 76. It is of the form $(x - y)$, where x, y are incommensurable in square, and $(x^2 + y^2)$ is 'rational,' while xy is 'medial.' As shown in the note on X. 76, the *minor* irrational straight line is of the form

$$\frac{\rho}{\sqrt{2}}\sqrt{1 + \frac{k}{\sqrt{1 + k^2}}} - \frac{\rho}{\sqrt{2}}\sqrt{1 - \frac{k}{\sqrt{1 + k^2}}}.$$

The proposition may be put as follows. $ABCDE$ being a regular pentagon inscribed in a circle, AG, BH the diameters through A, B meeting CD in L and AC in M respectively, FK is made equal to $\frac{1}{4}AF$.

Now, the radius $AF(r)$ being rational, so are FK, BK.

The arcs CG, GD are equal;

hence $\angle$s at L are right, and $CD = 2CL$.

Similarly $\angle$s at M are right, and $AC = 2CM$.

We have to prove

(1) that BM is an apotome,

(2) that BM is a fourth apotome,

(3) that BA is a minor irrational straight line.

Remembering that, if CA is divided in extreme and mean ratio, the greater segment is equal to the side of the pentagon [XIII. 8], and that accordingly [XIII. 1] $(CD + \frac{1}{2}CA)^2 = 5(\frac{1}{2}CA)^2$, we work towards a proportion containing the ratio $(CD + CM)^2 : CM^2$, thus.

The $\triangle$s ACL, AFM are equiangular and therefore similar.

Therefore $\quad LC : CA = MF : FA$,

and accordingly $\quad 2LC : CA = MF : \frac{1}{2}FA$;

thus $\quad 2LC : \frac{1}{2}CA = MF : \frac{1}{4}FA$,

or $\quad DC : CM = MF : FK$;

whence, *componendo*, and squaring,

$$(DC + CM)^2 : CM^2 = MK^2 : KF^2.$$

But $\quad (DC + CM)^2 = 5CM^2$;

therefore $\quad MK^2 = 5KF^2$.

[This means that $$MK^2 = \tfrac{5}{16} r^2,$$

or $$MK = \frac{\sqrt{5}}{4} r.]$$

It follows that, KF being rational, MK^2, and therefore MK, is rational.

(1) To prove that BM is an *apotome* and MK its *annex*.

We have $$BF = 4FK;$$

therefore $$BK = 5FK,$$
$$BK^2 = 25FK^2$$
$$= 5MK^2, \text{ from above;}$$

therefore BK^2 has not to MK^2 the ratio of a square number to a square number;

therefore BK, MK are incommensurable in length.

They are therefore rational and commensurable in square only;

accordingly BM is an *apotome*.

[$BK^2 = 5MK^2 = \frac{25}{16} r^2$, and $BK = \frac{5}{4} r$.

Consequently $$BK - MK = \left(\frac{5}{4} r - \frac{\sqrt{5}}{4} r\right).]$$

(2) To prove that BM is a *fourth apotome*.

First, since KF, FB are commensurable,

BK, BF are commensurable, i.e. BK is commensurable with BH, a given rational straight line.

Secondly, if $$N^2 = BK^2 - KM^2,$$

since $$BK^2 : KM^2 = 5 : 1,$$

it follows that $$BK^2 : N^2 = 5 : 4,$$

whence BK, N are incommensurable.

Therefore BM is a *fourth apotome*.

(3) To prove that BA is a *minor* irrational straight line.

If a fourth apotome form a rectangle with a rational straight line, the side of the square equivalent to the rectangle is *minor* [x. 94].

Now $$BA^2 = HB \,.\, BM,$$

HB is rational, and BM is a fourth apotome;

therefore BA is a *minor* irrational straight line.

$$[BA = r\sqrt{2} \,.\, \sqrt{\frac{5}{4} - \frac{\sqrt{5}}{4}} = \frac{r}{2}\sqrt{10 - 2\sqrt{5}}.$$

If this is separated into the difference between two straight lines, we have

$$BA = \frac{r}{2}\sqrt{5 + 2\sqrt{5}} - \frac{r}{2}\sqrt{5 - 2\sqrt{5}}.]$$

PROPOSITION 12.

If an equilateral triangle be inscribed in a circle, the square on the side of the triangle is triple of the square on the radius of the circle.

Let ABC be a circle,
and let the equilateral triangle ABC be inscribed in it;

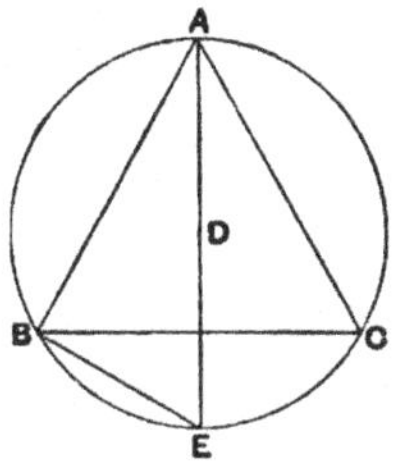

I say that the square on one side of the triangle ABC is triple of the square on the radius of the circle.

For let the centre D of the circle ABC be taken,
let AD be joined and carried through to E,
and let BE be joined.

Then, since the triangle ABC is equilateral,
therefore the circumference BEC is a third part of the circumference of the circle ABC.

Therefore the circumference BE is a sixth part of the circumference of the circle;
therefore the straight line BE belongs to a hexagon;
therefore it is equal to the radius DE. [IV. 15, Por.]

And, since AE is double of DE,
the square on AE is quadruple of the square on ED, that is, of the square on BE.

But the square on AE is equal to the squares on AB, BE; [III. 31, I. 47]
therefore the squares on AB, BE are quadruple of the square on BE.

Therefore, *separando*, the square on AB is triple of the square on BE.

But BE is equal to DE;
therefore the square on AB is triple of the square on DE.

Therefore the square on the side of the triangle is triple of the square on the radius.

Q. E. D.

Proposition 13.

To construct a pyramid, to comprehend it in a given sphere, and to prove that the square on the diameter of the sphere is one and a half times the square on the side of the pyramid.

Let the diameter AB of the given sphere be set out,
and let it be cut at the point C so that AC is double of CB;
let the semicircle ADB be described on AB,
let CD be drawn from the point C at right angles to AB,
and let DA be joined;
let the circle EFG which has its radius equal to DC be set out,
let the equilateral triangle EFG be inscribed in the circle EFG, [IV. 2]
let the centre of the circle, the point H, be taken, [III. 1]
let EH, HF, HG be joined;
from the point H let HK be set up at right angles to the plane of the circle EFG, [XI. 12]
let HK equal to the straight line AC be cut off from HK,
and let KE, KF, KG be joined.

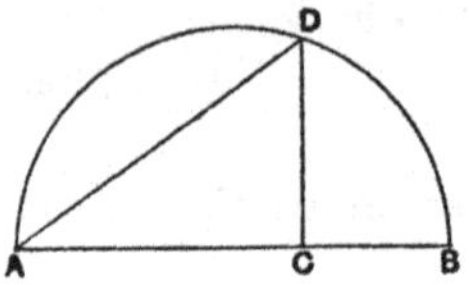

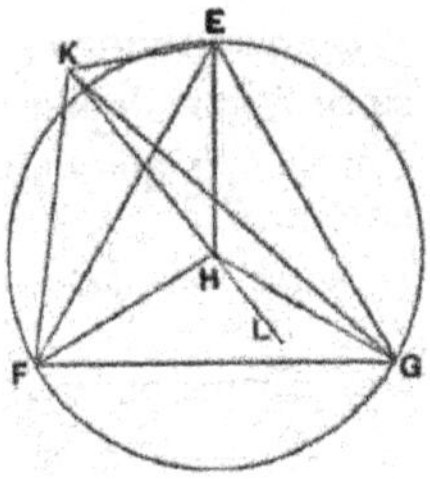

Now, since KH is at right angles to the plane of the circle EFG,
therefore it will also make right angles with all the straight lines which meet it and are in the plane of the circle EFG. [XI. Def. 3]

But each of the straight lines HE, HF, HG meets it:
therefore HK is at right angles to each of the straight lines HE, HF, HG.

And, since AC is equal to HK, and CD to HE,
and they contain right angles,
therefore the base DA is equal to the base KE. [I. 4]

For the same reason

each of the straight lines *KF*, *KG* is also equal to *DA*;

therefore the three straight lines *KE*, *KF*, *KG* are equal to one another.

And, since *AC* is double of *CB*,

therefore *AB* is triple of *BC*.

But, as *AB* is to *BC*, so is the square on *AD* to the square on *DC*, as will be proved afterwards.

Therefore the square on *AD* is triple of the square on *DC*.

But the square on *FE* is also triple of the square on *EH*, [XIII. 12]

and *DC* is equal to *EH*;

therefore *DA* is also equal to *EF*.

But *DA* was proved equal to each of the straight lines *KE*, *KF*, *KG*;

therefore each of the straight lines *EF*, *FG*, *GE* is also equal to each of the straight lines *KE*, *KF*, *KG*;

therefore the four triangles *EFG*, *KEF*, *KFG*, *KEG* are equilateral.

Therefore a pyramid has been constructed out of four equilateral triangles, the triangle *EFG* being its base and the point *K* its vertex.

It is next required to comprehend it in the given sphere and to prove that the square on the diameter of the sphere is one and a half times the square on the side of the pyramid.

For let the straight line *HL* be produced in a straight line with *KH*,

and let *HL* be made equal to *CB*.

Now, since, as *AC* is to *CD*, so is *CD* to *CB*, [VI. 8, Por.]

while *AC* is equal to *KH*, *CD* to *HE*, and *CB* to *HL*,

therefore, as *KH* is to *HE*, so is *EH* to *HL*;

therefore the rectangle *KH*, *HL* is equal to the square on *EH*. [VI. 17]

And each of the angles *KHE*, *EHL* is right;

therefore the semicircle described on *KL* will pass through *E* also. [cf. VI. 8, III. 31]

If then, KL remaining fixed, the semicircle be carried round and restored to the same position from which it began to be moved, it will also pass through the points F, G,

since, if FL, LG be joined, the angles at F, G similarly become right angles;

and the pyramid will be comprehended in the given sphere.

For KL, the diameter of the sphere, is equal to the diameter AB of the given sphere, inasmuch as KH was made equal to AC, and HL to CB.

I say next that the square on the diameter of the sphere is one and a half times the square on the side of the pyramid.

For, since AC is double of CB,

therefore AB is triple of BC;

and, *convertendo*, BA is one and a half times AC.

But, as BA is to AC, so is the square on BA to the square on AD.

Therefore the square on BA is also one and a half times the square on AD.

And BA is the diameter of the given sphere, and AD is equal to the side of the pyramid.

Therefore the square on the diameter of the sphere is one and a half times the square on the side of the pyramid.

Q. E. D.

LEMMA.

It is to be proved that, as AB is to BC, so is the square on AD to the square on DC.

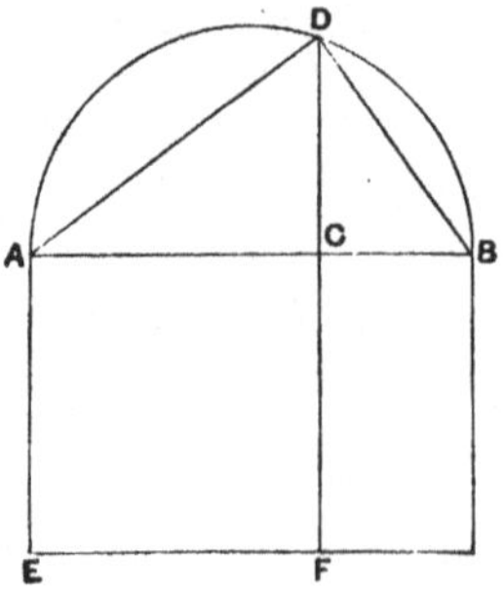

For let the figure of the semicircle be set out,

let DB be joined,

let the square EC be described on AC,

and let the parallelogram FB be completed.

Since then, because the triangle DAB is equiangular with the triangle DAC,

as BA is to AD, so is DA to AC, [VI. 8, VI. 4]

therefore the rectangle BA, AC is equal to the square on AD. [VI. 17]

And since, as AB is to BC, so is EB to BF, [VI. 1]
and EB is the rectangle BA, AC, for EA is equal to AC,
and BF is the rectangle AC, CB,
therefore, as AB is to BC, so is the rectangle BA, AC to the rectangle AC, CB.

And the rectangle BA, AC is equal to the square on AD, and the rectangle AC, CB to the square on DC,
for the perpendicular DC is a mean proportional between the segments AC, CB of the base, because the angle ADB is right. [VI. 8, Por.]

Therefore, as AB is to BC, so is the square on AD to the square on DC.

Q. E. D.

The Lemma is with reason suspected. Euclid commonly takes more difficult theorems for granted in the stereometrical Books. It is also clumsy in itself, while, from a gloss in the proposition rejected as an interpolation, it is clear that the interpolator of the gloss had not the Lemma. With the Lemma should disappear the words "as will be proved afterwards" (p. 469).

In the figure of the proposition, the semicircle really represents half of a section of the sphere through its centre and one edge of the inscribed tetrahedron (AD being the length of that edge).

The proof is in three parts, the object of which is to prove

(1) that $KEFG$ is a tetrahedron with all its edges equal to AD,

(2) that it is inscribable in a sphere of diameter equal to AB,

(3) that $$AB^2 = \tfrac{3}{2}AD^2.$$

To prove (1) we have to show

(*a*) that $$KE = KF = KG = AD,$$

(*b*) that $$AD = EF.$$

(*a*) Since $$HE = HF = HG = CD,$$
$$KH = AC,$$
and $$\angle\text{s } ACD, KHE, KHF, KHG \text{ are right},$$
$$\triangle\text{s } ACD, KHE, KHF, KHG \text{ are equal in all respects};$$
therefore $$KE = KF = KG = AD.$$

(*b*) Since $$AB = 3BC,$$
and $$AB : BC = AB.AC : AC.CB$$
$$= AD^2 : CD^2,$$
it follows that $$AD^2 = 3CD^2.$$

But [XIII. 12] $$EF^2 = 3EH^2;$$
and $EH = CD$, by construction.

Therefore $AD = EF$.

Thus $EFGK$ is a *regular* tetrahedron.

(2) We now observe the usefulness of Euclid's description of a sphere [in XI. Def. 14].

Producing $KH (= AC)$ to L so that $HL = CB$,

we have KL equal to AB;

thus KL is a diameter of the sphere which should circumscribe our tetrahedron,

and we have only to prove that E, F, G lie on semicircles described on KL as diameter.

E.g. for the point E,

since $$AC : CD = CD : CB,$$

while $$AC = KH,\ CD = HE,\ CB = HL,$$

we have $$KH : HE = HE : HL,$$

or $$KH . HL = HE^2,$$

whence, the angles KHE, EHL being right,

EKL is a triangle right-angled at E [cf. VI. 8].

Hence E lies on a semicircle on KL as diameter.

Similarly for F, G.

Thus a semicircle on KL as diameter revolving round KL passes successively through E, F, G.

(3) $$AB = 3BC;$$

therefore $$BA = \tfrac{3}{2} AC.$$

And $$BA : AC = BA^2 : BA . AC$$
$$= BA^2 : AD^2.$$

Therefore $$BA^2 = \tfrac{3}{2} AD^2.$$

If r be the radius of the circumscribed sphere,

$$(\text{edge of tetrahedron}) = \frac{2\sqrt{2}}{\sqrt{3}} . r = \tfrac{2}{3}\sqrt{6} . r.$$

It will be observed that, although in these cases Euclid's construction is equivalent to inscribing the particular regular solid in a given sphere, he does not actually construct the solid *in* the sphere but constructs a solid which a sphere *equal* to the given sphere will circumscribe. Pappus, on the other hand, in dealing with the same problems, actually constructs the respective solids in the given spheres. His method is to find circular sections in the given spheres containing a certain number of the angular points of the given solids. His solutions are interesting, although they require a knowledge of some properties of a sphere which are of course not found in the *Elements* but belonged to treatises such as the *Sphaerica* of Theodosius.

Pappus' solution of the problem of Eucl. XIII. 13.

In order to inscribe a regular pyramid or tetrahedron in a given sphere, Pappus (III. pp. 142—144) finds two circular sections equal and parallel to one another, each of which contains one of two opposite edges as its diameter. In this and the other similar problems he proceeds in the orthodox manner by analysis and synthesis. The following is a reproduction of his solution of this case.

Analysis.

Suppose the problem solved, A, B, C, D being the angular points of the required pyramid.

Through A draw EF parallel to CD; this will make equal angles with AC, AD; and, since AB does so too, EF is perpendicular to AB [Pappus has a lemma for this, p. 140, 12—24], and is therefore a tangent to the sphere (for EF is parallel to CD, the base of the triangle ACD, and therefore touches the circle circumscribing it, while it also touches the circular section AB made by the plane passing through AB and EF perpendicular to it).

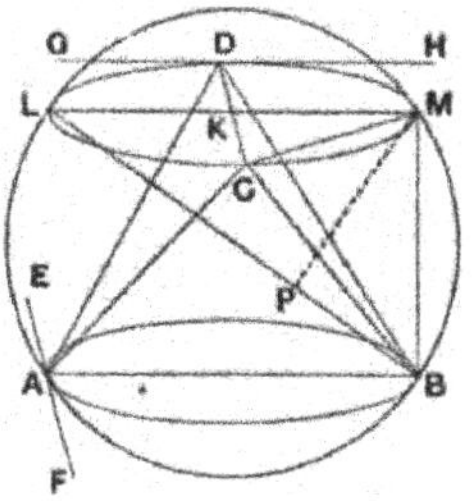

Similarly GH drawn through D parallel to AB touches the sphere.

And the plane through GH, CD makes a circular section equal and parallel to AB.

Through the centre K of that circular section, and in the plane of the section, draw LM perpendicular to CD and therefore parallel to AB. Join BL, BM.

BM is then perpendicular to AB, LM, and LB is a diameter of the sphere.

Join MC.

Then $$LM^2 = 2MC^2,$$
and $$BC = AB = LM,$$
so that $$BC^2 = 2MC^2.$$

And BM, being perpendicular to the plane of the circle LM, is perpendicular to CM,

whence $$BC^2 = BM^2 + MC^2,$$
so that $$BM = MC.$$
But $$BC = LM;$$
therefore $$LM^2 = 2BM^2.$$

And, since the angle LMB is right,
$$BL^2 = LM^2 + MB^2 = \tfrac{3}{2} LM^2.$$

Synthesis.

Draw two parallel circular sections of the sphere with diameter d', such that
$$d'^2 = \tfrac{2}{3} d^2,$$
where d is the diameter of the sphere.

[This is easily done by dividing BL, any diameter of the sphere, at P, so that $LP = 2PB$, and then drawing PM at right angles to LB meeting the great circle LMB of the sphere in M. Then $LM^2 : LB^2 = LP : LB = 2 : 3$.]

Draw sections through M, B perpendicular to MB, and in these sections respectively draw the parallel diameters LM, AB.

Lastly, in the section LM draw CD through the centre K perpendicular to LM.

$ABCD$ is then the required regular pyramid or tetrahedron.

Proposition 14.

To construct an octahedron and comprehend it in a sphere, as in the preceding case; and to prove that the square on the diameter of the sphere is double of the square on the side of the octahedron.

Let the diameter AB of the given sphere be set out,
and let it be bisected at C;
let the semicircle ADB be described on AB,
let CD be drawn from C at right angles to AB,
let DB be joined;
let the square $EFGH$, having each of its sides equal to DB, be set out,
let HF, EG be joined,
from the point K let the straight line KL be set up at right angles to the plane of the square $EFGH$ [XI. 12], and let it be carried through to the other side of the plane, as KM;
from the straight lines KL, KM let KL, KM be respectively cut off equal to one of the straight lines EK, FK, GK, HK,
and let LE, LF, LG, LH, ME, MF, MG, MH be joined.

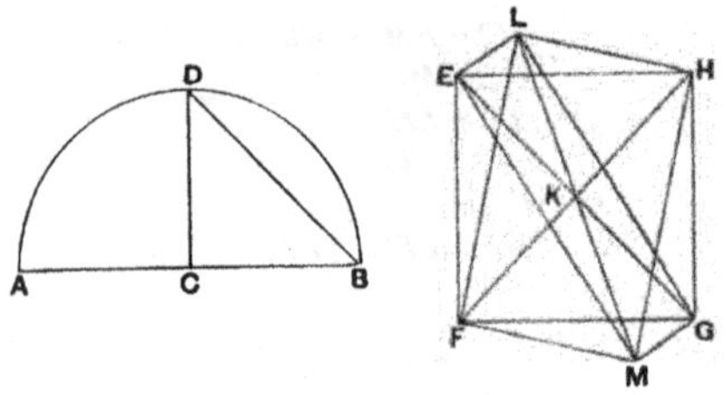

Then, since KE is equal to KH,
and the angle EKH is right,
therefore the square on HE is double of the square on EK. [I. 47]

Again, since LK is equal to KE,
and the angle LKE is right,
therefore the square on EL is double of the square on EK. [*id.*]

But the square on HE was also proved double of the square on EK;
therefore the square on LE is equal to the square on EH;
therefore LE is equal to EH.

For the same reason
LH is also equal to HE;
therefore the triangle LEH is equilateral.

Similarly we can prove that each of the remaining triangles of which the sides of the square $EFGH$ are the bases, and the points L, M the vertices, is equilateral;
therefore an octahedron has been constructed which is contained by eight equilateral triangles.

It is next required to comprehend it in the given sphere, and to prove that the square on the diameter of the sphere is double of the square on the side of the octahedron.

For, since the three straight lines LK, KM, KE are equal to one another,
therefore the semicircle described on LM will also pass through E.

And for the same reason,
if, LM remaining fixed, the semicircle be carried round and restored to the same position from which it began to be moved,
it will also pass through the points F, G, H,
and the octahedron will have been comprehended in a sphere.

I say next that it is also comprehended in the given sphere.

For, since LK is equal to KM,
while KE is common,
and they contain right angles,
therefore the base LE is equal to the base EM. [I. 4]

And, since the angle LEM is right, for it is in a semicircle, [III. 31]
therefore the square on LM is double of the square on LE. [I. 47]

Again, since AC is equal to CB,
AB is double of BC.

But, as AB is to BC, so is the square on AB to the square on BD;
therefore the square on AB is double of the square on BD.

But the square on LM was also proved double of the square on LE.

And the square on DB is equal to the square on LE, for EH was made equal to DB.

Therefore the square on AB is also equal to the square on LM;
therefore AB is equal to LM.

And AB is the diameter of the given sphere;
therefore LM is equal to the diameter of the given sphere.

Therefore the octahedron has been comprehended in the given sphere, and it has been demonstrated at the same time that the square on the diameter of the sphere is double of the square on the side of the octahedron.

Q. E. D.

I think the accompanying figure will perhaps be clearer than that in Euclid's text.

$EFGH$ being a square with side equal to BD, it follows that KE, KF, KG, KH are all equal to CB.

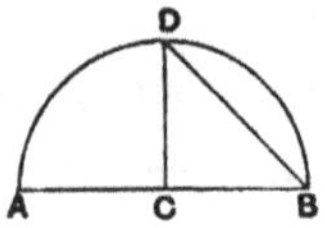

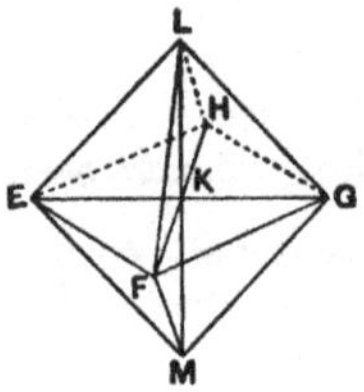

So are KL, KM, by construction;
hence LE, LF, LG, LH and ME, MF, MG, MH are all equal to EF or BD.

Thus (1) the figure is made up of eight equilateral triangles and is therefore a regular octahedron.

(2) Since $$KE = KL = KM,$$
the semicircle on LM in the plane LKE passes through E.

Similarly F, G, H lie on semicircles on LM as diameter.

Thus all the vertices of the tetrahedron lie on the sphere of which LM is a diameter.

(3) $$LE = EM = BD;$$
therefore $$LM^2 = 2EL^2 = 2BD^2$$
$$= AB^2,$$
or $$LM = AB.$$

(4) $$AB^2 = 2BD^2$$
$$= 2EF^2.$$

If r be the radius of the circumscribed sphere,

$$(\text{edge of octahedron}) = \sqrt{2} \,.\, r.$$

Pappus' method.

Pappus (III. pp. 148—150) finds the two equal and parallel sections of the sphere which circumscribe two opposite faces of the octahedron thus.

Analysis.

Suppose the octahedron inscribed, A, B, C; D, E, F being the vertices.

Through ABC, DEF describe planes making the circular sections ABC, DEF.

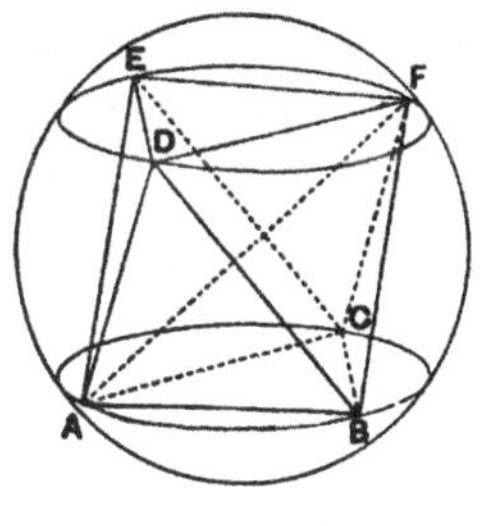

Since the straight lines DA, DB, DE, DF are equal, the points A, E, F, B lie on a circle of which D is the pole.

Again, since AB, BF, FE, EA are equal, $ABFE$ is a square inscribed in the said circle, and AB, EF are parallel.

Similarly DE is parallel to BC, and DF to AC.

Therefore the circles through D, E, F and A, B, C are parallel; and they are also equal because the equilateral triangles inscribed in them are equal.

Now, ABC, DEF being equal and parallel circular sections, and AB, EF equal and parallel chords "not on the same side of the centres,"

AF is a diameter of the sphere.

[Pappus has a lemma for this, pp. 136—138].

And $AE = EF$, so that $AF^2 = 2FE^2$.

But, if d' be the diameter of the circle DEF,

$$d'^2 = \tfrac{4}{3} EF^2. \qquad \text{[cf. XIII. 12]}$$

Therefore, if d be the diameter of the sphere,

$$d^2 : d'^2 = 3 : 2.$$

Now d is given, and therefore d' is given; hence the circles DEF, ABC are given.

Synthesis.

Draw two equal and parallel circular sections with diameter d', such that

$$d^2 = \tfrac{3}{2} d'^2,$$

where d is the diameter of the sphere.

Inscribe an equilateral triangle ABC in either circle (ABC).

In the other circle draw EF equal and parallel to AB but on the opposite side of the centre, and complete the inscribed equilateral triangle DEF.

$ABCDEF$ is the octahedron required.

It will be observed that, whereas in the problem of XIII. 13 Euclid first finds the circle circumscribing a face and Pappus first finds an edge, in this problem Euclid finds the edge first and Pappus the circle circumscribing a face.

PROPOSITION 15.

To construct a cube and comprehend it in a sphere, like the pyramid; and to prove that the square on the diameter of the sphere is triple of the square on the side of the cube.

Let the diameter *AB* of the given sphere be set out,
and let it be cut at *C* so that *AC* is double of *CB*;
let the semicircle *ADB* be described on *AB*,
let *CD* be drawn from *C* at right angles to *AB*,
and let *DB* be joined;
let the square *EFGH* having its side equal to *DB* be set out,
from *E*, *F*, *G*, *H* let *EK*, *FL*, *GM*, *HN* be drawn at right angles to the plane of the square *EFGH*,
from *EK*, *FL*, *GM*, *HN* let *EK*, *FL*, *GM*, *HN* respectively be cut off equal to one of the straight lines *EF*, *FG*, *GH*, *HE*,
and let *KL*, *LM*, *MN*, *NK* be joined;
therefore the cube *FN* has been constructed which is contained by six equal squares.

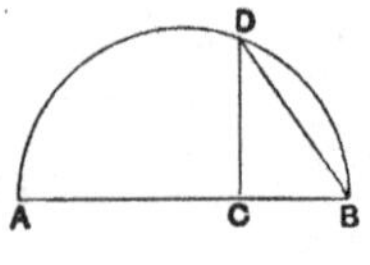

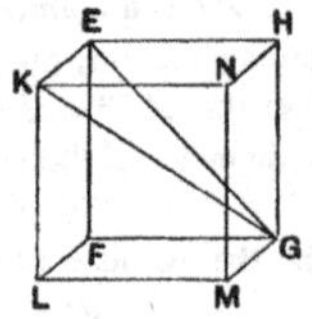

It is then required to comprehend it in the given sphere, and to prove that the square on the diameter of the sphere is triple of the square on the side of the cube.

For let *KG*, *EG* be joined.

Then, since the angle *KEG* is right, because *KE* is also at right angles to the plane *EG* and of course to the straight line *EG* also, [XI. Def. 3]
therefore the semicircle described on *KG* will also pass through the point *E*.

Again, since *GF* is at right angles to each of the straight lines *FL*, *FE*,
GF is also at right angles to the plane *FK*;
hence also, if we join *FK*, *GF* will be at right angles to *FK*;

and for this reason again the semicircle described on *GK* will also pass through *F*.

Similarly it will also pass through the remaining angular points of the cube.

If then, *KG* remaining fixed, the semicircle be carried round and restored to the same position from which it began to be moved,
the cube will be comprehended in a sphere.

I say next that it is also comprehended in the given sphere.

For, since *GF* is equal to *FE*,
and the angle at *F* is right,
therefore the square on *EG* is double of the square on *EF*.

But *EF* is equal to *EK*;
therefore the square on *EG* is double of the square on *EK*;
hence the squares on *GE*, *EK*, that is the square on *GK* [I. 47], is triple of the square on *EK*.

And, since *AB* is triple of *BC*,
while, as *AB* is to *BC*, so is the square on *AB* to the square on *BD*,
therefore the square on *AB* is triple of the square on *BD*.

But the square on *GK* was also proved triple of the square on *KE*.

And *KE* was made equal to *DB*;
therefore *KG* is also equal to *AB*.

And *AB* is the diameter of the given sphere;
therefore *KG* is also equal to the diameter of the given sphere.

Therefore the cube has been comprehended in the given sphere; and it has been demonstrated at the same time that the square on the diameter of the sphere is triple of the square on the side of the cube.

Q. E. D.

AB is divided so that $AC = 2CB$; *CD* is drawn at right angles to *AB*, and *BD* is joined.

KG is, by construction, a cube of side equal to *BD*.

To prove (1) that it is inscribable in a sphere.

Since *KE* is perpendicular to *EH*, *EF*,
KE is perpendicular to *EG*.

Thus, KEG being a right angle, E lies on a semicircle with diameter KG.

The same thing is proved in the same way of the other vertices F, H, L, M, N.

Thus the cube is inscribed in the sphere of which KG is a diameter.

(2) $$KG^2 = KE^2 + EG^2$$
$$= KE^2 + 2EF^2$$
$$= 3EK^2.$$

Also $$AB = 3BC,$$

while $$AB : BC = AB^2 : AB . BC$$
$$= AB^2 : BD^2;$$

therefore $$AB^2 = 3BD^2.$$

But $$BD = EK;$$

therefore $$KG = AB.$$

(3) $$AB^2 = 3BD^2$$
$$= 3KE^2.$$

If r be the radius of the circumscribed sphere,

$$(\text{edge of cube}) = \frac{2}{\sqrt{3}} . r = \tfrac{2}{3}\sqrt{3} . r.$$

Pappus' solution.

In this case too Pappus (III. pp. 144—148) gives the full analysis and synthesis.

Analysis.

Suppose the problem solved, and let the vertices of the cube be A, B, C, D, E, F, G, H.

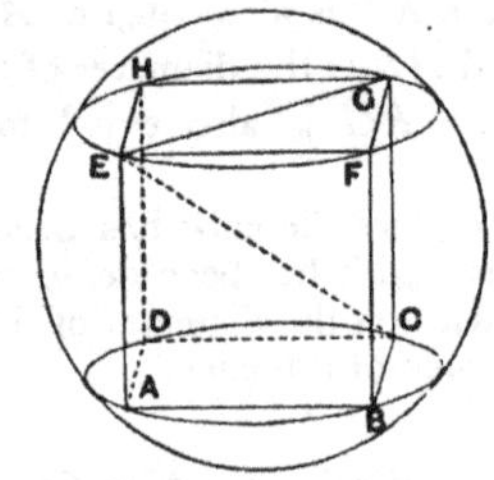

Draw planes through A, B, C, D and E, F, G, H respectively; these will produce parallel circular sections, which are also equal since the inscribed squares are equal.

And CE will be a diameter of the sphere.

Join EG.

Now, since $EG^2 = 2EH^2 = 2GC^2$, and the angle CGE is right,

$$CE^2 = GC^2 + EG^2 = \tfrac{3}{2}EG^2.$$

But CE^2 is given;

therefore EG^2 is given, so that the circles $EFGH$, $ABCD$, and the squares inscribed in them, are given.

Synthesis.

Draw two parallel circular sections with equal diameters d', such that

$$d^2 = \tfrac{3}{2}d'^2,$$

where d is the diameter of the given sphere.

Inscribe a square in one of the circles, as $ABCD$.

In the other circle draw FG equal and parallel to BC, and complete the square on FG inscribed in the circle $EFGH$.

The eight vertices of the required cube are thus determined.

Proposition 16.

To construct an icosahedron and comprehend it in a sphere, like the aforesaid figures; and to prove that the side of the icosahedron is the irrational straight line called minor.

Let the diameter AB of the given sphere be set out,
and let it be cut at C so that AC is quadruple of CB,
let the semicircle ADB be described on AB,
let the straight line CD be drawn from C at right angles to AB,
and let DB be joined;

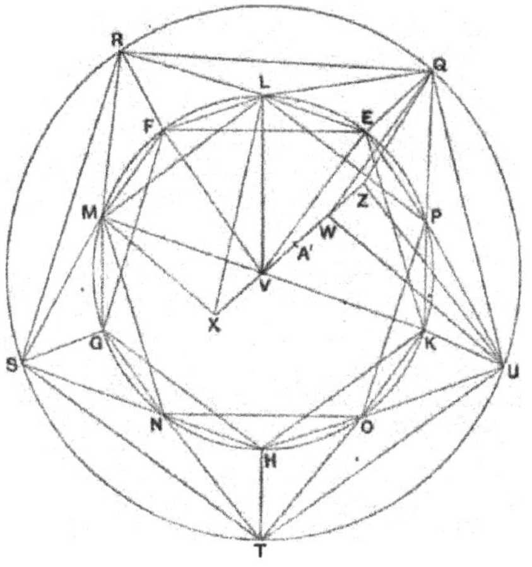

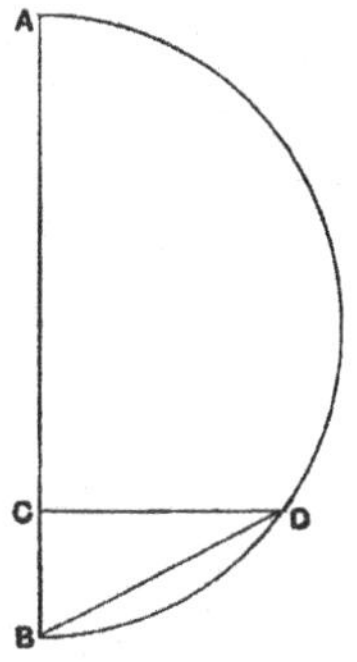

let the circle $EFGHK$ be set out and let its radius be equal to DB,
let the equilateral and equiangular pentagon $EFGHK$ be inscribed in the circle $EFGHK$,
let the circumferences EF, FG, GH, HK, KE be bisected at the points L, M, N, O, P,
and let LM, MN, NO, OP, PL, EP be joined.

Therefore the pentagon *LMNOP* is also equilateral, and the straight line *EP* belongs to a decagon.

Now from the points *E, F, G, H, K* let the straight lines *EQ, FR, GS, HT, KU* be set up at right angles to the plane of the circle, and let them be equal to the radius of the circle *EFGHK*,

let *QR, RS, ST, TU, UQ, QL, LR, RM, MS, SN, NT, TO, OU, UP, PQ* be joined.

Now, since each of the straight lines *EQ, KU* is at right angles to the same plane,

therefore *EQ* is parallel to *KU*. [XI. 6]

But it is also equal to it;

and the straight lines joining those extremities of equal and parallel straight lines which are in the same direction are equal and parallel. [I. 33]

Therefore *QU* is equal and parallel to *EK*.

But *EK* belongs to an equilateral pentagon;

therefore *QU* also belongs to the equilateral pentagon inscribed in the circle *EFGHK*.

For the same reason

each of the straight lines *QR, RS, ST, TU* also belongs to the equilateral pentagon inscribed in the circle *EFGHK*;

therefore the pentagon *QRSTU* is equilateral.

And, since *QE* belongs to a hexagon,

and *EP* to a decagon,

and the angle *QEP* is right,

therefore *QP* belongs to a pentagon;

for the square on the side of the pentagon is equal to the square on the side of the hexagon and the square on the side of the decagon inscribed in the same circle. [XIII. 10]

For the same reason

PU is also a side of a pentagon.

But *QU* also belongs to a pentagon;

therefore the triangle *QPU* is equilateral.

For the same reason

each of the triangles *QLR, RMS, SNT, TOU* is also equilateral.

And, since each of the straight lines *QL*, *QP* was proved to belong to a pentagon,
and *LP* also belongs to a pentagon,
therefore the triangle *QLP* is equilateral.

For the same reason
each of the triangles *LRM*, *MSN*, *NTO*, *OUP* is also equilateral.

Let the centre of the circle *EFGHK*, the point *V*, be taken;
from *V* let *VZ* be set up at right angles to the plane of the circle,
let it be produced in the other direction, as *VX*,
let there be cut off *VW*, the side of a hexagon, and each of the straight lines *VX*, *WZ*, being sides of a decagon,
and let *QZ*, *QW*, *UZ*, *EV*, *LV*, *LX*, *XM* be joined.

Now, since each of the straight lines *VW*, *QE* is at right angles to the plane of the circle,
therefore *VW* is parallel to *QE*. [XI. 6]
But they are also equal;
therefore *EV*, *QW* are also equal and parallel. [I. 33]
But *EV* belongs to a hexagon;
therefore *QW* also belongs to a hexagon.
And, since *QW* belongs to a hexagon,
and *WZ* to a decagon,
and the angle *QWZ* is right,
therefore *QZ* belongs to a pentagon. [XIII. 10]

For the same reason
UZ also belongs to a pentagon,
inasmuch as, if we join *VK*, *WU*, they will be equal and opposite, and *VK*, being a radius, belongs to a hexagon; [IV. 15, Por.]
therefore *WU* also belongs to a hexagon.
But *WZ* belongs to a decagon,
and the angle *UWZ* is right;
therefore *UZ* belongs to a pentagon. [XIII. 10]
But *QU* also belongs to a pentagon;
therefore the triangle *QUZ* is equilateral.

For the same reason
each of the remaining triangles of which the straight lines *QR*, *RS*, *ST*, *TU* are the bases, and the point *Z* the vertex, is also equilateral.

Again, since *VL* belongs to a hexagon,
and *VX* to a decagon,
and the angle *LVX* is right,
therefore *LX* belongs to a pentagon. [XIII. 10]

For the same reason,
if we join *MV*, which belongs to a hexagon,
MX is also inferred to belong to a pentagon.

But *LM* also belongs to a pentagon;
therefore the triangle *LMX* is equilateral.

Similarly it can be proved that each of the remaining triangles of which *MN*, *NO*, *OP*, *PL* are the bases, and the point *X* the vertex, is also equilateral.

Therefore an icosahedron has been constructed which is contained by twenty equilateral triangles.

It is next required to comprehend it in the given sphere, and to prove that the side of the icosahedron is the irrational straight line called minor.

For, since *VW* belongs to a hexagon,
and *WZ* to a decagon,
therefore *VZ* has been cut in extreme and mean ratio at *W*,
and *VW* is its greater segment; [XIII. 9]
therefore, as *ZV* is to *VW*, so is *VW* to *WZ*.

But *VW* is equal to *VE*, and *WZ* to *VX*;
therefore, as *ZV* is to *VE*, so is *EV* to *VX*.

And the angles *ZVE*, *EVX* are right;
therefore, if we join the straight line *EZ*, the angle *XEZ* will be right because of the similarity of the triangles *XEZ*, *VEZ*.

For the same reason,
since, as *ZV* is to *VW*, so is *VW* to *WZ*,
and *ZV* is equal to *XW*, and *VW* to *WQ*,
therefore, as *XW* is to *WQ*, so is *QW* to *WZ*.

And for this reason again,

if we join QX, the angle at Q will be right; [VI. 8]

therefore the semicircle described on XZ will also pass through Q. [III. 31]

And if, XZ remaining fixed, the semicircle be carried round and restored to the same position from which it began to be moved, it will also pass through Q and the remaining angular points of the icosahedron,

and the icosahedron will have been comprehended in a sphere.

I say next that it is also comprehended in the given sphere.

For let VW be bisected at A'.

Then, since the straight line VZ has been cut in extreme and mean ratio at W,

and ZW is its lesser segment,

therefore the square on ZW added to the half of the greater segment, that is WA', is five times the square on the half of the greater segment; [XIII. 3]

therefore the square on ZA' is five times the square on $A'W$.

And ZX is double of ZA', and VW double of $A'W$;

therefore the square on ZX is five times the square on WV.

And, since AC is quadruple of CB,

therefore AB is five times BC.

But, as AB is to BC, so is the square on AB to the square on BD; [VI. 8, V. Def. 9]

therefore the square on AB is five times the square on BD.

But the square on ZX was also proved to be five times the square on VW.

And DB is equal to VW,

for each of them is equal to the radius of the circle $EFGHK$;

therefore AB is also equal to XZ.

And AB is the diameter of the given sphere;

therefore XZ is also equal to the diameter of the given sphere.

Therefore the icosahedron has been comprehended in the given sphere.

I say next that the side of the icosahedron is the irrational straight line called minor.

For, since the diameter of the sphere is rational,
and the square on it is five times the square on the radius of the circle *EFGHK*,
therefore the radius of the circle *EFGHK* is also rational;
hence its diameter is also rational.

But, if an equilateral pentagon be inscribed in a circle which has its diameter rational, the side of the pentagon is the irrational straight line called minor. [XIII. 11]

And the side of the pentagon *EFGHK* is the side of the icosahedron.

Therefore the side of the icosahedron is the irrational straight line called minor.

PORISM. From this it is manifest that the square on the diameter of the sphere is five times the square on the radius of the circle from which the icosahedron has been described, and that the diameter of the sphere is composed of the side of the hexagon and two of the sides of the decagon inscribed in the same circle.

Q. E. D.

Euclid's method is

(1) to find the pentagons in the two parallel circular sections of the sphere, the sides of which form ten (five in each circle) of the edges of the icosahedron,

(2) to find the two points which are the poles of the two circular sections,

(3) to prove that the triangles formed by joining the angular points of the pentagons which are nearest to one another two and two are equilateral,

(4) to prove that the triangles of which the poles are the vertices and the sides of the pentagons the bases are also equilateral,

(5) that all the angular points other than the poles lie on a sphere the diameter of which is the straight line joining the poles,

(6) that this sphere is of the same size as the given sphere,

(7) that, if the diameter of the sphere is rational, the edge of the icosahedron is the *minor* irrational straight line.

I have drawn another figure which will perhaps show the pentagons, and the position of the poles with regard to them, more clearly than does Euclid's figure.

(1) If AB is the diameter of the given sphere, divide AB at C so that

$$AC = 4CB;$$

draw CD at right angles to AB meeting the semicircle on AB in D.

Join BD.

BD *is the radius of the circular sections containing the pentagons.*

[If r is the radius of the sphere,

since $$AB : BC = AB^2 : AB \,.\, BC$$
$$= AB^2 : BD^2,$$
while $$AB = 5BC,$$
it follows that $$AB^2 = 5BD^2,$$
or $$(\text{radius of section})^2 = \tfrac{4}{5}r^2.$$

Thus [XIII. 10, note] $$(\text{side of pentagon})^2 = \frac{r^2}{5}(10 - 2\sqrt{5}).]$$

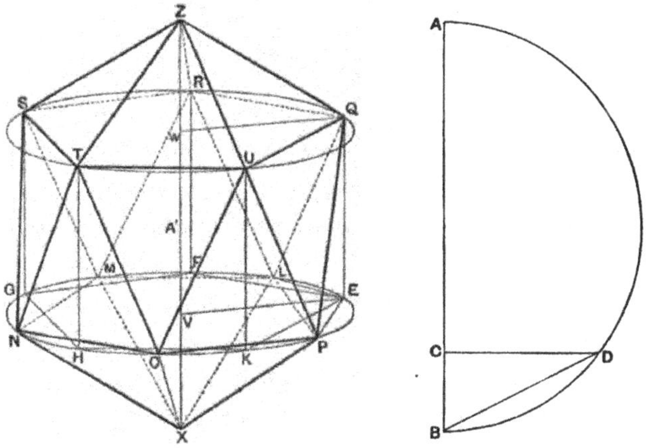

Inscribe the regular pentagon $EFGHK$ in the circle $EFGHK$ of radius equal to BD.

Bisect the arcs EF, FG, ..., so forming a decagon in the circle.

Joining successive points of bisection, we obtain another regular pentagon $LMNOP$.

LMNOP *is one of the pentagons containing five edges of the icosahedron.*

The other circle and inscribed pentagon are obtained by drawing perpendiculars from E, F, G, H, K to the plane of the circle, as EQ, FR, GS, HT, KU, and making each of these perpendiculars equal to the radius of the circle, or, as Euclid says, the side of the regular hexagon in it.

QRSTU *is the second pentagon* (of course equal to the first) *containing five edges of the icosahedron.*

Joining each angular point of one of the two pentagons to the two nearest angular points in the other pentagon, we complete ten triangles each of which has for one side a side of one or other of the two pentagons.

V, W are the centres of the two circles, and VW is of course perpendicular to the planes of both.

(2) Produce VW in both directions, making VX and WZ both equal to *a side of the regular decagon in the circle* (as EL).

Joining X, Z to the angular points of the corresponding pentagons, we have five more triangles formed with the sides of each pentagon as bases, ten more triangles in all.

Now we come to the proof.

(3) Taking two adjacent perpendiculars, EQ, KU, to the plane of the circle $EFGHK$, we see that they are parallel as well as equal;
therefore QU, EK are equal and parallel.

Similarly for QR, EF etc.

Thus the pentagons have their sides equal.

To prove that the triangles QPL etc., are equilateral, we have, e.g.

$$QL^2 = LE^2 + EQ^2$$
$$= (\text{side of decagon})^2 + (\text{side of hexagon})^2$$
$$= (\text{side of pentagon})^2, \qquad [\text{XIII. 10}]$$

i.e. $$QL = (\text{side of pentagon in circle})$$
$$= LP.$$

Similarly $$QP = LP,$$

and $$\triangle QPL \text{ is equilateral.}$$

So for the other triangles between the two pentagons.

(4) Since VW, EQ are equal and parallel,

VE, WQ are equal and parallel.

Thus WQ is equal to the side of a regular hexagon in the circles.

Now the angle ZWQ is right;

therefore $$ZQ^2 = ZW^2 + WQ^2$$
$$= (\text{side of decagon})^2 + (\text{side of hexagon})^2$$
$$= (\text{side of pentagon})^2. \qquad [\text{XIII. 10}]$$

Thus ZQ, ZR, ZS, ZT, ZU are all equal to QR, RS etc.; and the triangles with Z as vertex and bases QR, RS etc. are equilateral.

Similarly for the triangles with X as vertex and LM, MN etc. as bases.

Hence the figure is an icosahedron, being contained by twenty equal equilateral triangles.

(5) To prove that all the vertices of the icosahedron lie on the sphere which has XZ for diameter.

VW being equal to the side of a regular hexagon, and WZ to the side of a regular decagon inscribed in the same circle,

VZ is divided at W in extreme and mean ratio. [XIII. 9]

Therefore $$ZV : VW = VW : WZ,$$

or, since $$VW = VE, \quad WZ = VX,$$
$$ZV : VE = VE : VX.$$

Thus E lies on the semicircle on ZX as diameter. [VI. 8]

Similarly for all the other vertices of the icosahedron.

Hence the sphere with diameter XZ circumscribes it.

(6) To prove $XZ = AB$.

Since VZ is divided in extreme and mean ratio at W, and VW is bisected at A',

$$A'Z^2 = 5A'W^2. \qquad [\text{XIII. 3}]$$

Taking the doubles of $A'Z$, $A'W$, we have

$$XZ^2 = 5VW^2$$
$$= 5BD^2$$
$$= AB^2. \qquad [\text{see under (1) above}]$$

That is, $XZ = AB$.

[If r is the radius of the sphere,

$$VW = BD = \frac{2}{\sqrt{5}} r,$$

$$\begin{aligned} VX &= (\text{side of decagon in circle of radius } BD) \\ &= \frac{BD}{2}(\sqrt{5} - 1) \qquad \text{[XIII. 9, note]} \\ &= \frac{r}{\sqrt{5}}(\sqrt{5} - 1). \end{aligned}$$

Consequently

$$\begin{aligned} XZ &= VW + 2VX \\ &= \frac{2}{\sqrt{5}} r + \frac{2}{\sqrt{5}} r(\sqrt{5} - 1) \\ &= 2r.] \end{aligned}$$

(6) The radius of the circle $EFGHK$ is equal to $\frac{2}{\sqrt{5}} r$, and is therefore "rational" in Euclid's sense.

Hence the side of the inscribed pentagon is the irrational straight line called *minor*. [XIII. 11]

[The side of this pentagon is the edge of the icosahedron, and its value is (note on XIII. 10)

$$\begin{aligned} &\frac{BD}{2}\sqrt{10 - 2\sqrt{5}} \\ &= \frac{r}{\sqrt{5}}\sqrt{10 - 2\sqrt{5}} \\ &= \frac{r}{5}\sqrt{10(5 - \sqrt{5})}.] \end{aligned}$$

Pappus' solution.

This solution (Pappus, III. pp. 150—6) differs considerably from that of Euclid. Whereas Euclid uses *two* circular sections of the sphere (those circumscribing the pentagons of his construction), Pappus finds *four* parallel circular sections each passing through *three* of the vertices of the icosahedron; two of the circles are small circles circumscribing two opposite triangular faces respectively, and the other two circles are between these two circles, parallel to them and equal to one another.

Analysis.

Suppose the problem solved, the vertices of the icosahedron being A, B, C; D, E, F; G, H, K; L, M, N.

Since the straight lines BA, BC, BF, BG, BE drawn from B to the surface of the sphere are equal,

A, C, F, G, E are in one plane.

And AC, CF, FG, GE, EA are equal;

therefore $ACFGE$ is an equilateral and equiangular pentagon.

So are the figures $KEBCD$, $DHFBA$, $AKLGB$, $AKNHC$, and $CHMGB$.

Join EF, KH.

Now AC will be parallel to EF (in the pentagon $ACFGE$) and to KH (in the pentagon $AKNHC$), so that EF, KH are also parallel;

and further KH is parallel to LM (in the pentagon $LKDHM$).

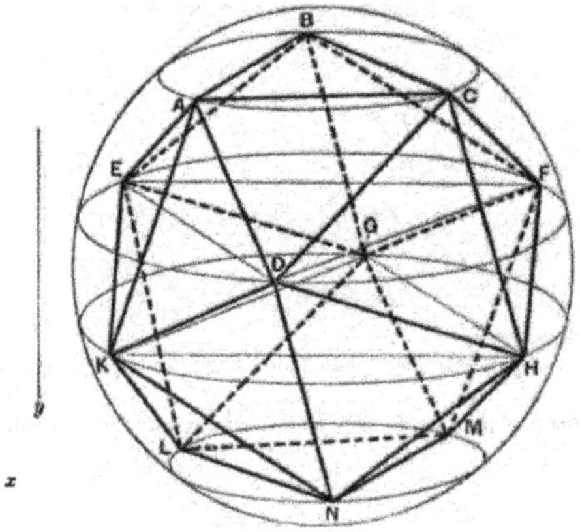

Similarly BC, ED, GH, LN are all parallel;

and likewise BA, FD, GK, MN are all parallel.

Since BC is equal and parallel to LN, and BA to MN, the circles ABC, LMN are equal and parallel.

Similarly the circles DEF, KGH are equal and parallel; for the triangles inscribed in them are equal (since each of the sides in both is the chord subtending an angle of equal pentagons), and their sides are parallel respectively.

Now in the equal and parallel circles DEF, KGH the chords EF, KH are equal and parallel, and on opposite sides of the centres;

therefore FK is a diameter of the sphere [Pappus' lemma, pp. 136—8], and the angle FEK is right [Pappus' lemma, p. 138, 20—26].

[The diameter FK is not actually drawn in the figure.]

In the pentagon $GEACF$, if EF be divided in extreme and mean ratio, the greater segment is equal to AC. [Eucl. XIII. 8]

Therefore $EF : AC =$ (side of hexagon) : (side of decagon in same circle). [XIII. 9]

And $$EF^2 + AC^2 = EF^2 + EK^2 = d^2,$$

where d is the diameter of the sphere.

Thus FK, EF, AC are as the sides of the pentagon, hexagon and decagon respectively inscribed in the same circle. [XIII. 10]

But FK, the diameter of the sphere, is given;

therefore EF, AC are given respectively;

thus the radii of the circles EFD, ACB are given (if r, r' are their radii, $r^2 = \frac{1}{3}EF^2$, $r'^2 = \frac{1}{3}AC^2$).

Hence the circles are given;
and so are the circles KHG, LMN which are equal and parallel to them respectively.

Synthesis.

If d be the diameter of the sphere, set out two straight lines x, y, such that d, x, y are in the ratio of the sides of the pentagon, hexagon and decagon respectively inscribed in one and the same circle.

Draw (1) two equal and parallel circular sections in the sphere, with radii equal to r, where $r^2 = \frac{1}{3}x^2$, as DEF, KGH,
and (2) two equal and parallel circular sections as ABC, LMN, with radius r' such that $r'^2 = \frac{1}{3}y^2$.

In the circles (1) draw EF, KH as sides of inscribed equilateral triangles, parallel to one another, and on opposite sides of the centres;
and in the circles (2) draw AC, LM as sides of inscribed equilateral triangles parallel to one another and to EF, KH, and so that AC, EF are on opposite sides of the centres, and likewise KH, LM.

Complete the figure.

The correctness of the construction is proved as in the analysis.

It follows also (says Pappus) that

$$(\text{diam. of sphere})^2 = 3\,(\text{side of pentagon in } DEF)^2.$$

For, by construction, $\quad KF : FE = p : h,$
where p, h are the sides of the pentagon and hexagon inscribed in the same circle DEF.

And $FE : h =$ the ratio of the side of an equilateral triangle to that of a hexagon inscribed in the same circle;

that is, $\quad FE : h = \sqrt{3} : 1,$

whence $\quad KF : p = \sqrt{3} : 1,$

or $\quad KF^2 = 3p^2.$

Another construction.

Mr H. M. Taylor has a neat construction for an icosahedron of edge a.

Let l be the length of the diagonal of a regular pentagon with side equal to a.

Then (figure of XIII. 8), by Ptolemy's theorem,

$$l^2 = la + a^2.$$

Construct a cube with edge equal to l.

Let O be the centre of the cube.

From O draw OL, OM, ON perpendicular to three adjacent faces, and in these draw PP', QQ', RR' parallel to AB, AD, AE respectively.

Make LP, LP', MQ, MQ', NR, NR' all equal to $\frac{1}{2}a$.

Let p, p', q, q', r, r' be the reflexes of P, P', Q, Q', R, R' respectively.

Then will P, P', Q, Q', R, R', p, p', q, q', r, r' be the vertices of a regular icosahedron.

The projections of PQ on AB, AD, AE are equal to $\frac{1}{2}(l-a)$, $\frac{1}{2}a$, $\frac{1}{2}l$ respectively.

Therefore

$$\begin{aligned} PQ^2 &= \tfrac{1}{4}(l-a)^2 + \tfrac{1}{4}a^2 + \tfrac{1}{4}l^2 \\ &= \tfrac{1}{2}(l^2 - al + a^2) \\ &= a^2. \end{aligned}$$

Therefore $PQ = a$.

Similarly it may be proved that every other edge is equal to a.

All the angular points lie on a sphere with radius OP, and

$$OP^2 = \tfrac{1}{4}(a^2 + l^2).$$

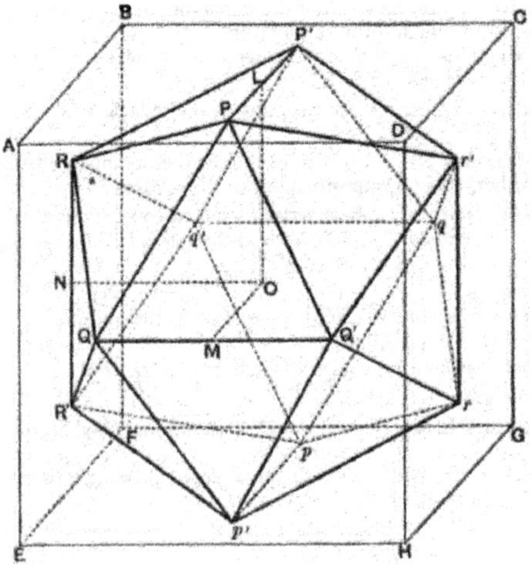

Each solid pentahedral angle is composed of five equal plane angles, each of which is the angle of an equilateral triangle.

Therefore the icosahedron is regular.

$$[a^2 = 4OP^2 - l^2.$$

And, from the equation $l^2 = la + a^2$, we derive

$$l = \frac{a}{2}(\sqrt{5} + 1).$$

Therefore, if r be the radius of the sphere,

$$a^2\left\{1 + \frac{(\sqrt{5}+1)^2}{4}\right\} = 4r^2,$$

whence
$$a = 4r/\sqrt{10 + 2\sqrt{5}}$$
$$= 4r\sqrt{10 - 2\sqrt{5}}/\sqrt{80}$$
$$= \frac{r}{\sqrt{5}}\sqrt{10 - 2\sqrt{5}}$$
$$= \frac{r}{5}\sqrt{10(5 - \sqrt{5})},$$

as above.]

PROPOSITION 17.

To construct a dodecahedron and comprehend it in a sphere, like the aforesaid figures, and to prove that the side of the dodecahedron is the irrational straight line called apotome.

Let *ABCD*, *CBEF*, two planes of the aforesaid cube at right angles to one another, be set out,

let the sides *AB*, *BC*, *CD*, *DA*, *EF*, *EB*, *FC* be bisected at *G*, *H*, *K*, *L*, *M*, *N*, *O* respectively,

let *GK*, *HL*, *MH*, *NO* be joined,

let the straight lines *NP*, *PO*, *HQ* be cut in extreme and mean ratio at the points *R*, *S*, *T* respectively,

and let *RP*, *PS*, *TQ* be their greater segments;

from the points *R*, *S*, *T* let *RU*, *SV*, *TW* be set up at right angles to the planes of the cube towards the outside of the cube,

let them be made equal to *RP*, *PS*, *TQ*,

and let *UB*, *BW*, *WC*, *CV*, *VU* be joined.

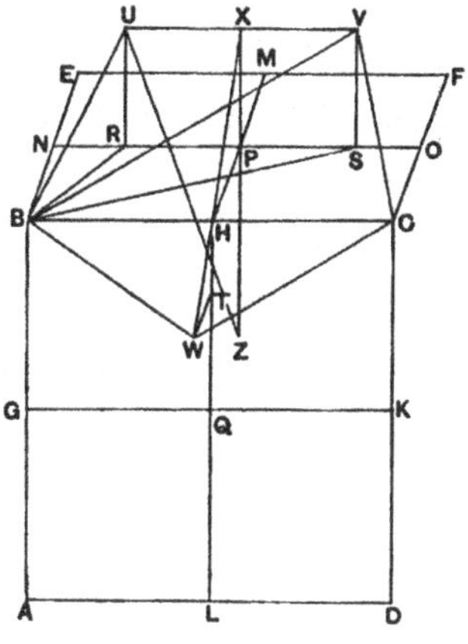

I say that the pentagon *UBWCV* is equilateral, and in one plane, and is further equiangular.

For let *RB*, *SB*, *VB* be joined.

Then, since the straight line *NP* has been cut in extreme and mean ratio at *R*,
and *RP* is the greater segment,
therefore the squares on *PN*, *NR* are triple of the square on *RP*. [XIII. 4]

But *PN* is equal to *NB*, and *PR* to *RU*;
therefore the squares on *BN*, *NR* are triple of the square on *RU*.

But the square on *BR* is equal to the squares on *BN*, *NR*; [I. 47]
therefore the square on *BR* is triple of the square on *RU*;
hence the squares on *BR*, *RU* are quadruple of the square on *RU*.

But the square on *BU* is equal to the squares on *BR*, *RU*;
therefore the square on *BU* is quadruple of the square on *RU*;
therefore *BU* is double of *RU*.

But *VU* is also double of *UR*,
inasmuch as *SR* is also double of *PR*, that is, of *RU*;
therefore *BU* is equal to *UV*.

Similarly it can be proved that each of the straight lines *BW*, *WC*, *CV* is also equal to each of the straight lines *BU*, *UV*.

Therefore the pentagon *BUVCW* is equilateral.

I say next that it is also in one plane.

For let *PX* be drawn from *P* parallel to each of the straight lines *RU*, *SV* and towards the outside of the cube, and let *XH*, *HW* be joined;
I say that *XHW* is a straight line.

For, since *HQ* has been cut in extreme and mean ratio at *T*, and *QT* is its greater segment,
therefore, as *HQ* is to *QT*, so is *QT* to *TH*.

But *HQ* is equal to *HP*, and *QT* to each of the straight lines *TW*, *PX*;
therefore, as *HP* is to *PX*, so is *WT* to *TH*.

And *HP* is parallel to *TW*,
for each of them is at right angles to the plane *BD*; [XI. 6]
and *TH* is parallel to *PX*,
for each of them is at right angles to the plane *BF*. [*id.*]

But if two triangles, as XPH, HTW, which have two sides proportional to two sides be placed together at one angle so that their corresponding sides are also parallel,
the remaining straight lines will be in a straight line; [VI. 32]
therefore XH is in a straight line with HW.

But every straight line is in one plane; [XI. 1]
therefore the pentagon $UBWCV$ is in one plane.

I say next that it is also equiangular.

For, since the straight line NP has been cut in extreme and mean ratio at R, and PR is the greater segment,
while PR is equal to PS,
therefore NS has also been cut in extreme and mean ratio at P,
and NP is the greater segment; [XIII. 5]
therefore the squares on NS, SP are triple of the square on NP. [XIII. 4]

But NP is equal to NB, and PS to SV;
therefore the squares on NS, SV are triple of the square on NB;
hence the squares on VS, SN, NB are quadruple of the square on NB.

But the square on SB is equal to the squares on SN, NB;
therefore the squares on BS, SV, that is, the square on BV—for the angle VSB is right—is quadruple of the square on NB;
therefore VB is double of BN.

But BC is also double of BN;
therefore BV is equal to BC.

And, since the two sides BU, UV are equal to the two sides BW, WC,
and the base BV is equal to the base BC,
therefore the angle BUV is equal to the angle BWC. [I. 8]

Similarly we can prove that the angle UVC is also equal to the angle BWC;
therefore the three angles BWC, BUV, UVC are equal to one another.

But if in an equilateral pentagon three angles are equal to one another, the pentagon will be equiangular, [XIII. 7]
therefore the pentagon *BUVCW* is equiangular.

And it was also proved equilateral;
therefore the pentagon *BUVCW* is equilateral and equiangular, and it is on one side *BC* of the cube.

Therefore, if we make the same construction in the case of each of the twelve sides of the cube,
a solid figure will have been constructed which is contained by twelve equilateral and equiangular pentagons, and which is called a dodecahedron.

It is then required to comprehend it in the given sphere, and to prove that the side of the dodecahedron is the irrational straight line called apotome.

For let *XP* be produced, and let the produced straight line be *XZ*;
therefore *PZ* meets the diameter of the cube, and they bisect one another,
for this has been proved in the last theorem but one of the eleventh book. [XI. 38]

Let them cut at *Z*;
therefore *Z* is the centre of the sphere which comprehends the cube,
and *ZP* is half of the side of the cube.

Let *UZ* be joined.

Now, since the straight line *NS* has been cut in extreme and mean ratio at *P*,
and *NP* is its greater segment,
therefore the squares on *NS*, *SP* are triple of the square on *NP*. [XIII. 4]

But *NS* is equal to *XZ*,
inasmuch as *NP* is also equal to *PZ*, and *XP* to *PS*.

But further *PS* is also equal to *XU*,
since it is also equal to *RP*;
therefore the squares on *ZX*, *XU* are triple of the square on *NP*.

But the square on *UZ* is equal to the squares on *ZX*, *XU*;
therefore the square on *UZ* is triple of the square on *NP*.

But the square on the radius of the sphere which comprehends the cube is also triple of the square on the half of the side of the cube,

for it has previously been shown how to construct a cube and comprehend it in a sphere, and to prove that the square on the diameter of the sphere is triple of the square on the side of the cube. [XIII. 15]

But, if whole is so related to whole, so is half to half also;

and *NP* is half of the side of the cube;

therefore *UZ* is equal to the radius of the sphere which comprehends the cube.

And *Z* is the centre of the sphere which comprehends the cube;

therefore the point *U* is on the surface of the sphere.

Similarly we can prove that each of the remaining angles of the dodecahedron is also on the surface of the sphere;

therefore the dodecahedron has been comprehended in the given sphere.

I say next that the side of the dodecahedron is the irrational straight line called apotome.

For since, when *NP* has been cut in extreme and mean ratio, *RP* is the greater segment,

and, when *PO* has been cut in extreme and mean ratio, *PS* is the greater segment,

therefore, when the whole *NO* is cut in extreme and mean ratio, *RS* is the greater segment.

[Thus, since, as *NP* is to *PR*, so is *PR* to *RN*,

the same is true of the doubles also,

for parts have the same ratio as their equimultiples; [V. 15]

therefore as *NO* is to *RS*, so is *RS* to the sum of *NR*, *SO*.

But *NO* is greater than *RS*;

therefore *RS* is also greater than the sum of *NR*, *SO*;

therefore *NO* has been cut in extreme and mean ratio,

and *RS* is its greater segment.]

But *RS* is equal to *UV*;

therefore, when *NO* is cut in extreme and mean ratio, *UV* is the greater segment.

And, since the diameter of the sphere is rational,
and the square on it is triple of the square on the side of the cube,
therefore NO, being a side of the cube, is rational.

[But if a rational line be cut in extreme and mean ratio, each of the segments is an irrational apotome.]

Therefore UV, being a side of the dodecahedron, is an irrational apotome. [XIII. 6]

PORISM. From this it is manifest that, when the side of the cube is cut in extreme and mean ratio, the greater segment is the side of the dodecahedron.

Q. E. D.

In this proposition we find Euclid using two propositions which precede but are used nowhere else, notably VI. 32, which some authors, in consequence of their having overlooked its use here, have been hard put to it to explain.

Euclid's construction in this case is really identical with that given by Mr H. M. Taylor, and also referred to by Henrici and Treutlein under "crystal-formation."

Euclid starts from the cube inscribed in a sphere, as in XIII. 15, and then finds the side of the regular pentagon in which the side of the cube is a diagonal.

Mr Taylor takes l to be the diagonal of a regular pentagon of side a, so that, by Ptolemy's theorem,

$$l^2 = al + a^2,$$

constructs a cube of which l is the edge, and gets the side of the pentagon by drawing ZX from Z, the centre of the cube, perpendicular to the face BF and equal to $\frac{1}{2}(l+a)$, then drawing UV through X parallel to BC, and making UX, XV both equal to $\frac{1}{2}a$.

Euclid finds UV thus.

Draw NO, MH bisecting pairs of opposite sides in the square BF and meeting in P.

Draw GK, HL bisecting pairs of opposite sides in the square BD and meeting in Q.

Divide PN, PO, QH respectively in extreme and mean ratio at R, S, T (PR, PS, QT being the greater segments); draw RU, SV, TW outwards perpendicular to the respective faces of the cube, and all equal in length to PR, PS, TQ.

Join BU, UV, VC, CW, WB.

Then *$BUVCW$ is one of the pentagonal faces of the dodecahedron*;
and the others can be constructed in the same way.

Euclid now proves

(1) that the pentagon $BUVCW$ is equilateral,

(2) that it is in one plane,

(3) that it is equiangular,

(4) that the vertex U is on the sphere which circumscribes the cube, and hence

(5) that all the other vertices lie on the same sphere,

and (6) that the side of the dodecahedron is an *apotome*.

(1) To prove that the pentagon $BUVCW$ is *equilateral*.

We have

$$\begin{aligned} BU^2 &= BR^2 + RU^2 \\ &= (BN^2 + NR^2) + RP^2 \\ &= (PN^2 + NR^2) + RP^2 \\ &= 3RP^2 + RP^2 \qquad \text{[XIII. 4]} \\ &= 4RP^2 \\ &= UV^2. \end{aligned}$$

Therefore $BU = UV$.

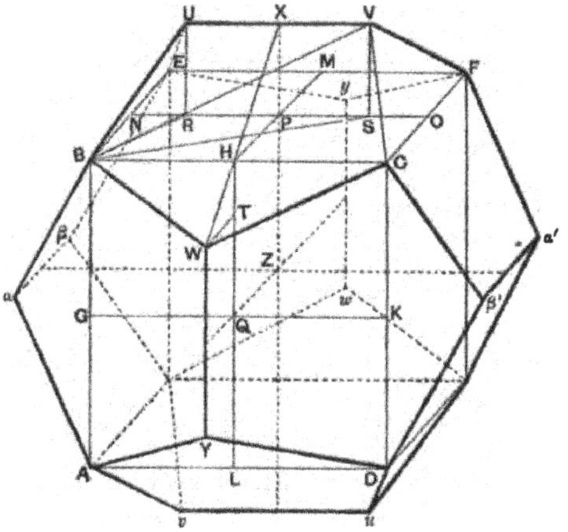

Similarly it may be proved that BW, WC, CV are all equal to UV or BU.

[Mr Taylor proceeds in this way. With his notation, the projections of BU on BA, BC, BE are respectively $\frac{1}{2}a$, $\frac{1}{2}(l-a)$, $\frac{1}{2}l$.

Therefore

$$\begin{aligned} BU^2 &= \tfrac{1}{4}a^2 + \tfrac{1}{4}(l-a)^2 + \tfrac{1}{4}l^2 \\ &= \tfrac{1}{2}(l^2 - al + a^2) \\ &= a^2. \end{aligned}$$

Similarly for BW, WC etc.]

(2) To prove that the pentagon $BUVCW$ is *in one plane*.

Draw PX parallel to RU or SV meeting UV in X.

Join XH, HW.

Then we have to prove that XH, HW are *in one straight line*.

Now HP, WT, being both perpendicular to the face BD, are parallel.

For the same reason XP, HT are parallel.

Also, since QH is divided at T in extreme and mean ratio,

$$QH : QT = QT : TH.$$

And $$QH = HP,\ QT = WT = PX.$$

Therefore $$HP : PX = WT : TH.$$

Consequently the triangles HPX, WTH *satisfy the conditions of* VI. 32; hence XHW is a straight line.

[Mr Taylor proves this as follows:

The projections of WH, WX on BE are $\frac{1}{2}a$ and $\frac{1}{2}(a+l)$, and the projections of WH, WX on BA are $\frac{1}{2}(l-a)$ and $\frac{1}{2}l$;

and $$a : (a+l) = (l-a) : l,$$

since $$al = l^2 - a^2.$$

Therefore WHX is a straight line.]

(3) To prove that the pentagon $BUVCW$ is *equiangular*.

We have
$$\begin{aligned} BV^2 &= BS^2 + SV^2 \\ &= (BN^2 + NS^2) + SP^2 \\ &= PN^2 + (NS^2 + SP^2) \\ &= PN^2 + 3PN^2, \end{aligned}$$

since NS is divided in extreme and mean ratio at P [XIII. 5], so that

$$NS^2 + SP^2 = 3PN^2. \qquad \text{[XIII. 4]}$$

Consequently
$$\begin{aligned} BV^2 &= 4PN^2 \\ &= BC^2, \end{aligned}$$

or $$BV = BC.$$

The $\triangle$s UBV, WBC are therefore equal in all respects,

and $$\angle BUV = \angle BWC.$$

Similarly $$\angle CVU = \angle BWC.$$

Therefore the pentagon is *equiangular*. [XIII. 7]

(4) To prove that the sphere which circumscribes the cube also circumscribes the dodecahedron we have only to prove that, if Z be the centre of the sphere, $ZU = ZB$, for example.

Now, by XI. 38, XP produced meets the diagonal of the cube, and the portion of XP produced which is within the cube and the diagonal bisect one another.

And
$$\begin{aligned} ZU^2 &= ZX^2 + XU^2 \\ &= NS^2 + PS^2 \\ &= 3PN^2, \end{aligned}$$

as before.

Also (cf. XIII. 15)
$$\begin{aligned} ZB^2 &= ZP^2 + PB^2 \\ &= ZP^2 + PN^2 + NB^2 \\ &= 3PN^2. \end{aligned}$$

Hence $$ZU = ZB.$$

(5) Similarly for ZV, ZW etc.

(6) Since PN is divided in extreme and mean ratio at R,

$$NP : PR = PR : RN.$$

Doubling the terms, we have

$$NO : RS = RS : (NR + SO),$$

so that, if NO is divided in extreme and mean ratio, the greater segment is equal to RS.

Now, since the diameter of the sphere is rational,

and $$(\text{diam. of sphere})^2 = 3\,(\text{edge of cube})^2,$$

the edge of the cube (i.e. NO) is rational.

Consequently RS is an *apotome*.

[This is proved in the spurious XIII. 6 above; Euclid assumes it, and the words purporting to quote the theorem are probably interpolated, like XIII. 6 itself.]

As a matter of fact, with Mr Taylor's notation,

$$l^2 = la + a^2,$$

and $$a = \frac{\sqrt{5} - 1}{2}\,l.$$

Since, if r is the radius of the circumscribing sphere, $r = \sqrt{3} \cdot \frac{l}{2}$,

$$a = \frac{r}{\sqrt{3}}(\sqrt{5} - 1) = \frac{r}{3}(\sqrt{15} - \sqrt{3}).$$

Pappus' solution.

Here too Pappus (III. pp. 156—162) finds four circular sections of the sphere all parallel to one another and all passing through five of the vertices of the dodecahedron.

Analysis.

Suppose (he says) the problem solved, and let the vertices of the dodecahedron be A, B, C, D, E; F, G, H, K, L; M, N, O, P, Q; R, S, T, U, V.

Then, as before, ED is parallel to FL, and AE to FG; therefore the planes $ABCDE$, $FGHKL$ are parallel.

But, since PA is parallel to BH, and BH to OC, PA is parallel to OC; and they are equal; therefore PO, AC are parallel, so that ST, ED are also parallel.

Similarly RS, DC are parallel, and likewise the pairs (TU, EA), (UV, AB), (VR, BC).

Therefore the planes $ABCDE$, $RSTUV$ are parallel; and the circles $ABCDE$, $RSTUV$ are equal, since the inscribed pentagons are equal.

Similarly the circles $FGHKL$, $MNOPQ$ are equal, since the pentagons inscribed in them are equal.

Now CL, OU are parallel because each is parallel to KN; therefore L, C, O, U are in one plane.

And LC, CO, OU, UL are all equal, since they subtend angles of equal pentagons.

Also L, C, O, U are on a plane section, i.e. a circle; therefore $LCOU$ is a square.

Therefore $$OL^2 = 2LC^2 = 2LF^2$$

(for LC, LF subtend angles of equal pentagons).

And the angle OLF is right; for PO, LF are equal and parallel chords in two equal and parallel circular sections of a sphere [Pappus' lemma, p. 138, 20—26].

Therefore $$OF^2 = OL^2 + FL^2 = 3FL^2. \quad \text{[from above]}$$

And OF is a diameter of the sphere; for PO, FL are on opposite sides of the centres of the circles in which they are [Pappus' lemma, pp. 136—8].

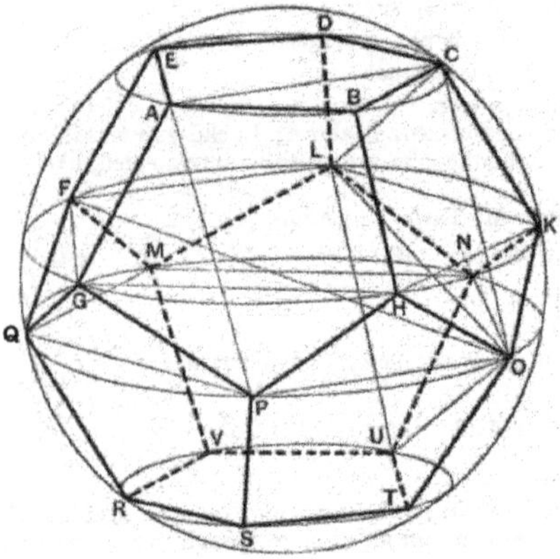

Now suppose p, t, h to be the sides of an equilateral pentagon, triangle and hexagon in the circle $FGHKL$, d the diameter of the sphere.

Then $$d : FL = \sqrt{3} : 1 \quad \text{[from above]}$$
$$= t : h; \quad \text{[Eucl. XIII. 12]}$$

and it follows *alternando* (since $FL = p$) that
$$d : t = p : h.$$

Now let d', p', h' be the sides of a regular decagon, pentagon and hexagon respectively inscribed in *any* one circle.

Since, if FL be divided in extreme and mean ratio, the greater segment is equal to ED, [XIII. 8]
$$FL : ED = h' : d'. \quad \text{[VI. Def. 3, XIII. 9]}$$

And $FL : ED$ is the ratio of the sides of the regular pentagons inscribed in the circles $FGHKL$, $ABCDE$, and is therefore equal to the ratio of the sides of the equilateral triangles inscribed in the same circles.

Therefore t : (side of △ in $ABCDE$) $= h' : d'$.

But $$d : t = p : h$$
$$= p' : h';$$

therefore, *ex aequali*, d : (side of △ in $ABCDE$) $= p' : d'$.

Now d is given;

therefore the sides of the equilateral triangles inscribed in the circles $ABCDE$, $FGHKL$ respectively are given, whence the radii of those circles are also given.

Thus the two circles are given, and so accordingly are the equal and parallel circular sections.

Synthesis.

Set out two straight lines x, y such that d, x, y are in the ratio of the sides of a regular pentagon, hexagon and decagon respectively inscribed in one and the same circle.

Find two circular sections of the sphere with radii r, r', where

$$r^2 = \tfrac{1}{3}x^2, \quad r'^2 = \tfrac{1}{3}y^2.$$

Let these be the circles *FGHKL*, *ABCDE* respectively, and draw the equal and parallel circles on the other side of the centre, namely *MNOPQ*, *RSTUV*.

In the first two circles inscribe regular pentagons with their sides respectively parallel, *ED* being parallel to *FL*.

Draw equal and parallel chords (on the other sides of the centres) in the other two circles, namely *ST* equal and parallel to *ED*, and *PO* equal and parallel to *FL*; and complete the regular pentagons on *ST*, *PO* inscribed in the circles.

Thus all the vertices of the dodecahedron are determined.

The proof of the correctness of the construction is clear from the analysis.

Pappus adds that the construction shows that the circles containing five vertices of the dodecahedron are the same respectively as those containing three vertices of the icosahedron, and that the same circle circumscribes the triangle of the icosahedron and the pentagonal face of the dodecahedron in the same sphere.

PROPOSITION 18.

To set out the sides of the five figures and to compare them with one another.

Let *AB*, the diameter of the given sphere, be set out,

and let it be cut at *C* so that *AC* is equal to *CB*, and at *D* so that *AD* is double of *DB*;

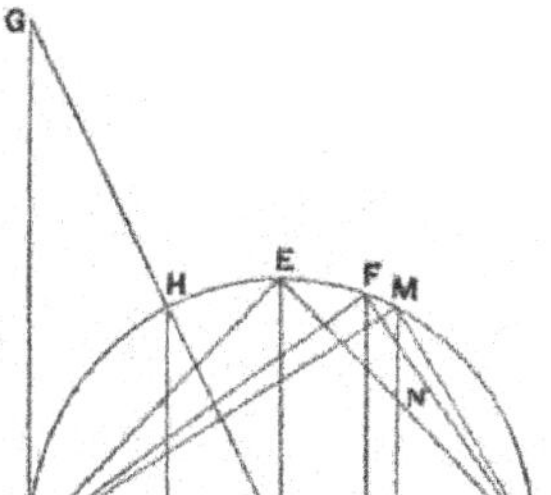

let the semicircle *AEB* be described on *AB*,

from *C*, *D* let *CE*, *DF* be drawn at right angles to *AB*,

and let *AF*, *FB*, *EB* be joined.

Then, since *AD* is double of *DB*,

therefore *AB* is triple of *BD*.

Convertendo, therefore, *BA* is one and a half times *AD*.

But, as BA is to AD, so is the square on BA to the square on AF, [v. Def. 9, vi. 8]
for the triangle AFB is equiangular with the triangle AFD; therefore the square on BA is one and a half times the square on AF.

But the square on the diameter of the sphere is also one and a half times the square on the side of the pyramid. [xiii. 13]

And AB is the diameter of the sphere;
therefore AF is equal to the side of the pyramid.

Again, since AD is double of DB,
therefore AB is triple of BD.

But, as AB is to BD, so is the square on AB to the square on BF; [vi. 8, v. Def. 9]
therefore the square on AB is triple of the square on BF.

But the square on the diameter of the sphere is also triple of the square on the side of the cube. [xiii. 15]

And AB is the diameter of the sphere;
therefore BF is the side of the cube.

And, since AC is equal to CB,
therefore AB is double of BC.

But, as AB is to BC, so is the square on AB to the square on BE;
therefore the square on AB is double of the square on BE.

But the square on the diameter of the sphere is also double of the square on the side of the octahedron. [xiii. 14]

And AB is the diameter of the given sphere;
therefore BE is the side of the octahedron.

Next, let AG be drawn from the point A at right angles to the straight line AB,
let AG be made equal to AB,
let GC be joined,
and from H let HK be drawn perpendicular to AB.

Then, since GA is double of AC,
for GA is equal to AB,
and, as GA is to AC, so is HK to KC,
therefore HK is also double of KC.

Therefore the square on HK is quadruple of the square on KC;

therefore the squares on HK, KC, that is, the square on HC, is five times the square on KC.

But HC is equal to CB;

therefore the square on BC is five times the square on CK.

And, since AB is double of CB,

and, in them, AD is double of DB,

therefore the remainder BD is double of the remainder DC.

Therefore BC is triple of CD;

therefore the square on BC is nine times the square on CD.

But the square on BC is five times the square on CK;

therefore the square on CK is greater than the square on CD;

therefore CK is greater than CD.

Let CL be made equal to CK,

from L let LM be drawn at right angles to AB,

and let MB be joined.

Now, since the square on BC is five times the square on CK,

and AB is double of BC, and KL double of CK,

therefore the square on AB is five times the square on KL.

But the square on the diameter of the sphere is also five times the square on the radius of the circle from which the icosahedron has been described. [XIII. 16, Por.]

And AB is the diameter of the sphere;

therefore KL is the radius of the circle from which the icosahedron has been described;

therefore KL is a side of the hexagon in the said circle. [IV. 15, Por.]

And, since the diameter of the sphere is made up of the side of the hexagon and two of the sides of the decagon inscribed in the same circle, [XIII. 16, Por.]

and AB is the diameter of the sphere,

while KL is a side of the hexagon,

and AK is equal to LB,

therefore each of the straight lines AK, LB is a side of the decagon inscribed in the circle from which the icosahedron has been described.

And, since LB belongs to a decagon, and ML to a hexagon,

for ML is equal to KL, since it is also equal to HK, being the same distance from the centre, and each of the straight lines HK, KL is double of KC,

therefore MB belongs to a pentagon. [XIII. 10]

But the side of the pentagon is the side of the icosahedron; [XIII. 16]

therefore MB belongs to the icosahedron.

Now, since FB is a side of the cube,

let it be cut in extreme and mean ratio at N,

and let NB be the greater segment;

therefore NB is a side of the dodecahedron. [XIII. 17, Por.]

And, since the square on the diameter of the sphere was proved to be one and a half times the square on the side AF of the pyramid, double of the square on the side BE of the octahedron and triple of the side FB of the cube,

therefore, of parts of which the square on the diameter of the sphere contains six, the square on the side of the pyramid contains four, the square on the side of the octahedron three, and the square on the side of the cube two.

Therefore the square on the side of the pyramid is four-thirds of the square on the side of the octahedron, and double of the square on the side of the cube;

and the square on the side of the octahedron is one and a half times the square on the side of the cube.

The said sides, therefore, of the three figures, I mean the pyramid, the octahedron and the cube, are to one another in rational ratios.

But the remaining two, I mean the side of the icosahedron and the side of the dodecahedron, are not in rational ratios either to one another or to the aforesaid sides;

for they are irrational, the one being minor [XIII. 16] and the other an apotome [XIII. 17].

That the side MB of the icosahedron is greater than the side NB of the dodecahedron we can prove thus.

For, since the triangle FDB is equiangular with the triangle FAB, [VI. 8]

proportionally, as DB is to BF, so is BF to BA. [VI. 4]

And, since the three straight lines are proportional, as the first is to the third, so is the square on the first to the square on the second; [V. Def. 9, VI. 20, Por.]

therefore, as *DB* is to *BA*, so is the square on *DB* to the square on *BF*;

therefore, inversely, as *AB* is to *BD*, so is the square on *FB* to the square on *BD*.

But *AB* is triple of *BD*;

therefore the square on *FB* is triple of the square on *BD*.

But the square on *AD* is also quadruple of the square on *DB*,

for *AD* is double of *DB*;

therefore the square on *AD* is greater than the square on *FB*;

therefore *AD* is greater than *FB*;

therefore *AL* is by far greater than *FB*.

And, when *AL* is cut in extreme and mean ratio, *KL* is the greater segment,

inasmuch as *LK* belongs to a hexagon, and *KA* to a decagon; [XIII. 9]

and, when *FB* is cut in extreme and mean ratio, *NB* is the greater segment;

therefore *KL* is greater than *NB*.

But *KL* is equal to *LM*;

therefore *LM* is greater than *NB*.

Therefore *MB*, which is a side of the icosahedron, is by far greater than *NB* which is a side of the dodecahedron.

Q. E. D.

I say next that ***no other figure, besides the said five figures, can be constructed which is contained by equilateral and equiangular figures equal to one another.***

For a solid angle cannot be constructed with two triangles, or indeed planes.

With three triangles the angle of the pyramid is constructed, with four the angle of the octahedron, and with five the angle of the icosahedron;

but a solid angle cannot be formed by six equilateral and equiangular triangles placed together at one point,

for, the angle of the equilateral triangle being two-thirds of a right angle, the six will be equal to four right angles:
which is impossible, for any solid angle is contained by angles less than four right angles. [XI. 21]

For the same reason, neither can a solid angle be constructed by more than six plane angles.

By three squares the angle of the cube is contained, but by four it is impossible for a solid angle to be contained,
for they will again be four right angles.

By three equilateral and equiangular pentagons the angle of the dodecahedron is contained;
but by four such it is impossible for any solid angle to be contained,
for, the angle of the equilateral pentagon being a right angle and a fifth, the four angles will be greater than four right angles:
which is impossible.

Neither again will a solid angle be contained by other polygonal figures by reason of the same absurdity.

Therefore etc.

Q. E. D.

Lemma.

But that *the angle of the equilateral and equiangular pentagon is a right angle and a fifth* we must prove thus.

Let *ABCDE* be an equilateral and equiangular pentagon,

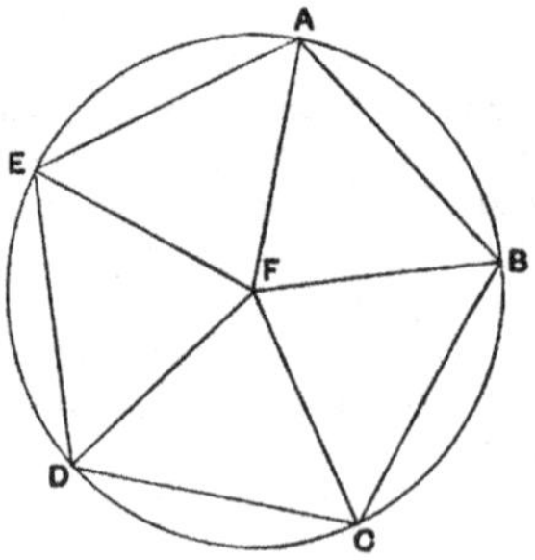

let the circle *ABCDE* be circumscribed about it,
let its centre *F* be taken,
and let *FA*, *FB*, *FC*, *FD*, *FE* be joined.

Therefore they bisect the angles of the pentagon at *A*, *B*, *C*, *D*, *E*.

And, since the angles at *F* are equal to four right angles and are equal,

therefore one of them, as the angle AFB, is one right angle less a fifth;
therefore the remaining angles FAB, ABF consist of one right angle and a fifth.

But the angle FAB is equal to the angle FBC;
therefore the whole angle ABC of the pentagon consists of one right angle and a fifth.

Q. E. D.

We have seen in the preceding notes that, if r be the radius of the sphere circumscribing the five solid figures,

$$\text{(edge of tetrahedron)} = \tfrac{2}{3}\sqrt{6}\,.\,r,$$
$$\text{(edge of octahedron)} = \sqrt{2}\,.\,r,$$
$$\text{(edge of cube)} = \tfrac{2}{3}\sqrt{3}\,.\,r,$$
$$\text{(edge of icosahedron)} = \frac{r}{5}\sqrt{10\,(5-\sqrt{5})},$$
$$\text{(edge of dodecahedron)} = \frac{r}{3}(\sqrt{15}-\sqrt{3}).$$

Euclid here exhibits the edges of all the five regular solids in one figure.

(1) Make AD equal to $2DB$.

Thus $BA = \tfrac{3}{2}AD,$

and $BA : AD = BA^2 : AF^2;$

therefore $BA^2 = \tfrac{3}{2}AF^2.$

Thus $AF = \sqrt{\tfrac{2}{3}}\,.\,2r = \tfrac{2}{3}\sqrt{6}\,.\,r = (\textit{edge of tetrahedron}).$

(2) $AB^2 : BF^2 = AB : BD$

$= 3 : 1.$

Therefore $BF^2 = \tfrac{1}{3}AB^2,$

or $BF = \dfrac{2}{\sqrt{3}}\,.\,r = \dfrac{2}{3}\sqrt{3}\,.\,r = (\textit{edge of cube}).$

(3) $AB^2 = 2BE^2.$

Therefore $BE = \sqrt{2}\,.\,r = (\textit{edge of octahedron}).$

(4) Draw AG perpendicular and equal to AB. Join GC, meeting the semicircle in H, and draw HK perpendicular to AB.

Then $GA = 2AC;$

therefore, by similar triangles, $HK = 2KC.$

Hence $HK^2 = 4KC^2,$

and therefore $5KC^2 = HK^2 + KC^2$

$= HC^2$

$= CB^2.$

Again, since $AB = 2CB$, and $AD = 2DB$,

by subtraction, $BD = 2DC,$

or $BC = 3DC.$

Therefore $9DC^2 = BC^2$
$= 5KC^2$.

Hence $KC > CD$.

Make CL equal to KC, draw LM at right angles to AB, and join AM, MB.

Since $CB^2 = 5KC^2$,
$AB^2 = 5KL^2$.

It follows that KL $(= \sqrt{\tfrac{4}{5}}\,.\,r)$ is the radius of, or the side of the regular hexagon in, the circle containing the pentagonal sections of the icosahedron. [XIII. 16]

And, since

$2r =$ (side of hexagon) + 2 (side of decagon in same circle) [XIII. 16, Por.]

$AK = LB =$ (side of decagon in the said circle).

But $LM = HK = KL =$ (side of hexagon in circle).

Therefore $LM^2 + LB^2$ $(= BM^2) =$ (side of pentagon in circle)^2 [XIII. 10]
$=$ (edge of icosahedron)^2,

and $BM =$ (*edge of icosahedron*).

[More shortly, $HK = 2KC$,

whence $HK^2 = 4KC^2$,

and $5KC^2 = HC^2 = r^2$.

Also $$AK = r - CK = r\left(1 - \frac{1}{\sqrt{5}}\right).$$

Thus
$$\begin{aligned} BM^2 &= HK^2 + AK^2 \\ &= \frac{4}{5}r^2 + r^2\left(1 - \frac{1}{\sqrt{5}}\right)^2 \\ &= r^2\left(\frac{10}{5} - \frac{2}{\sqrt{5}}\right) \\ &= \frac{r^2}{5}(10 - 2\sqrt{5}), \end{aligned}$$

and $$BM = \frac{r}{5}\sqrt{10(5 - \sqrt{5})} = (\textit{edge of icosahedron}).]$$

(5) Cut BF (the edge of the cube) in extreme and mean ratio at N. Then, if BN be the greater segment,

$BN =$ (*edge of dodecahedron*). [XIII. 17]

[Solving, we obtain
$$\begin{aligned} BN &= \frac{\sqrt{5} - 1}{2}\,.\,BF \\ &= \frac{\sqrt{5} - 1}{2}\,.\,\frac{2}{\sqrt{3}}\,.\,r \\ &= \frac{r}{3}(\sqrt{15} - \sqrt{3}) \\ &= (\textit{edge of dodecahedron}).] \end{aligned}$$

(6) If t, o, c are the edges of the tetrahedron, octahedron and cube respectively,

$$4r^2 = \tfrac{3}{2}t^2 = 2o^2 = 3c^2.$$

If each of these equals is put equal to X,

$$4r^2 = X,$$
$$t^2 = \tfrac{2}{3} . X,$$
$$o^2 = \tfrac{1}{2} . X,$$
$$c^2 = \tfrac{1}{3} . X,$$

whence $$4r^2 : t^2 : o^2 : c^2 = 6 : 4 : 3 : 2,$$

and the ratios between $2r$, t, o, c are all *rational* (in Euclid's sense).

The ratios between these and the edges of the icosahedron and the dodecahedron are *irrational*.

(7) To prove that

(edge of icosahedron) > (edge of dodecahedron),

i.e. that $$MB > NB.$$

By similar $\triangle$s FDB, AFB,

$$DB : BF = BF : BA,$$

or $$DB : BA = DB^2 : BF^2.$$

But $$3DB = BA;$$

therefore $$BF^2 = 3DB^2.$$

By hypothesis, $$AD^2 = 4DB^2;$$

therefore $$AD > BF,$$

and, *a fortiori*, $$AL > BF.$$

Now LK is the side of a hexagon, and AK the side of a decagon in the same circle;

therefore, when AL is divided in extreme and mean ratio, KL is the greater segment.

And, when BF is divided in extreme and mean ratio, BN is the greater segment.

Therefore, since $$AL > BF,$$
$$KL > BN,$$

or $$LM > BN.$$

And therefore, *a fortiori*, $$MB > BN.$$

APPENDIX.

I. THE CONTENTS OF THE SO-CALLED BOOK XIV. BY HYPSICLES.

This supplement to Euclid's Book XIII. is worth reproducing for the sake not only of the additional theorems proved in it but of the historical notices contained in the preface and in one or two later passages. Where I translate literally from the Greek text, I shall use inverted commas; except in such passages I reproduce the contents in briefer form.

I have already quoted from the Preface (Vol. I. pp. 5—6), but I will repeat it here.

"Basilides of Tyre, O Protarchus, when he came to Alexandria and met my father, spent the greater part of his sojourn with him on account of the bond between them due to their common interest in mathematics. And on one occasion, when looking into the tract written by Apollonius about the comparison of the dodecahedron and icosahedron inscribed in one and the same sphere, that is to say, on the question what ratio they bear to one another, they came to the conclusion that Apollonius' treatment of it in this book was not correct; accordingly, as I understood from my father, they proceeded to amend and rewrite it. But I myself afterwards came across another book published by Apollonius, containing a demonstration of the matter in question, and I was greatly attracted by his investigation of the problem. Now the book published by Apollonius is accessible to all; for it has a large circulation in a form which seems to have been the result of later careful elaboration.

"For my part, I determined to dedicate to you what I deem to be necessary by way of commentary, partly because you will be able, by reason of your proficiency in all mathematics and particularly in geometry, to pass an expert judgment upon what I am about to write, and partly because, on account of your intimacy with my father and your friendly feeling towards myself, you will lend a kindly ear to my disquisition. But it is time to have done with the preamble and to begin my treatise itself.

[**Prop. I.**] "*The perpendicular drawn from the centre of any circle to the side of the pentagon inscribed in the same circle is half the sum of the side of the hexagon and of the side of the decagon inscribed in the same circle.*"

Let ABC be a circle, and BC the side of the inscribed regular pentagon.

Take D the centre of the circle, draw DE from D perpendicular to BC, and produce DE both ways to meet the circle in F, A.

I say that DE is half the sum of the side of the hexagon and of the side of the decagon inscribed in the same circle.

Let DC, CF be joined; make GE equal to EF, and join GC.

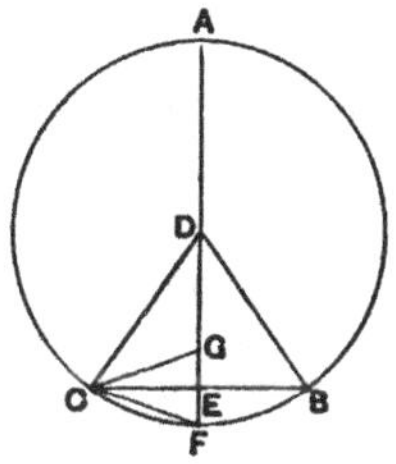

Since the circumference of the circle is five times the arc BFC,

and half the circumference of the circle is the arc ACF,

while the arc FC is half the arc BFC,

therefore (arc ACF) = 5 (arc FC)

or (arc AC) = 4 (arc CF).

Hence $\angle ADC = 4 \angle CDF$,

and therefore $\angle AFC = 2 \angle CDF$.

Thus $\angle CGF = \angle AFC = 2 \angle CDF$;

therefore [I. 32] $\angle CDG = \angle DCG$,

so that $DG = GC = CF$.

And $GE = EF$;

therefore $$DE = EF + FC.$$

Add DE to each;

therefore $$2DE = DF + FC.$$

And DF is the side of the regular hexagon, and FC the side of the regular decagon, inscribed in the same circle.

Therefore etc.

"Next it is manifest from the theorem [12] in Book XIII. that *the perpendicular drawn from the centre of the circle to the side of the equilateral triangle* [inscribed in it] *is half of the radius of the circle.*

[**Prop. 2.**] "*The same circle circumscribes both the pentagon of the dodecahedron and the triangle of the icosahedron inscribed in the same sphere.*

"This is proved by Aristaeus in his work entitled *Comparison of the five figures.* But Apollonius proves in the second edition of his comparison of the dodecahedron with the icosahedron that, as the surface of the dodecahedron is to the surface of the icosahedron, so also is the dodecahedron itself to the icosahedron, because the perpendicular from the centre of the sphere to the pentagon of the dodecahedron and to the triangle of the icosahedron is the same.

"But it is right that I too should prove that

[**Prop. 2**] *The same circle circumscribes both the pentagon of the dodecahedron and the triangle of the icosahedron inscribed in the same sphere.*

"For this I need the following

Lemma.

"*If an equilateral and equiangular pentagon be inscribed in a circle, the sum of the squares on the straight line subtending two sides and on the side of the pentagon is five times the square on the radius.*"

Let ABC be a circle, AC the side of the pentagon, D the centre; draw DF perpendicular to AC and produce it to B, E;

join AB, AE.

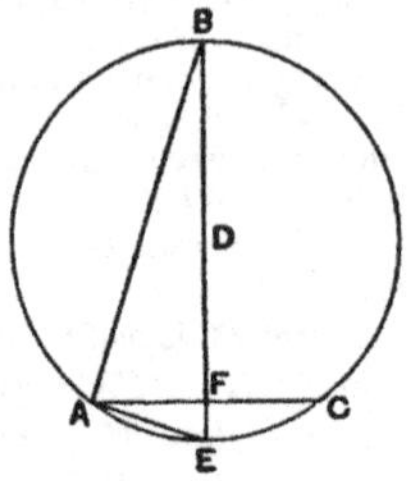

I say that

$$BA^2 + AC^2 = 5DE^2.$$

For, since $BE = 2ED$,

$$BE^2 = 4ED^2.$$

And $\qquad BE^2 = BA^2 + AE^2$;

therefore $\quad BA^2 + AE^2 + ED^2 = 5ED^2$.

But $\qquad AC^2 = DE^2 + EA^2$; [Eucl. XIII. 10]

therefore $\quad BA^2 + AC^2 = 5DE^2$.

"This being proved, it is required to prove that the same circle circumscribes both the pentagon of the dodecahedron and the triangle of the icosahedron inscribed in the same sphere."

Let AB be the diameter of the sphere, and let a dodecahedron and an icosahedron be inscribed.

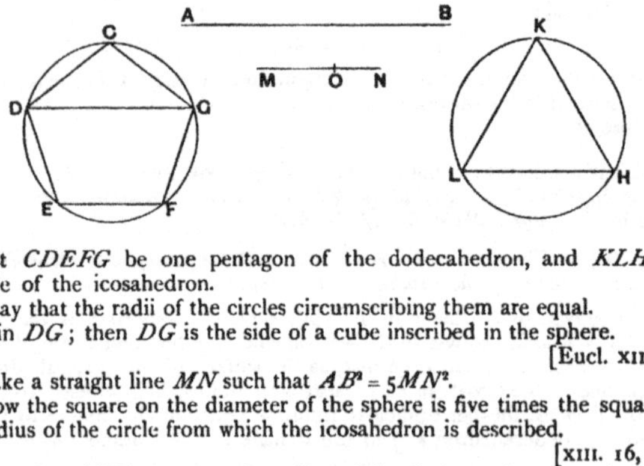

Let $CDEFG$ be one pentagon of the dodecahedron, and KLH one triangle of the icosahedron.

I say that the radii of the circles circumscribing them are equal.

Join DG; then DG is the side of a cube inscribed in the sphere. [Eucl. XIII. 17]

Take a straight line MN such that $AB^2 = 5MN^2$.

Now the square on the diameter of the sphere is five times the square on the radius of the circle from which the icosahedron is described. [XIII. 16, Por.]

Therefore MN is equal to the radius of the circle passing through the five vertices of the icosahedron which form a pentagon.

Cut MN in extreme and mean ratio at O, MO being the greater segment.

Therefore MO is the side of the decagon in the circle with radius MN. [XIII. 9 and 5, converse]

Now $\qquad 5MN^2 = AB^2 = 3DG^2$. [XIII. 15]

But $\qquad 3DG^2 : 3CG^2 = 5MN^2 : 5MO^2$

(since, if DG is cut in extreme and mean ratio, the greater segment is equal to CG, and, if two straight lines are cut in extreme and mean ratio, their segments are in the same ratio: see lemma later, pp. 518—9).

And $$5MO^2 + 5MN^2 = 5KL^2.$$

[This follows from XIII. 10, since KL is, by the construction of XIII. 16, the side of the regular pentagon in the circle with radius equal to MN, that is, the circle in which MN is the side of the inscribed hexagon and MO the side of the inscribed decagon.]

Therefore $$5KL^2 = 3CG^2 + 3DG^2.$$

But $$5KL^2 = 15 \text{ (radius of circle about } KLH)^2, \quad \text{[XIII. 12]}$$

and $$3DG^2 + 3CG^2 = 15 \text{ (radius of circle about } CDEFG)^2.$$

[Lemma above]

Therefore the radii of the two circles are equal.

Q. E. D.

[**Prop. 3.**] "*If there be an equilateral and equiangular pentagon and a circle circumscribed about it, and if a perpendicular be drawn from the centre to one side, then*

30 times the rectangle contained by the side and the perpendicular is equal to the surface of the dodecahedron."

Let $ABCDE$ be the pentagon, F the centre of the circle, FG the perpendicular on a side CD.

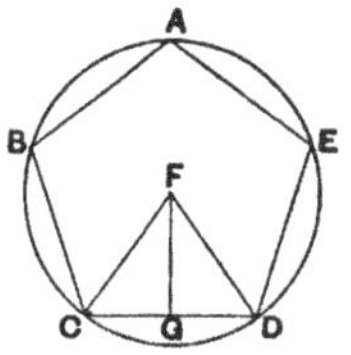

I say that

$$30 CD \,.\, FG = 12 \text{ (area of pentagon).}$$

Let CF, FD be joined.

Then, since

$$CD \,.\, FG = 2\,(\triangle CDF),$$

$$5 CD \,.\, FG = 10\,(\triangle CDF),$$

whence $$30 CD \,.\, FG = 12 \text{ (area of pentagon).}$$

Similarly we can prove that,

[**Prop. 4**] *If* ABC *be an equilateral triangle in a circle*, D *the centre, and* DE *perpendicular to* BC,

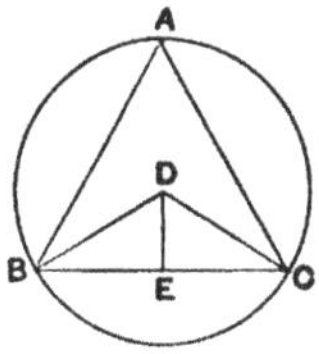

$$30 BC \,.\, DE = (\textit{surface of icosahedron}).$$

For $$DE \,.\, BC = 2\,(\triangle DBC);$$

therefore $$3 DE \,.\, BC = 6\,(\triangle DBC)$$
$$= 2\,(\triangle ABC),$$

whence $$30 DE \,.\, BC = 20\,(\triangle ABC).$$

It follows that [**Prop. 5**]

(*surface of dodecahedron*) : (*surface of icosahedron*)
= (*side of pentagon*) . (*its perpendicular*) : (*side of triangle*) . (*its perp.*).

"This being clear, we have next to prove that,

[**Prop. 6**] *As the surface of the dodecahedron is to the surface of the icosahedron, so is the side of the cube to the side of the icosahedron.*"

Let ABC be the circle circumscribing the pentagon of the dodecahedron and the triangle of the icosahedron, and let CD be the side of the triangle, AC that of the pentagon.

Let E be the centre, and EF, EG perpendiculars to CD, AC.

Produce EG to meet the circle in B and join BC.

Set out H equal to the side of the cube inscribed in the same sphere.

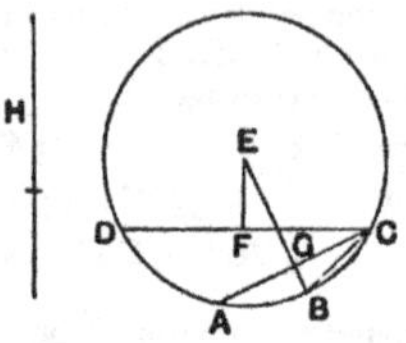

I say that

(surface of dodecahedron) : (surface of icosahedron) $= H : CD$.

For, since the sum of EB, BC is divided at B in extreme and mean ratio, and BE is the greater segment, [XIII. 9]
and $EG = \frac{1}{2}(EB + BC)$, [Prop. 1]
while $EF = \frac{1}{2}BE$, [see p. 513 above]
therefore, if EG is divided in extreme and mean ratio, the greater segment is equal to EF [that is to say, since EB is the greater segment of $EB + BC$ divided in extreme and mean ratio, $\frac{1}{2}EB$ is the greater segment of $\frac{1}{2}(EB + BC)$ similarly divided].

But, if H is also divided in extreme and mean ratio, the greater segment is equal to CA. [XIII. 17, Por.]

Therefore $H : CA = EG : EF$,
or $FE . H = CA . EG$.
And, since $H : CD = FE . H : FE . CD$,
and $FE . H = CA . EG$,
therefore $H : CD = CA . EG : FE . CD$
= (surface of dodecahedron) : (surf. of icos.). [Prop. 5]

Another proof of the same theorem.

Preliminary.

Let ABC be a circle and AB, AC sides of an inscribed regular pentagon.

Join BC; take D the centre of the circle, join AD and produce it to meet the circle at E. Join BD.

Let DF be made equal to $\frac{1}{2}AD$, and CH equal to $\frac{1}{3}CG$.

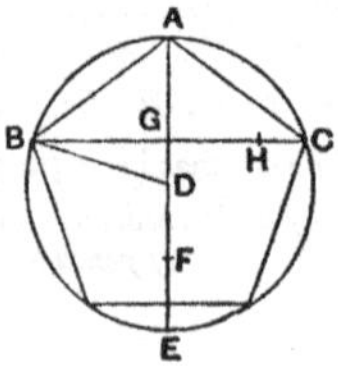

I say that

rect. $AF . BH =$ (area of pentagon).

For, since $AD = 2DF$,
$AF = \frac{3}{2}AD$.
And, since $GC = 3HC$,
$GC = \frac{3}{2}GH$.
Therefore $FA : AD = CG : GH$,
so that $AF . GH = AD . CG$
$= AD . BG$
$= 2(\triangle ABD)$.

Therefore

$$5AF \,.\, GH = 10\,(\triangle ABD) = 2 \text{ (area of pentagon)}.$$

And $GH = 2HC$;

therefore $5AF \,.\, HC =$ (area of pentagon),

or $AF \,.\, BH =$ (area of pentagon).

Proof of theorem.

This being clear, let the circle be set out which circumscribes the pentagon of the dodecahedron and the triangle of the icosahedron inscribed in the same sphere.

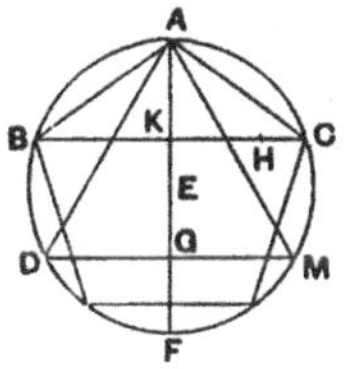

Let ABC be the circle, and AB, AC two sides of the pentagon; join BC.

Take E the centre of the circle, join AE and produce it to F.

Let $AE = 2EG$, $KC = 3CH$.

Through G draw DM at right angles to AF meeting the circle at D, M;

DM is then the side of the inscribed equilateral triangle.

Join AD, AM, which are equal to DM.

Now, since $AG \,.\, BH =$ (area of pentagon),

and $AG \,.\, GD =$ (area of triangle),

therefore $BH : GD =$ (area of pentagon) : (area of triangle),

and $12BH : 20GD =$ (surface of dod.) : (surface of icos.).

But $12BH = 10BC$, since $BH = 5HC$, and $BC = 6HC$;

and $20GD = 10DM$;

therefore (surface of dodecahedron) : (surface of icosahedron)

= (side of cube) : (side of icosahedron).

"Next we have to prove that,

[**Prop. 7**] *If any straight line whatever be cut in extreme aud mean ratio, then, as is* (1) *the straight line the square on which is equal to the sum of the squares on the whole line and on the greater segment to* (2) *the straight line the square on which is equal to the sum of the squares on the whole and on the lesser segment, so is* (3) *the side of the cube to* (4) *the side of the icosahedron.*"

Let AHB be the circle circumscribing both the pentagon of the dodecahedron and the triangle of the icosahedron inscribed in the same sphere, C the centre of the circle, and CB any radius divided at D in extreme and mean ratio, CD being the greater segment.

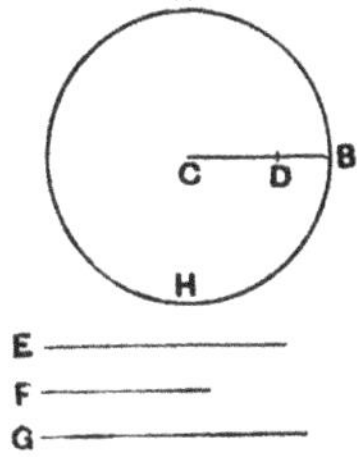

CD is then the side of the decagon inscribed in the circle. [XIII. 9 and 5, converse]

Let E be the side of the icosahedron, F that of the dodecahedron, and G that of the cube, inscribed in the sphere.

Then E, F are the sides of the equilateral triangle and pentagon inscribed in the circle, and, if G is divided in extreme and mean ratio, the greater segment is equal to F. [XIII. 17, Por.]

Thus $E^2 = 3BC^2$, [XIII. 12]

and $CB^2 + BD^2 = 3CD^2$. [XIII. 4]

Therefore $E^2 : CB^2 = (CB^2 + BD^2) : CD^2$,

or $E^2 : (CB^2 + BD^2) = CB^2 : CD^2$

$= G^2 : F^2$.

Therefore, alternately and inversely,

$$G^2 : E^2 = F^2 : (CB^2 + BD^2).$$

But $F^2 = BC^2 + CD^2$; for the square on the side of the pentagon is equal to the sum of the squares on the sides of the hexagon and decagon inscribed in the same circle. [XIII. 10]

Therefore $G^2 : E^2 = (BC^2 + CD^2) : (CB^2 + BD^2)$,

which is the result required.

It has now to be proved that

[**Prop. 8**] (*Side of cube*) : (*side of icosahedron*)

= (*content of dodecahedron*) : (*content of icosahedron*).

Since equal circles circumscribe the pentagon of the dodecahedron and the triangle of the icosahedron inscribed in the same sphere,

and in a sphere equal circular sections are equally distant from the centre,

the perpendiculars from the centre of the sphere to the faces of the two solids are equal;

in other words, the pyramids with the centre as vertex and the pentagons of the dodecahedron and the triangles of the icosahedron respectively as bases are of equal height.

Therefore the pyramids are to one another as their bases.

Thus (12 pentagons) : (20 triangles)

= (12 pyramids on pentagons) : (20 pyramids on triangles),

or (surface of dodecahedron) : (surface of icosahedron)

= (content of dod.) : (content of icos.).

Therefore

(content of dodecahedron) : (content of icosahedron)

= (side of cube) : (side of icosahedron). [Prop. 6]

Lemma.

If two straight lines be cut in extreme and mean ratio, the segments of both are in one and the same ratio.

Let AB be cut in extreme and mean ratio at C, AC being the greater segment;

and let DE be cut in extreme and mean ratio at F, DF being the greater segment.

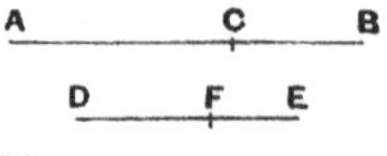

I say that $AB : AC = DE : DF$.

Since $AB \cdot BC = AC^2$,

and $DE \cdot EF = DF^2$,

$AB \cdot BC : AC^2 = DE \cdot EF : DF^2$,

and $4AB \cdot BC : AC^2 = 4DE \cdot EF : DF^2$.

Componendo,

$$(4AB \,.\, BC + AC^2) : AC^2 = (4DE \,.\, EF + DF^2) : DF^2,$$

or $$(AB + BC)^2 : AC^2 = (DE + EF)^2 : DF^2; \qquad [\text{II. } 8]$$

therefore $$(AB + BC) : AC = (DE + EF) : DF.$$

Componendo,

$$(AB + BC + AC) : AC = (DE + EF + DF) : DF,$$

or $$2AB : AC = 2DE : DF;$$

that is, $$AB : AC = DE : DF.$$

Summary of results.

If AB be *any* straight line divided at C in extreme and mean ratio, AC being the greater segment, and if we have a cube, a dodecahedron and an icosahedron inscribed in one and the same sphere, then:

(1) (side of cube) : (side of icosahedron) $= \sqrt{(AB^2 + AC^2)} : \sqrt{(AB^2 + BC^2)}$;

(2) (surface of dod.) : (surface of icos.)
= (side of cube) : (side of icosahedron);

(3) (content of dod.) : (content of icos.)
= (surface of dod.) : (surface of icos.);

and (4) (content of dodecahedron) : (content of icos.)
$= \sqrt{(AB^2 + AC^2)} : \sqrt{(AB^2 + BC^2)}$.

II. NOTE ON THE SO-CALLED "BOOK XV."

The second of the two Books added to the genuine thirteen is also supplementary to the discussion of the regular solids, but is much inferior to the first, "Book XIV." Its contents are of less interest and the exposition leaves much to be desired, being in some places obscure and in others actually inaccurate. It consists of three portions unequal in length. The first (Heiberg, Vol. v. pp. 40—48) shows how to inscribe certain of the regular solids in certain others, (*a*) a tetrahedron ("pyramid") in a cube, (*b*) an octahedron in a tetrahedron ("pyramid"), (*c*) an octahedron in a cube, (*d*) a cube in an octahedron and (*e*) a dodecahedron in an icosahedron. The second portion (pp. 48—50) explains how to calculate the number of edges and the number of solid angles in the five solids respectively. The third (pp. 50—66) shows how to determine the angle of inclination between faces meeting in an edge of any one of the solids. The method is to construct an isosceles triangle with vertical angle equal to the said angle of inclination; from the middle point of any edge two perpendiculars are drawn to it, one in each of the two faces intersecting in that edge; these perpendiculars (forming an angle which is the inclination of the two faces to one another) are used to determine the two equal sides of an isosceles triangle, and the base of the triangle is easily found from the known properties of the particular solid. The rules for drawing the respective isosceles triangles are first given all together in general terms (pp. 50—52); and the special interest of the passage consists in the fact that the rules are attributed to "Isidorus our great teacher." This Isidorus is no doubt Isidorus of Miletus, the architect of the Church of St Sophia at Constantinople (about 532 A.D.), whose pupil Eutocius also was; he is often referred to by Eutocius (*Comm. on Archimedes*) as ὁ Μιλήσιος μηχανικὸς Ἰσίδωρος ἡμέτερος διδάσκαλος. Thus the third portion of the Book at all events was written by a pupil of Isidorus in the sixth century. Kluge (*De Euclidis elementorum libris qui feruntur XIV et XV*, Leipzig, 1891) has closely examined the language and style of the three portions and conjectures that they may be the work of different authors; the first portion may, he thinks, date from the end of the third century (the time of Pappus), and the second portion too may be older than the third. Hultsch however (art. "Eukleides" in Pauly-Wissowa's *Real-Encyclopädie der classischen Altertumswissenschaft*, 1907) does not think his arguments convincing.

It may be worth while to set out the particulars of Isidorus' rules for constructing isosceles triangles with vertical angles equal respectively to the angles of inclination between faces meeting in an edge of the several regular solids. A certain base is taken, and then with its extremities as centres and a certain other straight line as radius two circles are drawn; their point of intersection determines the vertex of the particular isosceles triangle. In the case of the cube the triangle is of course right-angled; in the other cases the bases and the equal sides are as shown below.

	Base of isosceles triangle	*Equal sides of isosceles triangle*
For the tetrahedron	the side of a triangular face	the perpendicular from the vertex of a triangular face to its base
For the octahedron	the diagonal of the square on one side of a triangular face	ditto
For the icosahedron	the chord joining two non-consecutive angular points of the regular pentagon on an edge (the "pentagon of the icosahedron")	ditto
For the dodecahedron	the chord joining two non-consecutive angular points of a pentagonal face [*BC* in the figure of Eucl. XIII. 17]	the perpendicular from the middle point of the chord joining two non-consecutive angular points of a face to the parallel side of that face [*HX* in the figure of Eucl. XIII. 17]

ADDENDA ET CORRIGENDA.

Frontispiece. This is a facsimile of a page (fol. 45 verso) of the famous Bodleian MS. of the *Elements*, D'Orville 301 (formerly X. 1 inf. 2, 30), written in the year 888. The scholium in the margin, not very difficult to decipher, though some letters are almost rubbed out, is one of the scholia Vaticana given by Heiberg (Vol. V. p. 263) as III. No. 15: *Διὰ τοῦ κέντρου οὐσῶν οὐκ ἦν ζητήσεως ἄξιον, εἰ δίχα τέμνουσιν ἀλλήλας· τὸ γὰρ κέντρον αὐτῶν ἡ διχοτομία. ὁμοίως καὶ ἡ εἰ τῆς ἑτέρας διὰ τοῦ κέντρου οὔσης ἡ ἑτέρα μὴ διὰ τοῦ κέντρου εἴη, ὅτι οὐ δίχα τέμνεται ἡ διὰ τοῦ κέντρου.* The *ἡ* before *εἰ* in the last sentence should be omitted. PFVat. read *ἡ* without *εἰ*. The marginal references lower down are of course to propositions quoted, (1) *διὰ τὸ α′ τοῦ γ′*, "by III. 1," and (2) *διὰ τὸ γ′ τοῦ αὐτοῦ*, "by 3 of the same."

Vol. I. p. 20. I am aware that the assumption that the reference in the *Mechanics* (I. 24, p. 62, ed. Nix and Schmidt) is to Posidonius *of Rhodes* is disputed. It is pointed out that the context seems to show that the Posidonius referred to lived before Archimedes. Hoppe considers that the reference is to Posidonius *of Alexandria*, who was a pupil of Zeno the Stoic in the third century B.C. (cf. Meier, *De Heronis aetate*, pp. 19—21). The passage of the *Mechanics* in the German translation is as follows: "Posidonius, ein Stoiker, hat den Schwer- und Neigungspunkt in einer natürlichen (physikalischen?) Definition bestimmt und gesagt: der Schwer- oder Neigungspunkt ist ein solcher Punkt, dass, wenn die Last in demselben aufgehängt wird, sie in zwei gleiche Teile geteilt wird. Deshalb haben Archimedes und seine Anhänger in der Mechanik diesen Satz spezialisiert und einen Unterschied gemacht zwischen dem Aufhängepunkt und dem Schwerpunkt." This passage may certainly indicate that Posidonius' definition "represents a more imperfect standpoint than that of Archimedes" (Eneström in *Bibliotheca Mathematica* VIII_3, p. 177). But I do not feel certain that "deshalb" necessarily means so much as that it was the particular definition given by Posidonius *personally* which suggested to Archimedes the necessity for a distinction between the "Aufhängepunkt" and the "Schwerpunkt." I agree however with Meier (p. 21) that the doubt as to the reference makes it impossible to build upon the passage for the purpose of determining the date of Heron.

Vol. I. pp. 32—33. As bearing on the question whether Proclus continued his commentary beyond Book I., I should have referred to the scholium published by Heiberg in *Hermes* XXXVIII., 1903, p. 341, No. 17. It begins with the heading "Scholium on the scholium of Proclus on the 9th proposition where he says..." the words then quoted being taken from the last five lines of the long scholium x. No. 62 (Heiberg, Vol. v. pp. 450—2), one of the scholia Vaticana; and similar words lower down are accompanied by the parenthetical remark, "as the scholium of the divine Proclus says." If Proclus was really the author of the scholium, this is a point in favour of those who maintain that Proclus did write commentaries on the other Books (cf. Meier, *De Heronis aetate*, pp. 27—28). Heiberg however points out that, while the scholium shows that a Byzantine scholar took the collection of scholia Vaticana to be the work of Proclus, it does not prove more than this, and certainly it is not conclusive evidence that Proclus' commentaries covered all the Books. That this is *possible* cannot be denied; the scholia Vaticana to the other Books *may*, like those to Book I., have been extracted from Proclus, as also may the fragments which they contain of the commentary of Pappus, though it is not easy to explain why Proclus should have included extracts from Pappus which had already been put into the text by Theon. But it is much more probable, Heiberg thinks, that a Byzantine mathematician who had in his MS. of Euclid the collection of scholia Vaticana, and knew that those on Book I. came from Proclus, himself attached the name of Proclus to the rest of the collection; and this hypothesis seems to be confirmed by the fact that none of the other, older, sources of the scholia Vaticana have Proclus' name in x. No. 62.

Vol. I. pp. 64—66. Hultsch has some valuable remarks on the origin of the scholia (*Bibliotheca Mathematica* VIII_3, pp. 225 sqq. and art. "Eukleides" in Pauly-Wissowa's *Real-Encyclopädie der classischen Altertumswissenschaft*, 1907). Theodorus, Plato's teacher, is quoted in Plato's *Theaetetus* 147 D as having proved the irrationality of $\sqrt{3}$, $\sqrt{5}$ etc. up to $\sqrt{17}$; and the expressions used to describe such square roots, evidently Theodorus' own, are δύναμις ποδιαία, δύναμις τρίπους, δύναμις πεντάπους etc., the "square root" or "side" of "one, three, five etc. square feet." The same phraseology survives in the scholia x. Nos. 52, 94, 149, where we have the expressions ἡ τρίπους, ἡ τετράπους, ἡ πεντάπους, ἡ ἑξάπους, ἡ ἑπτάπους, ἡ ὀκτάπους, ἡ ἐννεάπους etc. Hultsch concludes that the sources go back as far as Theodorus. As regards the extracts from Geminus, Hultsch observes that the scholia to Book I. contain a considerable portion of Geminus' commentary on the definitions. They are specially valuable because they contain extracts from Geminus *only*, whereas Proclus, though drawing mainly upon him, quotes from others as well. On the postulates and axioms the scholia give more than is found in Proclus. Hultsch considers it probable that the scholium at the beginning of Book V. (No. 3) attributing the discovery of the theorems to Eudoxus but their arrangement to Euclid represents the tradition going back to Geminus; similarly he regards scholium XIII. No. 1 as having the same origin.

Vol. I. p. 71. The scholium numbered 17 on page 341 in *Hermes* XXXVIII. is taken from a MS. which was written in the 11th cent. Since the Arabic figures in it are in the first hand, it follows that the acquaintance of the Byzantines with these figures dates 100 years further back than the date given (12th cent.).

Vol. I. p. 71. In the numerical illustrations of Euclid's propositions sexagesimal fractions are often used; e.g. approximations to the values of surds are expressed as so many units, so many of the fractions $1/60$, so many of the fractions $1/60^2$ etc., going as far as "fourth-sixtieths" or the fractions $1/60^4$. Hultsch wrote a short paper on the sexagesimal fractions in the scholia to Book X (*Bibliotheca Mathematica* V_3, pp. 225—233). He shows that numbers expressed in these fractions are handled with skill and sometimes include results of surprising accuracy, as when $\sqrt{27}$ is given (allowing a slight correction of the last fraction by means of the context) as $5^\circ\ 11'\ 46''\ 10'''$, where $^\circ$ represents units and dashes the successive sexagesimal fractions, which gives for $\sqrt{3}$ the approximation $1^\circ\ 43'\ 55''\ 23'''$, being the same result as that given by Hipparchus in his tables of chords reproduced by Ptolemy and correct to the seventh decimal place. Similarly $\sqrt{8}$ is given as $2^\circ\ 49'\ 42''\ 20'''\ 10''''$, which is equivalent to $\sqrt{2} = 1 \cdot 4142135$. Hultsch gives instances of the various operations, addition, subtraction, multiplication and division, carried out in these fractions, and shows how the extraction of the square roots was effected, after the method which Theon of Alexandria in his commentary on Ptolemy's σύνταξις applies to the evaluation of $\sqrt{4500}$, and which evidently goes back to Hipparchus.

Vol. I. p. 101. In the *Bibliotheca Mathematica* IX_3, 1908, p. 76, A. Sturm notes that the preface to Camerarius' Euclid was not by Rhaeticus but by Camerarius himself, since the printer of the Steinmetz edition, Johann Steinmann, says, in a short preliminary notice, that Camerarius had written the preface 28 years before "sub alieno nomine."

Vol. I. p. 116. The date given for Eudoxus is that arrived at by Susemihl, "Die Lebenszeit des Eudoxos von Knidos" in *Rheinisches Museum für Philologie*, LIII., 1898, pp. 626—8. Hultsch however shows cause for rejecting this conjecture and for adhering to the earlier determination of the date as 408—355 B.C.

Vol. I. pp. 249, 370. The statement that Euclid does not use the expression αἱ ΒΑΓ, "the straight lines *BAC*," for "the straight lines *BA*, *AC*" is not accurate. Although I have not found it in the early Books, it is somewhat common in Books X, XI and XIII. Thus, e.g., in Book X "the rectangle (contained) by *BD*, *DC*" is often written τὸ ὑπὸ τῶν ΒΔΓ or τὸ ὑπὸ ΒΔΓ, and in one place (X. 59) we find τὸ συγκείμενον ἐκ τῶν ἀπὸ τῶν ΜΝΞ for "the sum of the *squares on MN*, *NO*." In Book XI the contracted form is used in expressions for the plane through two straight lines, e.g. τὸ διὰ τῶν ΒΔΑ ἐπίπεδον, "the plane through *BD*, *DA*." In XIII. 11 we have συναμφότερος ἡ ΔΓΜ for "the sum of the two straight lines *DC*, *CM*," where *DC*, *CM* form an angle.

Vol. I. pp. 343—4, 351; Vol. II. p. 97; Vol. III. pp. 1—3, etc. Heinrich Vogt's paper "Die Geometrie des Pythagoras" in the *Bibliotheca Mathematica* IX_3 (September, 1908), pp. 14—54, unfortunately appeared too late to be noticed in the proper places. I do not think it would have enabled me to modify greatly what I have written regarding the supposed discoveries of Pythagoras and the early Pythagoreans, because I have throughout endeavoured to give the traditions on the subject for what they are worth and no more, and not to build too much upon them. Vogt's paper is however a valuable piece of criticism, deserving of careful study; and it requires notice here so far as considerations of space allow. G. Junge had in his paper *Wann haben die Griechen das Irrationale entdeckt?* mentioned above (Vol. I. p. 351, Vol. III. p. 1 *n.*)

tried to prove that Pythagoras himself could not have discovered the irrational; and the object of Vogt's paper is to go further on the same lines and to maintain (1) that the theory of the irrational was first discovered by Theodorus, to whom Plato refers, and (2) that neither could Pythagoras himself have been the discoverer (*a*) of the theorem of Eucl. I. 47, or (*b*) of the construction of the five regular solids in the sense in which they are respectively constructed in Eucl. XIII., or (*c*) of the application of areas in its widest sense, equivalent to the solution of a quadratic equation in its most general form. Vogt's main argument as regards (*a*) the theorem of I. 47 is based on a new translation which he gives of the well-known passage of Proclus' note on the proposition (p. 426, 6—9), Τῶν μὲν ἱστορεῖν τὰ ἀρχαῖα βουλομένων ἀκούοντας τὸ θεώρημα τοῦτο εἰς Πυθαγόραν ἀναπεμπόντων ἐστὶν εὑρεῖν καὶ βουθύτην λεγόντων αὐτὸν ἐπὶ τῇ εὑρέσει. Vogt translates this as follows: "Unter denen, welche das Altertum erforschen wollen, kann man einige finden, welche denen Gehör geben, die dieses Theorem auf Pythagoras zurückführen und ihn als Stieropferer bei dieser Gelegenheit bezeichnen," "Among those who have a taste for research into antiquity, we can find some who give ear to those who refer this theorem to Pythagoras and describe him as sacrificing an ox on the strength of the discovery." According to this version the words τῶν... βουλομένων and the words ἀναπεμπόντων...καὶ...λεγόντων refer respectively to two different sets of persons, in fact two different generations; the latter are older authorities who are supposed to be cited by the former; the former are a later generation, perhaps contemporaries of Proclus, some of whom accepted the view of the older authorities while others did not. But this would have required the article τῶν before ἀναπεμπόντων, or some such expression as ἄλλων τινῶν οἳ ἀναπέμπουσι instead of ἀναπεμπόντων. Vogt's interpretation is therefore quite inadmissible. The persons denoted by ἀναπεμπόντων are *some of* the persons denoted by τῶν βουλομένων; hence Tannery's translation, to which mine (Vol. I. p. 350) is equivalent, is the only possible one, namely "Si l'on écoute ceux qui veulent raconter l'histoire des anciens temps, on peut en trouver qui attribuent ce théorème à Pythagore et lui font sacrifier un bœuf après sa découverte" (*La Géométrie grecque*, p. 103). ἀκούοντας agrees with the assumed *subject* of εὑρεῖν; ἀναπεμπόντων and λεγόντων should, strictly speaking, have been ἀναπέμποντας and λέγοντας agreeing with τινὰς (the direct *object* of εὑρεῖν) understood, but are simply attracted into the case of βουλομένων; the construction is quite intelligible. I agree with Vogt that Eudemus' history contained nothing attributing the theorem to Pythagoras. The words of Proclus imply this; but I do not think that they imply (as Vogt maintains) any pronouncement by Proclus himself *against* such attribution. In my opinion, Proclus is simply determined not to commit himself to any view; his way of evading a decision is the sentence following, ἐγὼ δὲ θαυμάζω μὲν καὶ τοὺς πρώτους ἐπιστάντας τῇ τοῦδε τοῦ θεωρήματος ἀληθείᾳ, μειζόνως δὲ ἄγαμαι τὸν στοιχειωτήν... ; the plural τοὺς πρώτους ἐπιστάντας is, I hold, used for the very purpose of making the statement as vague as possible; he will not even allow it to be inferred that he attributed the discovery to any single person. Returning to ἡ τῶν ἀλόγων πραγματεία (Proclus, p. 65, 19), we may concede that the imperfect (arithmetical) theory of proportion would probably be discovered earlier than the theory of the irrational; but we can hardly accept the reading ἀναλόγων or ἀναλογιῶν (instead of ἀλόγων) until it is confirmed by further investigation of the MSS. I do not agree in Vogt's contention that the theory of the irrational was first discovered by Theodorus. It seems to me that we have evidence to the contrary in the very passage of Plato referred to. Plato (*Theaetetus* 147 D) mentions $\sqrt{3}$, $\sqrt{5}$, ... up to $\sqrt{17}$ as dealt with by Theodorus, but *omits* $\sqrt{2}$. This fact, along with Plato's allusions elsewhere to the irrationality of $\sqrt{2}$, and to approximations to it, in the expressions ἄρρητος and ῥητὴ διάμετρος τῆς πεμπάδος, as if those expressions had a well-known signification, implies that the discovery of the irrationality of $\sqrt{2}$ had been made before the time of Theodorus. The words ἡ τῶν ἀλόγων πραγματεία might well be used even if the reference is only to $\sqrt{2}$, because the first step would be the most difficult, and πραγματεία need not mean the establishment of a complete theory or anything more than "investigation" of a subject. Coming now to (*b*) the construction of the cosmic figures, ἡ τῶν κοσμικῶν σχημάτων σύστασις (Proclus, p. 65, 20), I agree with Vogt to the following extent. It is unlikely that Pythagoras or even the early Pythagoreans "constructed" the five regular solids in the sense of a complete theoretical construction such as we find, say, in Eucl. XIII.; and it is possible that Theaetetus was the first to give these constructions, whether ἔγραψε in Suidas' notice, πρῶτος δὲ τὰ πέντε καλούμενα στερεὰ ἔγραψε, means "constructed" or "wrote upon." But σύστασις in the above phrase of Proclus may well mean something less than the theoretical constructions and proofs of Eucl. XIII.; it may mean, as Vogt says, simply the "putting together" of the figures in the same way as Plato puts them together in the *Timaeus*, i.e. by bringing a certain number of angles of equilateral triangles and of regular pentagons together at one point. There is no reason why the early Pythagoreans should not have "constructed" the five regular solids in this sense; in fact the supposition that they did so agrees well with what we know of their having put angles of certain regular figures together round a point (in connexion with the theorem of Eucl. I. 32) and shown that only three kinds of such angles would fill up the space *in one plane* round the point. But I do not agree in the apparent refusal of Vogt to credit the Pythagoreans with the knowledge of the theoretical construction of the regular pentagon as we find it in Eucl. IV. 10, 11. I do not know of any reason for rejecting the evidence of the Scholia IV. Nos. 2 and 4 which say categorically that "this Book" (Book IV) and "the whole of the theorems" in it (including therefore Props. 10, 11) are discoveries of the Pythagoreans. And the division of a straight line in extreme and mean ratio, on which the construction of the regular pentagon depends, comes in Eucl. Book II. (Prop. 11), while we have sufficient grounds for regarding the whole of the substance of Book II. as Pythagorean. I am sorry that, when I was writing on the subject of the "five bodies of the sphere" in the fragment of Philolaus (Vol. II. p. 97), my attention had not been called to the version of the passage in Diels' *Fragmente der Vorsokratiker* (Berlin 1903, p. 254, and 2nd ed. Berlin 1906, p. 244): καὶ τὰ μὲν τᾶς σφαίρας σώματα πέντε ἐντί, τὰ ἐν τᾶι σφαίραι πῦρ <καὶ> ὕδωρ καὶ γᾶ καὶ ἀήρ, καὶ ὁ τᾶς σφαίρας ὁλκάς, πέμπτον, "Und zwar gibt es fünf Elemente der Weltkugel: die *in* der Kugel befindlichen, Feuer, Wasser, Erde und Luft, und was der Kugel Lastschiff ist, das fünfte." If this version is right, there is (as Vogt points out) no allusion here to the five regular solids, and the fragment ceases to have any bearing on the present question. I will permit myself one more criticism out of many which Vogt's paper is sure to evoke. I think he bases too much on the fact that it was left for Oenopides (in the period from, say, 470 to 450 B.C.) to discover two elementary constructions (with ruler and compass only), namely that of a perpendicular to a straight line from an external point (Eucl. I. 12), and that of an angle equal to a given rectilineal angle (Eucl. I. 23). Vogt infers that geometry must have been in a very rudimentary condition at the time. I do not think this follows; the explanation would seem to be rather that, the restriction of the instruments used in constructions to the ruler and compass not having been definitely established before the time when Oenopides wrote, it had not previously occurred to anyone to substitute new constructions based on that principle for others previously in vogue. In the case of the perpendicular, for example, the construction would no doubt, in earlier days, have been made by means of a set square.

Vol. I. p. 411, column 2, line 12, for ἑκατέρᾳ ἑκατέρᾳ read ἑκατέρα ἑκατέρᾳ.

Vol. II. pp. 189—190. Hultsch (art. "Eukleides" in Pauly-Wissowa's *Real-Encyclopädie der classischen Altertumswissenschaft*) thinks that the definition of compound ratio (VI. Def. 5) is genuine. His grounds are (1) that it stood in the παλαιὰ ἔκδοσις represented by P (though P only has it in the margin) and (2) that some explanation on the subject must have been given by way of preparation for VI. 23, while there is nothing in the definition which is *inconsistent* with the mode of statement of VI. 23. If however the definition is after all genuine, I should be inclined to regard it as a mere survival from earlier text-books, like the first of the two alternative definitions of a solid angle (XI. Def. 11); for its form seems to suit the old theory of proportion applicable to commensurable quantities only better than the generalised theory due to Eudoxus.

Vol. II. pp. 424—5. I should have added to the note on "perfect numbers" the following references. Nicomachus (I. 16, 2—7) observes that perfect numbers are rare, there being only one among the units (6), one among the tens (28), one among the hundreds (496) and one among the thousands (8128), and that they end alternately in 6 and 8. Cf. Iamblichus, p. 33, 15—25.

Nesselmann (*Die Algebra der Griechen*, p. 164, note) gives a reference to a letter from Fermat to Mersenne (*Varia opera mathematica Petri de Fermat*, Tolosae, 1679, p. 177) in which Fermat enunciates three propositions which much facilitate the investigation whether a number of the form $2^n - 1$ is prime or not. If we write in one line the successive exponents 1, 2, 3, 4 etc. of the successive powers of 2 and underneath them respectively, in another line, the numbers representing the corresponding powers of 2 diminished by 1, thus,

1	2	3	4	5	6	7	8	9	10	11 ......n
1	3	7	15	31	63	127	255	511	1023	2047......$2^n - 1$

the following relations are found to subsist between the numbers in the first line and those directly below them in the second line.

1. If the exponent is not a prime number, the corresponding number is not a prime number either (since $a^{pq} - 1$ is always divisible by $a^p - 1$ as well as by $a^q - 1$).

2. If the exponent is a prime number, the corresponding number diminished by 1 is divisible by twice the exponent. $\left(\frac{2^n - 2}{2n} = \frac{2^{n-1} - 1}{n}\right.$; so that this is a special case of "Fermat's theorem" that, if p is a prime number and a is prime to p, then $a^{p-1} - 1$ is divisible by p.$\left.\right)$

3. If the exponent n is a prime number, the corresponding number is only divisible by numbers of the form $(2mn + 1)$. If therefore the corresponding number in the second line has no factor of this form, it has no integral factor.

The first and third of these propositions are those which are specially useful for the purpose in question. As usual, Fermat does not give his proofs but merely adds: "Voilà trois fort belles propositions que j'ay trouvées et prouvées non sans peine. Je les puis appeller les fondements de l'invention des nombres parfaits."

The first four perfect numbers, those mentioned above as given by Nicomachus, are

$$2(2^2-1)=6,\quad 2^2(2^3-1)=28,\quad 2^4(2^5-1)=496,\quad 2^6(2^7-1)=8128.$$

Hultsch investigated the next four, the fifth to the eighth (*Nachr. d. Gesellschaft d. Wissensch. zu Göttingen*, 1895, pp. 246 sqq.); the fifth is $2^{12}(2^{13}-1)=33{,}550{,}336$, the sixth $2^{16}(2^{17}-1)$, the seventh $2^{18}(2^{19}-1)$, and the eighth $2^{30}(2^{31}-1)$, which is greater than 2 trillions. The ninth, $2^{60}(2^{61}-1)$, was discovered by P. Seelhoff (*Zeitschrift für Math. u. Physik* XXXI., 1886, pp. 174—8) and verified by Lucas (*Mathésis* VII. pp. 45—46); Hultsch also wrote upon it (*Abhandlungen der Gesellschaft d. Wissensch. zu Göttingen*, 1897, pp. 47 sq.); it has 37 digits.

Loria (*Il periodo aureo della geometria greca*, p. 39) gives further references. He observes that the question of the existence of further prime numbers of the form $2^n - 1$ where $n > 61$ is not yet solved; it would be, however, if it were found possible to prove the *empirical* theorem of Catalan that, if $2^n - 1$ is a prime number $(=p)$, the numbers $p' = 2^p - 1$, $p'' = 2^{p'} - 1$ etc. will also be prime numbers ("Mélanges mathématiques" in *Mémoires de la Société de Liége*, 2e Série, XII. p. 376). There have also been attempts, so far unsuccessful, to solve the question whether there exist other "perfect numbers" than those of Euclid and, in particular, perfect numbers which are *odd* (cf. several notes by Sylvester in *Comptes rendus* CVI., 1888; Catalan, "Mélanges mathématiques" in *Mém. de la Soc. de Liége*, 2e Série, XV., 1888, pp. 205—7; C. Servais, in *Mathésis* VII. pp. 228—230 and VIII. pp. 92—93, 135; E. Cesàro in *Mathésis* VII. pp. 245—6; E. Lucas in *Mathésis* X. pp. 74—76).

GENERAL INDEX OF GREEK WORDS AND FORMS.

[The references are to volumes and pages.]

GENERAL INDEX.

[The references are to volumes and pages.]

CAMBRIDGE: PRINTED BY JOHN CLAY, M.A. AT THE UNIVERSITY PRESS.

Spotted something that isn't quite right?

Let us know and we'll fix it for future reprints. And as a thank you for your diligence, we'll send you a free Sothis book of your choice.

publisher@sothis.press

This page intentionally left blank

Except of course for that line above. And this one.
And whilst we have your attention, perhaps you'd
like to follow us on Instagram @sothis.press. Go on.

Made in the USA
Monee, IL
08 November 2026

68091704R00329